Real and Complex Geometry

Liviu Ornea
Editor

Real and Complex Geometry

In Honour of Paul Gauduchon

Springer

Editor
Liviu Ornea
Mathematics
University of Bucharest
Bucuresti, Romania

ISBN 978-3-031-92296-1 ISBN 978-3-031-92297-8 (eBook)
https://doi.org/10.1007/978-3-031-92297-8

This Springer imprint is published by the registered company Springer Nature Switzerland AG
The registered company address is: Gewerbestrasse 11, 6330 Cham, Switzerland

Contents

Introduction

It is difficult to open a modern paper on complex geometry without finding a mention of Paul Gauduchon. In fact, a search on MathSciNet returns 2284 entries where the term "Gauduchon" appears "Anywhere."

His name appears in titles as well, with numerous concepts bearing his mark: Gauduchon metric, k-Gauduchon metric, strongly Gauduchon metric, generalized Gauduchon metric, Gauduchon manifold, Gauduchon connection, Gauduchon (scalar) curvature, Gauduchon degree, k-th Gauduchon condition, and Gauduchon–Tod structure, among others.

Over the past 50 years, Riemannian and complex geometry—both algebraic and differential—have been profoundly shaped by his work, as evidenced by 2535 citations across 1812 publications. (We may not be fond of scientometrics, but still…)

Paul is undoubtedly one of the foremost figures in modern geometry. It is impossible to conceive of non-Kähler geometry without the metric he modestly called "standard," which has since been generalized in various ways. His contributions to extremal metrics theory, twistor theory, bi-Hermitian geometry, CR, and Sasakian geometry (to name just a few of his research interests) have been deeply influential. His writings—81 so far, with more to come—are models of mathematical exposition, characterized by rigor and clarity. What a pity that his book on Kähler geometry and extremal metrics remains unpublished and available only "on request."

But Paul is not only a brilliant mathematician—he is also generous with his ideas (as attested by his 20 coauthors and counting), eager to share, explain, and guide. Discussing mathematics with Paul is always a pleasure—and so is discussing music, literature, history… One never gets bored in his company.

The book you are about to read brings together contributions from those who, in one way or another, have benefited from his friendship, collaboration, and teaching. It is our way of expressing gratitude for his mathematics and for all he has given to us and to the community.

Bon anniversaire, cher ami !

March 23, 2025 Liviu Ornea

About the Editor

Liviu Ornea (born 1960) is a professor in the Department of Mathematics at the University of Bucharest and a senior researcher at the Institute of Mathematics "Simion Stoilow" in the Romanian Academy. His main fields of interest include differential geometry of complex manifolds, with a special focus on locally conformally Kähler manifolds. He is the (co)author of more than 90 published papers and two research monographs. He has visited MPIM Bonn, ICTP Trieste, IMPA (Rio de Janeiro), Ecole Polytechnique (Palaiseau), Ecole Polytechnique Fédérale de Lausanne, Tokyo Metropolitan University, Tokyo Institute for Technology, University of New Mexico (Albuquerque), Università di Firenze, Università di Roma "La Sapienza", etc.

From Kähler Ricci Solitons to Calabi-Yau Kähler Cones

Vestislav Apostolov, Abdellah Lahdili, and Eveline Legendre

Dedicated to Paul Gauduchon on the occasion of his 80th birthday, with respect and gratitude.

Abstract We show that if X is a smooth Fano manifold which carries a Kähler Ricci soliton, then the canonical cone of the product of X with a complex projective space of sufficiently large dimension is a Calabi–Yau cone, i.e. admits a Ricci-flat Kähler cone metric. This can be seen as an asymptotic version of a conjecture by Mabuchi and Nikagawa. This result is obtained by the relative openness of the set of weight functions v over the momentum polytope of a given smooth Fano manifold, for which a v-soliton exists. We discuss other ramifications of this approach, including a Licherowicz type obstruction to the existence of a Kähler Ricci soliton and a Fujita type volume bound for the existence of a v-soliton.

V.A. was supported in part by an NSERC Discovery Grant, an FRQNT Team Grant and a Connect Talent Grant of the Region de Pays de la Loire. He is grateful to the Institut Camille Jordan Lyon 1 for hospitality during the preparation of this work. A.L. was supported by the Villum Young Investigator grant 0019098 and UQAM. E.L. was supported in part by France ANR project BRIDGES No ANR-21-CE40-0017, she is grateful to the Simons foundation, CRM and FQRNT for offering her a visiting position during the preparation of this work. The authors thank the referees for their useful remarks.

V. Apostolov
Département de Mathématiques, UQAM, C.P. 8888, Montréal, QC, Canada

Institute of Mathematics and Informatics, Bulgarian Academy of Sciences, Sofia, Bulgaria
e-mail: apostolov.vestislav@uqam.ca

A. Lahdili
Département de Mathématiques, UQAM, C.P. 8888, Montréal, QC, Canada

E. Legendre (✉)
Université Claude Bernard Lyon 1, Institut Camille Jordan équipe AGL, Villeurbanne, France
e-mail: eveline.legendre@math.univ-lyon1.fr

L. Ornea (ed.), *Real and Complex Geometry*,
https://doi.org/10.1007/978-3-031-92297-8_1

Keywords Calabi-Yau · Sasaki-Einstein · v-Soliton · Weighted cscK · Kahler-Ricci solitons

1 Introduction

A number of different notions of *special* Kähler metrics have emerged in the last 20 years or so, in connection with Calabi's seminal program [14] of finding a canonical representative of a given de Rham class of Kähler metrics on a smooth compact Kähler manifold X. Perhaps the most studied notion of all, introduced by Calabi himself, is that of *constant scalar curvature* Kähler (cscK) metrics.

In the case of a smooth Fano variety X endowed with its anti-canonical de Rham Kähler class $2\pi c_1(X)$, a cscK metric in $2\pi c_1(X)$ is necessarily a Kähler–Einstein metric with scalar curvature equal to $2n$. The existence problem for such Kähler metrics is understood in terms of the Yau–Tian–Donaldson (YTD) conjecture [26, 54, 61] which states that X admits a Kähler-Einstein metric in $2\pi c_1(X)$ if and only if the anticanonical polarization (X, K_X^{-1}) is K-polystable. There are, by now, many different proofs of this conjecture [10, 19, 45, 55, 62], following the initial work of Chen–Donaldson–Sun [16–18] (who proved K-polystability implies existence) and Tian and Berman (who proved existence implies K-stability)[6, 54].

Beyond the study of Kähler-Einstein metrics on $(X, 2\pi c_1(X))$, other notions of canonical Kähler metrics have been considered. These allow to treat cases where a Kähler-Einstein metric do not exist due to the classical obstructions in terms of the automorphisms of X [29, 47, 51]. Tian-Zhu [56, 57] initiated a systematic study of the so-called *Kähler-Ricci solitons* (KRS) on $(X, 2\pi c_1(X))$, whereas Martelli–Sparks–Yau [50] developed the theory of Calabi–Yau cone structures (or, equivalently, Sasaki–Einstein structures) defined on the affine cone $K_X^\times$ associated to X. These works prompted separated investigations of the corresponding existence theories, and the formulation and proofs of appropriate modifications of the YTD conjecture in each case, see respectively [21, 23].

More recently, there have been developments providing a framework to treat the existence problems mentioned above all together. The unifying geometric object is that of a *v-soliton* Kähler metric $\omega \in 2\pi c_1(X)$ (called g-soliton in [36]), which was introduced in a foundational work by Mabuchi [48], followed by further comprehensive studies by Berman–Witt Nyström [9] and Han–Li [36]. A v-soliton is defined in terms of a fixed maximal compact torus $\mathbb{T} \subset \mathrm{Aut}(X)$ with associated canonical polytope $\mathrm{P}_X \subset (\mathrm{Lie}(\mathbb{T}))^*$, and a positive smooth function $v(x) > 0$ on P_X, via the equation

$$\mathrm{Ric}(\omega) - \omega = \frac{1}{2} dd^c \log v(\mu_\omega).$$

In the above formula, ω is a $\mathbb{T}$-invariant Kähler metric in $2\pi c_1(X)$, $\mathrm{Ric}(\omega) \in 2\pi c_1(X)$ is its Ricci form and $\mu_\omega : X \to \mathrm{P}_X$ is the canonically normalized

$\mathbb{T}$-momentum map. Thus, Kähler–Einstein metrics correspond to 1-solitons, KRS metrics to e^{ℓ}-solitons [57], and Calabi–Yau cone structures on $K_X^{\times}$ to $\ell^{-(n+2)}$-solitons [3], where $\ell(x)$ is a suitably defined (and in general different for each case) affine-linear function on P_X. As an outcome, the work [36] gives a YTD type correspondence for the existence of a v-soliton on $(X, \mathbb{T}, 2\pi c_1(X))$, expressed in terms of a suitable notion of *uniform v-weighted Ding stability* of $(X, \mathbb{T}, 2\pi c_1(X)))$ on $\mathbb{T}_{\mathbb{C}}$-equivariant test-configurations.

Despite the above remarkable progress, it remains a very challenging problem to use effectively the YTD correspondence in order to produce examples of v-solitons. A specific ramification in this general direction is the following

Problem 1.1 Suppose $(X, \mathbb{T}, 2\pi c_1(X))$ is a smooth Fano manifold. Describe the set $\mathcal{S}(X)$ of positive smooth functions v on P_X, such that $(X, \mathbb{T}, 2\pi c_1(X))$ admits a v-soliton.

By an idea going back to A. Futaki [29], it turns out that $\mathcal{S}(X) \subset \mathcal{F}(X)$ where $\mathcal{F}(X)$ is the subset

$$\mathcal{F}(X) := \left\{ v \in C^{\infty}_{>0}(\mathrm{P}_X) \mid \int_{\mathrm{P}_X} \langle \zeta, x \rangle v(x) d\mu_{\mathrm{DH}} = 0, \ \forall \zeta \in \mathrm{Lie}(\mathbb{T}) \right\}.$$

In the above formula, $d\mu_{\mathrm{DH}}$ denotes the Duistermaat–Heckman measure [27] on P_X induced by $2\pi c_1(X)$. Indeed, $\mathcal{F}(X)$ is identified with the set of weight functions v for which the corresponding v-Futaki invariant $\mathrm{Fut}_v : \mathrm{Lie}(\mathbb{T}) \to \mathbb{R}$ on X vanishes. Clearly, $\mathcal{F}(X)$ is a relatively closed convex cone in $C^{\infty}_{>0}(\mathrm{P}_X)$ (in the relative $C^0(\mathrm{P}_X)$-topology).

Problem 1.1 is particularly interesting when $\mathcal{S}(X)$ is a non-empty subset of $\mathcal{F}(X)$. We know in this case that the usual Calabi–Lichnerowicz–Matsushima obstruction vanishes, i.e. the connected component of the identity $\mathrm{Aut}_{\circ}(X)^{\mathbb{T}}$ of the group of complex automorphisms of X commuting with $\mathbb{T}$ is reductive (see. e.g. [43] for a general statement). In [36], it is shown that if X is a smooth *toric* Fano variety, one has the equality $\mathcal{S}(X) = \mathcal{F}(X)$. Similar *weight insensitive* highly symmetric Fano examples were recently found by L. Wang [59] and T. Delcroix [24]; these works also demonstrate examples satisfying $\mathcal{S}(X) \subsetneq \mathcal{F}(X)$.

Despite the above progress, a general understanding of $\mathcal{S}(X)$ remains an open problem. While we do not attempt in this work to obtain any deep structure result for $\mathcal{S}(X)$, we underline a basic property which provides a strong openness result à la LeBrun–Simanca [42]. To state it, we introduce the space of weights

$$\mathcal{D}(X) := \left\{ v \in C^{\infty}_{>0}(\mathrm{P}_X) \,\middle|\, \mathbf{D}_v \text{ is } \mathbb{T}_{\mathbb{C}} - \text{coercive} \right\},$$

where $\mathbf{D}_v$ denotes the v-weighted Ding functional (see Definitions 2.3 and 2.4).

Theorem 1.1 *$\mathcal{D}(X)$ is an open convex cone in $\mathcal{F}(X)$ with respect to the relative $C^0(\mathrm{P}_X)$-topology.*

By [36, Theorem 1.7], $\mathcal{S}(X) = \mathcal{D}(X)$. Thus, Theorem 1.1 yields an effective estimate for the radius of an open ball in $\mathcal{S}(X)$ around a given $v_0 \in \mathcal{S}(X)$, expressed in terms of the coercivity slope of the v_0-weighed Ding functional, see Corollary 3.1. Our main geometric applications below use the fact that $\mathcal{S}(X)$ is relatively open in $\mathcal{F}(X)$; to obtain these results, instead of Theorem 1.1 and [36], one can merely use a LeBrun–Simanca [42] type perturbation argument as in [41], which yields the relative openness of finite dimensional smooth families of weights in $\mathcal{S}(X)$. We present such a perturbation result adapted to our context in Appendix A.

Remark 1.1 In a somewhat similar vein, on a given Kähler cone Y endowed with a maximal compact torus $\hat{\mathbb{T}} \subset \mathrm{Aut}(Y)$, the authors study in [13] the cone $\hat{\mathfrak{t}}^{\mathrm{ext}} \subset \mathrm{Lie}(\hat{\mathbb{T}})$ of Sasaki–Reeb vector fields $\hat{\xi}$ such that Y admits an extremal Sasaki metric polarized by $\hat{\xi}$. Translated to the weighted Kähler terminology (see [1, 2, 41]), this is equivalent to study on a given quasi-regular quotient $(X, L, \mathbb{T})$ (endowed with a momentum polytope $\mathrm{P}_L \subset \mathfrak{t}^*$) the set $\mathfrak{t}^{\mathrm{ext}}$ of positive affine-linear functions ℓ on P_L such that X admits a $\mathbb{T}$-invariant $(\ell^{-n-1}, \ell_{\mathrm{ext}}\ell^{-n-3})$-cscK metric in $2\pi c_1(L)$. Here, ℓ is arbitrary positive affine linear function on P_L and ℓ_{ext} is uniquely determined by ℓ. In this setup, it is shown in [13] that $\mathfrak{t}^{\mathrm{ext}}$ is not convex in general, which contrasts with Theorem 1.1.

We now discuss geometric applications of Theorem 1.1. To this end, we consider the Sasaki geometry corresponding to the canonical bundle K_X of a Fano manifold X. It was shown in [3] that the existence of a Sasaki–Einstein structure (or equivalently, the existence of a Ricci-flat Kähler cone structure on the affine cone $K_X^\times$) is equivalent to the existence of a v-soliton in $(X, 2\pi c_1(X))$ for a weight function $v = \ell^{-(n+2)}$ where $\ell(x)$ is a positive affine-linear function on P_X.

Thus, Problem 1.1 incorporates as a special case answering the following Conjecture due to Mabuchi and Nakagawa:

Conjecture 1.1 ([49]) If the smooth Fano manifold X admits a KRS, then its canonical cone $Y := K_X^\times$ admits a Calabi–Yau cone structure.

We note that the recent work [38] shows that the above conjecture generally fails for Fano orbifolds.

Using Theorem 1.1, we make the following observation related to Conjecture 1.1.

Theorem 1.2 *Suppose X is a smooth Fano manifold which admits a Kähler–Ricci soliton. Then, there exists a non-negative integer k_0, such that for any $k \geq k_0$, the canonical cone $K_Z^\times$ of the Fano manifold*

$$Z := X \times \mathbb{P}^k_{\mathbb{C}}$$

admits a Ricci-flat Kähler cone metric.

Theorem 1.2 is obtained along the following lines: we construct a sequence of positive affine-linear functions $\ell_N(x)$ on P_X, satisfying that: (1) $(\ell_N(x))^{-N} \in \mathcal{F}(X)$, and (2) $(\ell_N(x))^{-N}$ converges in $C^0(\mathrm{P}_X)$ to the weight $v = e^{\langle \tau, x\rangle}$ corresponding to the KRS. By Theorem 1.1 and [36, Theorem 1.7] (or equivalently by Corollary A.1), for $N >> 1$ there exists an $(\ell_N(x))^{-N}$-soliton on X, and hence an $(\ell_N)^{-(\dim(Z)+2)}$-soliton on $Z = X \times \mathbb{P}_{\mathbb{C}}^{N-\dim(X)-2}$. The latter in turn defines a Calabi–Yau cone structure on $K_Y^{\times}$ by [3, Prop.2] (see also Remark 4.4 below).

Our proof only shows that a sufficiently large k_0 as above exists, but it does not yield a quantitative bound of k_0. Of course, Mabuchi–Nakagawa conjecture predicts that $k_0 = 0$. We also note that the k-dimensional complex projective space $\mathbb{P}_{\mathbb{C}}^k$ in Theorem 1.2 can be replaced with any other k-dimensional Kähler–Einstein Fano manifold.

We next extend the geometric setup by considering, more generally, Sasaki structures on the unitary bundle N in K_X, which are transversal Kähler–Ricci solitons in the sense of Futaki–Ono–Wang [31]. Equivalently, such Sasaki structures correspond to cone Kähler metrics on the canonical cone $K_X^{\times}$, polarized by vector fields in the Lie algebra of $\hat{\mathbb{T}} = \mathbb{T} \times \mathbb{S}^1$, which give rise to transversal Kähler–Ricci solitons. We show in Corollary 4.1 below that transversal KRS Sasaki structures on N correspond to v-solitons on X with $v = \ell_1^{-(n+2)} e^{\ell_2/\ell_1}$ for affine-linear functions $\ell_1 > 0$, ℓ_2 on P_X. The relative openness of $\mathcal{S}(X)$ and the fact that ℓ_1 and ℓ_2 are determined by the corresponding Sasaki–Reeb vector field (see Lemma 4.2 below) allow us to recover a result of D. Petrecca:

Theorem 1.3 ([53]) *Let X be a smooth Fano manifold with canonical cone $K_X^{\times}$ and maximal compact torus $\hat{\mathbb{T}} \subset \mathrm{Aut}(K_X^{\times})$. Suppose $K_X^{\times}$ admits a compatible cone Kähler metric, polarized by a Sasaki–Reeb vector field $\hat{\xi}_0 \in \mathrm{Lie}(\hat{\mathbb{T}})$, which is a transversal Kähler Ricci soliton. Then, for any Sasaki–Reeb vector field $\hat{\xi} \in \mathrm{Lie}(\hat{\mathbb{T}})$ which is sufficiently close to $\hat{\xi}_0$, $K_X^{\times}$ admits a compatible cone Kähler metric polarized by $\hat{\xi}$, which is transversal Kähler–Ricci soliton.*

We finally obtain some a priori constraints for the weight functions $v \in \mathcal{S}(X)$.

First, we give a uniform upper bound for the weight $v(x) = (\langle \xi, x \rangle + 1)^{-(n+2)} \exp\left(\frac{\langle \tau_\xi, x\rangle}{1+\langle \xi, x\rangle}\right)$ corresponding to a Sasaki transversal Kähler–Ricci soliton.

Theorem 1.4 *Let X be a smooth Fano manifold admitting a KRS with soliton vector field τ, invariant under a maximal torus $\mathbb{T} \subset \mathrm{Aut}(X)$. Then, on the canonical momentum polytope P_X,*

$$\langle \tau, x \rangle < n.$$

More generally, if $\hat{\xi} \in \mathrm{Lie}(\hat{\mathbb{T}})$ is a Sasaki–Reeb polarization of the cone $Y = K_X^{\times}$, normalized by the identity $\mathcal{L}_{-J\hat{\xi}}\Omega = \Omega$ where Ω is the Liouville holomorphic volume form of K_X, and if, moreover, $(Y, \hat{\xi})$ admits a compatible Kähler cone metric which is transversal KRS, then the projected vector field $\xi \in \mathrm{Lie}(\mathbb{T})$ and the corresponding soliton vector field $\tau_\xi \in \mathrm{Lie}(\mathbb{T})$ satisfy on P_X

$$\langle \tau_\xi, x \rangle < n\left(\langle \xi, x \rangle + 1\right).$$

Second, we derive a weighted version of Fujita'a inequality [28]:

Theorem 1.5 *Let X be a smooth Fano manifold admitting a v-soliton ω. Suppose without loss of generality that v is normalized so that*

$$\int_X v(\mu_\omega)\omega^{[n]} = \int_X \omega^{[n]}.$$

Then the first Chern number of X satisfies

$$c_1^n(X) \leq \left(\frac{n+1}{\inf_{\mathrm{P}_X} v}\right)^n.$$

The above inequality was established in the Kähler–Einstein case (i.e. with $v \equiv 1$) in a series of works by Berman-Berndtsson [7, 8] under some additional conditions, and by Fujita [28] in general. The latter proof uses the resolution of the YTD conjecture and a computation of the Donaldson–Futaki invariant of a certain family of test configurations. Our approach here is to adapt these computations to the weighted soliton case, using the (weighted) YTD correspondence and the notions of weighted beta invariant and weighted volume from Han-Li [36, 37]. Notice that Theorem 1.5 yields an a priori constraint for $v \in \mathcal{S}(X)$ on a given Fano manifold X.

The paper is organized as follows. In Sect. 2, we review the theory of Kähler-Ricci solitons and their generalization, the v-solitons. In Sect. 3, we establish Theorem 1.1 as an application of the results in [36]. In Sect. 4, we introduce the point of view of Sasaki geometry and the corresponding Kähler cones. We recall here the definition of transversal Kähler-Ricci soliton on the canonical cone $K_X^\times$ of a smooth Fano variety, and show in Corollary 4.1 that these structures give rise to special v-solitons on X. With this observation, we apply the openness Theorem 1.1 to derive the proofs of Theorems 1.2 and 1.3. In Sect. 5, we prove Theorem 1.4 and show how it can be seen as a variant of the so-called *Lichnerowicz obstruction* [33] which is a necessary condition for the existence of Calabi–Yau cone metrics on $K_X^\times$. The final Sect. 5, we prove Theorem 1.5. In the Appendix A, we recast in the setup of v-solitons studied in this paper a (weaker) openness result à la LeBrun–Simanca from [41], which gives an alternative tool for obtaining our main geometric applications, Theorems 1.2 and 1.3 above. In the final Appendix B, we gather some well-known curvature identities for gradient Ricci solitons and use them to observe, by a simple application of the maximum principle, that the transversal scalar curvature of a compact Sasaki transversal Kähler-Ricci soliton is positive. This is a key ingredient for our proof of Theorem 1.4.

2 Preliminaries

2.1 *Fano Manifolds: Notation and Normalization*

In what follows, X will denote a smooth compact complex manifold of complex dimension n, for which the anti-canonical bundle K_X^{-1} is ample. Such an X is called a *smooth Fano variety*. The Fano condition implies that X is projective, and that the de Rham class $\alpha = 2\pi c_1(X) = 2\pi c_1(K_X^{-1})$ contains Kähler metrics.

Any Kähler metric $\omega \in \alpha$ is de Rham cohomologous with the corresponding Ricci form $\mathrm{Ric}(\omega) \in 2\pi c_1(X)$, and thus we can write in this case

$$\mathrm{Ric}(\omega) - \omega = \frac{1}{2} dd^c h_\omega,$$

for a smooth function h_ω which (by the maximum principle) is unique up to an additive constant. Such a function will be referred to as a *Ricci potential* of ω; we can further fix the additive constant by requiring that

$$\int_X e^{\mathring{h}_\omega} \omega^{[n]} = \int_X \omega^{[n]} =: \boldsymbol{vol}(X), \tag{2.1}$$

where $\omega^{[n]} := \omega^n/n!$ stands for the Riemannian volume form of the Kähler metric ω. We shall then refer to this uniquely defined Ricci potential $\mathring{h}_\omega$ as the *normalized Ricci potential* of ω. In these terms, the *Kähler–Einstein* condition

$$\mathrm{Ric}(\omega) = \omega \tag{2.2}$$

is equivalent to $\mathring{h}_\omega = 0$.

We shall next fix once for all a maximal compact real torus $\mathbb{T}$ inside the connected component of identity $\mathrm{Aut}_\circ(X)$ of the group of complex automorphisms of X. The corresponding complex torus will be denoted by $\mathbb{T}_\mathbb{C}$. There is a canonical lift (still denoted by $\mathbb{T}$) of the action of $\mathbb{T}$ on X to an action on the canonical bundle K_X. The latter bundle has a further $\mathbb{S}^1$-extension of the lifted $\mathbb{T}$ action, given by fibre-wise multiplications with complex numbers $e^{i\theta} \in \mathbb{S}^1$. We denote by $\hat{\mathbb{T}} = \mathbb{T} \times \mathbb{S}^1$ the resulting $(\dim(\mathbb{T}) + 1)$-dimensional torus acting on K_X. We shall respectively denote by $\mathfrak{t}$ and $\hat{\mathfrak{t}}$ the Lie algebras of $\mathbb{T}$ and $\hat{\mathbb{T}}$.

We consider the space $\mathcal{K}_\alpha^\mathbb{T}(X)$ of $\mathbb{T}$-invariant Kähler metrics ω on X, belonging to α; by a standard averaging argument, $\mathcal{K}_\alpha^\mathbb{T}(X) \neq \emptyset$. Introducing a base-point $\omega_0 \in \mathcal{K}_\alpha^\mathbb{T}(X)$, we will identify $\mathcal{K}_\alpha^\mathbb{T}(X)$ with the Fréchet space $\mathcal{H}_{\omega_0}^\mathbb{T}(X)/\mathbb{R}$, where

$$\mathcal{H}_{\omega_0}^\mathbb{T}(X) := \left\{\varphi \in C^\infty(X)^\mathbb{T} \mid \omega_\varphi := \omega_0 + dd^c\varphi > 0\right\}$$

is the space of smooth $\mathbb{T}$-invariant Kähler potentials with respect to ω_0.

For each $\omega \in \mathcal{K}_\alpha^{\mathbb{T}}(X)$, we let H_ω denote the Hermitian metric on K_X whose Chern curvature is $R^{H_\omega} = -i\omega$, and by ∇^{H_ω} the H_ω-Chern connection on K_X. A basic fact in the theory is that any lift of the $\mathbb{T}$ action on K_X gives rise to a $\mathbb{T}$-momentum map μ_ω for $(X, \mathbb{T}, \omega)$, defined as follows: for any $\xi \in \mathfrak{t}$ and any smooth section $s \in C^\infty(X, -K_X)$,

$$\mathcal{L}_\xi s = \nabla_\xi^{H_\omega} s - i\mu_\omega^\xi s, \tag{2.3}$$

where μ_ω^ξ satisfies

$$\omega(\xi, \cdot) = -d\mu_\omega^\xi$$

see [32, Proposition 8.7.2]. Therefore, the canonical lift of $\mathbb{T}$ to $-K_X$ gives rise to a *canonically normalized* momentum map $\mu_\omega : X \to \mathfrak{t}^*$ whose image P_X is a compact convex polytope [4, 34]; one can further show (see e.g. [9, 41]) that $\mu_\omega(\mathrm{P}_X)$ is independent of the choice of $\omega \in \mathcal{K}_\alpha^{\mathbb{T}}(X)$. In this paper, we shall refer to P_X as the *canonical polytope* of $(X, \mathbb{T})$.

Remark 2.1 In general, a $\mathbb{T}$-momentum map $\mu_\omega : X \to \mathfrak{t}^*$ is defined only up to a translation with an element of $\mathfrak{t}^*$; the fact that in the Fano case there is a canonical normalization for μ_ω follows from the existence of a canonical lift of the $\mathbb{T}$-action on X to K_X. An alternative way to define the canonical normalization for μ_ω (see e.g. [3, 57]) is to require that for any $\zeta \in \mathfrak{t}$, the function $\mu_\omega^\zeta := \langle \mu_\omega, \zeta \rangle$ satisfies

$$\int_X \mu_\omega^\zeta e^{h_\omega} \omega^{[n]} = 0, \tag{2.4}$$

where h_ω is any Ricci potential of ω.

Once we have suitably normalized P_X, we can define the *Duistermaat-Heckman* measure $d\mu_{\mathrm{DH}}$ on P_X as the push-forward via μ_ω of the Riemannian measure of (X, ω): for any continuous function f on P_X, we let

$$\int_{\mathrm{P}_X} f(x) d\mu_{\mathrm{DM}} := \int_X f(\mu_\omega)\omega^{[n]}. \tag{2.5}$$

The fact that the LHS is independent of the choice of $\omega \in \mathcal{K}_\alpha^{\mathbb{T}}(X)$ follows for instance from the $\mathbb{T}$-equivariant Moser lemma (see e.g. [30]).

2.2 *Kähler–Ricci Solitons*

Following [57], a *Kähler Ricci soliton* (KRS for short) is a Kähler metric $\omega \in 2\pi c_1(X)$ which satisfies

$$\mathrm{Ric}(\omega) - \omega = -\frac{1}{2}\mathcal{L}_{J\tau}\omega, \tag{2.6}$$

where τ is a Killing vector field for the Kähler structure ω. In the case $\tau = 0$, (2.6) reduces to the Kähler–Einstein condition (2.2). Tian–Zhu [57] have extended the Matsushima's theorem [51] to the case of a KRS, which in turn yields that any Kähler metric satisfying (2.6) must be invariant by the action of a maximal torus in $\mathrm{Aut}_\circ(X)$, containing the flow of τ. Up to a pull-back by an element of $\mathrm{Aut}_0(X)$, we can and will assume that a KRS on X belongs to $\mathcal{K}_\alpha^{\mathbb{T}}(X)$ and $\tau \in \mathfrak{t}$. Thus, similarly to the Kähler–Einstein case, the KRS condition can be rewritten as

$$h_\omega = \mu_\omega^\tau \tag{2.7}$$

or, equivalently,

$$\mathrm{Ric}(\omega) - \omega = \frac{1}{2} dd^c \mu_\omega^\tau, \qquad \tau \in \mathfrak{t}. \tag{2.8}$$

By Remark 2.1, (2.7) and (2.5), if X admits a KRS in $\mathcal{K}_\alpha^{\mathbb{T}}(X)$, then for any $\zeta \in \mathfrak{t}$, we have

$$\int_{\mathrm{P}_X} \langle \zeta, x \rangle e^{\langle \tau, x \rangle} d\mu_{\mathrm{DH}} = 0.$$

The above condition means that τ is a critical point of the function $F : \mathfrak{t} \to \mathbb{R}$:

$$F(\zeta) := \int_{\mathrm{P}_X} e^{\langle \zeta, x \rangle} d\mu_{\mathrm{DH}}.$$

Tian–Zhu [57] further show that F admits a unique critical point, independent of the existence of a KRS on X. We shall refer to this $\tau \in \mathfrak{t}$ as the *KRS vector field* of $(X, \mathbb{T})$ and to the positive smooth function $v(x) := e^{\langle \tau, x \rangle}$ on $\mathfrak{t}^*$ as the *KRS weight function.*

2.3 v-Solitons

The notion of KRS extends to the following more general geometric situation, studied by Berman–Witt Nyström in [9] and, more recently, by Han–Li in [36]. We follow the notation of [3, Sect.2].

Definition 2.1 (v-Soliton) In the setup as above, let $v(x)$ be a given positive function defined on P_X. A Kähler metric $\omega \in \mathcal{K}_\alpha^{\mathbb{T}}(X)$ is called a *v-soliton* if it satisfies

$$\mathrm{Ric}(\omega) - \omega = \frac{1}{2} dd^c \log v(\mu_\omega). \tag{2.9}$$

Clearly, Kähler–Einstein metrics are 1-solitons whereas KRS are $v = e^{\langle \tau, x\rangle}$-solitons. Notice that if ω is a v-soliton it is also a λv-soliton for any $\lambda > 0$. To read off this ambiguity, we shall sometimes consider *normalized* weight functions $\mathring{v} := \frac{v}{\int_{\mathrm{P}_X} v d\mu_{\mathrm{DH}}}$, i.e.

$$\int_{\mathrm{P}_X} \mathring{v}(x) d\mu_{\mathrm{DH}} = 1. \tag{2.10}$$

We also notice that for any v-soliton, $h_\omega = \log(v(\mu_\omega))$, the linear function

$$\mathrm{Fut}_v : \mathfrak{t} \to \mathbb{R}, \qquad \mathrm{Fut}_v(\zeta) := \int_{\mathrm{P}_X} \langle \zeta, x\rangle v(x) d\mu_{\mathrm{DH}} \tag{2.11}$$

identically vanishes by Remark 2.1.

Definition 2.2 (v-Futaki Invariant) The linear function defined by (2.11) is called the v-*Futaki invariant* of $(X, \mathbb{T})$.

We next define a functional $\mathbf{I}_v$ on the space $\mathcal{H}^{\mathbb{T}}_{\omega_0}(X)$ of $\mathbb{T}$-invariant Kähler potentials (see [36, 41]:

$$d_\varphi \mathbf{I}_v(\dot\varphi) = \int_X v(\mu_{\omega_\varphi})\dot\varphi\omega_\varphi^{[n]}, \qquad \mathbf{I}_v(0) = 0.$$

Following [36], we introduce

Definition 2.3 (v-Ding Functional) The v-Ding functional is the map $\mathbf{D}_v : \mathcal{H}^{\mathbb{T}}_{\omega_0}(X) \to \mathbb{R}$ given by

$$\mathbf{D}_v(\varphi) := -\left(\frac{\mathbf{I}_v(\varphi)}{\boldsymbol{vol}_v(X)}\right) - \frac{1}{2}\log\left(\int_X e^{\mathring{h}_{\omega_0} - 2\varphi}\frac{\omega_0^{[n]}}{vol(X)}\right),$$

where we have set $vol_v(X) := \int_{\mathrm{P}_X} v(x) d\mu_{\mathrm{DH}} = \int_X v(\mu_\omega)\omega^{[n]}$ and $\mathring{h}_{\omega_0}$ stands for the normalized Ricci potential of the base point ω_0, see (2.1).

Notice that $\mathbf{D}_v$ does not change if we add a constant to φ, so it actually descends to a functional, denoted $\mathbf{D}_v(\omega_\varphi)$, on the space $\mathcal{H}^{\mathbb{T}}_{\omega_0}(X)/\mathbb{R} \cong \mathcal{K}^{\mathbb{T}}_\alpha(X)$.

It is not hard to see that the differential of $\mathbf{D}_v$ is given by

$$(d_{\omega_\varphi}\mathbf{D}_v)(\dot\varphi) = \int_X \dot\varphi\left(\frac{e^{\mathring{h}_{\omega_\varphi}}}{vol(X)} - \frac{v(\mu_{\omega_\varphi})}{vol_v(X)}\right)\omega_\varphi^{[n]}, \tag{2.12}$$

so that the critical points of $\mathbf{D}_v$ are precisely the Kähler metrics ω_φ for which

$$e^{\mathring{h}_{\omega_\varphi}} = vol(X)\mathring{v}(\mu_{\omega_\varphi})$$

i.e. the v-solitons.

Another consequence of the formula (2.12) is the following

Lemma 2.1 *The v-Ding functional is $\mathbb{T}_{\mathbb{C}}$-invariant, i.e. satisfies*

$$\mathbf{D}_v(\sigma^*(\omega)) = \mathbf{D}_v(\omega) \qquad \forall \sigma \in \mathbb{T}_{\mathbb{C}},$$

iff the v-Futaki invariant $\mathrm{Fut}_v \equiv 0$.

Proof $\mathbf{D}_v$ is clearly $\mathbb{T}$ invariant. For any $\zeta \in \mathfrak{t}$, we consider the flow of $-J\zeta$, say $\sigma_t \in \mathbb{T}_{\mathbb{C}}$, and take the derivative at $t = 0$ of $\mathbf{D}_v(\sigma_t^*(\omega))$. By (2.12)

$$\frac{d}{dt}_{|t=0} \mathbf{D}_v(\sigma_t^*(\omega)) = \int_X \mu_\omega^\zeta \left(\frac{e^{\mathring{h}_\omega}}{vol(X)} - \frac{v(\mu_\omega)}{vol_v(X)} \right) \omega^{[n]} = -\mathrm{Fut}_v(\zeta),$$

where we have used (2.4) for the canonically normalized momentum map μ_ω. The claim follows from the above by a standard argument. □

We end-up this section with stating one of the main results of [36], which gives an analytic criterion for the existence of a v-soliton on $(X, \mathbb{T})$ in terms of $\mathbf{D}_v$. To state it, we recall the definition [5] of the Aubin functional $\mathbf{J} : \mathcal{H}^{\mathbb{T}}_{\omega_0}(X) \to \mathbb{R}$:

$$\mathbf{J}(\varphi) := \int_X \varphi \omega_0^{[n]} - \mathbf{I}_1(\varphi),$$

which descends to $\mathcal{H}^{\mathbb{T}}(X)/\mathbb{R}$, and has the property that $\mathbf{J}(\omega_\varphi) \geq 0$ with $\mathbf{J}(\omega_\phi) = 0$ iff $\omega_\varphi = \omega_0$.

Definition 2.4 We say that $\mathbf{D}_v$ is coercive relative to $\mathbb{T}_{\mathbb{C}}$ if it is $\mathbb{T}_{\mathbb{C}}$-invariant and there are positive constants Λ, C such that

$$\mathbf{D}_v(\omega) \geq \Lambda \inf_{\sigma \in \mathbb{T}_{\mathbb{C}}} \mathbf{J}(\sigma^*(\omega)) - C. \tag{2.13}$$

The constant $\Lambda > 0$ is called the slope of coercivity of $\mathbf{D}_v$.

Theorem 2.1 ([36]) *Let X be Fano manifold, $\mathbb{T} \subset \mathrm{Aut}_0(X)$ a maximal compact torus with canonical momentum polytope $\mathrm{P}_X \subset \mathfrak{t}^*$. Let $v > 0$ be a positive smooth function on P_X. Then X admits a $\mathbb{T}$-invariant v-soliton in $2\pi c_1(X)$ if and only if the v-weighted Ding functional $\mathbf{D}_v$ is coercive with respect to the complex torus $\mathbb{T}_{\mathbb{C}}$.*

Versions of the above theorem have been known for KRS by the works of Cao–Tian–Zhu (see [15, Theorems 0.1 and 0.2]) and Darvas–Rubinstein (see [22, Theorem 2.11]).

3 A Quantitative Openness Result: Proof of Theorem 1.1

We have the following elementary

Proposition 3.1 *Suppose v_1, v_2 are positive weights with respective normalizations $\mathring{v}_1$ and $\mathring{v}_2$ such that*

$$\inf_{\mathrm{P}_X}(\mathring{v}_1 - \mathring{v}_2) = -\lambda_0, \qquad \lambda_0 > 0.$$

Then, there exists a constant C_{v_1,v_2} such that for any $\varphi \in \mathcal{H}^{\mathbb{T}}_{\omega_0}(X)$

$$\big(\mathbf{D}_{v_1}(\omega_\varphi) - \mathbf{D}_{v_2}(\omega_\varphi)\big) \geq -\lambda_0 \mathbf{J}(\omega_\varphi) + C_{v_1,v_2}.$$

Proof Using that $\mathbf{D}_v = \mathbf{D}_{\mathring{v}}$, we have

$$\begin{aligned}\big(\mathbf{D}_{v_1}(\omega_\varphi) - \mathbf{D}_{v_2}(\omega_\varphi)\big) &= -\big(\mathbf{I}_{\mathring{v}_1}(\varphi) - \mathbf{I}_{\mathring{v}_2}(\varphi)\big)\\ &= \mathbf{J}_{\mathring{v}_1}(\varphi) - \mathbf{J}_{\mathring{v}_2}(\varphi) - \int_X \varphi \mathring{v}_1(\mu_{\omega_0})\omega_0^{[n]} + \int_X \varphi \mathring{v}_2(\mu_{\omega_0})\omega_0^{[n]},\end{aligned}$$

where $\mathbf{J}_v$ denotes the v-weighted Aubin–Mabuchi functional defined by

$$\mathbf{J}_v(\varphi) := \int_X \varphi\, v(\mu_{\omega_0})\omega_0^{[n]} - \mathbf{I}_v(\varphi).$$

It is easy to check from the above formula that $\mathbf{J}_v(\varphi) = \mathbf{J}_v(\omega_\varphi)$ is a functional defined on the space of $\mathbb{T}$-invariant Kähler metrics $\omega_\varphi \in \alpha$, which is *linear* in v. Another key property established in [36] is that if $v > 0$ on P, then

$$\mathbf{J}_v(\omega_\varphi) \geq 0.$$

The above is actually true, by continuity in v (see [3, Lemma 6.5]), even if we merely assume $v \geq 0$ on P. We thus have, by the assumption in the proposition,

$$\mathbf{J}_{\mathring{v}_1}(\varphi) - \mathbf{J}_{\mathring{v}_2}(\varphi) + \lambda_0 \mathbf{J}(\varphi) = \mathbf{J}_{\mathring{v}_1 - \mathring{v}_2 + \lambda_0}(\varphi) \geq 0,$$

which leads to the inequality

$$\big(\mathbf{D}_{v_1}(\omega_\varphi) - \mathbf{D}_{v_2}(\omega_\varphi)\big) \geq -\lambda_0 \mathbf{J}(\omega_\varphi) + \int_X \varphi \mathring{v}_2(\mu_{\omega_0})\omega_0^{[n]} - \int_X \varphi \mathring{v}_1(\mu_{\omega_0})\omega_0^{[n]}.$$

Proposition 3.1 then follows from Lemma 3.1 below. □

Lemma 3.1 ([36, Lemma 13]) *Let $\mathring{v}$ be a normalized positive weigh-function on P_X. Then, there exists a uniform positive constant $C_{\mathring{v}} > 0$, such that for any $\varphi \in \mathcal{H}^{\mathbb{T}}_{\omega_0}(X)$,*

$$\sup_X \varphi - C_{\mathring{v}} \leq \int_X \varphi\, \mathring{v}(\mu_{\omega_0})\omega_0^{[n]} \leq \sup_X \varphi.$$

Proof The RHS of the inequality is obvious. To obtain the LHS we observe that the inequality is invariant under translations of φ with a constant, so we can assume without loss that $\sup_X \varphi = 0$. In this case, we need to prove

$$\int_X \varphi \mathring{v}(\mu_{\omega_0})\omega_0^{[n]} \geq -C_{\mathring{v}}.$$

The above is established for instance in [22, (27)] for $\mathring{v} = \mathring{1} = 1/\mathrm{Vol}(X)$. For a general weight function $\mathring{v} > 0$, we let

$$\sup_{\mathrm{P}_X} \mathring{v} = \lambda_{\mathring{v}} > 0.$$

Using $\sup_X \varphi = 0$ we then have

$$\int_X \varphi\, \mathring{v}(\mu_{\omega_0})\omega_0^{[n]} \geq \lambda_{\mathring{v}} \int_X \varphi\omega_0^{[n]} \geq -C_{\mathring{1}}\lambda_{\mathring{v}} =: -C_{\mathring{v}}.$$

□

Applying Theorem 2.1 and the previous observation we get the following quantitative openness result.

Corollary 3.1 *Let $\Lambda_0 > 0$ be such that $\mathbf{D}_{v_0}$ is coercive with respect to $\mathbb{T}_{\mathbb{C}}$ of slope Λ_0. Then for all $v \in \mathcal{F}(X)$ such that*

$$0 \leq -\inf_{\mathrm{P}_X}(\mathring{v} - \mathring{v}_0) = \lambda_0 < \Lambda_0$$

$\mathbf{D}_v$ is coercive relative to $\mathbb{T}_{\mathbb{C}}$ of slope $(\Lambda - \lambda_0)$.

Proof of Theorem 1.1 Let $v_0 > 0$ be a weight such that $\mathbf{D}_{v_0}$ is coercive relative to $\mathbb{T}_{\mathbb{C}}$ of slope $\Lambda_0 > 0$. Suppose $v > 0$ is a weight such that

$$\left\| v - v_0 \right\|_{C^0(\mathrm{P}_X)} = \varepsilon.$$

For ε small enough, the above inequality yields

$$\inf_{\mathrm{P}_X}(\mathring{v} - \mathring{v}_0) = -\lambda_0, \qquad 0 < \lambda_0 < \Lambda_0.$$

We also assume that $\mathrm{Fut}_v \equiv 0$. By Lemma 2.1, we know that the latter condition is equivalent to $\mathbf{D}_v$ being $\mathbb{T}_{\mathbb{C}}$-invariant. We let $\varphi \in \mathcal{H}^{\mathbb{T}}_{\omega_0}(X)$ and denote by ω^*_φ a Kähler metric in the $\mathbb{T}_{\mathbb{C}}$-orbit of ω_φ, such that

$$\inf_{\sigma \in \mathbb{T}_{\mathbb{C}}} \mathbf{J}(\sigma^* \omega_\varphi) = \mathbf{J}(\omega^*_\varphi).$$

Such a metric exists by [36, Lemma 29]. By the argument in the proof of Lemma 3.1 above and the $\mathbb{T}_{\mathbb{C}}$-invariance of $\mathbf{D}_v$ and $\mathbf{D}_{v_0}$, we have

$$\begin{aligned}\left(\mathbf{D}_v - \mathbf{D}_{v_0}\right)(\omega_\varphi) &= \left(\mathbf{D}_v - \mathbf{D}_{v_0}\right)(\omega^*_\varphi)\\ &\geq -\lambda_0 \mathbf{J}(\omega^*_\varphi) - C_v = -\lambda_0 \inf_{\sigma \in \mathbb{T}_{\mathbb{C}}} \mathbf{J}(\sigma^* \omega_\varphi) - C_v.\end{aligned}$$

As $\mathbf{D}_{v_0}$ is Λ_0-coercive relative to $\mathbb{T}_{\mathbb{C}}$ and $\Lambda_0 > \lambda_0$, we conclude that $\mathbf{D}_v$ is also coercive relative to $\mathbb{T}_{\mathbb{C}}$.

The convexity of $\mathcal{D}(X)$ follows from the fact that the subspace of normalized weight functions is linearly convex. □

4 Sasaki Geometry and Transversal Kähler–Ricci KRS

In this section, we recall a notion of special Sasaki geometry introduced in [31]. We use the point of view of [2], which will allow us to recast the possibly irregular transversal Kähler geometry of a Sasaki manifold in terms of the Kähler geometry of a given regular or quasi-regular quotient.

4.1 Sasaki Structures

We consider the following general set up: $(N, \mathcal{D}_0, J_0)$ is a compact $(2n+1)$-dimensional strictly pseudo-convex CR manifold invariant under the action of a compact torus $\hat{\mathbb{T}}$ whose Lie algebra is denoted by $\hat{\mathfrak{t}}$. We say that $\hat{\xi} \in \hat{\mathfrak{t}}$ is a *Sasaki–Reeb vector field* if $\hat{\xi}$ is transversal to $\mathcal{D}_0$ and the corresponding contact 1-form $\eta_0^{\hat{\xi}}$, which vanishes on $\mathcal{D}$ and is equal to 1 when evaluated at $\hat{\xi}$, defines a transversal Kähler form $d\eta_0^{\hat{\xi}}$ on $(\mathcal{D}_0, J_0)$, i.e. $(d\eta_0^{\hat{\xi}})_{|\mathcal{D}_0} > 0$. We denote by $\hat{\mathfrak{t}}_+(N) \subset \hat{\mathfrak{t}}$ the *Sasaki–Reeb* cone of Sasaki–Reeb vector fields, and assume that $\hat{\mathfrak{t}}_+(N)$ is non-empty. For any $\hat{\xi} \in \hat{\mathfrak{t}}_+(N)$, the data $(\hat{\xi}, \eta_0^{\hat{\xi}}, \mathcal{D}_0, J_0)$ is referred to as a *Sasaki structure* on N.

Example 4.1 (Regular Sasaki Structures) A basic example of Sasaki manifolds are the so-called *regular Sasaki* structures, described as follows. Let X be a smooth compact complex manifold X polarized by a line bundle L. Consider the unitary

$\mathbb{S}^1$-bundle $N_{\omega_0} \subset L^{-1}$ with respect to the Hermitian metric H_{ω_0} on L^{-1} whose curvature is $-i\omega_0$. Here, $\omega_0 \in \mathcal{K}^{\mathbb{T}}_{2\pi c_1(L)}(X)$, where $\mathbb{T}$ is a maximal torus in the group of reduced automorphisms of X, corresponding to a maximal torus $\hat{\mathbb{T}} \subset \mathrm{Aut}(X, L)$. Clearly, $\hat{\mathbb{T}}$ acts on N_{ω_0} preserving the induced CR structures $(\mathcal{D}_0, J_0)$. If $\hat{\chi} \in \hat{\mathfrak{t}}$ denotes the generator of the $\mathbb{S}^1$-action on the fibres, then $\hat{\chi} \in \hat{\mathfrak{t}}_+(N)$. The corresponding 1-form $\eta_0^{\hat{\chi}}$ is the unique connection 1-form on N_{ω_0} with curvature $d\eta_0^{\hat{\chi}} = \pi^*\omega_0$ whereas $\mathcal{D}_0$ becomes the horizontal distribution of $\eta_0^{\hat{\chi}}$. The transversal Kähler structure in this case is just the pull-back of ω_0 to $(\mathcal{D}_0, J_0)$. Note that $X = N_{\omega_0}/\mathbb{S}^1_{\hat{\chi}}$, which allows us to recover the Kähler structure ω_0 on X from the corresponding Sasaki structure $(\hat{\chi}, \eta_0^{\hat{\chi}}, \mathcal{D}_0, J_0)$ on N_{ω_0}. This correspondence also applies to periodic Sasaki-Reeb vector fields $\hat{\xi} \in \hat{\mathfrak{t}}_+(N)$, in which case $(N, \hat{\xi}, \mathcal{D}_0, J_0)$ is the (smooth) total space of an $\mathbb{S}^1$-orbibundle over the Kähler orbifold $(X := N/\mathbb{S}^1_{\hat{\xi}}, \omega_0)$. This situation is referred to in the literature as a *quasi-regular* Sasaki structure.

We now introduce the variation spaces of Sasaki structures: For $\hat{\xi} \in \hat{\mathfrak{t}}_+(N)$ on $(N, \mathcal{D}_0, J_0)$ and $\eta_0^{\hat{\xi}}$ the associated contact form, we let

$$\begin{aligned}&\Xi^{\hat{\mathbb{T}}}_{\hat{\xi},\eta_0^{\hat{\xi}},J^{\hat{\xi}}}(N)\\&:= \left\{\varphi \in C^\infty(N)^{\hat{\mathbb{T}}} \mid \eta_\varphi^{\hat{\xi}} := \eta_0^{\hat{\xi}} + d^c_{\hat{\xi}}\varphi \text{ satisfies } d\eta_\varphi^{\hat{\xi}} > 0 \text{ on } \mathcal{D}_\varphi := \ker(\eta_\varphi^{\hat{\xi}})\right\},\end{aligned}$$

where $J^{\hat{\xi}} \in \mathrm{End}(TN)$ extends $J_0 \in \mathrm{End}(\mathcal{D}_0)$ by letting $J^{\hat{\xi}}(\hat{\xi}) = 0$, $d^c_{\hat{\xi}}\varphi := -d\varphi \circ J^{\hat{\xi}}$, and the positivity of $d\eta_\varphi^{\hat{\xi}}$ on $\mathcal{D}_\varphi$ is defined with respect to $J^{\hat{\xi}}_{|\mathcal{D}_\varphi}$. We refer to [2, §2] for the basic properties of this space. In particular, for any $\varphi \in \Xi^{\hat{\mathbb{T}}}_{\hat{\xi},\eta_0^{\hat{\xi}},J^{\hat{\xi}}}(N)$, $(\hat{\xi}, \eta_\varphi^{\hat{\xi}}, \mathcal{D}_\varphi, J^{\hat{\xi}}_{|\mathcal{D}_\varphi})$ is a Sasaki structure and the space $\Xi^{\hat{\mathbb{T}}}_{\hat{\xi},\eta_0^{\hat{\xi}},J^{\hat{\xi}}}(N)$ is called a *slice of* $(\hat{\xi}, J^{\hat{\xi}})$*-compatible* $\hat{\mathbb{T}}$*-invariant Sasaki potentials*. We shall use through this paper the following identification established in [2, 39].

Proposition 4.1 *For any Sasaki–Reeb vector fields* $\hat{\chi}, \hat{\xi} \in \hat{\mathfrak{t}}_+(N)$, *there exists a natural bijection*

$$\Theta_{\hat{\chi},\hat{\xi}} : \Xi^{\hat{\mathbb{T}}}_{\hat{\xi},\eta_0^{\hat{\xi}},J^{\hat{\xi}}}(N) \cong \Xi^{\hat{\mathbb{T}}}_{\hat{\chi},\eta_0^{\hat{\chi}},J^{\hat{\chi}}}(N).$$

Remark 4.1 For a (quasi) regular Sasaki structure $(N, \hat{\chi}, \eta_0^{\hat{\chi}}, \mathcal{D}_0, J_0)$ (see Example 4.1), we have

$$\Xi^{\hat{\mathbb{T}}}_{\hat{\chi},\eta_0^{\hat{\chi}},J^{\hat{\chi}}}(N) \equiv \mathcal{H}^{\mathbb{T}}_{\omega_0}(X),$$

which, together with Proposition 4.1, allows one to study $(\hat{\xi}, J^{\hat{\xi}})$-compatible $\hat{\mathbb{T}}$-invariant Sasaki structures on N via the induced Kähler geometry of $(M, \mathbb{T}, \alpha = 2\pi c_1(L))$.

4.2 *Kähler Cones vs Sasaki Structures*

A smooth complex cone [2, 20] $(Y, J, \hat{\xi})$ is a non-compact $(n+1)$-dimensional complex manifold endowed with a free holomorphic $\mathbb{R}^+$-action (sometimes denoted by $\mathbb{R}^+_{-J\hat{\xi}}$) generated by the flow of a real holomorphic vector field $-J\hat{\xi}$. We shall further assume that $N = Y/\mathbb{R}^+_{-J\hat{\xi}}$ is a compact $(2n+1)$-dimensional manifold N. A typical example is $Y = L^\times$, where L is a holomorphic line bundle over X and $L^\times$ denotes the total space of L with its zero section removed. In this case, we can take $\hat{\chi}$ be the generator of the natural fiberwise $\mathbb{S}^1$-action on L, so that $L^\times/\mathbb{R}^+_{-J\hat{\chi}} = N$ as in Example 4.1.

We shall further fix a maximal compact torus $\hat{\mathbb{T}} \subset \mathrm{Aut}(Y)$ and consider free $\mathbb{R}^+_{-J\hat{\xi}}$-actions for $\hat{\xi} \in \hat{\mathfrak{t}}$, where $\hat{\mathfrak{t}} = \mathrm{Lie}(\hat{\mathbb{T}})$. For any such $\hat{\xi} \in \hat{\mathfrak{t}}$, we consider the space of functions

$$\mathcal{R}^{\hat{\mathbb{T}}}_{\hat{\xi}}(Y) := \left\{ r \in C^\infty(Y, \mathbb{R}_{>0})^{\hat{\mathbb{T}}} \mid \mathcal{L}_{-J\hat{\xi}} r = r,\ \hat{\omega}_r := \frac{1}{4} dd^c r^2 > 0 \right\},$$

and assume that $\mathcal{R}^{\hat{\mathbb{T}}}_{\hat{\xi}}(Y) \neq \emptyset$ for at least some $\hat{\xi} \in \hat{\mathfrak{t}}$; clearly, for such a $\hat{\xi}$, any $r \in \mathcal{R}^{\hat{\mathbb{T}}}_{\hat{\xi}}(Y)$ gives rise to a Kähler cone metric $\hat{\omega}_r$, i.e. satisfying $\mathcal{L}_{-J\hat{\xi}} \hat{\omega}_r = 2\hat{\omega}_r$ with respect to the $\mathbb{R}^+$-action generated by $-J\hat{\xi}$. We shall refer to the pair $(Y, \hat{\xi})$ as a *polarized Kähler cone* and to $\hat{\xi}$ as a *polarization*. We denote by $\hat{\mathfrak{t}}_+(Y) \subset \hat{\mathfrak{t}}$ the affine cone of polarizations of Y.

Example 4.2 (Regular Kähler Cones) An important example is the case of *regular Kähler cones*. In this case, $Y = (L^{-1})^\times$ for a polarization L of X, and $\hat{\chi}$ is the generator of the $\mathbb{S}^1$ fiberwise action on L^{-1}. In this case, the fibre-norm $r_\omega = ||\cdot||_{H_\omega}$ with respect the Hermitian metric H_ω on L^{-1} with curvature $-i\omega$ where $\omega \in \mathcal{K}_{2\pi c_1(L)}(X)$ gives rise to a Kähler cone structure on Y polarized by $\hat{\chi}$, compare with Example 4.1. Furthermore, if we fix a maximal torus $\hat{\mathbb{T}} \subset \mathrm{Aut}_0(X, L)$, which projects to a maximal torus $\mathbb{T} \subset \mathrm{Aut}_{\mathrm{red}}(X)$ in the reduced automorphisms group of X, then $r_\omega \in \mathcal{R}^{\hat{\mathbb{T}}}_{\hat{\chi}}(Y)$ provided that $\omega \in \mathcal{K}^{\mathbb{T}}_{2\pi c_1(L)}(X)$.

Any $r_0 \in \mathcal{R}^{\hat{\mathbb{T}}}_{\hat{\xi}}(Y)$ gives rise to a Sasaki structure on $N = Y/\mathbb{R}^+_{-J\hat{\xi}}$ as follows: The torus action of $\hat{\mathbb{T}}$ descends to N, and $\hat{\xi} \in \hat{\mathfrak{t}}$ defines a vector field on N. The 1-form $\eta_0^{\hat{\xi}} := d^c \log r_0$ on Y is $-J\hat{\xi}$-basic, so we can view it as a 1-form on N; furthermore, the distribution $\mathcal{D}_0 := \ker(\eta_0^{\hat{\xi}}) \subset TN$ inherits a CR structure J_0, defined through the almost complex structure on $\langle \hat{\xi}, J\hat{\xi} \rangle^{\perp_{\omega_{r_0}}} \subset TY$. The data $(\hat{\xi}, \eta_0^{\hat{\xi}}, \mathcal{D}_0, J_0, \hat{\mathbb{T}})$ is a Sasaki structure on N, as defined in the previous subsection. Furthermore, for any other $r \in \mathcal{R}_{\hat{\xi}}^{\hat{\mathbb{T}}}(Y)$, the smooth function $\varphi := \log(r/r_0)$ is $(-J\hat{\xi})$-invariant and defines an element of $\Xi^{\hat{\mathbb{T}}}_{\hat{\xi}, \eta_0^{\hat{\xi}}, J^{\hat{\xi}}}(N)$. It turns out that

$$\hat{\mathfrak{t}}_+(Y) = \hat{\mathfrak{t}}_+(N) =: \hat{\mathfrak{t}}_+, \qquad \mathcal{R}_{\hat{\xi}}^{\hat{\mathbb{T}}}(Y) \cong \Xi^{\hat{\mathbb{T}}}_{\hat{\xi}, \eta_0^{\hat{\xi}}, J^{\hat{\xi}}}(N), \qquad \forall \hat{\xi} \in \hat{\mathfrak{t}}_+.$$

These identifications are discussed at length for example in [2, §1, §2]. In the special case when $(Y, J, \hat{\chi}) = (L^{-1})^{\times}$ for a polarization L of X, our discussion here and Proposition 4.1 provide the following sequence of identifications:

$$\mathcal{R}_{\hat{\xi}}^{\hat{\mathbb{T}}}(Y) \cong \Xi^{\hat{\mathbb{T}}}_{\hat{\xi}, \eta_0^{\hat{\xi}}, J^{\hat{\xi}}}(N) \cong \Xi^{\hat{\mathbb{T}}}_{\hat{\chi}, \eta_0^{\hat{\chi}}, J^{\hat{\chi}}}(N) \equiv \mathcal{H}^{\mathbb{T}}_{\omega_0}(X), \qquad \forall \hat{\xi} \in \hat{\mathfrak{t}}_+. \tag{4.1}$$

4.3 Kähler Cones with Trivial Canonical Bundle

We now consider a compact Sasaki manifold $(N, \hat{\xi}, \eta_0^{\hat{\xi}}, \mathcal{D}_0, J_0, \hat{\mathbb{T}})$ which is *transversally Fano*, that is it satisfies the following additional condition: the $\hat{\xi}$-basic first Chern class $c_1^{\hat{\xi}}(\mathcal{D}_0, J_0)$ verifies

$$c_1^{\hat{\xi}}(\mathcal{D}_0, J_0) = \lambda_{\hat{\xi}} [d\eta_0^{\hat{\xi}}]_{B, \hat{\xi}}, \tag{4.2}$$

for some $\lambda_{\hat{\xi}} > 0$. This is equivalent to the conditions that $c_1(\mathcal{D}_0, J_0) = 0$ and the $\hat{\xi}$-basic first Chern class of $(\mathcal{D}_0, J_0)$ is positive (cf. [31, Prop.4.3]). It follows that (4.2) holds true for any other Sasaki-Reeb field $\hat{\chi} \in \hat{\mathfrak{t}}_+$, but the corresponding positive constants $\lambda_{\hat{\xi}}$ and $\lambda_{\hat{\chi}}$ are in general different.[1]

By the work of Martelli–Sparks–Yau [50] (see also [58, Proposition 2.5]), in the case when $\lambda_{\hat{\xi}} > 0$ and $\hat{\xi}$ is quasi-regular, the condition (4.2) is equivalent to the existence of a (unique up to a multiplicative constant) holomorphic volume form Ω on the corresponding polarized Kähler cone $(Y, \hat{\xi}, \hat{\mathbb{T}})$, such that

$$\mathcal{L}_{-J\hat{\xi}} \Omega = \langle \gamma_Y, \hat{\xi} \rangle \Omega, \tag{4.3}$$

where the multiplicative constant $\langle \gamma_Y, \hat{\xi} \rangle \in \mathbb{R}_{>0}$ is canonically determined by $(Y, \hat{\xi})$. The holomorphic volume form Ω on Y satisfying (4.3) depends a priori on

[1] If (4.2) holds with $\lambda_{\hat{\xi}} \leq 0$ then the Sasaki-Reeb cone $\hat{\mathfrak{t}}_+$ is one dimensional.

the quasi-regular Sasaki–Reeb vector field $\hat{\xi}$ we have chosen, but the uniqueness yields that it can be chosen so that (4.3) holds for any other quasi-regular element of $\hat{\Sigma}^+$. We can further extend the validity of (4.3) to irregular elements of $\hat{\Sigma}_+$ by density.

As the left hand side of (4.3) is linear with respect to $\hat{\xi} \in \hat{\mathfrak{t}}$, this defines $\gamma_Y \in \hat{\mathfrak{t}}^*$ by[2]

$$\langle \gamma_Y, \hat{\chi} \rangle := \frac{\mathcal{L}_{-J\hat{\chi}}\Omega}{\Omega}. \tag{4.4}$$

Definition 4.1 Let $(Y, \hat{\xi}, \hat{\mathbb{T}})$ be a polarized Kähler cone which admits a global holomorphic volume form Ω satisfying (4.3). We consider the sub-algebra

$$\mathfrak{t} := \left\{\zeta \in \hat{\mathfrak{t}} \,|\, \langle \gamma_Y, \zeta \rangle = 0\right\},$$

and the convex open polytope of *normalized* Reeb vector fields

$$\hat{\Sigma}_+ := \left\{\hat{\xi} \in \hat{\mathfrak{t}}_+ \mid \langle \gamma_Y, \hat{\xi} \rangle = 1.\right\}.$$

Both $\lambda_{\hat{\xi}}$ from (4.2) and $\langle \gamma_Y, \hat{\xi} \rangle$ are linear functions on $\hat{\mathfrak{t}}_+$ along the ray $\mathbb{R}_{>0} \cdot \hat{\xi}$ and agree at one point (the point at which we have $\lambda_{\hat{\xi}} = n + 1$, see [58, Proposition 2.5]). Thus, we get

Lemma 4.1 *For any $\hat{\xi} \in \hat{\mathfrak{t}}_+$, we have $\lambda_{\hat{\xi}} = \langle \gamma_Y, \hat{\xi} \rangle$. In particular,*

$$\hat{\Sigma}_+ = \left\{\hat{\xi} \in \hat{\mathfrak{t}}_+ \mid (4.2) \textit{ holds with } \lambda_{\hat{\xi}} = 1.\right\}.$$

Remark 4.2 One can derive a more general relationship between the constants $\lambda_{\hat{\xi}} \in \mathbb{R}_{>0}$ and $\gamma_Y \in \hat{\mathfrak{t}}^*$. To this end, let $\eta := \eta_0^{\hat{\xi}}$ be the contact form of $\hat{\xi}$ on $(N, \mathcal{D}_0, J)$ and

$$\mu^\eta : N \to \hat{\mathfrak{t}}^*, \qquad \langle \mu^\eta, a \rangle := \eta(a)$$

the induced contact $\hat{\mathbb{T}}$-momentum map on (N, η). Denote by $\omega = d\eta$ the transversal Kähler form of $(N, \mathcal{D}, J, \hat{\xi})$ and by ρ_ω the corresponding transversal Ricci form. The equality (4.2) tells us that there exists a unique up to additive constant $\hat{\mathbb{T}}$-invariant function h on N such that

$$\rho_\omega - \lambda_{\hat{\xi}}\omega = \frac{1}{2} dd^c_{\hat{\xi}} h, \tag{4.5}$$

[2] In [44, §2.2] and [46], the vector $\gamma_Y \in \hat{\mathfrak{t}}^*$ appears as the restriction of the log discrepancy seen as a map on $\hat{\mathfrak{t}}_+$, itself identified with a subspace of valuations.

where d^c_ξ is the ξ-basic d^c-operator. Contracting (4.5) with an element of $\mathfrak{t}^*$ then yields

$$\gamma_Y = \lambda_\xi \mu^\eta - \frac{1}{2}\Delta_\omega \mu^\eta + \frac{1}{2} d^c_\xi h \tag{4.6}$$

where Δ_ω is the ξ-basic Laplace operator associated to the transversal Kähler structure ω. (Equivalently, Δ_ω coincides with the Riemannian Laplace operator of (N, g) acting on $\hat{\mathbb{T}}$-invariant smooth functions, where g is the Sasaki Riemannian metric $g = \eta^2 + d\eta(\cdot, J_0\cdot)$.)

Example 4.3 (The Canonical Cone of a Fano Manifold) The condition (4.2) is clearly fulfilled on the regular Sasaki manifold $(N, \hat{\chi}, \eta_0^{\hat{\chi}}, \mathcal{D}_0, J_0, \hat{\mathbb{T}})$ of Example 4.1, provided that X is a Fano manifold polarized by its anti-canonical bundle, i.e. $L = K_X^{-1}$. In this case $\lambda_{\hat{\chi}} = 1$. Notice that the corresponding Kähler cone is then the canonical cone $Y = K_X^\times$ whereas Ω is the Liouville holomorphic volume form on K_X. Furthermore, $\mathfrak{t} \subset \hat{\mathfrak{t}}$ correspond to the canonical lift of $\mathrm{Aut}(X)$ to $\mathrm{Aut}(K_X)$ and $\hat{\Sigma}_+$ to the dual of the momentum cone via the identification $\mathfrak{t} \ni \xi \mapsto \langle \xi, \cdot \rangle + 1$.

4.4 *Transversal Kähler–Ricci Solitons: Definition and Normalization*

Definition 4.2 (See e.g. [31]) A (compact) Sasaki manifold $(N, \hat{\xi}, \eta^{\hat{\xi}}, \mathcal{D}, J, \hat{\mathbb{T}})$ is a transversal KRS if there exists a $\hat{\tau} \in \hat{\mathfrak{t}}$ such that

$$\rho_{\eta^{\hat{\xi}}} - \lambda_{\hat{\xi}} d\eta^{\hat{\xi}} = \frac{1}{2} dd^c_{\hat{\xi}}\, \eta^{\hat{\xi}}(\hat{\tau}), \tag{4.7}$$

where $\rho_{\eta^{\hat{\xi}}}$ is the transversal Ricci curvature of the Sasaki structure, and $\lambda_{\hat{\xi}} > 0$ is a positive constant.

Some immediate observations are in order.

(i) For any transversal KRS the condition (4.2) is satisfied and $\lambda_{\hat{\xi}} = \langle \gamma_Y, \hat{\xi} \rangle$ by Lemma 4.1.

(ii) If $(\hat{\xi}, \eta^{\hat{\xi}}, \mathcal{D}, J)$ is a transversal KRS for $\hat{\tau} \in \hat{\mathfrak{t}}$ and $\lambda_{\hat{\xi}} > 0$ it is also a transversal KRS for $\hat{\tau} + c\hat{\xi}$ and the same constant $\lambda_{\hat{\xi}} > 0$. Thus, there exists a *unique* such $\hat{\tau}$ satisfying the condition $\hat{\tau} \in \mathfrak{t}$ see Definition 4.1. We shall denote by $\tau_{\hat{\xi}}$ this normalized element.

(iii) If $(\hat{\xi}, \eta^{\hat{\xi}}, \mathcal{D}, J)$ is a transversal KRS with respect to $\tau_{\hat{\xi}} \in \mathfrak{t}$ and $\lambda_{\hat{\xi}} > 0$, then for any $\lambda > 0$ $(\lambda\hat{\xi}, \eta^{\lambda\hat{\xi}}, \mathcal{D}, J)$ is a transversal KRS satisfying $\tau_{\lambda\hat{\xi}} = \tau_{\hat{\xi}}$ and $\lambda_{\lambda\hat{\xi}} = \lambda\lambda_{\hat{\xi}}$. We can then assume $\hat{\xi} \in \hat{\Sigma}_+$ in which case $\lambda_{\hat{\xi}} = 1$.

(iv) The case when $(\hat{\xi}, \eta^{\hat{\xi}}, \mathcal{D}, J)$ is a transversal KRS with $\tau_{\hat{\xi}} = 0$ corresponds to a positive Sasaki–Einstein structure (up to scale of $\hat{\xi}$, see Remark 4.3 below) or, equivalently, to a Ricci-flat Kähler cone metric $\hat{\omega} = \frac{1}{4}dd^c r^2$ on $(Y, \hat{\xi})$.

Assumption 4.1 *By (i), (ii) and (iii) above, without loss of generality we may and we shall from now on consider transversal KRS* $(N, \hat{\xi}, \eta^{\hat{\xi}}, \mathcal{D}, J, \hat{\mathbb{T}})$ *such that* $\hat{\xi} \in \Sigma_+$ *and* $\tau_{\hat{\xi}} \in \mathfrak{t}$*, see Definition 4.1.*

Suppose $(\hat{\xi}, \eta, \mathcal{D}, J)$ is a transversal KRS with $\hat{\xi} \in \Sigma_+$ and $\tau \in \mathfrak{t}$. In [31], Futaki–Ono–Wang extended to the Sasaki setting the Tian–Zhou Futaki-type obstruction [57], by showing that τ satisfies

$$\int_N \eta(a)\, e^{\eta(\tau)} \eta \wedge (d\eta)^n = 0, \qquad \forall a \in \mathfrak{t}. \tag{4.8}$$

This follows from (4.6) where we take $h = \eta(\tau)$ and observe that

$$\frac{1}{2}\Delta_\omega \eta(a) - \mathcal{L}_{-J^{\hat{\xi}}a}\eta(\tau) = \frac{1}{2}\Delta_\omega \eta(a) - \mathcal{L}_{-J^{\hat{\xi}}\tau}\eta(a)$$

is self-dual with respect to the volume form $e^{\eta(\tau)}\eta \wedge (d\eta)^n$.

By the arguments of [31, 57] and [50], we have

Lemma 4.2 *Let* $(N, \hat{\chi}, \mathcal{D}_0, J_0, \hat{\mathbb{T}})$ *be a compact Sasaki manifold satisfying* (4.2). *Then*

- *For any* $\hat{\xi} \in \hat{\Sigma}_+$ *there exists a unique* $\tau_{\hat{\xi}} \in \mathfrak{t}$ *such that* (4.8) *holds.*
- *There exists a unique* $\hat{\xi} \in \hat{\Sigma}_+$ *such that* $\tau_{\hat{\xi}} = 0$.

Proof For any $\hat{\xi} \in \hat{\Sigma}_+$, pick $\eta = \eta_\varphi^{\hat{\xi}}$ for $\varphi \in \Xi(\hat{\xi}, \eta_0, J^{\hat{\xi}})^{\hat{\mathbb{T}}}$ the functional $V_{\hat{\xi}} : \mathfrak{t} \to \mathbb{R}$ defined by

$$V_{\hat{\xi}}(a) := \int_N e^{\eta(a)} \eta \wedge (d\eta)^n, \quad a \in \mathfrak{t}, \tag{4.9}$$

is convex and does not depend on the choice of η within $\Xi(\hat{\xi}, \eta_0^{\hat{\xi}}, J^{\hat{\xi}})^{\hat{\mathbb{T}}}$. Therefore, there exists a unique $\tau \in \mathfrak{t}$, the critical point of $V_{\hat{\xi}}$, satisfying (4.8).

For the second part, by [50], there exists a unique $\hat{\xi}_0 \in \Sigma_+$ such that the corresponding contact 1-form η_0 satisfies

$$\int_N \eta_0(a)\eta_0 \wedge (d\eta_0)^n = 0 \qquad \forall a \in \mathfrak{t}.$$

This in turn is equivalent to (4.8) for $\hat{\xi} \in \Sigma_+$ and $\tau_{\hat{\xi}} = 0$. □

Similarly, we can introduce transversal KRS Kähler cones.

Definition 4.3 We say that a polarized Kähler cone $(Y, J, \hat{\xi})$ admits a *compatible transversal KRS cone Kähler metric* $\hat{\omega} = \frac{1}{4} dd^c r^2$ with $r \in \mathcal{R}_{\hat{\xi}}(Y)$ if the associated Sasaki structure on $N = r^{-1}(1)$ (see Sect. 4.2) is a transversal KRS.

Remark 4.3 A polarized Kähler cone $(Y, J, \hat{\xi}, \hat{\omega})$ with $\hat{\omega} = \frac{1}{4} dd^c r^2$ for a radial potential function $r \in \mathcal{R}_{\hat{\xi}}(Y)$ is Kähler Ricci-flat if and only if the associated Sasaki manifold $(N, \hat{\xi}, \eta^{\hat{\xi}}, \mathcal{D}, J, \hat{\mathbb{T}})$ (see Sect. 4.2) is transversally Kähler–Einstein with transversal scalar curvature equal to $2n(n+1)$. To see this, let $\omega = d\eta^{\hat{\xi}}$ and ρ_ω denote the corresponding transversal Kähler and Ricci forms on the Sasaki manifold $N = r^{-1}(1)$. Then we have (see e.g. [50])

$$\rho_{\hat{\omega}} = \rho_\omega - (n+1)\omega. \tag{4.10}$$

4.5 Transversal KRS as v-Solitons

We start with a technical result obtained in the following general set up: $(N, \mathcal{D}, J)$ is a fixed strictly pseudo-convex CR manifold and $(\hat{\xi}, \hat{\chi})$ are commuting Sasaki–Reeb vector fields. We denote by $\omega_{\hat{\xi}} := d\eta^{\hat{\xi}}$ and $\omega_{\hat{\chi}} := d\eta^{\hat{\chi}}$ the corresponding transversal Kähler forms and by $g_{\hat{\xi}}$ and $g_{\hat{\chi}}$ the corresponding transversal Riemannian metrics. We further suppose that $(N, \mathcal{D}, J, \hat{\xi})$ (with $\hat{\xi} \in \hat{\Sigma}_+$) is transversally Fano, i.e. the transversal Ricci form $\rho_{\omega_{\hat{\xi}}}$ satisfies (see (4.5))

$$\rho_{\omega_{\hat{\xi}}} - \omega_{\hat{\xi}} = \frac{1}{2} dd^c_{\hat{\xi}} h,$$

where h is a $(\hat{\xi}, \hat{\chi})$-invariant smooth function.

Lemma 4.3 *Let $\lambda_{\hat{\chi}}$ be the constant introduced in Sect. 4.3. Then the transversal Ricci form $\rho_{\omega_{\hat{\chi}}}$ of the transversal Kähler structure $(g_{\hat{\chi}}, \omega_{\hat{\chi}})$ satisfies*

$$\rho_{\omega_{\hat{\chi}}} - \lambda_{\hat{\chi}} \omega_{\hat{\chi}} = \frac{1}{2} dd^c_{\hat{\chi}} h - \frac{(n+2)}{2} dd^c_{\hat{\chi}} \log f, \qquad f := \eta^{\hat{\chi}}(\hat{\xi}).$$

Proof Let $f := \eta^{\hat{\chi}}(\hat{\xi})$ be the positive Killing potential of $(g_{\hat{\chi}}, \omega_{\hat{\chi}})$. We use the following general relation between the transversal Ricci curvatures (see e.g. [40]):

$$\rho_{\omega_{\hat{\xi}}} - \rho_{\omega_{\hat{\chi}}} = \frac{(n+2)}{2} dd^c_{\hat{\chi}} \log f - \frac{1}{2} d\left(\left(\frac{\Delta_{g_{\hat{\chi}}} f}{f} + (n+2)\frac{|df|^2_{g_{\hat{\chi}}}}{f^2}\right)\eta^{\hat{\chi}}\right). \tag{4.11}$$

Furthermore, using that h is $\hat{\chi}$-invariant, $\eta^{\hat{\chi}} = f\eta^{\hat{\xi}}$, and (see [1, Lemma 1])

$$\hat{\xi} = f\hat{\chi} - \omega_{\hat{\chi}}^{-1}(df_{|\mathcal{D}}),$$

we get

$$d^c_{\hat{\xi}}h = d^c_{\hat{\chi}}h - \langle dh, df\rangle_{g_{\hat{\chi}}}\,\eta^{\hat{\chi}}. \tag{4.12}$$

Finally, using (4.11), (4.12) and

$$\omega_{\hat{\xi}} = d\eta^{\hat{\xi}}, \qquad \omega^{\hat{\chi}} = d\eta^{\hat{\chi}}, \qquad \eta_{\hat{\xi}} = f\eta^{\hat{\chi}},$$

we get

$$\begin{aligned} 0 =& \rho_{\omega_{\hat{\xi}}} - \omega_{\hat{\xi}} - \frac{1}{2}dd^c_{\hat{\xi}}h \\ =& \rho_{\omega_{\hat{\chi}}} + \frac{(n+2)}{2}dd^c_{\hat{\chi}}\log f - \frac{1}{2}dd^c_{\hat{\chi}}h \\ &- d\left[f + \frac{1}{2}\left(\frac{\Delta_{g_{\hat{\chi}}}f}{f} + (n+2)\frac{|df|^2_{g_{\hat{\chi}}}}{f^2}\right) - \frac{1}{2}\langle dh, df\rangle_{g_{\hat{\chi}}}\right]\eta^{\hat{\chi}} \end{aligned}$$

Evaluating the latter on the vector field $\hat{\chi}$, we get that

$$\left[f + \frac{1}{2}\left(\frac{\Delta_{g_{\hat{\chi}}}f}{f} + (n+2)\frac{|df|^2_{g_{\hat{\chi}}}}{f^2}\right) - \frac{1}{2}\langle dh, df\rangle_{g_{\hat{\chi}}}\right] = \lambda$$

is a constant. The first claim follows easily from the above. This constant must be positive when N is compact (which follows by considering a point of maximum of $f > 0$) and is then also determined from the basic cohomology of $\hat{\chi}$ by the relation $[c_1(\mathcal{D})]_{B,\hat{\chi}} = \lambda[\omega_{\hat{\chi}}]$. Thus, $\lambda = \lambda_{\hat{\chi}}$, see Sect. 4.3. □

We thus get the following generalization of [3, Prop.2].

Corollary 4.1 *Suppose $\hat{\mathfrak{t}}$ is a maximal abelian subalgebra of $\mathfrak{cr}(N, J, \mathcal{D})$, and $\hat{\xi}, \hat{\chi} \in \hat{\mathfrak{t}}$. Then $(\hat{\xi}, \mathcal{D}, J)$ is a transversal KRS with soliton vector $\tau_{\hat{\xi}} \in \mathfrak{t}$ iff $(\hat{\chi}, \mathcal{D}, J)$ is a transversal $v_{\hat{\xi}}$-soliton for the weigh function*

$$v_{\hat{\xi}} := e^{\left(\frac{\eta^{\hat{\chi}}(\tau_{\hat{\xi}})}{\eta^{\hat{\chi}}(\hat{\xi})}\right)}\left(\eta^{\hat{\chi}}(\hat{\xi})\right)^{-(n+2)}. \tag{4.13}$$

4.6 Transversal KRS's from a Fixed Regular Quotient

By virtue of Corollary 4.1, we now suppose $(X, \mathbb{T})$ is a Fano manifold with $\mathbb{T} \subset \mathrm{Aut}_r(X)$ a maximal torus with canonically normalized momentum polytope P_X. Any $\mathbb{T}$-invariant Kähler metric $\omega \in 2\pi c_1(X)$ can be viewed as the reduction of a Sasaki manifold $(N, J, \mathfrak{D})$ by a regular Sasaki Reeb vector field $\hat{\chi} \in \hat{\mathfrak{t}}$, where $\hat{T} = \mathbb{S}^1_{\hat{\chi}} \times \mathbb{T}$ is defined on N via the natural lift of $\mathbb{T}$ to K_X. Furthermore, the Lie algebra $\mathfrak{t} = \mathrm{Lie}(\mathbb{T})$ is naturally identified with the subspace $\mathfrak{t} \subset \hat{\mathfrak{t}}$ defined in Definition 4.1.

The polytope $\Sigma_+ \subset \hat{\mathfrak{t}}_+$ can be equivalently characterized in terms of $(X, \mathbb{T}, \omega)$ as the set of elements $\zeta \in \mathfrak{t}$ such that the affine linear function $\ell_\zeta(x) := \langle x, \zeta \rangle + 1$ is strictly positive on P_X. Indeed, any affine-linear function ℓ_ζ as above gives rise to an element $\hat{\zeta} \in \Sigma_+$ given by the horizontal lift of ζ to $\mathfrak{D}$ plus $\ell_\zeta(\mu_\omega)\hat{\chi}$; conversely, the quantity $\eta_{\hat{\chi}}(\hat{\xi})$ defines an affine-linear function on P_X of the form $\langle \zeta, x \rangle + 1$. Thus, we have an identification of Σ_+ with the dual (open) polytope $\mathring{\mathrm{P}}^*_X$ of P_X:

$$\Sigma_+ \cong \mathring{\mathrm{P}}^*_X, \qquad \mathring{\mathrm{P}}^*_X := \left\{ \zeta \in \mathfrak{t} \,|\, \ell_\zeta(x) := \langle \zeta, x \rangle + 1 > 0 \ \forall x \in \mathrm{P}_X \right\}.$$

By Lemma 4.2, there exists a smooth function

$$\tau : \mathring{\mathrm{P}}^*_X \to \mathfrak{t}$$

such that

$$\forall \xi \in \mathring{\mathrm{P}}^*_X, \quad v_\xi := \exp\left(\frac{\langle \mu, \tau(\xi) \rangle}{\ell_\xi} \right) (\ell_\xi)^{-(n+2)} \text{ satisfies } \mathrm{Fut}_{v_\xi} \equiv 0, \qquad (4.14)$$

where Fut_v is the weighted Futaki invariant defined in (2.11).

Remark 4.4 A special case of Lemma 4.2 appears when $\xi_0 \in \mathring{\mathrm{P}}^*_X$ is such that $\tau(\xi_0) = 0$. Equivalently, $v_{\xi_0} = \ell_{\xi_0}^{-(n+2)}$ where ξ_0 is uniquely determined (see [50]) by the property $\mathrm{Fut}_{\ell_{\xi_0}^{-(n+2)}} \equiv 0$ or, equivalently,

$$(dV_{\xi_0})_0(\dot{\tau}) = \int_X \langle \mu_\omega, \dot{\tau} \rangle \frac{\omega^{[n]}}{\ell_{\xi_0}(\mu_\omega)^{n+2}} = \mathrm{Fut}_{\ell_{\xi_0}^{-(n+2)}}(\dot{\tau}) = 0.$$

The corresponding v_{ξ_0}-soliton (if it exists) then corresponds, up to a transversal homothety (see Remark 4.3), to a Calabi–Yau Kähler cone structure on $K_X^\times$, compare with [3, Prop.2]

4.7 From Kähler–Ricci Solitons to Calabi–Yau Cones: Proof on Theorem 1.2

Proof of Theorem 1.2 Let $(X, \mathbb{T})$ be a smooth Fano variety admitting a KRS. We consider the sequence of rational functions on $\mathfrak{t} \times \mathfrak{t}^*$

$$q_N(\xi, x) := \left(1 - \frac{\langle x, \xi\rangle}{N}\right)^{-N},$$

which for $N >> 1$ is well-defined and converges uniformly on a compact subset of $\mathfrak{t} \times \mathfrak{t}^*$ to the function $e^{\langle \mu, \xi\rangle}$.

Let $\mathrm{P}_X^* \subset \mathfrak{t}$ denote the dual polytope of P_X, i.e.

$$\mathrm{P}_X^* := \{\xi \in \mathfrak{t} \mid \langle \xi, x\rangle + 1 \geq 0 \; \forall x \in \mathrm{P}_X\},$$

and $\mathring{\mathrm{P}}_X^*$ its interior. Thus, for any $\xi \in -N\mathring{\mathrm{P}}_X^*$, $q_N(\xi, x)$ is a positive function on P_X and we consider the volume functional

$$V_N(\xi) := \int_X \left(1 - \frac{\langle \mu_\omega, \xi\rangle}{N}\right)^{-N+1} \omega^{[n]}, \qquad \omega \in \mathcal{K}^{\mathbb{T}}_{2\pi c_1(X)}$$

which is well-defined, convex and proper on $-N\mathring{\mathrm{P}}_X^*$. Let $\xi_N \in -N\mathring{\mathrm{P}}_X^*$ be the unique minimizer of V_N: ξ_N then satisfies

$$\int_X \mu_\omega^\zeta \left(1 - \frac{\langle \mu_\omega, \xi_N\rangle}{N}\right)^{-N} \omega^{[n]} = 0, \qquad \forall \zeta \in \mathfrak{t}, \; \forall \omega \in \mathcal{K}^{\mathbb{T}}_{2\pi c_1(X)}. \tag{4.15}$$

i.e. $v_N(x) := q_N(\xi_N, x) > 0$ on P_X and $\mathrm{Fut}_{v_N}(\zeta) = 0$ for any $\zeta \in \mathfrak{t}$.

We now show that if $\tau \in \mathfrak{t}$ is the Tian–Zhu KRS vector field (so that $(X, \mathbb{T})$ admits an $e^{\langle \tau, x\rangle}$-soliton by the hypothesis) then

$$\lim_{N \to \infty} \xi_N = \tau.$$

To see this, we can use for instance the implicit function theorem. Indeed, consider the function

$$F(a, t) := e^{\left(\frac{a+1}{a}\right)\log(1+at)}, \qquad (a, t) \in \mathbb{R}^2 : \; 0 < 1 + at < 2,$$

which admits an analytic extension at $a = 0$ with $F(0, t) = e^t$. Notice that $F(-\frac{1}{N}, t) = (1 - \frac{t}{N})^{-N+1}$ and, for any fixed a, the function $t \to F(a, t)$ is strictly convex as soon as $a > -1$. Consider the function

$$W(a, \xi) := \int_X F(a, \langle \xi, \mu_\omega\rangle)\omega^{[n]},$$

defined on the domain

$$U := \{(a, \xi) \in \mathbb{R} \times \mathfrak{t} \,|\, a > -1, \ -1 < a\langle \xi, x \rangle < 1 \ \forall x \in \mathrm{P}_X\}.$$

By the properties of $F(a, t)$ mentioned above, $W(-\frac{1}{N}, \xi) = V_N(\xi)$ and, for any fixed $a > -1$, the map $\xi \to W(a, \xi)$ is strictly convex on the domain

$$U_a := \{\xi \in \mathfrak{t} \mid -1 < a\langle \xi, x \rangle < 1\}.$$

Applying the implicit function theorem to

$$\Psi : U \to \mathfrak{t}^*, \qquad \Psi(a, \xi) := (d_\xi W)(a, \xi),$$

there exists a smooth path $(a, \tau_a) \in U$, defined for $|a| < \varepsilon$, such that $\Psi(a, \tau_a) = 0$ and $\tau_0 = \tau$. Clearly, for $N >> 1$, $\xi_N = \tau_{-\frac{1}{N}}$ by the uniqueness of the critical point, and whence $\lim_{N\to\infty} \xi_N = \tau$.

Using $\lim_{N\to\infty} \xi_N = \tau$, we have $\lim_{N\to\infty} v_N(x) = e^{\langle \tau, x \rangle}$ in $C^0(\mathrm{P}_X)$. As X admits an $e^{\langle \tau, x \rangle}$-soliton by assumption, and $v_N(x)$ tends uniformly on P_X to $e^{\langle \tau, x \rangle}$ satisfying $\mathrm{Fut}_{v_N} = 0$, we can apply Theorems 1.1 and 2.1 to conclude that there are v_N-solitons for any $N >> 1$. Alternatively, we can appeal to Corollary A.1 applied to the smooth family of weights $v_a(x) := e^{\frac{1}{a}\log(1+a\langle \tau_a, x \rangle)} \in \mathcal{F}(X)$.

Taking the product of a v_N-soliton on X with the Fubini–Study metric on $\mathbb{P}^{N-n-2}_{\mathbb{C}}$ gives rise to a v_N-soliton on $Z = X \times \mathbb{P}^{N-n-2}_{\mathbb{C}}$. As Z is $(N-2)$-dimensional, this in turn defines a Calabi–Yau cone structure on $K_Z^\times$ by Remark 4.4. □

4.8 Openness of Transversal KRS: Proof of Theorem 1.3

Proof of Theorem 1.3 We consider the smooth family of weights v_ξ, $\xi \in \mathring{\mathrm{P}}^*_X$ defined by (4.14). By Corollary 4.1, Proposition 4.1 and Remark 4.1, the existence of a $\mathbb{T}$-invariant v_ξ-soliton in $2\pi c_1(X)$ is equivalent with the existence of a transversal KRS soliton in $\Xi^{\hat{\mathbb{T}}}_{\hat{\xi}, \eta_0^{\hat{\xi}}, J^{\hat{\xi}}}(N)$. By the discussion in Sect. 4.2, this is also equivalent to the existence of a $\hat{\xi}$-polarized transversal KRS cone metric on $Y = K_X^\times$. Theorem 1.3 then follows from Theorems 1.1 and 2.1, or by Corollary A.1 applied to v_ξ. □

Remark 4.5 The assumption in the above proof that $K_X^\times$ is a cone of *smooth* Fano manifold, or, equivalently, that the Sasaki manifold $(N, \mathcal{D}, J, \hat{\mathbb{T}})$ admits a regular Sasaki-Reeb vector field can be removed by considering instead a quasi-regular Sasaki–Reeb field $\hat{\chi} \in \hat{\mathfrak{t}}_+$. In this case, the quotient $X := N/\mathbb{S}^1_{\hat{\chi}}$ is a Kähler Fano orbifold. We can still apply Corollary 4.1 which hold for orbifolds. Instead of Theorem 1.1, which uses the results in [36] on a smooth Fano manifold, one can use an openness result a la LeBrun–Simanca (which we recall in Appendix A and whose proof can be adapted to orbifolds) in order to establish the existence of a transversal KRS in $\Xi^{\hat{\mathbb{T}}}_{\xi,\eta_0^{\hat{\xi}},J^{\hat{\xi}}}(N)$ for all Sasaki–Reeb vector fields near $\hat{\xi}_0$. This is closer to the original approach in [53].

5 A Lichnerowicz Type Obstruction for Transversal KRS Soliton: Proof of Theorem 1.4

We discuss in this subsection a necessary condition in terms of the function $\tau(\xi)$ for the weight function v_ξ to belong to $\mathcal{S}$, i.e $K_X^\times$ to admit a compatible transversal KRS polarized by $\hat{\xi}$.

5.1 *Proof of Theorem 1.4*

Proposition 5.1 *Let $(Y, \hat{\xi})$ be a smooth polarized complex cone endowed with a holomorphic volume form Ω, satisfying $\mathcal{L}_{-J\hat{\xi}}\Omega = \Omega$. Let $\hat{\mathbb{T}}$ be a maximal torus in* Aut(Y) *such that $\hat{\xi} \in \hat{\mathfrak{t}} =$* Lie$(\hat{\mathbb{T}})$. *Suppose that $(Y, \hat{\xi})$ admits a compatible transversal Kähler–Ricci soliton, $\hat{\omega} = \frac{1}{4}dd^c r$, $r \in \mathcal{R}^{\hat{\mathbb{T}}}_{\hat{\xi}}(Y)$, with corresponding soliton vector field $\tau_{\hat{\xi}} \in \hat{\mathfrak{t}}$ normalized by $\mathcal{L}_{-J\tau_{\hat{\xi}}}\Omega = 0$. Then*

$$\hat{\xi} - \frac{1}{n}\tau_{\hat{\xi}}$$

is a Sasaki–Reeb polarization on Y.

Proof In the notation of the previous subsections, we have $\hat{\xi} \in \hat{\Sigma}_+$ and $\tau_{\hat{\xi}} \in \mathfrak{t}$. We shall work on the corresponding Sasaki manifold $(N := r^{-1}(1), \mathcal{D}, J, \hat{\xi}, \eta^{\hat{\xi}})$, which is a transversal KRS by assumption. We want to show that $\hat{\xi} - \frac{1}{n}\tau_{\hat{\xi}} \in \hat{\mathfrak{t}}_+(N)$ or, equivalently, (see [2, 12, 20])

$$\eta^{\hat{\xi}}\left(\hat{\xi} - \frac{1}{n}\tau_{\hat{\xi}}\right) = 1 - \frac{1}{n}\eta^{\hat{\xi}}(\tau_{\hat{\xi}}) > 0. \tag{5.1}$$

Contracting the transversal KRS equation (4.7) with the transversal Kähler form $\omega := d\eta^{\hat{\xi}}$ gives for the transversal scalar curvature $\mathrm{Scal}(\omega)$

$$\mathrm{Scal}(\omega) = 2n - \Delta_\omega f, \qquad f := \eta^{\hat{\xi}}(\tau_{\hat{\xi}}).$$

By (4.6) (with $h = f$ and contracting $\tau_{\hat{\xi}}$), we also get

$$f - \frac{1}{2}\Delta_\omega f + \frac{1}{2}\langle d^c_{\hat{\xi}} f, \tau_{\hat{\xi}}\rangle = 0, \tag{5.2}$$

so we obtain

$$\mathrm{Scal}(\omega) = 2n - 2f - \langle d^c_{\hat{\xi}} f, \tau_{\hat{\xi}}\rangle.$$

As $\mathrm{Scal}(\omega) > 0$ everywhere on N by Corollary B.1 established in the Appendix B, evaluating the above equality at a point of global maximum of f yields (5.1). □

Proof of Theorem 1.4 We can realize X as the regular Kähler quotient of a polarized cone $(Y = K_X^\times, \hat{\xi})$ endowed with a compatible transversal Kähler–Ricci soliton with vector field $\hat{\tau}$. The lifts $\hat{\xi}$ and $\hat{\tau}$ of the vector fields 0 and τ on X are normalized respectively by $\mathcal{L}_{-J\hat{\xi}}\Omega = \Omega$ and $\mathcal{L}_{-J\hat{\tau}}\Omega = 0$. Thus, by Proposition 5.1, the vector field $n\hat{\xi} - \hat{\tau}$ defines a Sasaki–Reeb polarization on Y. In terms of X, this means that $n - \mu^\tau_\omega > 0$. The second statement follows similarly. □

5.2 Theorem 1.4 as a Lichnerowicz Type Condition

It is well-known that the normalized characters for the linear action of $\hat{\mathbb{T}}$ on the spaces $H^0(X, -mK_X)$ of m-plurianticanonical sections of X are rational lattice points inside P_X (see e.g. [41, Lemma 13]). From this point of view, the first inequality in Theorem 1.4 yields an a priori bound on the spectrum of the operator $\mathcal{L}_\tau$ acting on $H^0(X, -mK_X)$:

Corollary 5.1 *If $(X, \mathbb{T})$ admits a $\mathbb{T}$-invariant KRS with soliton vector field τ and $s \in H^0(X, -mK_X)$ is a non-trivial m-plurianticanonical holomorphic section such that $\frac{1}{m}\mathcal{L}_\tau s = i\lambda s$, $\lambda \in \mathbb{R}$, then $(n - \lambda) > 0$.*

On the other hand, plurianticanonical sections of X give rise to certain holomorphic functions on the corresponding cone $Y = K_X^\times$. Thus, Corollary 5.1 should also lead to an a priori bound on the eigenvalues (also called *charges* in [33]) for the infinitesimal action of the soliton vector field τ on the space of holomorphic functions on Y. This is reminiscent to but different from the bounds on the charges of holomorphic functions of Y under the action of the Reeb field of a Calabi–Yau cone structure on Y, obtained in [33].

We provide below a direct argument for this bound, independent of the proof of Theorem 1.4.

Lemma 5.1 *Let* $(Y, \hat{\xi})$ *be a smooth polarized complex cone satisfying the conditions of Proposition 5.1 with respect to a maximal torus of automorphisms* $\hat{\mathbb{T}} \subset \mathrm{Aut}(Y)$. *If* $\varphi : Y \to \mathbb{C}$ *is a non-trivial holomorphic function on* Y *satisfying* $\mathcal{L}_{\hat{\xi}}\varphi = i\kappa\varphi$ *and* $\mathcal{L}_{\tau_{\hat{\xi}}}\varphi = i\kappa\lambda\varphi$ *for some* $\lambda \in \mathbb{R}$ *and* $\kappa \in \mathbb{R}_{>0}$, *then*

$$(n - \lambda) > 0.$$

Proof Let $r \in \mathcal{R}^{\hat{\mathbb{T}}}_{\hat{\xi}}(Y)$ be the potential of a compatible transversal KRS cone metric on $(Y, \hat{\xi})$, with soliton vector field $\tau = \tau_{\hat{\xi}} \in \mathfrak{t}$. As in the proof of Proposition 5.1, we shall work on the link $N := r^{-1}(1)$, with the induced transversal KRS Sasaki structure $(\mathcal{D}, J, \hat{\xi}, \eta^{\hat{\xi}})$. We still denote by $\omega := d\eta^{\hat{\xi}}$ the transversal Kähler structure and let $f := \eta^{\hat{\xi}}(\tau_{\hat{\xi}}) = \mu^{\tau}_{\omega}$ be the transversal Killing potential of $\tau_{\hat{\xi}}$.

We will first prove that, when restricted on N, φ satisfies the inequality

$$\Delta_{\omega,f}(e^{-\kappa f}|\varphi|^2) \leq 2\kappa(n - \lambda)e^{-\kappa f}|\varphi|^2, \tag{5.3}$$

where $\Delta_{\omega,f}\psi := \Delta_\omega\psi + \langle d\psi, df\rangle_\omega$ is the twisted $\hat{\xi}$-basic Laplacian (recall that Δ_ω denotes the $\hat{\xi}$-basic Laplacian).

As φ is holomorphic, the assumptions equivalently read as

$$\mathcal{L}_{-J\hat{\xi}}\varphi = \kappa\varphi, \qquad \mathcal{L}_{-J\tau_{\hat{\xi}}}\varphi = \kappa\lambda\varphi,$$

and hence

$$\mathcal{L}_{\hat{\xi}}|\varphi|^2 = 0, \qquad \mathcal{L}_{-J\hat{\xi}}|\varphi|^2 = 2\kappa|\varphi|^2, \qquad \mathcal{L}_{-J\tau_{\hat{\xi}}}|\varphi|^2 = 2\kappa\lambda|\varphi|^2. \tag{5.4}$$

At any point of N, we can write (see e.g. [1, Lemma 1])

$$\tau_{\hat{\xi}} = f\hat{\xi} - \omega^{-1}(df), \qquad J\tau_{\hat{\xi}} = fJ\hat{\xi} - g_\omega^{-1}(df). \tag{5.5}$$

We get from the above

$$\langle d|\varphi|^2, df\rangle_\omega = 2\kappa(\lambda - f)|\varphi|^2, \tag{5.6}$$

where $\langle\cdot,\cdot\rangle_\omega$ stands for the inner product induced by the transversal Kähler structure ω on $\hat{\xi}$-basic tensors. Together with (5.2), the latter yields

$$\left\langle d(e^{-\kappa f}|\varphi|^2), df\right\rangle_\omega = 2\kappa e^{-\kappa f}|\varphi|^2\left(\lambda - \frac{1}{2}\Delta_\omega f\right).$$

A direct computation using (5.6) shows that

$$\Delta_\omega(e^{-\kappa f}|\varphi|^2) = e^{-\kappa f}|\varphi|^2\left(-\kappa\Delta_\omega f - \kappa^2||df||^2_\omega + 4\kappa^2\lambda - 4\kappa^2 f\right) + e^{-\kappa f}\Delta_\omega|\varphi|^2,$$

so that, combining with (5.2), we obtain

$$\Delta_{\omega,f}(e^{-\kappa f}|\varphi|^2) = e^{-\kappa f}|\varphi|^2\left(-\kappa^2||df||^2_\omega + 2\kappa(2\kappa-1)\lambda - 4\kappa^2 f\right) + e^{-\kappa f}\Delta_\omega|\varphi|^2. \tag{5.7}$$

We next develop the term $\Delta_\omega|\varphi|^2$. To this end, recall that any holomorphic function φ on Y satisfies (on the open dense subset where $\varphi \neq 0$)

$$dd^c|\varphi|^2 = 2i d\varphi \wedge d\bar{\varphi} = \frac{1}{|\varphi|^2}\left(d|\varphi|^2 \wedge d^c|\varphi|^2\right). \tag{5.8}$$

Furthermore, if $\iota : N \hookrightarrow Y$ denotes the inclusion and ψ is any smooth function on Y, then we have on N

$$d^c_\xi(\iota^*\psi) = \iota^*(d^c\psi) + (\mathcal{L}_{J\xi}\psi)\eta^\xi, \qquad dd^c_\xi\psi = \iota^*(dd^c\psi) - d\left((\mathcal{L}_{-J\xi}\psi)\eta^\xi\right).$$

We use the previous two identities and (5.4) to compute

$$\Delta_\omega|\varphi|^2 = -tr_\omega dd^c_\xi|\varphi|^2 = -tr_\omega(2i d\varphi \wedge d\bar{\varphi}) + 2n\kappa|\varphi|^2.$$

Substituting back in (5.7) and regrouping the terms yields

$$\begin{aligned}\Delta_{\omega,f}(e^{-\kappa f}|\varphi|^2) =& 2\kappa e^{-\kappa f}|\varphi|^2(n-\lambda)\\ &+ 2\kappa^2 e^{-\kappa f}|\varphi|^2\left(-\frac{||df||^2_\omega}{2} + 2\lambda - 2f\right)\\ &- e^{-\kappa f}tr_\omega(2i d\varphi \wedge d\bar{\varphi})\end{aligned} \tag{5.9}$$

Working outside the zero locus of φ for a moment, we compute

$$||d(e^{-\frac{\kappa f}{2}}|\varphi|)||^2_\omega = \frac{\kappa^2}{4}|\varphi|^2 e^{-\kappa f}(||df||^2_\omega) - \kappa|\varphi|e^{-\kappa f}\langle df, d|\varphi|\rangle_\omega + e^{-\kappa f}||d|\varphi|||^2_\omega.$$

Using (5.2), the second term of the right hand side of the above identity may be rewritten and we get

$$||d(e^{-\frac{\kappa f}{2}}|\varphi|)||^2_\omega = \frac{\kappa^2}{4}|\varphi|^2 e^{-\kappa f}(||df||^2_\omega) - \kappa^2|\varphi|^2 e^{-\kappa f}(\lambda - f) + e^{-\kappa f}||d|\varphi|||^2_\omega. \tag{5.10}$$

The second line of (5.9) is exactly $-4||d(e^{-\frac{\kappa f}{2}}|\varphi|)||^2_\omega + 4e^{-\kappa f}||d|\varphi|||^2_\omega$, so that

$$\begin{aligned}\Delta_{\omega,f}(e^{-\kappa f}|\varphi|^2) =& 2\kappa e^{-\kappa f}|\varphi|^2(n-\lambda)\\ &-4||d(e^{-\frac{\kappa f}{2}}|\varphi|)||^2_\omega + 4e^{-\kappa f}||d|\varphi|||^2_\omega\\ &-e^{-\kappa f}\, tr_\omega(2id\varphi\wedge d\bar{\varphi}).\end{aligned}$$

Using the second equality in (5.8) we finally obtain

$$\Delta_{\omega,f}(e^{-\kappa f}|\varphi|^2) = 2\kappa e^{-\kappa f}|\varphi|^2(n-\lambda) - 4||d(e^{-\frac{\kappa f}{2}}|\varphi|)||^2_\omega, \tag{5.11}$$

which yields the inequality (5.3) everywhere on Y.

From (5.3), we conclude that $(n-\lambda) \geq 0$ by the maximum principle applied on a local quotient of N by $\hat{\xi}$, near a point of global maximum of $e^{-kf}|\varphi|^2$ on N. (Notice that $|\varphi|^2$ cannot be identically zero on N as φ is a non-trivial holomorphic function on Y by assumption.) If $\lambda = n$, then the maximum principle tells us that $e^{-\kappa f}|\varphi|^2 = C$ for some constant C on N. By (5.4), we have on Y

$$|\varphi|^2 = Cr^{2\kappa}e^{\kappa f}.$$

Taking Lie derivative in the direction of $-J\tau_\xi$ in both sides of the last equation (and using (5.4) and (5.5)) gives

$$\lambda = C(|df|^2 + f),$$

which in turn implies that f is a constant and thus $\tau_\xi = 0$. In that case $\lambda = 0$, a contradiction. □

6 A Weighted Version of the Fujita Volume Bound: Proof of Theorem 1.5

6.1 The Weighted β-Invariant of a Polarized Manifold and Fujita's Volume Bound

Here we recall the setup in [36, 37] of valuative characterization of weighted Ding stability of smooth Fano varieties.

Let $\pi : L \to X$ be a holomorphic line bundle endowed with the action of a real torus $\mathbb{T} \subset \mathrm{Aut}(X, L)$, covering $\mathbb{T} \subset \mathrm{Aut}(X)$. Let $W_L \subset \mathfrak{t}^*$ be the convex hull of normalized weights

$$\left\{\frac{\alpha}{k} \in \mathfrak{t}^* \mid k \in \mathbb{N}^*, H^0(X, kL)_\alpha \neq 0\right\}$$

of the linear action of $\mathbb{T}$ on the vector spaces $H^0(X, kL)$. When L is ample, similarly to (2.3), one can associate to the chosen lift $\mathbb{T} \subset \mathrm{Aut}(X, L)$ a momentum polytope $P_L \subset \mathfrak{t}^*$ and it is well-known that in this case $P_L = W_L$, see for example [60, Chapter 8].

For a smooth weight function $v \in C^\infty(W_L, \mathbb{R}_{>0})$, Han–Li introduced in [36] the (algebraic) v-weighted volume of L as the limit

$$Vol_v(L) := \lim_{k \to +\infty} \frac{n!}{k^n} \sum_{\alpha \in \hat{\mathfrak{t}}^*} v\left(\frac{\alpha}{k}\right) \dim \left(H^0(X, kL)\right)_\alpha,$$

the existence of which is justified in [36, p.40]. We set

$$Vol(L) := Vol_1(L).$$

In the case when L is ample, we have (see e.g. [41, Lemma 14])

$$Vol_v(L) = \frac{n!}{(2\pi)^n} \int_{\mathrm{P}_X} v(x) d\mu_{\mathrm{DH}},$$

where $d\mu_{\mathrm{DH}}$ is introduced through (2.5) for a $\omega \in 2\pi c_1(L)$. It thus follows that in this case

$$Vol(L) = c_1(L)^n. \tag{6.1}$$

Let now X be a smooth Fano variety and $\mathbb{T} \subset \mathrm{Aut}(X)$ a maximal torus as in the setup of Sect. 2.1. The torus $\mathbb{T}$ admits a canonical lift on $L = -K_X$, and by the above discussion, $W_{-K_X} = \mathrm{P}_X$, where P_X is the canonically normalized polytope of $(X, \mathbb{T})$. Let $\hat{X}$ be the blow-up of X at a point $p \in X$ fixed by $\mathbb{T}$. We thus have a $\mathbb{T}$-equivariant birational morphism $b : \hat{X} \to X$, and a $\mathbb{T}$-stable exceptional divisor $D \subset \hat{X}$. Notice that D is, by definition, an instance of a *prime divisor* over $(X, \mathbb{T})$, so we can use the theory of weighted β-invariant [36, 37], which we recast to our setup below.

Given $x \in \mathbb{Q}$, there exists $m_x \in \mathbb{N}^*$ such that $m_x(-b^*K_X - x[D])$ is a holomorphic line bundle on $\hat{X}$. Notice that $-b^*m_xK_X - xm_x[D]$ comes with an induced $\mathbb{T}$-action through the blowing-up construction, and we have a $\mathbb{T}$- equivariant exact sequence of sheaves

$$\begin{aligned} 0 &\longrightarrow \mathcal{O}_{\hat{X}}(-b^*m_xK_X - xm_x[D]) \\ &\longrightarrow \mathcal{O}_{\hat{X}}(-b^*m_xK_X) \longrightarrow \mathcal{O}_D(-xm_x[D]) \longrightarrow 0, \end{aligned}$$

showing the inclusion $\left(H^0(X, m_x(-b^*K_X - x[D]))\right)_\alpha \subset \left(H^0(X, -m_xK_X)\right)_\alpha$ for each weight α. It follows that $W_{m_x(-b^*K_X - x[D])} \subset W_{m_x(-K_X)} = m_xP_X$. One can therefore define the v-weighted volume $Vol_v(m_x(-b^*K_X - x[D]))$ of $m_x(-b^*K_X - x[D])$ over $\hat{X}$ as

$$Vol_v(m_x(-b^*K_X - x[D])) := \lim_{k\to+\infty} \frac{n!}{k^n} \sum_{\alpha\in\hat{\mathfrak{t}}^*} v\left(\frac{\alpha}{m_x k}\right) \dim\left(H^0(\hat{X}, km_x(-b^*K_X - x[D]))\right)_\alpha .$$

According to [36], this weighted volume is well-defined, continuous in x, and homogeneous of order n in m_x. We can thus set

$$Vol_v((-b^*K_X - x[D])) := \frac{1}{m_x^n}\left(\lim_{k\to+\infty} \frac{n!}{k^n} \sum_{\alpha\in\hat{\mathfrak{t}}^*} v\left(\frac{\alpha}{m_x k}\right) \dim\left(H^0(\hat{X}, km_x(-b^*K_X - x[D]))\right)_\alpha\right),$$

which is independent of m_x. The above definition yields the monotonicity property

$$Vol_{v_1}(-b^*K_X - x[D]) \leq Vol_{v_2}(-b^*K_X - x[D]) \qquad \forall\, v_1 \leq v_2 \text{ on } \mathrm{P}_X, \tag{6.2}$$

which we shall use below.

The weighted β invariant of $D \subset \hat{X}$ is then introduced by

$$\beta_v(D) := A_X(D)Vol_v(-K_X) - \int_0^\infty Vol_v(b^*(-K_X) - x[D])dx, \tag{6.3}$$

where $A_X(D)$ stands for the log discrepancy of the prime divisor D, see [25, Definition 2.5]. In the case when $D \subset \hat{X}$ is the exceptional divisor as above, $A_X(D) = n$ (see e.g. [28]).

Proof of Theorem 1.5 From [36, Theorems 1.7 & 5.18] it follows that under the hypothesis of Theorem 1.5, we have $\beta_v(D) \geq 0$, i.e.

$$\frac{1}{n}A_X(D)Vol_v(-K_X) = Vol_v(-K_X) \geq \frac{1}{n}\int_0^\infty Vol_v(-b^*K_X - x[D]).$$

Let $m_v := \min_{\mu\in\mathrm{P}_X} v(\mu) > 0$. By (6.2) we have

$$Vol_v(-b^*K_X - x[D]) \geq m_v Vol(-b^*K_X - x[D])$$

whereas by [28, Theorem 2.3]

$$Vol(b^*K_X - x[D]) \geq Vol(-K_X) - x^n.$$

The above inequalities give

$$
\begin{aligned}
Vol_v(-K_X) \geq & \frac{1}{n}\int_0^\infty Vol_v(-b^*K_X - x[D])dx \\
\geq & \frac{m_v}{n}\int_0^\infty Vol(-b^*K_X - x[D])dx \\
\geq & \frac{m_v}{n}\int_0^{(Vol(-K_X))^{1/n}} Vol(-b^*K_X - x[D])dx \\
\geq & \frac{m_v}{n}\int_0^{(Vol(-K_X))^{1/n}} \left(Vol(-K_X) - x^n\right)dx \\
= & \frac{m_v}{n}\left((Vol(-K_X))^{1/n} Vol(-K_X) - \frac{1}{n+1}(Vol(-K_X))^{(n+1)/n}\right) \\
= & \frac{m_v}{n+1}(Vol(-K_X))^{(n+1)/n}
\end{aligned}
$$

Under the normalization $Vol_v(-K_X) = Vol(-K_X)$ for the weight v, we obtain

$$
Vol(-K_X) \leq \left(\frac{n+1}{m_v}\right)^n.
$$

The claim follows by (6.1). □

Remark 6.1 More generally, let $(X, L, \mathbb{T})$ be a polarized projective manifold and $\mathbb{T} \subset \mathrm{Aut}(X, L)$ a lift of the action of a real torus in $\mathrm{Aut}(X)$, with corresponding momentum polytope $\mathrm{P}_L \subset \mathfrak{t}^*$. For any two weight functions $v > 0$, w on P_L, and a $\mathbb{T}$-stable prime divisor D of $(X, \mathbb{T})$, one can associate the (v, w)-weighted β-invariant of D by the formula

$$
\begin{aligned}
\beta_{v,w}(D) := A_X(D)Vol_v(L) + & \frac{1}{2}\int_0^\infty Vol_w(L - x[D])dx \\
& + \int_0^\infty Vol_v'(L - x[D]) \cdot [K_X]dx,
\end{aligned} \tag{6.4}
$$

where, as in [25], $Vol_v'(L - x[D]) \cdot [K_X]dx$ stands for the (formal) derivative of Vol_v in the direction of the canonical bundle K_X. The above formula extends the "unweighted" situation with $v = 1$, $w = 2n\left(\frac{c_1(L)\cdot L^{n-1}}{L^n}\right)$ studied in [25]. In general, in (6.4), we assume the differentiability of the extension of Vol_v on $\mathbb{R}$-line bundles in each direction, a fact established in the case $v = 1$ in [11] but which is open in general.

Similarly to [25, Corollary 3.11], using integration by parts, one can rewrite (6.4) as

$$\beta_{v,w}(D) = A_X(D)Vol_v(L) + \frac{1}{2}\int_0^\infty Vol_{w-(\tilde{v}+v)}(L - x[D])dx$$
$$+ \int_0^\infty Vol_v'(L - x[D]) \cdot [L + K_X]dx.$$

When X is Fano, $L = -K_X$ and $w = \tilde{v} := 2(n + d\log v)v$, the above reduces to (6.3), showing the consistency of (6.4). It would be interesting to show that, similarly to the unweighted case [25], the value $\beta_{v,w}(D)$ is related to the (v, w)-weighted Donaldson-Futaki invariant [41] of a certain test configuration.

Appendix A: Smooth Deformations of Weighted cscK Metrics

Let X be a compact Kähler manifold endowed with a Kähler class $\alpha \in H^{1,1}(X, \mathbb{R})$ and $\mathbb{T} \subset \mathrm{Aut}_r(X)$ a maximal torus with momentum polytope $\mathrm{P} \subset \mathfrak{t}^*$. Let $v, w \in C^\infty(\mathrm{P})$, $v > 0$ be a pair of weight functions on P. Recall that (v, w)-weighted Futaki invariant (see [41]) vanishes if for some (and hence any) $\mathbb{T}$-invariant Kähler metric $\omega \in \alpha$ and for any affine-linear function $\ell(x)$ on $\mathfrak{t}^*$,

$$\mathrm{Fut}_{v,w}(\ell) := \int_X \left(\mathrm{Scal}_v(\omega) - w(\mu_{\omega_\omega})\right) \ell(\mu_\omega)\omega^{[n]} = 0,$$

where $\mathrm{Scal}_v(\omega)$ stands for the v-scalar curvature of ω given by

$$\mathrm{Scal}_v(\omega) := v(\mu_\omega)\mathrm{Scal}(\omega) + 2\Delta_\omega v(\mu_\omega) + \left\langle g_\omega, \mu_\omega^* (\mathrm{Hess}(v)) \right\rangle,$$

with $\mathrm{Scal}(\omega)$ being the usual scalar curvature of the Riemannian metric g_ω associated to ω, Δ_ω standing for the Laplace operator of g_ω, and $\langle \cdot, \cdot \rangle$ denoting the contraction between the smooth $\mathfrak{t}^* \otimes \mathfrak{t}^*$-valued function g_ω on X (the restriction of the Riemannian metric g_ω to $\mathfrak{t} \subset C^\infty(X, TX)$) and the smooth $\mathfrak{t}\otimes\mathfrak{t}$-valued function ${\mu_\omega}^*$ (Hess(v)) on X (given by the pull-back by μ_ω of Hess$(v) \in C^\infty(\mathrm{P}, \mathfrak{t} \otimes \mathfrak{t})$).

The following result is a slight modification of [41, Theorem B2] (see also [35, Theorem 6.1.2] for the special case when w is fixed), where the weight functions (v_t, w_t) are supposed to be positive on P.

Lemma A.1 *Let $\{(v_t, w_t),\ v_t, w_t \in C^\infty(\mathrm{P}),\ v_t > 0,\ t \in U \subset \mathbb{R}^k\}$ be a finite dimensional smooth family of weight functions on* P*, parametrized by a neighbourhood U of* 0 *and satisfying* $\mathrm{Fut}_{v_t,w_t} \equiv 0$ *for all $t \in U$. Suppose ω_0 is a $\mathbb{T}$-invariant Kähler metric in α such that*

$$\mathrm{Scal}_{v_0}(\omega_0) = w_0(\mu_{\omega_0}).$$

Then, there exists $\epsilon > 0$ *and a differentiable family of smooth* $\mathbb{T}$*-invariant Kähler metrics* $\omega_t \in \alpha$, *such that for* $||t|| < \epsilon$

$$\mathrm{Scal}_{v_t}(\omega_t) = w_t(\mu_{\omega_t}),$$

where $\mu_{\omega_t} : X \to \mathrm{P}$ *is the* ω_t*-momentum map of* $\mathbb{T}$ *normalized by* $\mu_{\omega_t}(X) = \mathrm{P}$.

Proof The proof is a standard adaptation of the arguments in [42], using the computations in [41].

Consider the Fréchet space $C^\infty(X)^{\mathbb{T}}$ of smooth $\mathbb{T}$-invariant functions on X and denote by $\mathcal{K}^{\mathbb{T}}_{\omega_0}(X)$ the subset of $\mathbb{T}$-invariant ω_0-relative Kähler potentials; for any $\varphi \in \mathcal{K}^{\mathbb{T}}_{\omega_0}(X)$, we let $\omega_\varphi := \omega_0 + dd^c\varphi > 0$ be the corresponding Kähler structure on X, $\mu_\varphi := \mu_{\omega_0} + d^c\varphi$ be the normalized ω_φ momentum map, and we let

$$\mathcal{P}_{\omega_\varphi} := \{\ell(\mu_\varphi),\ \ell \in \mathrm{Aff}(\mathrm{P})\}$$

denote the space of all ω_φ-Hamiltonians of elements of $\mathrm{Lie}(\mathbb{T})$. We denote by

$$\Pi_{\omega_\varphi} : C^\infty(X)^{\mathbb{T}} \to \mathcal{P}_{\omega_\varphi}$$

the orthogonal projection with respect to the global L^2 product defined by $\omega_\varphi^{[n]}$ and let

$$C^\infty_\perp(X)^{\mathbb{T}} := (1 - \Pi_{\omega_0})(C^\infty(X)^{\mathbb{T}})$$

be the space of smooth $\mathbb{T}$-invariant functions on X, which are L^2-orthogonal to the finite dimensional space $\mathcal{P}_{\omega_0}$.

We consider the map $\Psi : U \times \left(\mathcal{K}^{\mathbb{T}}_{\omega_0}(X) \cap C^\infty_\perp(X)^{\mathbb{T}}\right) \to \mathbb{R}^k \times C^\infty_\perp(X)^{\mathbb{T}}$ defined by

$$\Psi(t, \varphi) := \Big(t, (1 - \Pi_{\omega_0})\left(\mathrm{Scal}_{v_t}(\omega_\varphi) - w_t(\mu_\varphi)\right)\Big).$$

By the computations in [41] and using that $\mathrm{Scal}_{v_0}(\omega_0) = w_0(\mu_{\omega_0})$, the differential $(T_{(0,0)}\Psi)$ of Ψ at $(0, 0)$ in the direction of φ is given by

$$(T_{(0,0)}\Psi)(0, \dot\varphi) = \left(0, -2\left((1 - \Pi_{\omega_0}\right)\left(\delta_{\omega_0}\delta_{\omega_0}(v_0(\mu_{\omega_0})(Dd\dot\varphi)^-\right)\right),$$

where $\delta_{\omega_0} T := -\sum_{i=1}^{2n}(D_{e_i}T)(e_i, \cdot)$ denotes the codifferential of a $(p, 0)$-tensor T with respect to ω_0, D stands for the Levi-Civita connection of ω_0 and $(Dd\dot\varphi)^-$ is the $(2, 0) + (0, 2)$ part of the Hessian of $\dot\varphi \in C^\infty(X)^{\mathbb{T}}$.

Notice that the weighted Lichnerowicz operator $\mathbb{L}_{\omega_0}(\dot\varphi) := \delta_{\omega_0}\delta_{\omega_0}\left(v_0(\mu_{\omega_0})(Dd\dot\varphi)^-\right)$ is a linear self-adjoint operator with respect to the L^2 inner product of ω_0, whose kernel in $C^\infty(X)^{\mathbb{T}}$ consists of all $\mathbb{T}$-invariant Killing potentials of ω_0. As $\mathbb{T} \subset \mathrm{Aut}_r(X)$ is a maximal torus, we conclude that the kernel of $\mathbb{L}_{\omega_0}$ is $\mathcal{P}_{\omega_0}$ and its image is in $C^\infty_\perp(X)^\mathbb{T}$. Thus, by standard linear elliptic theory, $\mathbb{L}_{\omega_0}$ (and hence $\dot{\varphi} \to (T_{0,0}\Psi)(0, \dot{\varphi})$) is an isomorphism when restricted to $C^\infty_\perp(X)^\mathbb{T}$, and hence also when viewed as a linear map between the Banach spaces $W^{k+4}_\perp(X)^\mathbb{T}$ and $W^k_\perp(X)^\mathbb{T}$, where for any positive integer k, $W^k(X)^\mathbb{T}$ denotes the completion in $L^2(X, \omega_0)$ of $C^\infty(X)^\mathbb{T}$ with respect to the Sobolev norm involving derivatives up to order k, and $W^k_\perp(X)^\mathbb{T} := (1 - \Pi_{\omega_0})(W^k(X)^\mathbb{T})$. By Sobolev's embedding theorem, for k large enough, we have $W^{k+4}(X)^\mathbb{T} \subset C^4(X)^\mathbb{T}$. As $\mathrm{Scal}_{v_t}(\omega_\varphi)$ is a 4th order quasilinear operator in φ, $\Psi(t, \varphi)$ admits an extension (still denoted by Ψ) to a C^1 map $\Psi : U \times \left(\mathcal{K}^\mathbb{T}_{\omega_0}(X) \cap W^{k+4}_\perp(X)^\mathbb{T}\right) \to \mathbb{R}^k \times W^k_\perp(X)^\mathbb{T}$, where the open subspace $\mathcal{K}^\mathbb{T}_{\omega_0}(X) \cap W^{k+4}_\perp(X)^\mathbb{T}$ is of C^2-Kähler metrics in α. Applying the implicit function theorem to Ψ and using bootstraping elliptic regularity for the map $\varphi \to \mathrm{Scal}_v(\omega_\varphi)$, we conclude that we can find a differentiable path of potentials $\varphi_t \in \mathcal{K}^\mathbb{T}_{\omega_0}(X) \cap C^\infty_\perp(X)^\mathbb{T}$, $|t| < \epsilon_0$ with $\varphi_0 \equiv 0$, such that

$$\mathrm{Scal}_{v_t}(\omega_{\varphi_t}) - w_t(\mu_{\varphi_t}) = \Pi_{\omega_0}\left(\mathrm{Scal}_{v_t}(\omega_{\varphi_t}) - w_t(\mu_{\varphi_t})\right).$$

Now, by the assumption $\mathrm{Fut}_{v_t, w_t} = 0$, $\Pi_{\omega_{\varphi_t}}\left(\mathrm{Scal}_{v_t}(\omega_{\varphi_t}) - w_t(\mu_{\varphi_t})\right) = 0$. Thus, for $||t|| < \epsilon$ small enough, one also has $\Pi_{\omega_0}(\mathrm{Scal}_{v_t}(\omega_{\varphi_t}) - w_t(\mu_{\varphi_t}) = 0$, which finishes the proof. □

The above deformation result applies in particular in the case of a v_0-soliton and a smooth family weights $v_t \in \mathcal{F}(X)$.

Corollary A.1 *Let* $(X, \mathbb{T})$ *be a smooth Fano manifold with canonically normalized polytope* P_X *and* $\{v_t \in C^\infty(\mathrm{P}_X),\ t \in U \subset \mathbb{R}^k\}$ *be a finite dimensional smooth family of positive weight functions, parameterized by a neighbourhood* $0 \in U \subset \mathbb{R}^k$*, which satisfy* $\mathrm{Fut}_{v_t} \equiv 0$ *for all* $t \in U$*. Suppose* $v_0 \in \mathcal{S}(X)$*, i.e.* X *admits a* $\mathbb{T}$*-invariant* v_0*-soliton. Then there exits* $\varepsilon > 0$ *such that for all* $||t|| < \varepsilon$, $v_t \in \mathcal{S}(X)$.

Proof By [3, Prop.1], a v-soliton ω on $(X, \mathbb{T}, 2\pi c_1(X))$ can be characterized by the property

$$\mathrm{Scal}_v(\omega) = \tilde{v}(\mu_\omega),$$

where $\tilde{v}(x) := 2(n + \langle d \log v, x\rangle)v(x)$. Furthermore, by [3, Lemma B1] we have

$$\mathrm{Fut}_v \equiv 0 \Leftrightarrow \mathrm{Fut}_{v,\tilde{v}} \equiv 0.$$

The claim then follows from Lemma A.1. □

Appendix B: Curvature Identities for KRS

Lemma B.1 ([52]) *Suppose (g, J, ω) is a $\mathbb{T}$-invariant Kähler metric which is a KRS, i.e. satisfies*

$$\mathrm{Ric}(\omega) - \omega = \frac{1}{2} dd^c f, \tag{B.1}$$

where f is the potential of a real Killing and holomorphic vector field $\tau = Jgrad_g f$. Then the scalar curvature $\mathrm{Scal}(\omega)$ *satisfies the following identity*

$$\frac{1}{2}\Delta \mathrm{Scal}(\omega) - \langle d\mathrm{Scal}(\omega), df\rangle_g + \mathrm{Scal}(\omega) = ||\mathrm{Ric}_g||^2. \tag{B.2}$$

Proof This is the forth formula in [52, Lemma 2.5] (which generally holds for any gradient Ricci soliton). We reproduce its proof in our Kähler situation for convenience of the Reader.

We denote by Ric_g the (symmetric) Ricci tensor of g, so that the (Kähler) Ricci form is $\mathrm{Ric}(\omega)(\cdot,\cdot) = \langle \mathrm{Ric}_g J\cdot,\cdot\rangle$. As $\tau := J\mathrm{grad}_g f$ is a Killing vector field, we have

$$\nabla d^c f = \frac{1}{2} dd^c f,$$

and, by Bochner's identity, $\Delta(d^c f) = 2\mathrm{Ric}_g(d^c f)$ or, equivalently,

$$-d(\Delta f) = 2\mathrm{Ric}(\omega)(\tau). \tag{B.3}$$

Taking trace with respect to ω in the soliton equation (B.1) yields

$$\mathrm{Scal}(\omega) - 2n = -\Delta f,$$

so that we have

$$d\mathrm{Scal}(\omega) = -\Delta df = 2\mathrm{Ric}(\omega)(\tau), \qquad \Delta \mathrm{Scal}(\omega) = -\Delta\Delta f. \tag{B.4}$$

Applying the covariant derivative ∇_X in (B.1) yields

$$\nabla_X (\mathrm{Ric}(\omega)) = \frac{1}{2}\nabla_X(dd^c f). \tag{B.5}$$

Taking interior product with ξ in (B.5) and using (B.4) and (B.1) again gives

$$\begin{aligned}\frac{1}{2}(\nabla_X dd^c f)(\tau) &= (\nabla_X \mathrm{Ric}(\omega))(\tau) = \nabla_X(\mathrm{Ric}(\omega)(\tau)) - \mathrm{Ric}(\omega)(\nabla_X \tau)\\ &= \frac{1}{2}\nabla_X d\mathrm{Scal}(\omega) - \mathrm{Ric}(\omega)(\mathrm{Ric}_g(JX) - JX)\\ &= \frac{1}{2}\nabla_X d\mathrm{Scal}(\omega) + \mathrm{Ric}_g(\mathrm{Ric}_g(X) - X).\end{aligned}$$

Taking trace gives

$$\frac{1}{2}(\Delta d^c f)(\tau) = -\frac{1}{2}\Delta\mathrm{Scal}(\omega) + ||\mathrm{Ric}_g||^2 - \mathrm{Scal}(\omega).$$

By (B.4) we have $\frac{1}{2}(\Delta d^c f)(\tau) = -\frac{1}{2}\langle d\mathrm{Scal}, df\rangle$, so we finally get (B.2). □

As an immediate corollary, we get the following fact, which we presume is well-known to experts, but which we were not able to trace in the literature.

Corollary B.1 *Suppose $(N, \mathcal{D}, J, \hat{\xi}, \eta_{\hat{\xi}})$ is a compact Sasaki manifold which is a transversal KRS. Then the transversal scalar curvature is everywhere positive.*

Proof We denote by Scal(ω) the transversal scalar curvature and consider a local $\hat{\xi}$-quotient (U, J, ω, g) around a point of global minimum of Scal(ω) on N. Notice that (U, J) is a (open) complex manifold and ω descends to define a KRS on U. By Lemma B.1, the smooth function Scal(ω) satisfies on U the identity (B.2). As Scal(ω) achieves its minimum in the interior of U, we get by the strong maximum principle $\mathrm{Scal}(\omega) > 0$ on U, and hence on N. □

References

1. V. Apostolov, D.J. Calderbank, The CR geometry of weighted extremal Kahler and Sasaki metrics. Math. Ann. **379**, 1047–1088 (2021)
2. V. Apostolov, D.J. Calderbank, E. Legendre, Weighted K-stability of polarized varieties and extremality of Sasaki manifolds. Adv. Math. **391** (2021). Paper No 107969
3. V. Apostolov, S. Jubert, A. Lahdili, Weighted K-stability and coercivity with applications to extremal Kähler and Sasaki metrics. Geom. Topol. **8**, 3229–3302 (2023)
4. M.F. Atiyah, Convexity and commuting Hamiltonians. Bull. Lond. Math. Soc. **14**, 1–15 (1982)
5. Th. Aubin, Réduction du cas positif de l'équation de Monge-Ampèere sur les variétés kählériennes compactes à la démonstration d'une inégalité . J. Funct. Anal. **57**, 143–153 (1984)
6. R. Berman, K-polystability of Q-Fano varieties admitting Kähler-Einstein metrics. Invent. Math. **203**, 973–1025 (2016)
7. R. Berman, B. Berndtsson, The projective space has maximal volume among all toric Kähler-Einstein manifolds. arXiv:1112.4445
8. R. Berman, B. Berndtsson, The volume of Kähler–Einstein Fano varieties and convex bodies. J. Reine Angew. Math. **723**, 127–152 (2017)
9. R.J. Berman, D. Witt-Nyström, Complex optimal transport and the pluripotential theory of Kähler-Ricci solitons. arXiv:1401.8264

10. R. Berman, S. Boucksom, M. Jonsson, A variational approach to the Yau-Tian-Donaldson conjecture. J. Am. Math. Soc. **34**, 605–652 (2021)
11. S. Boucksom, C. Favre, M. Jonsson, Differentiability of volumes of divisors and a problem of Teissier. J. Algebraic Geom. **18**, 279–308 (2009)
12. C.P. Boyer, C. van Coevering, Relative K-stability and extremal Sasaki metrics. Math. Res. Lett. **28**, 1–19 (2018)
13. C. Boyer, H. Huang, E. Legendre, C.W. Tønnesen-Friedman, Existence and non-existence of constant scalar curvature and extremal Sasaki metrics. Math. Z. **304** (2023), Article No 61
14. E. Calabi, On Kähler manifolds with vanishing canonical class, in *Algebraic Geometry and Topology: A Symposium in Honour of Lefschetz* (Princeton University Press, Princeton, 1955), pp. 78–89
15. H.-D. Cao, G. Tian, X. Zhu, Kähler–Ricci solitons on compact complex manifolds with $c_1(M) > 0$. Geom. Funct. Anal. **15**, 697–719 (2005)
16. X.X. Chen, S. Donaldson, S. Sun, Kähler–Einstein metrics on Fano manifolds, I: approximation of metrics with cone singularities. J. Am. Math. Soc. **28**, 183–197 (2015)
17. X.X. Chen, S. Donaldson, S. Sun, Kähler–Einstein metrics on Fano manifolds, II: limits with cone angle less than 2π. J. Am. Math. Soc. **28**, 199–234 (2015)
18. X.X. Chen, S. Donaldson, S. Sun, Kähler–Einstein metrics on Fano manifolds, III: limits as cone angle approaches 2π and completion of the main proof. J. Am. Math. Soc. **28**, 235–278 (2015)
19. X.X. Chen, S. Sun, B. Wang, Kähler-Ricci flow, Kähler-Einstein metric, and K-stability. Geom. Topol. **22**, 3145–3173 (2018)
20. T. Collins, G. Székelyhidi, K-semistability for irregular Sasaki manifolds. J. Differ. Geom. **109**, 81–109 (2018)
21. T. Collins, G. Székelyhidi, Sasaki-Einstein metrics and K-stability. Geom. Topol. **23**, 1339–1413 (2019)
22. T. Darvas, Y. Rubinstein, Tian's properness conjectures and Finsler geometry of the space of Kähler metrics. J. Am. Math. Soc. **30**, 347–387 (2017)
23. V. Datar, G. Székelyhidi, Kähler-Einstein metrics along the smooth continuity method. Geom. Funct. Anal. **26**, 975–1010 (2016)
24. T. Delcroix, Weight sensibility in K-stability of Fano varieties. arXiv:2411.07864
25. R. Dervan, E. Legendre, Valuative stability of polarized varieties. Math. Ann. **385**, 357–391 (2023)
26. S. Donaldson, Scalar curvature and stability of toric varieties. J. Differ. Geom. **62**, 289–349 (2002)
27. J. Duistermaat, G.J. Heckman, On the variation in the cohomology of the symplectic form of the reduced phase space. Invent. Math. **69**, 259–268 (1982)
28. K. Fujita, Optimal bounds for the volumes of Kähler-Einstein Fano manifolds. Am. J. Math. **140**, 391–414 (2018)
29. A. Futaki, An obstruction to the existence of Einstein Kähler metrics. Invent. Math. **73**, 437–443 (1983)
30. A. Futaki, T. Mabuchi, Bilinear forms and extremal Kähler vector fields associated with Kähler classes. Math. Ann. **301**, 199–210 (1995)
31. A. Futaki, H. Ono, G. Wang, Transverse Kähler geometry of Sasaki manifolds and toric Sasaki–Einstein manifolds. J. Differ. Geom. **83**, 585–635 (2009)
32. P. Gauduchon, *Calabi's extremal metrics: an elementary introduction.* Lecture notes available upon request
33. J.P. Gauntlett, D. Martelli, J. Sparks, S.-T. Yau, Obstructions to the existence of Sasaki–Einstein metrics. Commun. Math. Phys. **273**, 803–827 (2007)
34. V. Guillemin, S. Sternberg, Convexity properties of the moment mapping. Invent. Math. **67**, 491–513 (1982)
35. M. Hallam, The geometry and stability of fibrations. Oxford Ph.D. thesis (2022). Available at https://ora.ox.ac.uk/objects/uuid:941370e7-4267-40ce-ae45-8b009c168130
36. J. Han, C. Li, On the Yau-Tian-Donaldson conjecture for generalized Kähler-Ricci soliton equations. Commun. Pure Appl. Math. **76**, 1793–1867 (2023)

37. J. Han, C. Li, Algebraic uniqueness of Kähler-Ricci flow limits and optimal degenerations of Fano varieties. Geom. Topol. **28**, 539–592 (2024)
38. D. Hättig, J. Hausen, H. Süss, Log DelPezzo $\mathbb{C}^*$-surfaces, Kähler–Einstein metrics, Kähler–Ricci solitons and sasaki–Einstein metrics. Michigan Math. J. (to appear). arXiv:2306.03796v1
39. W. He, S. Sun, Frankel conjecture and Sasaki geometry. Adv. Math. **291**, 912–960 (2016)
40. D. Jerison, J.M. Lee, Extremals for the Sobolev inequalitiy for the Heisenberg group and the CR Yamabe problem. J. Am. Math. Soc. **1**, 1–13 (1988)
41. A. Lahdili, Kähler metrics with constant weighted scalar curvature and weighted K-stability. Proc. Lond. Math. Soc. **119**, 1065–1114 (2019)
42. C.R. LeBrun, S. Simanca, Extremal Kähler metrics and complex deformation theory. Geom. Funct. Anal. **4**, 298–336 (1994)
43. K.L. Lee, J. Sturm, X. Wang, Moment map, convex function and extremal point. arXiv:2208.03724
44. C. Li, Minimizing normalized volumes of valuations. Math. Z. **289**, 491–513 (2018)
45. C. Li, G-uniform stability and Kähler-Einstein metrics on singular Fano varieties. Invent. Math. **227**, 661–744 (2022)
46. C. Li, Y. Liu, C. Xu, A guided tour to normalized volume, in *Geometric Analysis*, vol. 333. Progr. Math. (Birkauser/Springer, Cham, 2020), pp. 167–219
47. A. Lichnerowicz, Géométrie des groupes de transformation. Travaux et Recherches Mathématiques 3, Dunod (1958)
48. T. Mabuchi, Multiplier Hermitian structures on Kähler manifolds. Nagoya J. Math. **170**, 73–115 (2003)
49. T. Mabuchi, Y. Nakagawa, New examples of Sasaki–Einstein manifolds. Tohoku Math. J. **65**, 243–252 (2013)
50. D. Martelli, J. Sparks, S.-T. Yau, Sasaki-Einstein manifolds and volume minimisation. Commun. Math. Phys. **280**, 611–673 (2008)
51. Y. Matsushima, Sur la structure du groupe d'homéomorphismes analytiques d'une certaine variété kählérienne. Nagoya Math. J. **11**, 145–150 (1957)
52. P. Petersen, W. Wylie, Rigidity of gradient Ricci solitons. Pacific J. Math. **241**, 329–345 (2009)
53. D. Petrecca, On Sasaki–Ricci solitons and their deformations. Adv. Geom. **16**, 57–70 (2016)
54. G. Tian, Kähler–Einstein metrics with positive scalar curvature. Invent. Math. **130**, 1–37 (1997)
55. G. Tian, K-stability and Kähler–Einstein metrics. Commun. Pure Appl. Math. **68**, 1085–1156 (2015); Corrigendum: **68**, 2082–2083 (2015)
56. G. Tian, X. Zhu, Uniqueness of Kähler-Ricci solitons. Acta Math. **184**, 271–305 (2000)
57. G. Tian, X. Zhu, A new holomorphic invariant and uniqueness of Kähler-Ricci solitons. Comment. Math. Helv. **77**, 297–325 (2002)
58. C. van Coevering, Examples of asymptotically conical Ricci-flat Kähler manifolds. Math. Z. **267**, 465–496 (2011)
59. L. Wang, A valuative criterion of K-polystability. arXiv:2406.06176
60. C. Woodward, Moment maps and geometric invariant theory. Les cours du CIRM, **Tome 1**(1), 55–98 (2010)
61. S.T. Yau, Open problems in geometry, in *Differential Geometry: Partial Differential Equations on Manifolds*. Proc. Sympos. Pure Math., vol. 54 (AMS, Los Angeles, 1990), pp. 1–28
62. K. Zhang, A quantization proof of the uniform Yau–Tian–Donaldson conjecture. J. Eur. Math. Soc. (2023). Published online: https://doi.org/10.4171/JEMS/1373

Projective Structures on Curves and Conformal Geometry

Florin Belgun and Andrei Moroianu

The authors are grateful to Paul Gauduchon, whose work (especially the unpublished work [5]) motivated us to develop the theory contained in this paper, which we dedicate to his 80th anniversary.

Abstract Projective structures on curves appear naturally in many areas of mathematics, from extrinsic conformal geometry to analysis, where the main problem is to find qualitative information about the solutions of Hill equations. In this paper, we describe in detail the correspondence between different equivalent definitions of projective structures and their isomorphism classes, correcting long-standing inexactitudes in the literature. As an application, we show that the *Yamabe problem for curves* in a conformal/Möbius ambient space has no solutions in general.

Keywords Conformal structure · Projective structure · Laplace structure · Möbius structure

2010 Mathematics Subject Classification Primary 53A20, 53A30, 53A55

This work was partly supported by the PNRR-III-C9-2023-I8 grant CF 149/31.07.2023 *Conformal Aspects of Geometry and Dynamics*.

F. Belgun
Institute of Mathematics "Simion Stoilow" of the Romanian Academy, Bucharest, Romania
e-mail: fbelgun@imar.ro

A. Moroianu (✉)
Université Paris-Saclay, CNRS, Laboratoire de mathématiques d'Orsay, Orsay, France

Institute of Mathematics "Simion Stoilow" of the Romanian Academy, Bucharest, Romania
e-mail: andrei.moroianu@math.cnrs.fr

L. Ornea (ed.), *Real and Complex Geometry*,
https://doi.org/10.1007/978-3-031-92297-8_2

1 Introduction

There are not many structures one can define on a 1-dimensional manifold, and all we can think of are *flat*, i.e. they do not admit local invariants.

For example, a conformal structure on a curve is simply its smooth structure, and all curves are orientable, even parallelizable. A Riemannian structure on a connected curve is characterized globally by its total length, and, for the ones of infinite length, whether they are complete or not.

There is, however, a quite interesting type of structures on a curve, and the moduli space of such structures—in case of closed curves—turns out to be really intricate. We are talking here about *projective structures*.

Historically, projective structures have been studied for a long time as particular cases of a (G, X)-geometry (a viewpoint first introduced by Felix Klein in his Erlangen Program, then formalized by Thurston). For the 1-dimensional case, Kuiper gave a classification in 1954 [11], but there are some flaws as we will see in the present paper.

Projective structures on curves are tightly linked with the vast domain of analysis focusing on *Hill's equation*

$$x'' + Fx = 0, \ \ F : \mathbb{R} \to \mathbb{R}, \ \ F(t + T_0) = F(t), \ \ \forall t \in \mathbb{R}.$$

Indeed, two independent solutions (x_1, x_2) of the above equation define a map

$$\mathbb{R} \ni t \mapsto [x_1(t) : x_2(t)] \in \mathbb{RP}^1$$

that provides local charts forming a projective atlas as a $(\mathrm{PSL}(2, \mathbb{R}), \mathbb{RP}^1)$-geometry, [13].

The latter analytic approach can be stated in an invariant way as a *Laplace structure* on a curve C, which is a second-order linear differential operator, with no first-order term (there is indeed an invariant way to state this, see Proposition 3.3) on a certain *weight bundle* on C (cf. Definition 3.4).

Our interest is motivated by the existence of various structures induced on a curve C embedded in a conformal manifold M by the conformal structure of the ambient space (more precisely, the ambient space M needs to be *conformal* if $\dim M \geq 3$, and furthermore *Möbius* if $\dim M = 2$, see Sect. 2 and [2]). Every such curve carries special parametrizations, for which the *conformal acceleration* vanishes, and a canonically induced *Laplace structure*. We show in Proposition 3.9 that these notions are strongly related.

The induced Laplace structure on a curve in a conformal/Möbius ambient space is, in fact, a particular case of the *conformal Laplacian*, or *Yamabe operator*, that is, a second order linear differential operator on a weight bundle, with symbol equal to the conformal contraction and with vanishing first order term, determined by the conformal structure alone. For all dimensions $n \geq 3$, the *Yamabe problem* consists in searching for a metric with constant scalar curvature in the conformal class (which turns out to be, up to a constant factor, the zero order term of the above Laplace structure, viewed in an appropriate gauge).

In the 1-dimensional case, Calderbank and Burstall formulated in [3] the *Yamabe problem for curves* in a conformal/Möbius ambient space: is there a metric in the conformal class for which the induced Laplace structure has constant zero order term?

In order to give a (negative) answer to this question, we classify the projective structures on connected curves up to isomorphism. This endeavor is not new, we could find many references dealing with this topic, starting with the classification of these structures by Kuiper [11], the classification (up to diffeomorphism) of vector fields preserving a projective structure on a circle [9], and many other authors interested in the qualitative behavior of solutions of Hill's equation [8].

Nevertheless, we consider that our differential-geometric approach brings more clarity and situates the study of projective structures in a broader context of differential geometry, like Lie group actions and differential operators associated to a given G-structure (in the sense of Kobayashi [10]).

We prove several versions of the equivalence between a Laplace structure on a curve and a projective structure, which in turn can be characterized by a *developing map* and its *holonomy*, [7], see Theorem 3.6.

The developing map is the main tool to characterize a projective structure on an open curve (diffeomorphic to $\mathbb{R}$), cf. Theorem 4.15, while the holonomy turns out to determine the isomorphism class of a projective structure on a closed curve (diffeomorphic to S^1), cf. Theorem 4.5.

For open curves, the developing map $D : C \to \widetilde{\mathbb{RP}^1}$ defines an embedding of the curve in the universal cover of $\mathbb{RP}^1$, and we define the *winding number* of the curve as the "average" number of points in the preimage in C of the points of $\mathbb{RP}^1$ (if this "average"is not integer, we "round it up" to a number in $\frac{1}{2} + \mathbb{N}$). In Proposition 4.15 and in Corollary 4.16 we retrieve the classification of open projective curves, first obtained in [11], see also [7].

Using some technical results about $\widetilde{\mathrm{SL}}(2, \mathbb{R})$, we approach in Sect. 5 the classification of projective structures on closed curves, in which the main role is played by the conjugacy class of a generator of the *holonomy group*, which acts freely on the image of the developing map. In fact, there exists a canonical generator (for a given orientation on the curve) of the holonomy group, and its conjugacy class determines the projective structure on the curve up to isomorphism, Proposition 5.4 and Corollary 5.5.

We complete the theory by Theorem 5.7, which produces a natural, continuous and bijective map from the set of Laplace structures on S^1, up to diffeomorphism, and the set of conjugacy classes of *positive generators* of the holonomy groups of projective structures on S^1. We use here the classification of conjugacy classes of $\widetilde{\mathrm{SL}}(2, \mathbb{R})$, Theorem 5.1, which seems to be a classical result but was only implicitly (and inexactly) used in [11], however it is only very recently that a complete study of the conjugacy classes of elements of $\widetilde{\mathrm{SL}}(2, \mathbb{R})$ has been published [14].

Finally, we describe in Proposition 5.8, and Proposition 5.9 some geometric properties that characterize each projective class, and in Proposition 5.12 we determine the automorphism group of each projective class. For the impatient reader, the table at the end of Sect. 5 provides most of the relevant information.

We conclude the paper by the proof of the fact that the zero order term of a Laplace structure can not always be constant, giving therefore a negative answer to the Yamabe problem for curves in a conformal/Möbius ambient space, cf. Proposition 6.1.

2 Curves in Conformal Geometry

Definition 2.1 On a n-dimensional manifold M, denote by L the (oriented) line bundle associated to the representation

$$\mathrm{GL}(n,\mathbb{R}) \ni A \mapsto |\det A|^{1/n} \in \mathbb{R}^*_+$$

and we refer to it as the *weight bundle of weight 1*.

Analogously, using other powers of the absolute value of the determinant, one can consider any real power L^k of the weight bundle L. We can easily notice that L^0 is canonically isomorphic to the trivial line bundle $M \times \mathbb{R}$, L^{-1} is isomorphic with the dual of L, and L^k, with $k \in N$, is isomorphic to the k-th tensorial power $L^{\otimes k}$ of L. Clearly, L^k are all oriented (we know what a *positive* section of these bundles is), but there is no canonical trivialization. However, a trivialization of L^k, for $k \neq 0$, induces a trivialization of all nonzero powers of L, including $L^{-n} \simeq |\Lambda^n M|$, the bundle of *densities* on M.

More generally, a tensor bundle on M (i.e., a bundle associated to the frame bundle of M via a representation $\rho : \mathrm{GL}(n,\mathbb{R}) \to \mathrm{GL}(m,\mathbb{R})$) is said to have *intrinsic weight* $k \in \mathbb{R}$ if and only if $\rho(a\mathrm{Id}_n) = a^k \mathrm{Id}_m$ for every $\alpha \in \mathbb{R}^*_+$. The bundles L^k make no exception to this rule, and k is indeed the *intrinsic weight* of L^k for every $k \in \mathbb{R}$.

Definition 2.2 A *conformal structure* on a manifold M is a symmetric, non-degenerate bilinear form c on TM with values in L^2, which is normalized, in the sense that the induced bilinear form on $|\Lambda^n M| \simeq L^{-n}$ with values in L^{-2n} is the identity of the square of the latter bundle.

In this paper we will only discuss *Riemannian* conformal structures, i.e. such that c is *positive definite*, which is a notion that makes sense because the target space of c is L^2, a canonically oriented bundle.

Notice that all bundle homomorphisms induced by c (including c itself, as a bundle map from $TM \otimes TM$ to L^2) are between spaces with the same intrinsic weight. In particular, the so-called *rising/lowering of indices* from T$^*M$ to TM

must be adjusted to satisfy this "weight conservation" principle. Thus, c defines an isomorphism between T^*M and $\mathrm{T}M \otimes L^{-2}$

The weight bundles L^k, for $k \in \mathbb{R}$, are also important in the case of a 1-dimensional manifold, a curve. Here, of course, a conformal structure c is nothing else but the very smooth structure of the curve.

For further notions and results about conformal geometry, we refer to [2]. Let us briefly summarize some of them:

Definition 2.3 A *Weyl structure* is a torsion-free conformal connection on (M, c). It induces connections on all bundles associated to the frame bundle of M.

Two Weyl structures ∇ and ∇' are related by

$$\nabla'_X Y - \nabla_X Y = \tilde{\theta}_X Y := \theta(X)Y + \theta(Y)X - c(X, Y)\theta, \text{ for } X, Y \in \mathrm{T}M, \tag{1}$$

where θ is a 1-form. Note that $c(X, Y)\theta \in L^2 \otimes \mathrm{T}^*M \simeq \mathrm{T}M$.

A remark (for example, from [2]) is that for every curve in a conformal manifold, the latter carries *adapted* Weyl structures (associated to parametrizations $\gamma : I \to C \subset M$ of the curve), for which the parametrized curve is *geodesic*, i.e. $\nabla_{\dot\gamma}\dot\gamma = 0$. In order to single out a particular class of parametrizations or even a class of immersed curves, one needs to introduce further objects induced by c.

The *Schouten-Weyl tensor* h^∇ of a Weyl structure ∇ is a bilinear form on $\mathrm{T}M$ and it is computed, for $n \geq 3$, from the curvature of ∇ [5, 6], see also [2, Definition 2.9]. The Schouten-Weyl tensor satisfies the following transformation rule when the Weyl structures are changed via (1), as shown in [5, 6]:

$$h^{\nabla'} = h^\nabla - \nabla\theta + \theta \otimes \theta - \frac{1}{2}c(\theta, \theta)c. \tag{2}$$

The formula from [2, Definition 2.9] cannot be used for $n = 2$, but one can define a Schouten-Weyl-like tensor on a *Möbius* surface [4], see also [2, Definition 3.2] (which is a structure that "rigidifies" the conformal structure of a 2-dimensional conformal manifold), and this tensor satisfies the gauge transformation law (2) as well. We use here the notion of a Möbius structure merely to be able to have a 2-dimensional ambient space and still get the same results as for higher-dimensional conformal spaces.

We make the following convention of naming a *conformal/Möbius* manifold (M, c): if $\dim M \geq 3$ this will simply mean (M, c) as a conformal manifold, if $n = 2$ we will assume M is endowed with a Möbius structure that is compatible with c. The consequence in both cases is that, for every Weyl structure ∇, there is an associated Schouten-Weyl tensor h^∇, which changes with the Weyl structure according to (2).

We are interested in the structure induced by a conformal/Möbius ambient space on a curve.

Let us first recall the definition of the *conformal acceleration* of a parametrized curve in conformal/Möbius ambient space [2, Definition 4.1 and Lemma 4.8]:

Definition 2.4 ([2]) The *conformal acceleration* $a_c(\gamma)$ of a curve $\gamma : I \to \mathcal{M}$ is equal to the vector field $c(\dot\gamma, \dot\gamma)h^{\nabla}(\dot\gamma)$ where ∇ is any Weyl structure adapted to γ.

As before, the factor $c(\dot\gamma, \dot\gamma) \in L^2$ ensures that the conformal acceleration is a vector field on $C := \gamma(I)$. This vector field has a *normal* and a *tangential* component

$$a_c(\gamma) = a_c^N(\gamma) + a_c^T(\gamma),$$

where $a_c^N \perp \dot\gamma$ and $a_c^T(c) \in TC$ are both independent of the choice of the Weyl structure ∇, and $a_c^N(\gamma)$ is furthermore independent also of the parametrization of C [2, Lemma 4.2].

We define the following class of parametrizations of any embedded curve $C \subset (M, c)$ [1]:

Definition 2.5 A (local) parametrization of a curve C in a conformal/Möbius manifold is *special* if and only if the tangential part of its conformal acceleration vanishes identically.

Such parametrizations always exist, as solutions of an ODE. An equivalent description of special parametrizations will be given in Sect. 3.2, Proposition 3.9.

3 Laplace Structures on Curves, Projective Parametrizations and Hill's Equation

We review here two main descriptions of projective structures on curves and describe the equivalence of these approaches. More precisely, a projective structure on a curve is defined by:

Definition 3.1 A *projective structure* on a given curve C is an atlas $\mathcal{A}$ of charts from open subsets of C with values in $\mathbb{R}$, such that the transition maps consist of restrictions of maps of the form

$$f_A(x) := \frac{ax+b}{cx+d}, \text{ for } A := \begin{pmatrix} a & b \\ c & d \end{pmatrix} \in \mathrm{GL}(2, \mathbb{R}),$$

called *homographies*. A *projective curve* is a curve C together with such an atlas $\mathcal{A}$, and a *projective map* $f : C_1 \to C_2$ between projective curves is a map that is induced, in corresponding projective charts of $\mathcal{A}_1$, respectively $\mathcal{A}_2$, by homographies.

A projective map is thus a local diffeomorphism. Conversely, if f is an immersion $f : C_1 \to C_2$ of curves, and C_2 has a projective structure, then f induces a projective structure on C_1.

A classical result states that the homographies f_A, as described in Definition 3.1, satisfy the equation $S(f_A) = 0$, where S is the *Schwarzian derivative*:

$$S(\varphi) := \frac{\varphi'''}{\varphi'} - \frac{3}{2}\left(\frac{\varphi''}{\varphi'}\right)^2, \tag{3}$$

defined on local diffeomorphisms $\varphi : I_1 \to I_2$ between intervals in $\mathbb{R}$.

Conversely, for every local diffeomorphism φ satisfying $S(\varphi) = 0$ there exists a matrix A such that φ coincides on an open set with f_A.

In the next section we will investigate in full detail the projective structures on curves from the viewpoint of (G, X)-structures (in our case, $(\mathrm{PGL}(2, \mathbb{R}), \mathbb{RP}^1)$-structures), but in the present section we focus on a class of differential-geometric structures on a curve, the *Laplace structures*, that we will further show to be equivalent to the projective structures from Definition 3.1.

3.1 Laplace Structures on Curves

Let C be a smooth curve. A connection ∇ on $\mathrm{T}C$ induces a connection on the canonical weight bundle $L := |\mathrm{T}C|$ and on every real power of L.

Remark 3.2 $L = |\mathrm{T}C|$ is the oriented line bundle associated to $\mathrm{T}C$; every orientation of C induces a canonical isomorphism of L with $\mathrm{T}C$. The bundle L being oriented (hence topologically trivial), all real powers of L are well defined and trivial. In the sequel, if X is a non-vanishing vector field along C, e.g., the speed vector field $\dot{\gamma}$ induced by some parametrization γ, then $|X|^k$ will denote the corresponding section of L^k.

Every parametrization γ of C induces a unique connection ∇^γ defined by requiring $\dot{\gamma}$ to be ∇^γ-parallel. Conversely, every connection admits compatible *local* (but not necessarily global) parametrizations. For example, the tautological connection on the half-line $(0, +\infty)$ induces a connection on the closed curve $C := (0, +\infty)/\sim$, where $t \sim 2t$, $\forall t > 0$, which carries no global parallel sections.

If ∇ is any connection on C, the corresponding connection on L^k defines a *Hessian* $\mathrm{Hess}^\nabla : \mathcal{C}^\infty(L^k) \to \mathcal{C}^\infty(\mathrm{T}^*C \otimes \mathrm{T}^*C \otimes L^k) \simeq \mathcal{C}^\infty(L^{k-2})$, acting on a section l on L^k by

$$\mathrm{Hess}^\nabla l(X, Y) := \nabla_X \nabla_Y l - \nabla_{\nabla_X Y} l.$$

When comparing the Hessians induced by two connections on L^k, [2, Proposition 3.7] (see also [9]) shows:

Lemma 3.3 *The difference between the Hessians on L^k of two arbitrary connections ∇, ∇' on a curve C is a zero-order operator if and only if $k = 1/2$.*

Proof The difference $\nabla' - \nabla$ is a 1–form $\theta \in \mathcal{C}^\infty(\mathrm{T}^*C)$. For every section l of L^k one thus has $\nabla'_X l - \nabla_X l = k\theta(X)l$. As the Hessians $\mathrm{Hess}^{\nabla'} l$ and $\mathrm{Hess}^{\nabla} l$ are bilinear forms on $\mathrm{T}C$ with values in L^k, we can choose the arguments $X, Y \in \mathrm{T}C$ to be non-zero, and such that $\nabla_X Y = 0$ at the point where we make the computation. This implies

$$\begin{aligned}
(\mathrm{Hess}^{\nabla'} l)(X, Y) &= \nabla'_X \nabla'_Y l - \nabla'_{\nabla'_X Y} l = \nabla'_X(\nabla_Y l + k\theta(Y)l) - \nabla'_{(\theta(X)Y)} l \\
&= \nabla_X \nabla_Y l + k(\nabla_X\theta)(Y)l + k\theta(Y)\nabla_X l + k\theta(X)\nabla_Y l \\
&\quad + k^2\theta(X)\theta(Y)l - \theta(X)\nabla_Y l - k\theta(X)\theta(Y)l \\
&= (\mathrm{Hess}^{\nabla} l)(X, Y) + (2k-1)\theta(X)\nabla_Y l \\
&\quad + k\left(\nabla_X\theta(Y) + (k-1)\theta(X)\theta(Y)\right) l,
\end{aligned}$$

where in the last equality we use the fact that $\theta(X)\nabla_Y l = \theta(Y)\nabla_X l$ since X and Y are collinear. This concludes the proof. □

The reason for which we included here the proof instead of merely citing [2], is to obtain the explicit relation between Hessians on $L^{1/2}$:

$$\mathrm{Hess}^{\nabla'} - \mathrm{Hess}^{\nabla} = \tfrac{1}{2}\left(\nabla\theta - \tfrac{1}{2}\theta \otimes \theta\right). \tag{4}$$

If ∇^{γ_1} and ∇^{γ_2} are the connections induced by two parametrizations γ_1, γ_2 of C, then the above formula reads

$$\mathrm{Hess}^{\gamma_2} - \mathrm{Hess}^{\gamma_1} = \tfrac{1}{2}\left(\nabla^{\gamma_1}\theta - \tfrac{1}{2}\theta \otimes \theta\right), \tag{5}$$

where $\theta := \nabla^{\gamma_2} - \nabla^{\gamma_1} \in \mathcal{C}^\infty(\mathrm{T}^*C)$.

Definition 3.4 (See also [2, Definition 3.8]) A *Laplace structure* $\mathcal{L}$ on a smooth curve C is a second order linear differential operator on $L^{1/2}$ with values in $\mathrm{T}^*C \otimes \mathrm{T}^*C \otimes L^{1/2} \simeq L^{-3/2}$, which differs from some (hence any) Hessian Hess^{∇} by a zero-order term.

Note that some authors, (e.g. Hitchin [9] or Guha [8]) use the term *projective connection* to denote what we call here a Laplace structure. Their main reason is that, although the operator has second order, it shares some features with linear connections: its symbol, as a bundle map from $\mathrm{Sym}^2(\mathrm{T}^*C) \otimes L^{1/2} \simeq L^{-3/2}$ to $L^{-3/2}$, is the identity, and (precisely because it contains no first-order derivatives), it also satisfies a relation similar to the Leibnitz rule. However, we prefer the terminology *Laplace structure* since it refers to a second-order operator (which moreover can also be defined on higher-dimensional manifolds [2]).

Recall (see Remark 3.2) that any (local) parametrization $\gamma : I \to C$ of C defines privileged local sections $|\dot\gamma|^k \circ \gamma^{-1} \in \mathcal{C}^\infty(L^k)$ for every k. A Laplace structure $\mathcal{L}$ on C thus induces, for each parametrization γ, a second order operator $\mathcal{L}^\gamma$ acting on functions $x : I \to \mathbb{R}$ as follows:

$$|\dot{\gamma}|^{-3/2}\mathcal{L}^{\gamma}(x) \circ \gamma^{-1} := \mathcal{L}((x \circ \gamma^{-1})|\dot{\gamma}|^{1/2}) \tag{6}$$

(this is an equality between sections of $\mathcal{L}^{-3/2}$ defined along the open set $\gamma(I) \subset C$). Lemma 3.3 shows that the operator $\mathcal{L}^{\gamma}$ differs from the second derivative on functions by a zero-order term (i.e., it does not involve any first-order derivative):

$$\mathcal{L}^{\gamma}x = x'' + F^{\gamma}x. \tag{7}$$

Proposition 3.5 *A Laplace structure $\mathcal{L}$ on a curve C determines a map*

$$\mathcal{F}: \{\text{local parametrizations } \gamma : I \to C\} \longrightarrow \{\text{functions on } I\}, \qquad \gamma \mapsto F^{\gamma}.$$

such that the dependence of F^{γ} on γ satisfies the following rule with respect to gauge transformations: If $\gamma_1 : I_1 \to C$ is a local parametrization and $\varphi : I_2 \to I_1$ is a diffeomorphism, then the function F^{γ_2} associated to the local parametrization $\gamma_2 := \gamma_1 \circ \varphi$ reads

$$F^{\gamma_2} = (\varphi')^2 F^{\gamma_1} \circ \varphi + \tfrac{1}{2}S(\varphi), \tag{8}$$

where $S(\varphi)$ is the Schwarzian derivative of φ (3).

Proof Let $\mathcal{L}$ be a Laplace structure on C, let γ_1 and $\gamma_2 := \gamma_1 \circ \varphi$ be two local parametrizations of C and let $x_1 : I_1 \to \mathbb{R}$ be any smooth function. We denote by $x_2 := x_1 \circ \varphi$. From (6) and (7) it follows that

$$x_1'' + F^{\gamma_1}x_1 = (|\dot{\gamma}_1|^{3/2}\mathcal{L}((x_1 \circ \gamma_1^{-1})|\dot{\gamma}_1|^{1/2})) \circ \gamma_1 \tag{9}$$

and

$$x_2'' + F^{\gamma_2}x_2 = (|\dot{\gamma}_2|^{3/2}\mathcal{L}((x_2 \circ \gamma_2^{-1})|\dot{\gamma}_2|^{1/2})) \circ \gamma_2. \tag{10}$$

On the other hand, the local vector fields $\dot{\gamma}_1$ and $\dot{\gamma}_2$ along C are related by $\dot{\gamma}_2 = \varphi' \circ \gamma_2^{-1}\dot{\gamma}_1$. Using that $x_1 \circ \gamma_1^{-1} = x_2 \circ \gamma_2^{-1}$, we can write the argument of the right hand side of (9) as $(x_1 \circ \gamma_1^{-1})|\dot{\gamma}_1|^{1/2} = (x_2(\varphi')^{-1/2}) \circ \gamma_2^{-1})|\dot{\gamma}_2|^{1/2}$. Applying (10) to $(x_2(\varphi')^{-1/2}$ instead of x_2 yields

$$(x_2(\varphi')^{-1/2})'' + F^{\gamma_2}(x_2(\varphi')^{-1/2}) = (|\dot{\gamma}_2|^{3/2}\mathcal{L}((x_1 \circ \gamma_1^{-1})|\dot{\gamma}_1|^{1/2})) \circ \gamma_2. \tag{11}$$

Comparing (9) and (11) we thus obtain

$$(x_2(\varphi')^{-1/2})'' + F^{\gamma_2}(x_2(\varphi')^{-1/2}) = (\varphi')^{3/2}(x_1'' + F^{\gamma_1}x_1) \circ \varphi. \tag{12}$$

Since $x_2 = x_1 \circ \varphi$, we have $x_2' = \varphi' x_1' \circ \varphi$ and $x_2'' = (\varphi')^2 x_1'' \circ \varphi + \varphi'' x_1' \circ \varphi$, whence

$$\begin{aligned}(x_2(\varphi')^{-1/2})'' &= x_2''(\varphi')^{-1/2} + 2x_2'((\varphi')^{-1/2})' + x_2((\varphi')^{-1/2})'' \\ &= (\varphi')^{3/2} x_1'' \circ \varphi + (\varphi')^{-1/2} \varphi'' x_1' \circ \varphi \\ &\quad - (\varphi')^{-1/2} \varphi'' x_1' \circ \varphi + x_2((\varphi')^{-1/2})'' \\ &= (\varphi')^{3/2} x_1'' \circ \varphi + x_2((\varphi')^{-1/2})''.\end{aligned}$$

From (12) we thus get

$$(\varphi')^{-1/2} F^{\gamma_2} + ((\varphi')^{-1/2})'' = (\varphi')^{3/2} F^{\gamma_1} \circ \varphi,$$

which is equivalent to (8). □

Conversely, every such map $\mathcal{F}$ satisfying the above gauge transformation rule defines a Laplace structure on C. Indeed, for every local parametrization $\gamma : I \to C$, the function F^γ on I defines by (7) a second order differential operator $\mathcal{L}^\gamma$ on functions, which in turn defines a second order differential operator $\mathcal{L}$ on $L^{1/2}$ (along the image of γ). The fact that this operator does not depend on the choice of γ follows, by reversing the previous argument, from the transformation rule (8).

We state now the main result of this section:

Theorem 3.6 *Let C be a closed curve. Then there is a canonical equivalence between:*

(1) Projective structures on C;
(2) Laplace structures $\mathcal{L} : C^\infty(L^{1/2}) \to C^\infty(L^{-3/2})$.

Moreover, if $\mathcal{A}$ is the projective atlas corresponding to $\mathcal{L}$, we have:

(a) For any local parametrization $\gamma : I \to C$, the induced operator $\mathcal{L}^\gamma$ acting on functions on I is the second derivative (i.e., $F^\gamma \equiv 0$) if and only if γ is locally the inverse of a projective chart.
(b) For any choice of an orientation of C (and thus of an isomorphism $L \simeq TC$) a local non-vanishing section l of $L^{1/2}$ satisfies the Laplace equation

$$\mathcal{L} l = 0, \tag{13}$$

if and only if there exists a projective chart compatible with the orientation whose inverse γ satisfies $\dot{\gamma} = l^2$.
(c) For any pair of linearly independent solutions l_1, l_2 of Eq. (13), the map

$$l_{1,2} : C \to \mathbb{R}^*, \ l_{1,2}(p) := \frac{l_1(p)}{l_2(p)}, \tag{14}$$

defined where $l_2 \neq 0$, is a projective chart.

Proof Let $\mathcal{A}$ be a projective structure on C as in Definition 3.1. The charts of $\mathcal{A}$ will be referred to as *projective charts*. The inverse, $\gamma_1 : I_1 \to \gamma_1(I_1) \subset C$ of a projective chart is a local parametrization of C. We define a Laplace structure along $\gamma_1(I_1)$ by

$$\mathcal{L}_1 := \mathrm{Hess}^{\gamma_1} \text{ on sections of } L^{1/2}.$$

This is equivalent to say that the function F^{γ_1} defined by $\mathcal{L}_1$ in Proposition 3.5 vanishes. If $\gamma_2 : I_2 \to C$ is the inverse of another projective chart, one can write on $\gamma_2^{-1}(\gamma_1(I_1))$ the relation $\gamma_2 = \gamma_1 \circ \varphi$ for some homography $\varphi : \gamma_2^{-1}(\gamma_1(I_1)) \to \gamma_1^{-1}(\gamma_2(I_2))$. Because of (8) and by the well-known fact that homographies have vanishing Schwarzian derivative, it turns out that $\mathcal{L}_1$ and $\mathcal{L}_2$ coincide on their common domain of definition. Consequently, the local Laplace structures defined in this way glue together to define a global Laplace structure $\mathcal{L}^{\mathcal{A}}$ on C.

Conversely, suppose $\mathcal{L}$ is a Laplace structure on C. We claim that around any point there exist local parametrizations γ such that $F^\gamma = 0$. Indeed, pick any local parametrization γ_0 of C and look for a re-parametrization $\gamma := \gamma_0 \circ \varphi$ such that $F^\gamma = 0$. From (8), this is a third order non-linear differential equation for φ:

$$S(\varphi) + 2(\varphi')^2 F^{\gamma_0} \circ \varphi = 0,$$

which obviously has local solutions φ around each point.

If γ_1 and γ_2 are local parametrizations with $F^{\gamma_1} = F^{\gamma_2} = 0$, and p is any point in $\mathrm{Im}(\gamma_1) \cap \mathrm{Im}(\gamma_2)$, then the local inverses of γ_1 and γ_2 near p are real-valued charts on C whose coordinate change $\varphi := \gamma_1^{-1} \circ \gamma_2$ is a homography because by (8) it satisfies the equation $S(\varphi) = 0$. The set of charts constructed in this way defines a projective atlas $\mathcal{A}^{\mathcal{L}}$. The correspondences $\mathcal{A} \mapsto \mathcal{L}^{\mathcal{A}}$ and $\mathcal{L} \mapsto \mathcal{A}^{\mathcal{L}}$ are clearly inverse to each other.

We now turn to the second part of the statement. The point (a) follows directly from (7). To prove (b), note that every non-vanishing section l of $L^{1/2}$, together with the choice of an orientation of C (inducing an isomorphism of TC with L) defines local parametrizations γ (unique up to composition with a translation) such that $\dot\gamma = l^2$. If $\mathcal{L}l = 0$ then (6) and (7) applied to the constant function $x \equiv 1$ yield $F^\gamma \equiv 0$, so γ is by (a) the inverse of a projective chart. Conversely, if γ is the inverse of a projective chart then $F^\gamma = 0$ and from (6) and (7) it follows that $l := |\dot\gamma|^{1/2}$ satisfies $\mathcal{L}l = 0$.

For (c), we choose a local projective parametrization (i.e. inverse of projective chart) $\gamma : I \to C$. By (a), we have $F^\gamma = 0$. We write (for $i = 1, 2$) $l_i = (x_i \circ \gamma^{-1})|\dot\gamma|^{1/2}$, where $x_i : I \to \mathbb{R}$ are smooth functions. Then by (6) and (7) we get $x_i'' = 0$, so $f := x_1/x_2$ is a homography. Consequently $l_1/l_2 = f \circ \gamma^{-1}$ is the composition of a homography with a projective chart, i.e. it is a projective chart as well. □

Remark 3.7 The map $C \ni x \mapsto l_1(x)/l_2(x) \in \mathbb{R}$ from the point (c) in Theorem 3.6 is defined only on the open set where the solution l_2 of the Laplace equation does not vanish. Instead we can consider, for any pair l_1, l_2 of linearly independent solutions of (13) (note that they are defined globally on $\tilde{C} \simeq \mathbb{R}$ as solutions to a linear ODE), the map

$$[l_1 : l_2] : \tilde{C} \to \mathbb{RP}^1, \ x \mapsto [l_1(p(x) : l_2(p(x)], \tag{15}$$

which is well-defined because the isomorphisms of $L^k \oplus L^k$ with $\mathbb{R}^2$ induced by any trivialization of L all induce the same identification between $\mathbb{P}(L^k \oplus L^k)$ and $\mathbb{P}(\mathbb{R}^2) = \mathbb{RP}^1$. As shown in Theorem 3.6, the map $[l_1 : l_2]$ is locally a projective chart, hence any lift:

$$D_{l_1,l_2} : \tilde{C} \to \widetilde{\mathbb{RP}^1} \tag{16}$$

to $\widetilde{\mathbb{RP}^1}$ of this map is a *developing map* for the induced projective structure—see Definition 4.3 below.

3.2 Laplace Structures on Curves in Conformal/Möbius Geometry

A *Laplace structure* [2, Definition 3.8] on a conformal manifold (of any dimension $n \geq 1$) is defined as a second-order linear differential operator, with vanishing first order term and symbol equal to the conformal contraction ($c : \mathrm{T}^*M \otimes \mathrm{T}^*M \to L^{-2}$) on sections of $L^{1-n/2}$, cf. [2]. For $n = 1$, we get Definition 3.4 above.

Moreover, if M is conformal manifold of dimension $n \geq 3$ or a Möbius surface, then any curve $C \subset M$ turns out to inherit a canonical Laplace structure induced by the ambient space [2, Proposition 4.25]:

Proposition 3.8 *For a curve C in a conformal/Möbius manifold (M, c), the following operator is a Laplace structure which only depends on the embedding and the structure of the ambient space:*

$$\mathcal{L}^c : C^\infty(L^{1/2}) \to C^\infty(L^{-3/2}), \ \mathcal{L}^c l := \mathrm{Hess}^\nabla l + \frac{1}{2}\frac{h^\nabla(\dot\gamma, \dot\gamma)}{c(\dot\gamma, \dot\gamma))}, \tag{17}$$

where γ is a local parametrization and ∇ an adapted Weyl structure.

As a consequence of Theorem 3.6, any curve in a conformal/Möbius ambient space inherits a projective structure, and the local projective parametrizations γ for this structure are such that $h^\nabla(\dot\gamma, \dot\gamma) = 0$ for the corresponding adapted Weyl structure. We can therefore make the connection with the *special parametrizations* defined in Definition 2.5:

Proposition 3.9 *For a curve in a conformal/Möbius manifold, the* special *parametrizations (for which the tangential conformal acceleration vanishes) are exactly the projective parametrizations for the induced Laplace structure.*

Therefore, the projective structure on a curve in a conformal/Möbius ambient space can be either seen via the projective charts from Theorem 3.6 (c) (for the induced Laplace structure) or via the parametrizations that have vanishing tangential part of the conformal acceleration.

4 Projective Structures on Curves

In this section we discuss the projective structures on open and closed curves from the classical viewpoint of Felix Klein's Erlangen Program, and present a complete classification of these structures up to isomorphism, see also [7, 9, 11].

A homography f_A as in Definition 3.1 actually only depends on the image $\hat{A} \in \mathrm{PGL}(2, \mathbb{R})$ of the matrix A, so we may also think of a projective structure as of a (G, X)-structure in the sense of Thurston (see also the works of Kuiper, Kobayashi, Ehresmann, Klein), where $G := \mathrm{PGL}(2, \mathbb{R})$ and $X = \mathbb{RP}^1$. We recall:

Definition 4.1 Let G be a Lie group acting effectively and transitively on the manifold X. A (G, X)*-structure* on a manifold M is an atlas of diffeomorphisms $\varphi : U \to \varphi(U) \subset X$ such that, for any two charts φ_1, φ_2, on each connected component of its domain of definition, the transition function $\varphi_2 \circ \varphi_1^{-1}$ coincides with the action on that connected component of an element of G.

A map $f : M \to N$ between two (G, X)-manifolds is called a *G-map* if, for every charts $\varphi : U \to \varphi(U) \subset X$, $\psi : V \to \psi(V) \subset X$ on M, resp. N, with U connected and small enough such that $f(U) \subset V$, the composition $\psi \circ f \circ \varphi^{-1}$ is given by the action of an element $g \in G$.

4.1 The Developing Map and Holonomy of a Closed Curve

The general theory of (G, X)-structures [7] on a manifold M states that such a structure is equivalent with a pair (D, ρ) of a *developing map* $D : \tilde{M} \to X$ which is ρ-equivariant, $\rho : \pi_1(M) \to G$ being the *holonomy* homomorphism, up to equivalence:

Theorem 4.2 ([7]) *Let M be a (G, X)-manifold. Then there exists a G-map $D : \tilde{M} \to X$, and a group homomorphism $\rho : \pi_1(M) \to G$ such that*

$$\tilde{D}(\gamma x) = \tilde{\rho}(\gamma)\tilde{D}(x), \ \forall x \in \tilde{M}, \ \forall \gamma \in \pi_1(M). \tag{18}$$

The set of pairs (D, ρ) *as above has an equivalence relation:*

$$(D, \rho) \sim (D', \rho') \Longleftrightarrow \exists g \in G \; : \; D'(.) = gD(.) \textit{ and } \rho' = g\rho g^{-1}, \tag{19}$$

and the equivalence classes are in 1–1 correspondence to the isomorphism classes of (G, X)*-structures on* M.

Definition 4.3 D is called a *developing map* and ρ its *holonomy*.

Instead of a full proof for the classical result above (for which we refer to [7]), we recall here briefly the way the holonomy homomorphism appears in the general theory of (G, X)-structures:

A developing map is obtained by gluing (G, X)-charts on M, using elements of G on the intersections, and this gluing turns out to yield global maps from $\tilde{M}$ to X. The fundamental group $\pi_1(M)$ acts on $\tilde{M}$ by (G, X)-maps, which necessarily correspond to elements of G, and that defines the group homomorphism ρ.

If G is connected and its universal cover $\tilde{G}$ acts effectively on the universal cover $\tilde{X}$ of X, a (G, X)-structure induces a $(\tilde{G}, \tilde{X})$-structure and conversely.

In our case, as all curves are orientable, there is no loss of generality to first consider *oriented projective structures* on C compatible with a given orientation (i.e., an atlas of charts such that the transition functions are orientation-preserving homographies of type f_A with $A \in \mathrm{SL}(2, \mathbb{R})$) and then study orientation-reversing projective isomorphisms.

An *oriented projective* structure on C is thus a $(\mathrm{PSL}(2, \mathbb{R}), \mathbb{RP}^1)$-structure or, equivalently, a $(\widetilde{\mathrm{SL}}(2, \mathbb{R}), \mathbb{RP}^1)$-structure. The advantage of this latter description is that both the model space and the structure group are simply connected; its drawback is that $\widetilde{\mathrm{SL}}(2, \mathbb{R})$ is not a linear group.

From now on, we will study *oriented projective structures* on C to be a $(\widetilde{\mathrm{SL}}(2, \mathbb{R}), \widetilde{\mathbb{RP}^1})$-structure, whose isomorphism class is in turn characterized by a pair (D, ρ), where $D : \tilde{C} \to \widetilde{\mathbb{RP}^1}$ is the developing map and $\rho : \pi_1(M) \to \widetilde{\mathrm{SL}}(2, \mathbb{R})$ is its holonomy, up to the equivalence relation (19).

For a projective structure on C defined by an atlas $\mathcal{A}$ like in Definition 3.1 and a fixed orientation on C, we define an oriented projectve structure by the atlas $\mathcal{A}_+$ of orientation-preserving charts. Conversely, such an atlas $\mathcal{A}_+$ may be completed with orientation-reversing projective charts $\mathcal{A}_-$ (that also define an oriented projective structure, compatible with the other orientation on C). The only question is whether the same curve C, the projective structures defined by $(C, \mathcal{A}_+)$ and by $(C, \mathcal{A}_-)$ are equivalent *as oriented projective structures*, i.e. whether there exists an orientation-reversing automorphism τ of $(C, \mathcal{A}_+)$. We will answer this question separately for open, resp. closed curves at the end of this Section, resp in Proposition 5.6.

The developing map D is identified (up to a diffeomorphism of C) with its image $U \subset \widetilde{\mathbb{RP}^1}$. For ρ, there are two possible cases: C is an open curve (in which case ρ is trivial), and C is a closed curve, when ρ is identified with its image in $\widetilde{\mathrm{SL}}(2, \mathbb{R})$, an infinite cyclic group acting freely and properly on U.

In the latter case, one can further choose from the two possible generators of $\rho(\pi_1(C))$:

Definition 4.4 For a closed projective curve, with developing map $D : \tilde{C} \to \widetilde{\mathbb{RP}^1}$ and holonomy $\rho : \pi_1(C) \to \widetilde{\mathrm{SL}}(2, \mathbb{R})$, the generator of $\rho(\pi_1(C))$ such that

$$\tilde{A}(x) > x, \ \forall x \in U = D(\tilde{C}), \tag{20}$$

is called *the positive generator* of the holonomy.

We retrieve the following result (used implicitly in [7] and [11]):

Theorem 4.5 *The oriented projective structures on $C \simeq S^1$, up to diffeomorphism, are in 1–1 correspondence with the equivalence classes of pairs $(U, \tilde{A})$ where $U \subset \widetilde{\mathbb{RP}^1}$ is an interval and $\tilde{A} \in \widetilde{\mathrm{SL}}(2, \mathbb{R})$ such that $\tilde{A}(U) = U$ and* (20) *holds, modulo the equivalence relation*

$$(U, \tilde{A}) \sim (U', \tilde{A}') \iff \exists \tilde{B} \in \widetilde{\mathrm{SL}}(2, \mathbb{R}) : \ U' = \tilde{B}(U) \text{ and } \tilde{A}' = \tilde{B}\tilde{A}\tilde{B}^{-1}. \tag{21}$$

To complete the classification we will need to study in detail the action of $\widetilde{\mathrm{SL}}(2, \mathbb{R})$ on $\widetilde{\mathbb{RP}^1}$, starting from the action of $\mathrm{SL}(2, \mathbb{R})$ on $\mathbb{RP}^1$.

4.2 The Action of $\widetilde{\mathrm{SL}}(2, \mathbb{R})$ on $\widetilde{\mathbb{RP}^1}$

Denote by $\mathbf{p} : \mathbb{R} \to \mathbb{RP}^1$, $\mathbf{p}(t) := [\cos t : \sin t]$. Clearly, this is a realization of the universal covering of $\mathbb{RP}^1$. In order to underline the projective structure of the domain of definition of $\mathbf{p}$, we will always denote it by $\widetilde{\mathbb{RP}^1}$ instead of $\mathbb{R}$. Note that the affine chart $\{[x : 1] \ , \ x \in \mathbb{R}\} \subset \mathbb{RP}^1$, which is the complement of the "point at infinity" $[1 : 0] \in \mathbb{RP}^1$, is isomorphic via $\mathbf{p}$ with each of the intervals $(k\pi, (k+1)\pi) \subset \widetilde{\mathbb{RP}^1}$, $k \in \mathbb{Z}$.

The action of $\mathrm{PSL}(2, \mathbb{R})$ on $\mathbb{RP}^1$ defined by the homographies lifts to a faithful, transitive action of $\widetilde{\mathrm{SL}}(2, \mathbb{R})$ on $\widetilde{\mathbb{RP}^1}$. The center of $\widetilde{\mathrm{SL}}(2, \mathbb{R})$ is isomorphic to $\mathbb{Z}$ and coincides with the kernel of the projection $\Phi : \widetilde{\mathrm{SL}}(2, \mathbb{R}) \to \mathrm{PSL}(2, \mathbb{R})$, and also with the automorphism group of the Galois covering $\mathbf{p} : \widetilde{\mathbb{RP}^1} \to \mathbb{RP}^1$. We clearly have

$$\mathbf{p}(\tilde{A}(x)) = \Phi(\tilde{A})(\mathbf{p}(x)), \ \forall x \in \widetilde{\mathbb{RP}^1}, \ \forall \tilde{A} \in \widetilde{\mathrm{SL}}(2, \mathbb{R}). \tag{22}$$

We will also consider the "intermediate" covering $\bar{\mathbf{p}} : \widetilde{\mathbb{RP}^1} \to S^1 \subset \mathbb{R}^2$, $\bar{\mathbf{p}}(t) := (\cos t, \sin t)$, because through it we can also see the intermediate projection $\bar{\Phi} : \widetilde{\mathrm{SL}}(2, \mathbb{R}) \to \mathrm{SL}(2, \mathbb{R})$, which is the double cover of $\mathrm{PSL}(2, \mathbb{R})$: indeed, an element $\tilde{A} \in \widetilde{\mathrm{SL}}(2, \mathbb{R})$ (which therefore acts on $\widetilde{\mathbb{RP}^1}$) is mapped to a matrix $A := \bar{\Phi}(\tilde{A}) \in \mathrm{SL}(2, \mathbb{R})$ such that $\bar{\mathbf{p}} \circ \tilde{A} = A \circ \bar{\mathbf{p}}$, where $A \in \mathrm{SL}(2, \mathbb{R})$ acts on S^1. Now, the action of $\mathrm{SL}(2,\mathbb{R})$ on S^1 is given by

$$\mathrm{SL}(2,\mathbb{R}) \times S^1 \ni (A,x) \mapsto \frac{Ax}{\|Ax\|} \in S^1,$$

where $\|x\|$ is the Euclidean norm of a vector in $\mathbb{R}^2$. The expression is nonlinear unless $A \in \mathrm{SO}(2)$, the maximal compact subgroup of $\mathrm{SL}(2,\mathbb{R})$:

Proposition 4.6 *The translations $\tilde{E}_\alpha : \widetilde{\mathbb{RP}^1} \to \widetilde{\mathbb{RP}^1}$ with $\alpha \in \mathbb{R}$, defined by $\tilde{E}_\alpha(x) := x + \alpha$, are exactly the elements in $\widetilde{\mathrm{SL}}(2,\mathbb{R})$ that project over the rotation matrices $E_\alpha := \begin{pmatrix} \cos\alpha & -\sin\alpha \\ \sin\alpha & \cos\alpha \end{pmatrix}$ acting on S^1, in the sense that $\bar{\mathbf{p}} \circ \tilde{E}_\alpha = E_\alpha \circ \bar{\mathbf{p}}$. The center of $\widetilde{\mathrm{SL}}(2,\mathbb{R})$ is equal to $\{\tilde{E}_{k\pi}\ ,\ k \in \mathbb{Z}\} = \ker(\Phi)$.*

It is well-known that $\mathrm{SL}(2,\mathbb{R})$ admits the partition $\mathcal{H} \cup \mathcal{P} \cup \mathcal{E}^0 \cup \mathcal{Z}$ defined by

$$\mathcal{H} := \{A \in \mathrm{SL}(2,\mathbb{R})\ ,\ |\mathrm{tr} A| > 2\},\ \mathcal{E}^0 := \{A \in \mathrm{SL}(2,\mathbb{R})\ ,\ |\mathrm{tr} A| < 2\},$$

which are both open subsets formed of *hyperbolic*, resp. *properly elliptic* elements, $\mathcal{Z} = \{\pm I_2\}$ is the center of $\mathrm{SL}(2,\mathbb{R})$ (called *central*, or *degenerate elliptic* elements of $\mathrm{SL}(2,\mathbb{R})$, cf. [12]) and $\mathcal{P} := \{A \in \mathrm{SL}(2,\mathbb{R}) \setminus \mathcal{Z}\ ,\ |\mathrm{tr} A| = 2\}$ is the set of *parabolic* elements.

Let us give a classical result about the set of representatives of the conjugacy classes in $\mathrm{SL}(2,\mathbb{R})$:

Proposition 4.7 *Every matrix $A \in \mathrm{SL}(2,\mathbb{R})$ is conjugated up to sign to one of the following:*

(1) $H(\lambda) := \begin{pmatrix} \lambda & 0 \\ 0 & \lambda^{-1} \end{pmatrix}$, $\lambda \in (0,1)$;
(2) $P^\pm := \begin{pmatrix} 1 & \pm 1 \\ 0 & 1 \end{pmatrix}$;
(3) E_α, $\alpha \in (0,\pi)$;
(4) id.

Moreover, the above matrices and their opposites represent different conjugacy classes.

Proof In the hyperbolic, resp. properly elliptic cases, the matrix A has two distinct eigenvalues, therefore it is either real-diagonalizable, hence conjugated to a matrix of the form $H(\lambda)$, with $\lambda \notin \{-1,0,1\}$, or it is complex-diagonalizable, hence conjugated in $\mathrm{SL}(2,\mathbb{R})$ to a matrix of the form E_α, $\alpha \in (-\pi,0) \cup (0,\pi)$.

In the hyperbolic case, we have $JH(\lambda)J^{-1} = H(\lambda^{-1})$, where $J := E_{\pi/2}$, which is not surprising as it has the same eigenvalues as $H(\lambda)$.

For $|\mathrm{tr}A| = 2$, we have either a diagonalizable matrix, hence $A = \pm\mathrm{id}$, or A is conjugated to $\pm P(x)$, which is defined as

$$P(x) := \begin{pmatrix} 1 & x \\ 0 & 1 \end{pmatrix}, \quad x \in \mathbb{R}. \tag{23}$$

On the other hand $H(\lambda)P(x)H(\lambda^{-1}) = P(\lambda^{-2}x)$, so by choosing $\lambda = \sqrt{|x|}$ we obtain that A is conjugated to $\pm P^{\pm}$.

We need to show that two different representatives in the above list are not conjugated to each other. First, as the trace doesn't change by conjugation, we only need to prove that matrices of the same type (hyperbolic, parabolic, properly elliptic and central) are not conjugated to one another.

For hyperbolic matrices this is clear as $\mathrm{tr}A$ determines uniquely λ such that $|\lambda| \in (0, 1)$ and $\lambda + \lambda^{-1} = \mathrm{tr}A$.

For elliptic matrices we only need to show that E_α is not conjugated to $E_{-\alpha}$, for $\alpha \in (0, \pi)$. Suppose $B \in \mathrm{SL}(2, \mathbb{R})$ such that $BE_\alpha B^{-1} = E_{-a}$. For a non-zero vector $X \in \mathbb{R}^2$, we can write

$$X \wedge E_{-\alpha}X = BB^{-1}X \wedge BE_\alpha B^{-1}X = \det B(B^{-1}X \wedge E_\alpha B^{-1}X) = Y \wedge E_\alpha Y,$$

for $Y := B^{-1}X$. This is a contradiction since for any non-zero vector $X \in \mathbb{R}^2$, the bi-vector $X \wedge E_\alpha X$ is a positive multiple of the volume element $e_1 \wedge e_2$, whereas $X \wedge E_{-\alpha}X$ is a negative multiple of $e_1 \wedge e_2$.

For parabolic matrices we need to show that P^+ and P^- are not conjugated in $\mathrm{SL}(2, \mathbb{R})$. If this were possible, then there would be a basis X_1, X_2 of $\mathbb{R}^2$ such that $P^- X_1 = X_1$ (this already implies $X_1 = \begin{pmatrix} a \\ 0 \end{pmatrix}$, with $a \in \mathbb{R}^*$), and $P^- X_2 = X_1 + X_2$. Denoting $X_2 = \begin{pmatrix} b \\ c \end{pmatrix}$, with $c \neq 0$, this last equation reads

$$\begin{pmatrix} b - c \\ c \end{pmatrix} = \begin{pmatrix} a + b \\ c \end{pmatrix},$$

thus $a = -c$ which implies that the basis X_1, X_2 is not positive, i.e., the basis change is made by a matrix with negative determinant, contradiction. □

The sign ambiguities are removed when considering the homographies associated to matrices in $\mathrm{SL}(2, \mathbb{R})$, i.e. the conjugacy classes in $\mathrm{PSL}(2, \mathbb{R})$:

Proposition 4.8 *The group* $\mathrm{PSL}(2, \mathbb{R})$ *admits the following partition;* $\mathrm{PSL}(2, \mathbb{R}) = \hat{\mathcal{H}} \cup \hat{\mathcal{P}} \cup \hat{\mathcal{E}}^0 \cup \{\mathrm{id}\}$ *invariant by conjugation, and a full set of representatives is given by, respectively:*

(1) $\hat{H}(\lambda)$, $\lambda \in (0, 1)$;
(2) $\hat{P}^{\pm}$;

(3) $\hat{E}_\alpha$, $\alpha \in (0, \pi)$;
(4) id.

Remark 4.9 A fixed point for the action of $A \in \mathrm{SL}(2, \mathbb{R})$ on $\mathbb{RP}^1$ corresponds to a real eigenspace of A as a matrix, hence the partitions above can also be retrieved by looking at the number of fixed points in $\mathbb{RP}^1$ of the respective elements: two for hyperbolic, one for parabolic, and none for properly elliptic elements. Central elements act trivially on $\mathbb{RP}^1$.

Pulling back through ϕ the above partition, we get

$$\widetilde{\mathrm{SL}}(2, \mathbb{R}) = \tilde{\mathcal{H}} \cup \tilde{\mathcal{P}} \cup \tilde{\mathcal{E}}^0 \cup \tilde{\mathcal{Z}}, \tag{24}$$

a partition in hyperbolic, parabolic, properly elliptic and central elements (which, together, form the elliptic elements) of $\widetilde{\mathrm{SL}}(2, \mathbb{R})$.

In order to give in $\widetilde{\mathrm{SL}}(2, \mathbb{R})$ a full set of representatives of the conjugacy classes, we need the following

Lemma 4.10 *Let* $\mathbf{p} : \tilde{G} \to G$ *be the universal cover of a Lie group* G*, and denote by* $Z := \ker \mathbf{p}$*, consisting of central elements in* $\tilde{G}$*. There exists a 1–1 correspondence between conjugacy classes in* $\tilde{G}$ *and pairs* $([g], z)$*, with* $g \in G$*,* $z \in Z$*, where* $[g]$ *is the conjugacy class of* g *in* G*.*

Proof Let $\tilde{g}_2 = \tilde{h}\tilde{g}_1\tilde{h}^{-1}$, for $\tilde{g}_{1,2}, \tilde{h} \in \tilde{G}$ and set $g_{1,2} := \mathbf{p}(\tilde{g}_{1,2})$, $h := \mathbf{p}(\tilde{h})$. Clearly g_1 is conjugated to g_2 in G. Conversely, if we have $hg_1h^{-1} = g_2$, the element $\tilde{h}\tilde{g}_1\tilde{h}^{-1}$ is the same for every lift $\tilde{h}$ of h via $\mathbf{p}$, because two such lifts differ by a central element. Therefore there exists $z \in Z$ such that $\tilde{g}_2 = z\tilde{h}\tilde{g}_1\tilde{h}^{-1}$.

This means that a conjugacy class in $\tilde{G}$ is determined by a conjugacy class in G and an element in Z. Note that in order to associate to $[\tilde{g}]$ a pair $([g], z)$ we need to fix a (non-canonic) lift of each representative $g \in [g] \subset G$ to an element $\tilde{g} \in \mathbf{p}^{-1}(g)$. □

We consider now lifts of the representatives from Proposition 4.8:

The translations in $\mathbb{R} = \widetilde{\mathbb{RP}^1}$ are properly elliptic if and only if $\alpha \notin \pi\mathbb{Z}$, they are central otherwise, cf. Proposition 4.6. From the Lemma above we conclude that $\{\tilde{E}_\alpha\ ,\ \alpha \in \mathbb{R}\}$ is a full set of representatives of conjugacy classes of elliptic elements $\tilde{\mathcal{E}}^0 \cup \tilde{\mathcal{Z}}$.

For $\tilde{A} \in \tilde{\mathcal{H}} \cup \tilde{\mathcal{P}}$, $\Phi(\tilde{A}) = \hat{A} \in \hat{\mathcal{H}} \cup \hat{\mathcal{P}}$ has two, resp. a fixed point on $\mathbb{RP}^1$ and the following can be said about $\tilde{A}$:

Lemma 4.11 *Let* $\tilde{A} \in \widetilde{\mathrm{SL}}(2, \mathbb{R})$ *and* $\hat{A} := \Phi(\tilde{A}) \in \mathrm{PSL}(2, \mathbb{R})$ *be its projection. If* $\tilde{A}$ *fixes* $x \in \widetilde{\mathbb{RP}^1}$*, then it fixes all points in* $\mathbf{p}^{-1}(\mathbf{p}(x)) = \{x + k\pi\ ,\ k \in \mathbb{Z}\}$*. If* $\hat{A}$ *has, in addition to* $\mathbf{p}(x)$*, also another fixed point* $z \in \mathbb{RP}^1$*, then* $\tilde{A}(y) = y$ *for all* $y \in \mathbf{p}^{-1}(z)$*.*

Conversely, if $\hat{A} \in \mathrm{SL}(2, \mathbb{R})$ *has a fixed point in* $\mathbb{RP}^1$*, then there exists a unique element* $\tilde{A}_0 \in \widetilde{\mathrm{SL}}(2, \mathbb{R})$ *such that* $\Phi(\tilde{A}_0) = A$ *and* $\tilde{A}_0$ *fixes one (hence all) points in* $\mathbf{p}^{-1}(z)$*, for any fixed point* z *of* $\hat{A}$*. Moreover,* $\Phi^{-1}(\hat{A}) = \{\tilde{A}_k := \tilde{A}_0\tilde{E}_{k\pi}\ ,\ k \in \mathbb{Z}\}$*.*

Proof Let $x + k\pi = \tilde{E}_{k\pi}(x)$ be any element of $\mathbf{p}^{-1}(\mathbf{p}(x))$. Then $\tilde{A}(x + k\pi) = \tilde{A}(\tilde{E}_{k\pi}(x)) = \tilde{E}_{k\pi}(\tilde{A}(x)) = \tilde{E}_{k\pi}x = x + k\pi$ for any $k \in \mathbb{Z}$, because $\tilde{E}_{k\pi}$ are in the center of $\widetilde{\mathrm{SL}}(2, \mathbb{R})$. This proves the first claim.

For the second claim, note that $\mathbf{p}^{-1}(z)$ is a $\tilde{A}$-invariant subset of $\widetilde{\mathbb{RP}^1}$, and it consists of one point y_k for each interval $(x + k\pi, x + (k + 1)\pi) \subset \widetilde{\mathbb{RP}^1}$. The endpoints of this interval are fixed points of $\tilde{A}$, which is also a monotone increasing map from $\widetilde{\mathbb{RP}^1} = \mathbb{R}$ to itself. That means that $\tilde{A}(y_k)$ also belongs to the same interval, hence $\tilde{A}(y_k) = y_k$.

Let us now consider $\hat{A} \in \mathrm{PSL}(2, \mathbb{R})$ that has a fixed point $z \in \mathbb{RP}^1$. Then any $\tilde{A} \in \Phi^{-1}(\hat{A})$ maps $\mathbf{p}^{-1}(z)$ to itself, in particular, for every $x \in \mathbf{p}^{-1}(z)$ there exists an integer k (which is actually independent of x) such that $\tilde{A}(x) = x + k\pi = \tilde{E}_{k\pi}(x)$. Therefore $\tilde{A} = \tilde{A}_k$ as in the statement. □

Remark 4.12 For an element $A \in \mathrm{SL}(2, \mathbb{R})$ acting (as $\hat{A} \in \mathrm{PSL}(2, \mathbb{R})$) with fixed points on $\mathbb{RP}^1$, we call $\tilde{A}_0$ as in Lemma 4.11 the *canonical lift* of A, resp. $\hat{A}$ in $\widetilde{\mathrm{SL}}(2, \mathbb{R})$. Note that the elements $\tilde{A}_k$, $k \in \mathbb{Z}^*$, have no fixed points on $\widetilde{\mathbb{RP}^1}$.

Remark 4.13 The action of the connected group $\widetilde{\mathrm{SL}}(2, \mathbb{R})$ on $\widetilde{\mathbb{RP}^1}$ is by monotone increasing diffeomorphisms. The orientation-reversing diffeomorphism of $\widetilde{\mathbb{RP}^1} \equiv \mathbb{R}$, defined by $\tilde{\tau}(t) := -t$ also projects via $\mathbf{p}$ to a diffeomorphism of $\mathbb{RP}^1$, namely the map $[x : y] \mapsto [x : -y]$ which actually is the homography f_T, defined by $T := \begin{pmatrix} 1 & 0 \\ 0 & -1 \end{pmatrix} \in \mathrm{GL}(2, \mathbb{R})$, hence τ is an orientation-reversing projective automorphism of $\widetilde{\mathbb{RP}^1}$. Note that, if $\tilde{A}_0(0) = 0$, then $\tilde{\tau} \circ \tilde{A}_0 \circ \tilde{\tau}$ is again an orientation-preserving diffeomorphism on $\widetilde{\mathbb{RP}^1}$ that fixes 0, thus it is equal to some $\tilde{B}_0$ and $Fix(\tilde{B}_0) = -Fix(\tilde{A}_0) \subset \widetilde{\mathbb{RP}^1} = \mathbb{R}$. Consequently, conjugation with $\tilde{\tau}$ maps $\tilde{\mathcal{H}}$, $\tilde{\mathcal{P}}$, $\tilde{\mathcal{E}}^0$ and $\tilde{\mathcal{Z}}$ to themselves. Here $Fix(F)$ denotes the set of fixed points of a diffeomorphism $F : \widetilde{\mathbb{RP}^1} \to \widetilde{\mathbb{RP}^1}$.

4.3 *Open Projective Curves*

An open, oriented, projective curve C can be seen, via a developing map, as an interval in $\mathbb{R} = \widetilde{\mathbb{RP}^1}$, and we need to classify them up to an isomorphism in $\widetilde{\mathrm{SL}}(2, \mathbb{R})$. There are three types of open intervals:

(1) $\widetilde{\mathbb{RP}^1}$ itself;
(2) a half-line;
(3) a bounded interval (α, β).

Obviously, any element $\tilde{A} \in \widetilde{\mathrm{SL}}(2, \mathbb{R})$ acts by bijections of $\widetilde{\mathbb{RP}^1}$, thus $\widetilde{\mathbb{RP}^1}$ can not be isomorphic, as a projective curve, to a half-line or some bounded interval. On the other hand, because $\tilde{A} : \widetilde{\mathbb{RP}^1} \to \widetilde{\mathbb{RP}^1}$ is monotone increasing (see Remark 4.13), we have

$$\tilde{A}\left((-\infty, p_0)\right) = (-\infty, \tilde{A}(p_0)),$$

hence, on one hand, such a *left* half-line is not isomorphic to any *right* half-line or to any bounded interval and, on the other hand, any two left (or right) half-lines are isomorphic (because $\widetilde{\mathrm{SL}}(2, \mathbb{R})$ acts transitively on $\widetilde{\mathbb{RP}^1}$). Thus, $\widetilde{\mathbb{RP}^1}$, left and right half-lines are three distinct isomorphism classes of oriented open projective curves.

Let us consider the remaining type (bounded intervals).

Definition 4.14 Let $U := (a, b) \subset \widetilde{\mathbb{RP}^1}$, with $a, b \in \mathbb{R}$, $a < b$.

(1) If $b = a + k\pi$, $k \in \mathbb{N}$, or, equivalently $b = \tilde{E}_{k\pi}(a)$, we say that the interval U has *winding number* $k \in \mathbb{N}$, and write $W(U) = k$.
(2) If $a + k\pi < b < a + (k+1)\pi$, (or equivalently $\tilde{E}_{k\pi}(a) < b < \tilde{E}_{(k+1)\pi}(a)$) with $k \in \mathbb{N}$, we say that the interval U has *winding number* $k + \frac{1}{2}$, $k \in \mathbb{N}$ and write $W(U) = n + \frac{1}{2}$.

For example, an interval projecting via $\mathbf{p}$ on a maximal affine chart, or $\mathbb{RP}^1$ without a point, e.g. the interval $(0, \pi)$, has winding number 1, and an interval projecting on an affine half-line inside an affine chart, e.g. the interval $(0, \frac{\pi}{2}) \subset \widetilde{\mathbb{RP}^1}$, has, by the above convention, winding number $\frac{1}{2}$.

In general the winding number of an interval $U \subset \widetilde{\mathbb{RP}^1}$ is related to the number in $U \cap \mathbf{p}^{-1}(x)$ (how many times U is *wrapped* around $\mathbb{RP}^1$), as follows; $\{\sharp(I \cap \mathbf{p}^{-1}(x))\ ,\ x \in \mathbb{RP}^1\}$ always consists of two non-negative integers $k, k+1$, the distinction is made by how many points $x \in \mathbb{RP}^1$ attain the lower number above: if there is only one point, then $W(U) = k+1$, if there are at least two (hence infinitely many), then $W(U) = k + \frac{1}{2}$.

We can now prove the following:

Proposition 4.15 *The winding number (with values in $\frac{1}{2}\mathbb{N}$) is the unique invariant of a bounded interval of $\widetilde{\mathbb{RP}^1}$ under the action of $\widetilde{\mathrm{SL}}(2, \mathbb{R})$: For U and U' bounded intervals of $\widetilde{\mathbb{RP}^1}$, there exists an element $\tilde{A} \in \widetilde{\mathrm{SL}}(2, \mathbb{R})$ such that $\tilde{A}(U) = U'$ and if and only if these intervals have the same winding numbers.*

Proof By mapping U, U' by appropriate elements of $\widetilde{\mathrm{SL}}(2, \mathbb{R})$, we can assume $U = \widetilde{(0, p)}$ and $U' = (0, p')$. Any further action on U needs to be by $\tilde{A}$ that fixes $0 \in \widetilde{\mathbb{RP}^1}$.

Assume that there exists $\tilde{A} \in \widetilde{\mathrm{SL}}(2, \mathbb{R})$ such that $\tilde{A}(0) = 0$ and $\tilde{A}(p) = p'$. If $W(0, p) = k \in \mathbb{N}$ then $p = k\pi$ and $p' = \tilde{A}(k\pi) = k\pi = p$ as implied by Lemma 4.11.

If the winding number of $(0, p)$ is $k + \frac{1}{2}$ then $p \in (k\pi, (k+1)\pi)$ and $p' = \tilde{A}(p)$ lies in the same interval (see the proof of Lemma 4.11), hence $W(0, p') = k + \frac{1}{2}$ as claimed.

Conversely, assume that $(0, p)$ and $(0, p')$ have the same winding numbers. If this number is $k \in \mathbb{N}$ then $p = p' = k\pi$ and we may even take $\tilde{A}$ to be the identity $\tilde{E}_0 \in \widetilde{\mathrm{SL}}(2, \mathbb{R})$.

If $W(0, p) = W(0, p') = k + \frac{1}{2}$, then $p, p' \in (k\pi, (k+1)\pi)$. Their projections $\mathbf{p}(p), \mathbf{p}(p') \in \mathbb{RP}^1$ are equal to $[x : 1]$, resp $[x' : 1]$, simply because they are both different from $\mathbf{p}(k\pi) = \mathbf{p}(0) = [1 : 0]$. We have $[x' : 1] = \hat{P}(x' - x)([x : 1])$, where $P(x' - x)$ is the parabolic matrix defined in (23). Lemma 4.11 implies that $\tilde{P}_0(x' - x)(0) = 0$ and $\tilde{P}_0(x' - x)(p) = p'$ for $\tilde{P}_0(x' - x)$ the canonical lift of $P(x' - x)$, see Remark 4.12.

Therefore $\tilde{P}_0(x' - x)$ maps $(0, p)$ onto $(0, p')$ as claimed. □

We see that the isomorphism class of an open, oriented projective curve is determined by its winding number or, if the latter is infinite, there are three possible isomorphism classes: $(0, \infty)$, $(-\infty, 0)$, $\mathbb{R} \subset \mathbb{RP}^1$.

For non-oriented open projective curves, note that the map $s \mapsto -s$ on $\widetilde{\mathbb{RP}}^1$ (which induces an orientation-reversing projective automorphism of $\widetilde{\mathbb{RP}}^1$) does not change the winding number, but establishes an isomorphism between a left- and a right-half-line. Therefore, we have (see also [7, 11]):

Corollary 4.16 *The moduli space $\mathcal{M}^o$ of isomorphism classes of connected, open projective curves (disregarding the orientation) is countable:*

$$\mathcal{M}^o = \{\infty, H\} \cup \tfrac{1}{2}\mathbb{N}^*.$$

The point ∞ corresponds to $\widetilde{\mathbb{RP}}^1$, H represents the isomorphism class of any half-line in $\widetilde{\mathbb{RP}}^1$, and any number x in $\frac{1}{2}\mathbb{N}$ represents the isomorphism class of a bounded interval in $\widetilde{\mathbb{RP}}^1$ of winding number x.

5 The Isomorphism Classes of Closed Projective Curves

In order to classify, up to isomorphism, the projective structures on a closed curve, we need to refine Theorem 4.5 using our knowledge of the action of $\widetilde{\mathrm{SL}}(2, \mathbb{R})$ on $\widetilde{\mathbb{RP}}^1$.

5.1 *The Conjugacy Classes in* $\widetilde{\mathrm{SL}}(2, \mathbb{R})$

Using Lemma 4.10, we refine the partition (24) by decomposing it further in conjugacy classes, and give for each of them a standard representative:

Theorem 5.1 *The conjugacy classes of elements in $\widetilde{\mathrm{SL}}(2, \mathbb{R})$ are the following, with respective representatives:*

(1) $\tilde{\mathcal{H}} = \bigcup_{k\in\mathbb{Z},\ \lambda\in(0,1)} \tilde{\mathcal{H}}_k(\lambda)$, *represented by* $\tilde{H}_k(\lambda) := \tilde{H}_0(\lambda)\tilde{E}_{k\pi} \in \tilde{\mathcal{H}}_k(\lambda)$, *where* $\tilde{H}_0(\lambda)$ *is the canonical lift of* $\hat{H}(\lambda) \in \mathrm{PSL}(2, \mathbb{R})$*;*

(2) $\tilde{\mathcal{P}} = \bigcup_{k\in\mathbb{Z}} (\mathcal{P}_k^+ \cup \mathcal{P}_k^-)$, *represented by* $\tilde{P}_k^\pm := \tilde{P}_0^\pm \tilde{E}_{k\pi} \in \tilde{\mathcal{P}}_k^\pm$, *where* $\tilde{P}_0^\pm$ *is the canonical lift of* $\hat{P}^\pm \in \mathrm{PSL}(2, \mathbb{R})$*;*

(3) $\tilde{\mathcal{E}}^0 = \bigcup_{\alpha\in\mathbb{R}\setminus\pi\mathbb{Z}} \tilde{\mathcal{E}}_\alpha$, *represented by* $\tilde{E}_\alpha \in \tilde{\mathcal{E}}_\alpha$*;*

(4) $\tilde{\mathcal{Z}} = \{\tilde{E}_{k\pi}\ ,\ k \in \mathbb{Z}\}$*, formed by conjugacy classes containing only one element each.*

Remark 5.2 The smooth map $\mathrm{SL}(2, \mathbb{R}) \ni A \mapsto \mathrm{tr}A \in \mathbb{R}$ is a submersion when restricted to $\mathrm{SL}(2, \mathbb{R}) \setminus \mathcal{Z}$, and the connected components of the level sets (which are, therefore, closed surfaces in $\mathrm{SL}(2, \mathbb{R}) \setminus \mathcal{Z}$) are exactly the conjugacy classes. We can define the map $\kappa : \widetilde{\mathrm{SL}}(2, \mathbb{R}) \to \mathbb{R}$, $\kappa(\tilde{A}); = \mathrm{tr}(\mathbf{p}(\tilde{A}))$, which is also a submersion on $\widetilde{\mathrm{SL}}(2, \mathbb{R}) \setminus \tilde{\mathcal{Z}}$, and its level sets are (grouped pairwise by the notation $\kappa^{-1}(\pm a) := \kappa^{-1}(a) \cup \kappa^{-1}(-a)$, for $a > 2$, $0 \leq a < 2$ and resp. $a = 2$):

$$\kappa^{-1}\left(\pm\tfrac{\lambda^2+1}{\lambda}\right) = \bigcup_{k\in\mathbb{Z}} \tilde{\mathcal{H}}_k(\lambda), \quad \kappa^{-1}(\pm 2\cos\alpha) = \bigcup_{k\in\mathbb{Z}} \tilde{\mathcal{E}}_{\alpha+k\pi}, \quad \kappa^{-1}(\pm 2) = \tilde{\mathcal{P}}.$$

Here, in order to write each component of a level set exactly once, we make the convention that $\alpha \in (0, \pi)$ and $\lambda \in (0, 1)$.

5.2 *Classification of Projective Structures on Closed Curves (With or Without a Fixed Orientation)*

We will simplify the data associated with a projective structure, replacing a pair $(U, \tilde{A})$ as in Theorem 4.5 with $\tilde{A}$ alone. As $\tilde{A}$ from the above mentioned theorem satisfies the inequality (20), we define a subset of $\widetilde{\mathrm{SL}}(2, \mathbb{R})$ where $\tilde{A}$ is contained:

Definition 5.3 The subset $\widetilde{\mathrm{SL}}(2, \mathbb{R})_+$ is the set of elements $\tilde{A}$ such that there exists at least one point $x \in \widetilde{\mathbb{RP}^1}$ such that $\tilde{A}(x) > x$.

Clearly, all elements conjugated to $\tilde{A} \in \widetilde{\mathrm{SL}}(2, \mathbb{R})_+$ are also in $\widetilde{\mathrm{SL}}(2, \mathbb{R})_+$. For every $\tilde{A} \in \widetilde{\mathrm{SL}}(2, \mathbb{R})$, $\tilde{A}$ or its inverse are in $\widetilde{\mathrm{SL}}(2, \mathbb{R})_+$ (sometimes both, see below), and if $\tilde{A}$ has no fixed points on $\widetilde{\mathbb{RP}^1}$, then exactly one of $\tilde{A}$ and $\tilde{A}^{-1}$ belongs to $\widetilde{\mathrm{SL}}(2, \mathbb{R})_+$. The element $\tilde{H}_0(\lambda)$ admits points on $\widetilde{\mathbb{RP}^1}$ that satisfy $\tilde{H}_0(\lambda)(x) > x$ and also points satisfying $\tilde{H}_0(\lambda)(y) < y$. On the other hand, we have $\tilde{P}_0^+(x) \leq x$ and $\tilde{P}_0^-(x) \geq x$ for all $x \in \widetilde{\mathbb{RP}^1}$. We conclude:

$$\widetilde{\mathrm{SL}}(2, \mathbb{R})_+ = \bigcup_{k\in\mathbb{N},\ \lambda\in(0,1)} \tilde{\mathcal{H}}_k(\lambda) \cup \bigcup_{\alpha>0} \tilde{\mathcal{E}}_\alpha \cup \bigcup_{k\in\mathbb{N}} \tilde{\mathcal{P}}_k^- \cup \bigcup_{k\in\mathbb{N}^*} \tilde{\mathcal{P}}_k^+. \tag{25}$$

We show now that an isomorphism class of an oriented closed projective curve C is encoded in the positive generator $\tilde{A}$ of its holonomy group, cf Definition 4.4. Note that if $\tilde{A}$ has no fixed points, then there is nothing to show since in that case $C = \widetilde{\mathbb{RP}^1}/\langle\tilde{A}\rangle$.

Proposition 5.4 *Let $\tilde{A} \in \widetilde{\mathrm{SL}}(2,\mathbb{R})_+$ and let $a < b$ be fixed points of $\tilde{A}$ such that $\tilde{A}(x) > x$, $\forall x \in U := (a,b) \subset \widetilde{\mathbb{RP}^1}$. Then, every other interval U' with same property is of the form $U' = (a + k\pi, b + k\pi)$ for some $k \in \mathbb{Z}$. Moreover, the projective curves $U/\langle\tilde{A}\rangle$ and $U'/\langle\tilde{A}\rangle$ are isomorphic.*

Proof The statement is invariant by conjugation in $\widetilde{\mathrm{SL}}(2,\mathbb{R})$ since the translation by $k\pi$ in $\mathbb{R}$ is given by the central element $\tilde{E}_{k\pi}$, and conjugation in $\widetilde{\mathrm{SL}}(2,\mathbb{R})$ preserves the inequalities in $\mathbb{R} = \widetilde{\mathbb{RP}^1}$ (in the sense that if $\tilde{A} = \tilde{B}\tilde{A}'\tilde{B}^{-1}$ and $A(x) > x$ on some interval U, then $A'(x) > x$ on $\tilde{B}(U)$.

If $\tilde{A} \in \widetilde{\mathrm{SL}}(2,\mathbb{R})_+$ has fixed points, we can thus assume up to conjugation that it is either equal to $\tilde{P}_0^-$, whose fixed points are $\pi\mathbb{Z}$, or to some $\tilde{H}_0(\lambda)$, $\lambda \in (0,1)$, whose fixed points are $\frac{\pi}{2}\mathbb{Z}$.

In the first case, the interval U is necessarily of the form $(k\pi, (k+1)\pi)$, which trivially implies our claim.

In the second case, since $H(\lambda)$ maps an element $[\cos\alpha : \sin\alpha] \in \mathbb{RP}$ onto $[\lambda\cos\alpha : \lambda^{-1}\sin\alpha]$, whose argument is larger than α for $\alpha \in (0, \frac{\pi}{2})$ and smaller than α on $(\frac{\pi}{2}, \pi)$, we obtain that U is necessarily of the form $(k\pi, k\pi + \frac{\pi}{2})$, and again the claim follows.

The last statement follows directly from the fact that the projective structures associated to the pairs $(U, \tilde{A})$ and $(\tilde{E}_{k\pi}U, \tilde{A})$ are isomorphic by taking $\tilde{B} = \tilde{E}_{k\pi}$ in Theorem 4.5. □

Corollary 5.5 *The moduli space of oriented projective structures on closed curves is the set of conjugacy classes of elements of $\widetilde{\mathrm{SL}}(2,\mathbb{R})_+$.*

We show now that we can drop the word "oriented" from the above result:

Proposition 5.6 *All closed projective curves admit orientation-reversing projective automorphisms. Therefore, the set of oriented projective structures on a closed curve C coincides with the set of projective structures on C.*

Proof It is enough to prove the statement for particular representatives $\tilde{A}$ of each conjugacy class in $\widetilde{\mathrm{SL}}(2,\mathbb{R})_+$. If $\tilde{A}$ has no fixed points on $\widetilde{\mathbb{RP}^1}$, it suffices to construct an orientation-reversing map on $\widetilde{\mathbb{RP}^1}$ (that induces on $\mathbb{RP}^1$ an orientation-reversing homography) such that the holonomy group $\langle\tilde{A}\rangle$ is preserved. In all cases we set $\tilde{\tau} : \widetilde{\mathbb{RP}^1} \to \widetilde{\mathbb{RP}^1}$ by $\tilde{\tau}(x) := -x$ (which induces the matrix $T \in \mathrm{GL}(2,\mathbb{R})$, see Remark 4.13, and we claim that

$$\tau\tilde{E}_\alpha\tau = \tilde{E}_{-\alpha}, \quad (\tau\tilde{J})\tilde{H}(\lambda)(\tilde{J}^{-1}\tilde{\tau}) = \tau\tilde{H}_n(\lambda^{-1})\tau = \tilde{H}_n(\lambda^{-1}),$$
$$\tau\tilde{P}_n^{\pm}\tau = P_{-n}^{\mp}, \ \forall n \in \mathbb{Z}, \tag{26}$$

where juxtaposition means composition of projective maps on $\widetilde{\mathbb{RP}^1}$. If $\tilde{A}$ is one of these elements, its conjugate by τ, resp. $(\tau\tilde{J})$, would be exactly its inverse, thus the group $\langle\tilde{A}\rangle$ would be preserved.

To prove (26) we proceed case by case: for $\tilde{A} = \tilde{E}_\alpha$ this is an identity involving affine transformations on $\mathbb{R} = \widetilde{\mathbb{RP}^1}$:

$$\tau\tilde{E}_\alpha\tau(x) = -(-x+\alpha) = x - \alpha = \tilde{E}_{-\alpha}(x), \ \forall x \in \widetilde{\mathbb{RP}^1}.$$

For the hyperbolic and parabolic cases we first prove the identities for $k = 0$, i.e., $\tilde{A} = \tilde{A}_0$ is the canonical lift of a matrix $A \in \mathrm{SL}(2,\mathbb{R})$ of hyperbolic, resp. parabolic type. It is straightforward that the right hand sides of (26) also have fixed points on $\mathbb{RP}^1$, hence the identities are implied by the corresponding identities of matrices, i.e.

$$TJ^{-1}H(\lambda)JT = TH(\lambda^{-1})T = H(\lambda^{-1}),$$

for which the first identity has already appeared in the proof of Proposition 4.7 and the second follows because T and $H(\lambda^{-1})$ are both diagonal, hence they commute, and

$$TP(\pm 1)T = P(\mp 1),$$

which is again a very simple computation.

In general $\tilde{A} = \tilde{A}_0\tilde{E}_{k\pi}$, $k \in \mathbb{Z}$, so we apply the already proved identities for $\tilde{E}_{k\pi}$ and for $\tilde{A}_0$, thus finishing the proof of (26). □

5.3 *Classification of Laplace Structures on Closed Curves*

The classification of Laplace structures on closed curves, that can be also obtained by following Theorems 3.6, 4.5 and Proposition 5.4, can be stated by exhibiting a natural map that behaves well under deformations:

Theorem 5.7 *There exists a natural map Ξ from the set of equivalence classes of real-valued, 1-periodic functions $F : \mathbb{R} \to \mathbb{R}$, up to the action of the diffeomorphism group of $\mathbb{R}/\mathbb{Z}$ described in* (8), *to the conjugacy classes of elements in $\widetilde{\mathrm{SL}}(2,\mathbb{R})_+$. This map is defined as $F \mapsto \tilde{R}(1)$, where R is the solution of the first order differential system $R' = R\begin{pmatrix} 0 & -F \\ 1 & 0 \end{pmatrix}$ with the initial condition $R(0) = I_2$, and $\tilde{R} : [0,1] \to \widetilde{\mathrm{SL}}(2,\mathbb{R})$ is the lift of the path $R : [0,1] \to \mathrm{SL}(2,\mathbb{R})$ with $\tilde{\mathbb{R}}(0) = \tilde{E}_0$.*

The map Ξ is bijective and continuous for the standard topologies of the corresponding quotient spaces.

Proof The system

$$\begin{cases} R' = RQ \\ R(0) = I_2 \end{cases} \qquad \text{where } Q := \begin{pmatrix} 0 & -F \\ 1 & 0 \end{pmatrix}, \tag{27}$$

is the matrix version of the same first order linear ODE associated to Hill's equation $x'' + Fx = 0$, i.e. for a solution $x(t)$ of Hill's equation, the (row) vector-valued map $X(t) := (x(t)\ x'(t)) \in \mathbb{R}^2$ (seen as row matrices) satisfies $X' = XQ$, and conversely, the second entry of a solution to $X' = XQ$ is the derivative of the first entry x, which satisfies the Hill's equation. So the first column of R consists of two solutions x_1, x_2 of $x'' + Fx = 0$, and the second column of their derivatives, thus the determinant of R is equal to the *Wronskian* $Wr(x_2, x_1) := x_2' x_1 - x_1' x_2$ which is constant, thus equal to 1 (the value of $\det R(0)$). Thus $R : \mathbb{R} \to \mathrm{SL}(2, \mathbb{R})$ and it admits a well-defined lift to $\widetilde{\mathrm{SL}}(2, \mathbb{R})$, denoted with $\tilde{R}$. From the standard theory of ODE's, the map

$$F \mapsto \tilde{R}(1) \in \widetilde{\mathrm{SL}}(2, \mathbb{R})$$

is well-defined and continuous. Clearly, the first column

$$\begin{pmatrix} x_1 \\ x_2 \end{pmatrix} = R \begin{pmatrix} 1 \\ 0 \end{pmatrix} \tag{28}$$

of R determines a map $\psi : \mathbb{R} = \tilde{S^1} \to \mathbb{RP}^1$ such that $\psi(0) = [1 : 0]$, that lifts to a map $\tilde{\psi} : \mathbb{R} \to \widetilde{\mathbb{RP}^1}$, such that $\tilde{\psi}(0) = 0$. The fact that the Wronskian $Wr(x_2, x_1)$ is positive implies that the map $\tilde{\psi}$ has positive derivative. In particular, $\tilde{\psi}(1) > \tilde{\psi}(0) = 0$. The map $\tilde{\psi} : \mathbb{R} \to \widetilde{\mathbb{RP}^1}$ is thus not only an immersion, but a monotone increasing one, and this is a developing map for the associated projective structure, cf. Theorem 3.6 and Remark 3.7.

We claim that the holonomy corresponding to the developing map $\tilde{\psi}$ is generated by $\tilde{R}(1)$: indeed, we need to prove that $\tilde{\psi}(t+1) = \tilde{R}(1)(\tilde{\psi}(t))$ or, equivalently, that $\psi(t+1) = \hat{R}(1)\psi(t)$ for all $t \in \mathbb{R}$. This follows from the fact that $t \mapsto R(1)R(t)$ and $t \mapsto R(t+1)$ satisfy the same equation $R' = RQ$ with the same initial condition at $t = 0$.

From (28) we have that $\tilde{\psi}(1) = \tilde{R}(1)(0)$ (the element $\tilde{R}(1)$ of $\widetilde{\mathrm{SL}}(2, \mathbb{R})$ applied to $0 \in \widetilde{\mathbb{RP}^1}$), thus $\tilde{R}(1) \in \widetilde{\mathrm{SL}}(2, \mathbb{R})_+$. Consequently, we get a continuous map

$$\Xi : \{F : \mathbb{R} \to \mathbb{R}\,,\ F \text{ smooth and } F(t+1) = F(t),\ \forall t \in \mathbb{R}\} \longrightarrow \widetilde{\mathrm{SL}}(2, \mathbb{R})_+ \tag{29}$$

which is equivariant for the action of the diffeomorphism group of $\mathbb{R}/\mathbb{Z}$ on the left side, resp. conjugation in $\widetilde{\mathrm{SL}}(2, \mathbb{R})$ on the right side, cf. Theorem 3.6.

Note that Ξ is a composition of the correspondence between Laplace and projective structures from Theorem 3.6 with the one from Theorem 4.5 (which associates to a projective structure an equivalence class of a pair $(U, \tilde{A})$) and with the one from Proposition 5.4 (which drops the interval U that turns out to be determined modulo $\pi\mathbb{Z}$ by $\tilde{A}$). Because we know from that the correspondences above are all 1–1, we conclude that Ξ is bijective as well. □

5.4 Geometric Properties of Closed Projective Curves

We will describe some of the geometric properties of a closed projective curve that allow us to distinguish various isomorphism classes. The first property is whether the developing map is surjective or not. Recall that by Corollary 5.5, a projective curve is identified to a conjugacy class of the adjoint action of $\widetilde{\mathrm{SL}}(2, \mathbb{R})$ on $\widetilde{\mathrm{SL}}(2, \mathbb{R})_+$, and those corresponding to non-surjective developing maps are $\tilde{\mathcal{P}}_0^- \cup \tilde{\mathcal{H}}_0(\lambda)$, $\lambda \in (0, 1)$.

Proposition 5.8 *The curves from $\tilde{\mathcal{P}}_0^- \cup \tilde{\mathcal{H}}_0(\lambda)$, $\lambda \in (0, 1)$, are the projective curves that are also* affine, *i.e. one can extract from the atlas of projective charts a smaller atlas of affine charts or, equivalently, it admits a global connection or which the corresponding Laplace operator is the Hessian. The distinction between $\tilde{\mathcal{P}}_0^-$ and $\tilde{\mathcal{H}}_0(\lambda)$ is that, in the first case, the above mentioned connection is unique and admits a global parallel section, and in the second case there are two such connections and none admits a parallel section.*

Proof First note that if there exists ∇ on C such that $\mathcal{L} = \mathrm{Hess}^\nabla$ on C, then every *local* ∇-parallel nontrivial section of TC induces a *local* parametrization that is projective, so there exists a sub-atlas formed by those parametrizations whose speed vector field is ∇-parallel, see Theorem 3.6. The transition maps are affine, because their second derivative vanishes (which of course implies the vanishing of the Schwarzian). These parametrizations can be globally defined on $\tilde{C}$ which is therefore sent through the developing map to an interval that projects diffeomorphically through $\mathbf{p}$ onto its image in an affine chart of $\mathbb{RP}^1$. This implies that the projective structure is of type $\tilde{\mathcal{P}}_0^-$ or $\tilde{\mathcal{H}}_0(\lambda)$, $\lambda \in (0, 1)$.

For $\tilde{A} = \tilde{P}_0^-$ we get that $(0, \pi)/\langle \tilde{P}_0^- \rangle$ is isomorphic with $\mathbb{R}/\mathbb{Z}$, as the element $\tilde{P}_0^-$ induces on $\{[x : 1]\,,\ x \in \mathbb{R}\} \simeq \mathbb{R}$ the transformation $[t : 1] \mapsto [t + 1 : 1]$. For $\tilde{A} = \tilde{H}_0(\lambda)$, its action on $(0, \pi/2)$ induces the transformation $[x : 1] \mapsto [\lambda^2 x : 1]$, for $x > 0$ (the image $\mathbf{p}((0, \pi/2)) = \{[x : 1]\,,\ x \in (0, \infty)\}$).

Note that for $\tilde{A} = \tilde{P}_0^-$, the image of D is contained in a single affine chart in $\mathbb{RP}^1$, hence the uniqueness of the affine connection ∇, which is (after identifying $\tilde{C}$ with $\mathbb{R}$, the trivial connection for which the derivative of the identity map (which is the affine parametrization) is the ∇-parallel (constant) vector field 1. As noticed before, $\tilde{P}_0^-$ acts on $\tilde{C} \simeq \mathbb{R}$ by translations.

For $\tilde{A} = \tilde{H}_0(\lambda)$, the developing map has values in $(0, \pi/2)$ and its projection through $\mathbf{p}$ on $\mathbb{RP}^1$ is contained in an infinity of affine charts, however only for two of them does $\tilde{A}$ act by affine transformations: on those affine charts which contain at most (in fact, exactly) one of the two fixed points of $\hat{A} \in \mathrm{PSL}(2, \mathbb{R})$ on $\mathbb{RP}^1$, namely on $\{[1 : x]\ ,\ x \in \mathbb{R}\} \supset \mathbf{p}(D(\tilde{C})) = \{[1 : x]\ ,\ x > 0\}$ or on $\{[x : 1]\ ,\ x \in \mathbb{R}\} \supset \mathbf{p}(D(\tilde{C})) = \{[x : 1]\ ,\ x > 0\}$. On each of these charts the holonomy is generated by $x \mapsto \lambda^2 x$, resp $x \mapsto \lambda^{-2} x$, so the ∇-parallel (constant) functions/vector fields are not $\tilde{A}$-invariant, hence not defined globally on C. □

The next invariant of an isomorphism class of closed projective curves is the set of winding numbers of the fundamental domains of $U/\langle\tilde{A}\rangle$:

Proposition 5.9 *If $I := [a, b)$ is a fundamental domain of $\widetilde{\mathbb{RP}^1}/\langle\tilde{A}\rangle$ (thus $b = \tilde{A}(a)$), then the winding number of I depends on $\tilde{A}$ as follows:*

(1) If $\tilde{A}$ is conjugated to $\tilde{E}_\alpha$, $\alpha > 0$, then $W(I) = n$ if and only if $\alpha = n\pi$, otherwise $W(I) = n + \frac{1}{2}$, where $n\pi < \alpha < (n+1)\pi$;
(2) If $\tilde{A}$ is conjugated to $\tilde{P}_n^+$ or $\tilde{P}_n^-$, for $n \in \mathbb{N}^*$, then $n + \frac{1}{2} \geq W(I) \geq n$, respectively $n - \frac{1}{2} \leq W(I) \leq n$;
(3) If $\tilde{A}$ is conjugated to $\tilde{H}_n(\lambda)$, $\lambda > 1$, $n \in \mathbb{N}^*$, then $n - \frac{1}{2} \leq W(I) \leq n + \frac{1}{2}$.

In cases (2) and (3), the winding number of (a, b) is integer if and only if the projection $\mathbf{p}(a) \in \mathbb{RP}^1$ is a fixed point of the projection $\hat{A} \in \mathrm{PSL}(2, \mathbb{R})$ of $\tilde{A}$. Moreover, the limits in the inequalities above are attained.

Proof As the claims are invariant to conjugation, we will assume $\tilde{A} \in \widetilde{\mathrm{SL}}(2, \mathbb{R})_+$ is one of the representatives of the elliptic, parabolic or hyperbolic conjugacy classes.

Note that if $W(a, \tilde{A}(a)) := W([a, \tilde{A}(a)))$ is an integer, then $\mathbf{p}(a) = \mathbf{p}(\tilde{A}(a))$ is a fixed point for $\hat{A}$ on $\mathbb{RP}^1$.

In case (1), $\tilde{A}$ is the translation with α, hence $W(x, x + \alpha) = n$ if and only if $\alpha = n\pi, n \in \mathbb{N}^*$, and $W(x, x + \alpha) = n + \frac{1}{2}$ if $\alpha \in (n\pi, (n+1)\pi)$, as claimed.

In case (2), $\tilde{A}(x) = \tilde{P}_n \pm = \tilde{P}_0^{\pm}(x) + n\pi, n \in \mathbb{N}^*$ or $\tilde{A} = \tilde{P}_0^-$. Clearly, $\tilde{P}_0^+(x) \leq x$ for all $x \in \widetilde{\mathbb{RP}^1}$, with equality for $x \in \pi\mathbb{Z}$, and $\tilde{P}_0^-(x) \geq x$ for all $x \in \widetilde{\mathbb{RP}^1}$, with equality for $x \in \pi\mathbb{Z}$. In both cases $\tilde{P}_0^{\pm}(x) - x \in (-\pi, \pi)$ (because all intervals of the form $(k\pi, (k+1)\pi)$ are mapped to themselves), hence $W(x, \tilde{P}_0^{\pm}(x) + n\pi) = n \mp \frac{1}{2}$ if $x \in \widetilde{\mathbb{RP}^1} \setminus \pi\mathbb{Z}$ and $W(x, \tilde{P}_0^{\pm}(x) + n\pi) = n$ if $x \in \pi\mathbb{Z}$, as claimed.

We argue similarly in case (3): $\tilde{A}(x) = \tilde{H}_0(\lambda)(x) + n\pi$, and the difference $\eta(x) := \tilde{H}_0(\lambda)(x) - x \in (-\pi, \pi)$, vanishes exactly for $x \in \frac{\pi}{2}\mathbb{Z} \subset \widetilde{\mathbb{RP}^1}$. For $k \in \mathbb{Z}$ and $x \in (k\pi, (k + \frac{1}{2})\pi)$, $\eta(x) > 0$ and for $x \in ((k - \frac{1}{2})\pi, k\pi)$, $\eta(x) < 0$. The corresponding winding numbers $W(x, \tilde{A}(x))$ for these values of x are therefore $n, n + \frac{1}{2}$, resp. $n - \frac{1}{2}$. □

Definition 5.10 The *winding number* $W(C)$ of a closed curve $C \simeq \widetilde{\mathbb{RP}^1}/\langle\tilde{A}\rangle$ is $n \in \mathbb{N}^*$ if there exist $a \neq b := \tilde{A}(a) \in \widetilde{\mathbb{RP}^1}$ such that $W([a, b)) = n$. Otherwise

$W(C) := W(I) \in \frac{1}{2}\mathbb{N}^* \setminus \mathbb{N}^*$ for any fundamental domain I of C as in the proposition above.

Remark 5.11 In [11], and also in [9], the authors did not distinguish between the classes $\tilde{P}_n^+$ and $\tilde{P}_n^-$ for the same $n \in \mathbb{N}^*$. Proposition 5.9 produces a geometric invariant (see also the table below) that distinguishes these classes.

Another invariant of an isomorphism class of closed projective curves is its group of global projective automorphisms:

Proposition 5.12 *The orientation-preserving automorphism group* $\mathrm{Aut}^+(C)$ *of an oriented closed projective curve* $C \simeq \widetilde{\mathbb{RP}^1}/\langle\tilde{A}\rangle$ *is 3-dimensional (equal to* $\widetilde{\mathrm{SL}}(2,\mathbb{R})/\langle\tilde{E}_{n\pi}\rangle$*) if and only if the curve is of class* $\mathcal{E}_{n\pi}$, $n \in \mathbb{N}^*$. *Otherwise,* $\mathrm{Aut}(C)$ *is a 1-dimensional Lie group and:*

(1) $\mathrm{Aut}^+(C)$ *is connected and compact (i.e., isomorphic to* S^1*) if and only if*

$$C \in \tilde{\mathcal{E}}^0 \cup \tilde{\mathcal{H}}_0 \cup \tilde{\mathcal{P}}_0^-;$$

(2) $\mathrm{Aut}^+(C)$ *is non-connected and non-compact if and only if*

$$C \in (\tilde{\mathcal{H}} \setminus \tilde{\mathcal{H}}_0) \cup (\tilde{\mathcal{P}} \setminus \tilde{\mathcal{P}}_0^-).$$

In the first case, C is a homogeneous *projective curve. In the second case, C is non-homogeneous and* $\mathrm{Aut}_0(C)$*, the connected component of the identity in* $\mathrm{Aut}^+(C)$*, has $n = W(C)$ fixed points if $C \in \mathcal{P}_n^\pm$, respectively $2n = 2W(C)$ fixed points if $C \in \mathcal{H}_n(\lambda)$. In these cases, the index of* $\mathrm{Aut}_0(C)$ *in* $\mathrm{Aut}^+(C)$ *is $W(C)$.*

Proof In general, $\mathrm{Aut}^+(C)$ is equal to $N(\langle\tilde{A}\rangle)/\langle\tilde{A}\rangle$, where $N(\langle\tilde{A}\rangle)$ denotes the normalizer of the discrete group $\langle\tilde{A}\rangle$ in $\mathrm{Aut}(\tilde{C})$ (this latter group being equal to $\widetilde{\mathrm{SL}}(2,\mathbb{R})$ if $\tilde{C} \simeq \widetilde{\mathbb{RP}^1}$, and an affine group otherwise). Moreover, $\mathrm{Aut}_0(C)$ is equal to the connected component of the identity in the centralizer of $\tilde{A}$. Because the group $\langle\tilde{A}\rangle$ is infinite cyclic, an element $\tilde{B}$ lies in its normalizer if an only if B is in its centralizer or $\tilde{B}\tilde{A}\tilde{B}^{-1} = \tilde{A}^{-1}$. For $\tilde{A} \in \widetilde{\mathrm{SL}}(2,\mathbb{R})_+$ a standard representative of its conjugacy class, the latter is only possible if $\tilde{A} = \tilde{H}_0(\lambda)$.

If $\tilde{A} \in \tilde{\mathcal{Z}} \cap \widetilde{\mathrm{SL}}(2,\mathbb{R})_+$ (i.e., $\tilde{A} = \tilde{E}_{n\pi}$, $n \in \mathbb{N}^*$), then the whole $\widetilde{\mathrm{SL}}(2,\mathbb{R})$ is the centralizer of $\tilde{A}$, and therefore $\mathrm{Aut}^+(\widetilde{\mathbb{RP}^1}/\langle\tilde{A}\rangle) = \widetilde{\mathrm{SL}}(2,\mathbb{R})/\langle\tilde{E}_{n\pi}\rangle$, and it is connected.

If $\tilde{A} = E_\alpha$, $\alpha \in \mathbb{R} \setminus \pi\mathbb{Z}$, then $\bar{\mathbf{p}}(\tilde{A}) = A = E_\alpha$, $\alpha \in (0,\pi) \cup (-\pi,0)$, has complex, non-real eigenvectors that are shared by any matrix $B \in \mathrm{SL}(2,\mathbb{R})$ that commutes with A, i.e., $B = E_\beta$, With $\beta \in (-\pi,\pi]$. The centralizer of $\tilde{A}$ is then $\{\tilde{E}_\beta\ ,\ \beta \in \mathbb{R}\}$. Moreover, $\tilde{E}_\alpha$, $\alpha > 0$, is not conjugated in $\widetilde{\mathrm{SL}}(2,\mathbb{R})$ with $\tilde{E}_{-\alpha}$, thus the normalizer of $\tilde{E}_\alpha$ is the same as its centralizer. We conclude $\mathrm{Aut}^+(\widetilde{\mathbb{RP}^1}/\langle\tilde{E}_\alpha\rangle) = \{\tilde{E}_\beta\ ,\ \beta \in \mathbb{R}\}/\langle\tilde{E}_\alpha\rangle \simeq \mathbb{R}/\alpha\mathbb{Z} \simeq S^1$, for all $\alpha \in \mathbb{R} \setminus \pi\mathbb{Z}$.

For $\tilde{A} = \tilde{P}_n^\pm$, $n \in \mathbb{N}$, then again the normalizer of $\langle\tilde{A}\rangle$ and its centralizer coincides, as $\tilde{A}$ is not conjugated in $\widetilde{\mathrm{SL}}(2,\mathbb{R})$ with $\tilde{A}^{-1}$. But $\hat{A} := \mathbf{p}(\tilde{A}) \in$

$\mathrm{PSL}(2,\mathbb{R})$ has a unique fixed point $[1:0] \in \mathbb{RP}^1$, hence every element $\hat{B}$ of $\mathrm{PSL}(2,\mathbb{R})$ that commutes with $\hat{A}$ must fix this point as well (and no other, unless $\hat{B} = \mathrm{id}$). This means that $\hat{B} = \hat{P}(t)$, $t \in \mathbb{R}$, where the parabolic matrix $P(t)$ is defined in (23). The canonical lift $\tilde{P}_0(t)$ commutes with $\tilde{A}_0$, the canonical lift of $\hat{A}$, and because $\tilde{A} = \tilde{A}_n = \tilde{A}_0 E_{n\pi}$ and $\tilde{B}_m = \tilde{B}_0 E_{m\pi}$ also commute, we conclude

$$\mathrm{Aut}^+(\widetilde{\mathbb{RP}^1}/\langle \tilde{P}_n^{\pm}\rangle) = \{\tilde{P}_m(t)\ ,\ m \in \mathbb{Z},\ t \in \mathbb{R}\}/\langle \tilde{P}_n(\pm 1)\rangle,$$

which is a semidirect product of $\mathrm{Aut}_0(\widetilde{\mathbb{RP}^1}/\langle \tilde{P}_n^{\pm}\rangle) \simeq \mathbb{R}$ with $\mathbb{Z}/n\mathbb{Z}$. $\mathrm{Aut}_0\widetilde{\mathbb{RP}^1}/\langle \tilde{P}_n^{\pm}\rangle)$ consists of the canonical lifts $\tilde{P}_0(t)$, $t \in \mathbb{R}$, hence all its elements fix all fixed points of $\tilde{A}_0$. There are exactly n such points in $C = \widetilde{\mathbb{RP}^1}/\langle \tilde{P}_n^{\pm}\rangle$.

For $\tilde{A} = \tilde{\mathbb{H}}_n(\lambda)$, $n \in \mathbb{N}^*$, $\tilde{A}$ is not conjugated in $\widetilde{\mathrm{SL}}(2,\mathbb{R})$ to $\tilde{A}^{-1}$ hence we again need to determine the centralizer of $\tilde{A}$. As in the elliptic case, if $\tilde{B}$ commutes with $\tilde{A}$, then $B := \bar{\mathbf{p}}(\tilde{B}) \in \mathrm{SL}(2,\mathbb{R})$ commutes with $A := \bar{\mathbf{p}}(\tilde{A})$ which has two real eigenvectors $(1,0)$ and $(0,1)$ in $\mathbb{R}^2$, and this means that B has the same real eigenvectors, hence $B = H(\mu)$, $\mu \neq 0$. As in the parabolic case, the canonical lift of B to $\widetilde{\mathrm{SL}}(2,\mathbb{R})$ commutes with $\tilde{A}_0$ and in fact all lifts $\tilde{B}_n = \tilde{B}_0\tilde{E}_{m\pi}$, $\mathrm{m} \in \mathbb{Z}$, of B commute with $\tilde{A}$. We conclude

$$\mathrm{Aut}^+(\widetilde{\mathbb{RP}^1}/\langle \tilde{H}_n(\lambda)\rangle) = \{\tilde{H}_m(\mu)\ ,\ m \in \mathbb{Z},\ \mu > 0\}/\langle \tilde{H}_n(\lambda)\rangle,$$

which is again a semidirect product of $\mathrm{Aut}_0(\widetilde{\mathbb{RP}^1}/\langle \tilde{H}_n(\lambda)\rangle) \simeq (0,\infty)$ with $\mathbb{Z}/n\mathbb{Z}$.

As $\mathrm{Aut}_0(\widetilde{\mathbb{RP}^1}/\langle \tilde{H}_n(\lambda)\rangle)$ consists of $\tilde{H}_0(\mu)$, $\mu > 0$, which all fix the set $\frac{\pi}{2}\mathbb{Z}$ of fixed points of $\tilde{A}_0$ in $\widetilde{\mathbb{RP}^1}$, we have that the set of $2n$ points on $C = \widetilde{\mathbb{RP}^1}/\langle \tilde{H}_n(\lambda)\rangle$ coming from $\frac{\pi}{2}\mathbb{Z}$ are fixed by all elements of $\mathrm{Aut}_0(\widetilde{\mathbb{RP}^1}/\langle \tilde{H}_n(\lambda)\rangle)$.

Finally, for $\tilde{A} = \tilde{P}_0^-$ or $\tilde{H}_0(\lambda)$, $\mathrm{Aut}^+(C)$ is a quotient by $\langle \tilde{A}\rangle$ of the automorphism group of the image of the developing map, which is $(0,\pi)$ in the first case and $(0,\pi/2)$ in the second case. Now, $\mathrm{Aut}^+(0,\pi) = \{\tilde{P}_0(t)\ ,\ t \in \mathbb{R}\} \simeq \mathbb{R}$ hence $\mathrm{Aut}^+(C) \simeq \mathbb{R}/\mathbb{Z} \simeq S^1$ and, in the second case, $\mathrm{Aut}^+(0,\pi/2) = \{\tilde{H}_0(\mu)\ ,\ \mu > 0\} \simeq (0,\infty)$, thus $\mathrm{Aut}^+(C) \simeq S^1$ as well. In both cases the action of the (connected) group of oriented projective automorphisms is free. □

Definition 5.13 A point on a closed projective curve c which is fixed by all elements of $\mathrm{Aut}_0(C)$ is called a *resonance point* of C.

In the classes $\tilde{\mathcal{P}}_n^{\pm}$ or $\tilde{\mathcal{H}}_n(\lambda)$, $n \in \mathbb{N}^*$, there are n, resp. $2n$ resonance points, moreover the set of these points is preserved by all projective automorphisms (preserving the orientation or not, see below), hence the action of the full automorphism group is not transitive; the projective curve is not *globally* homogeneous.

Recall, however, that all projective curves are *locally* homogeneous manifolds, there is no way to distinguish two points of C by looking at a neighborhood of each point; the existence, on some projective curves C, of points that are fixed (or simply permuted) by all projective automorphisms seems to be rather a *resonance* phenomenon, hence the use of this term in Definition 5.13.

Remark 5.14 The infinitesimal automorphisms of a non-homogeneous projective curve are *projective vector fields* (in fact, there is only one up to scale) that vanish at the finite set of resonance points. This vanishing is at order 1 for the classes $\tilde{\mathcal{H}}_n(\lambda)$, $n \in \mathbb{N}^*$, and at order 2 for the classes $\tilde{\mathcal{P}}_n^{\pm}$, $n \in \mathbb{N}^*$. On the projective classes $\tilde{\mathcal{P}}_0^-$, $\tilde{\mathcal{H}}_0(\lambda)$ and $\tilde{\mathcal{E}}^0$ there is again only one projective vector field (up to scale), and it is nowhere vanishing. On a curve of class $\tilde{\mathcal{E}}_{n\pi}$, i.e. on the n-fold cover of $\mathbb{RP}^1$, there are projective vector fields of all the kinds described above (if such a vector field vanishes, then it either vanishes at order 2 at n points, or it vanishes at order 1 at $2n$ points), but the zeros of such a vector field V vary depending on V, cf. [9].

Remark 5.15 As we have seen in Proposition 5.6, all closed curves admit orientation-reversing automorphisms, so $\mathrm{Aut}(C)$ is a semidirect product of $\mathrm{Aut}^+(C)$ with $\mathbb{Z}/2\mathbb{Z}$.

We see thus that all homogeneous projective curves (i.e, the classes $\tilde{\mathcal{E}} \cup \tilde{\mathcal{H}}_0 \cup \tilde{\mathcal{P}}_0^-$) have connected orientation-preserving automorphism group (isomorphic to S^1 except for the classes in $\tilde{\mathcal{E}}_{n\pi}$, $n \in \mathbb{N}^*$, where we get a n-fold Galois covering of $\mathrm{PSL}_2(\mathbb{R})$).

We summarize in the following table the geometric properties of each projective type of a given curve C:

- the winding number $W(\tilde{C})$ of its universal cover $\tilde{C} \subset \widetilde{\mathbb{RP}^1}$;
- the winding number $W(C)$ of C;
- the winding number of the fundamental domains of C (see Proposition 5.9)
- the connected component of the identity in the automorphism group of C;
- the number of resonance points when C is non-homogeneous / the compatible global affine connections (when they exist);
- (up to conjugation and sign) the projection in $\mathrm{SL}(2, \mathbb{R})$ of the positive generator of the holonomy of C.

Projective classes of curves and their geometric properties

Projective type of C	$W(\tilde{C})$	$W(C)$	$W(U)$	$\mathrm{Aut}_0(C)$	# dist. points/ affine	$\tilde{\mathbf{p}}(\tilde{A}) \in \mathrm{SL}(2,\mathbb{R})$ $\tilde{A}$ positive generator of the holonomy
$\tilde{\mathcal{P}}_0^-$	1	$\frac{1}{2}$	$\frac{1}{2}$	$\mathbb{R}/\mathbb{Z}$	∇^0	$P^- = \begin{pmatrix} 1 & -1 \\ 0 & 1 \end{pmatrix}$
$\tilde{\mathcal{P}}_n^+$, $n \in \mathbb{N}^*$	∞	n	$n - \frac{1}{2}, n^*$	$\mathbb{R}$	n	$P^+ = \begin{pmatrix} 1 & 1 \\ 0 & 1 \end{pmatrix}$
$\tilde{\mathcal{P}}_n^-$, $n \in \mathbb{N}^*$	∞	n	$n^*, n + \frac{1}{2}$	$\mathbb{R}$	n	$P^- = \begin{pmatrix} 1 & -1 \\ 0 & 1 \end{pmatrix}$
$\tilde{\mathcal{H}}_0(\lambda)$	$\frac{1}{2}$	$\frac{1}{2}$	$\frac{1}{2}$	$\mathbb{R}/\mathbb{Z}$	∇^+, ∇^-	$\tilde{H}(\lambda) = \begin{pmatrix} \lambda & 0 \\ 0 & \lambda^{-1} \end{pmatrix}$
$\tilde{\mathcal{H}}_n(\lambda)$, $n \in \mathbb{N}^*$	∞	n	$n - \frac{1}{2}, n^*, n + \frac{1}{2}$	$\mathbb{R}$	$2n$	$H(\lambda) = \begin{pmatrix} \lambda & 0 \\ 0 & \lambda^{-1} \end{pmatrix}$
$\tilde{\mathcal{E}}_\alpha$, $\alpha \notin \pi\mathbb{N}$	∞	$[\frac{\alpha}{\pi}] + \frac{1}{2}$	$[\frac{\alpha}{\pi}] + \frac{1}{2}$	$\mathbb{R}/\mathbb{Z}$		$E_\alpha = \begin{pmatrix} \cos\alpha & -\sin\alpha \\ \sin\alpha & \cos\alpha \end{pmatrix}$
$\tilde{\mathcal{E}}_{n\pi}$, $n \in \mathbb{N}^*$	∞	n	n	$\widetilde{\mathrm{SL}}(2,\mathbb{R})/\langle \tilde{E}_{n\pi} \rangle$		$\mathrm{id} = \begin{pmatrix} 1 & 0 \\ 0 & 1 \end{pmatrix}$

Here the asterisk denotes the winding number of the (countably many) particular choices of connected fundamental domains U of $\tilde{C} \subset \widetilde{\mathbb{RP}^1}$ starting (and ending) at some preimage through $\mathbf{p}$ of one of the resonance points of C, as it follows from Proposition 5.9.

6 The Yamabe Problem for Curves in Conformal Geometry

For a compact conformal manifold (M, c) of dimension at least 2, the Yamabe problem asks for a metric in the conformal class that has constant scalar curvature. Notably, the solution to this problem involves the *Yamabe operator*, which is precisely the canonical Laplace structure $\mathcal{L}$ induced on the manifold itself by its conformal structure [2, Proposition 3.7]. This Yamabe operator, or *conformal Laplacian*, is a second-order linear differential operator on some (rank 1) weight bundle whose symbol is the contraction with the conformal structure, there is no first order term, and the zero order term, written in a gauge defined by a metric g compatible with c, is, up to a constant factor, exactly the scalar curvature Scal_g.

One can reformulate the Yamabe problem (which does admit solutions) as a search for a metric gauge in which the zero order term of the operator $\mathcal{L}$ is constant.

In view of the above, D. Calderbank and F. Burstall formulate in [3] the *Yamabe problem for curves* in a conformal/Möbius ambient space:

Let C be a curve in a conformal/Möbius manifold (M, c). Is it possible to find a metric in the conformal class of the ambient space such that the zero order term of the induced Laplace structure on C, seen in an arc-length parametrization, is constant?

As we announced in the introduction, the answer is *No!*

Proposition 6.1 *There exist curves C in ambient conformal/Möbius spaces (M, c) for which the Yamabe problem has no solution.*

Proof In our notation from (7), F^γ should be constant for γ an arc-length parametrization of C. If this is possible, then the projective structure of C has to be homogeneous (the parametrization can be composed with translations, still remain arc-length, and the resulting zero order term is unchanged, see Proposition 3.5).

But we know that there exist (a lot of) projective structures on a closed curve C that are not homogeneous. On the other hand, [2, Theorem 4.22] states, for the particular case of a curve:

Proposition 6.2 *Let (C, g) be a Riemannian curve and $\mathcal{L}$ a Laplace structure on C. Let ν be a vector bundle over C. Then there exists a metric g^ν on the total space M of ν (and additionally a Möbius structure on $(M, [g])$ if $\dim M = 2$) such that $\mathcal{L}$ is the induced Laplace structure on C induced by the ambient space M.*

That means, every Laplace structure (hence, every projective structure, including non-homogeneous ones) on C can be realized as one induced by the embedding of C in a conformal/Möbius manifold M.

Such a non-homogeneous Laplace structure can not induce, in any parametrization, a Laplace structure with constant zero order term. □

References

1. T. Bailey, M. Eastwood, Conformal circles and parametrizations of curves in conformal manifolds. Proc. Am. Math. Soc. **108**(1), 215–221 (1990)
2. F. Belgun, Geodesics and submanifold structures in conformal geometry. J. Geom. Phys. **91**, 172–191 (2015)
3. F. Burstall, D. Calderbank, Conformal submanifold geometry I-III. arXiv:1006.5700
4. D. Calderbank, Möbius structures and two-dimensional Einstein–Weyl geometry. J. Reine Angew. Math. **504**, 37–53 (1998)
5. P. Gauduchon, Connexion canonique et structures de Weyl en géométrie conforme (1990, unpublished)
6. P. Gauduchon, Structures de Weyl-Einstein, espaces de twisteurs et variétés de type $S^1 \times S^3$. J. Reine Angew. Math. **469**, 1–50 (1995)
7. W.M. Goldman, *Geometric Structures on Manifolds*. Grad. Stud. Math., vol. 227 (American Mathematical Society, Providence, 2022)
8. P. Guha, Diffeomorphisms on S^1, projective structures and integrable systems. ANZIAM J. **44**, 169–180 (2002)

9. N.J. Hitchin, Vector fields on the circle, in *Mechanics, Analysis and Geometry: 200 Years After Lagrange*, ed. by M. Francaviglia. North-Holland Delta Ser. (1991), pp. 359–378
10. S. Kobayashi, *Transformations Groups in Riemannian Geometry* (Springer, Berlin, 1972)
11. N. Kuiper, Locally projective spaces of dimension one. Michigan Math. J. **2**, 95–97 (1954)
12. V.F. Lazutkin, M.F. Pankratova, Normal forms and versal deformations for Hill's equation. Funct. Anal. Appl. **9**(4), 306–311 (1975; 1976)
13. W. Magnus, S. Winkler, *Hill's Equation*. Interscience Tracts in Pure and Applied Mathematics, vol. 20 (Interscience Publishers John Wiley & Sons, New York-London-Sydney, 1966)
14. C. Táfula, Classification of the conjugacy classes of $\widetilde{\mathrm{SL}}(2, \mathbb{R})$. Expo. Math. **42**(6) (2024), Paper No. 125626, 13 pp.

Constant Scalar Curvature Sasaki Metrics

Charles P. Boyer and Christina W. Tønnesen-Friedman

Dedicated to our good friend and colleague Paul Gauduchon on the occasion of his 80th birthday

Abstract In this mostly survey paper we will consider Sasaki manifolds M whose regular quotient is of the form $N = \mathbb{P}(E_0 \oplus E_\infty) \to S$, where E_0 and E_∞ are both projectively flat hermitian holomorphic vector bundles over a compact Kähler manifold S. While N usually does not admit a constant scalar curvature (CSC) Kähler metric, we will see that for an appropriate choice of polarization of N, there always exists a ray of CSC Sasaki metrics somewhere in the Sasaki cone of M. In particular we obtain CSC Sasaki metrics on certain Yamazaki fiber joins over a compact Riemann surface.

Keywords Constant scalar curvature Sasaki metrics · Yamazaki fiber joins

1 Introduction

It is a distinct honor and privilege to dedicate this survey of our recent work to our good friend and colleague, Paul Gauduchon. As described in our previous

---

The first author "Charles P. Boyer" was partially supported by a grant (#245002) from the Simons Foundation.

C. P. Boyer
Department of Mathematics and Statistics, University of New Mexico, Albuquerque, NM, USA
e-mail: cboyer@unm.edu

C. W. Tønnesen-Friedman (✉)
Department of Mathematics, Union College, Schenectady, NY, USA
e-mail: tonnesec@union.edu

L. Ornea (ed.), *Real and Complex Geometry*,
https://doi.org/10.1007/978-3-031-92297-8_3

manuscript in honor of Paul [5], his influence on our work is absolutely apparent and cannot be overestimated.

The most obvious way to obtain constant scalar curvature Sasaki metrics is as an S^1 bundle (or orbi-bundle) over a constant scalar curvature Kähler metric. However, even if the natural Sasaki metric over a (Hodge) Kähler manifold or orbifold is not of constant curvature, there can be a constant scalar curvature Sasaki metric in the corresponding Sasaki cone. It is an important problem to determine necessary and sufficient conditions such that such constant scalar curvature Sasaki metrics exist. One approach to this problem has been given recently by the authors in [8].

Consider the natural Sasakian S^1-bundle over a Kähler orbifold, viz $S^1 \longrightarrow M \longrightarrow N$. This Sasakian structure has an underlying contact CR structure $(\mathcal{D}, J)$ and the Sasaki cone $\mathfrak{t}^+$ can be thought of as the moduli space of Sasakian structures with respect to $(\mathcal{D}, J)$. If M has dimension $2n+1$, then the dimension k of $\mathfrak{t}^+$ satisfies $1 \leq k \leq n+1$.

In this paper we will focus on N being a particular kind of Kähler manifold, namely (the total space of) a so-called *admissible projective bundle*. In full generality, these are described in [2]. Here we shall work with a simplified version that only has "one piece in the base."

Definition 1.1 A complex manifold N is a (total space of) an *admissible projective bundle* provided it is a projective bundle of the form

$$N = \mathbb{P}(E_0 \oplus E_\infty) \longrightarrow S$$

that satisfy the following conditions:

- S is a compact Kähler manifold (S, ω_1, g_1) of complex dimension d_1. We assume g_1 has CSC with scalar curvature $2d_1 s_1$.
- E_0 and E_∞ are holomorphic projectively flat Hermitian vector bundles over S of ranks (d_0+1) and $(d_\infty+1)$, respectively, which satisfy

$$\frac{c_1(E_\infty)}{(d_\infty+1)} - \frac{c_1(E_0)}{(d_0+1)} = \frac{1}{2\pi}[\omega_1].$$

In the case where S is a compact Riemann surface $\Sigma_{\mathfrak{g}}$ of genus $\mathfrak{g}$, we have from Theorem 5 in [2] that N does not admit any constant scalar curvature Kähler metrics. However, for any dimension of S, we can (when appropriate) search for constant scalar curvature Sasaki metrics in the Sasaki cone. In order to do so we will utilize the concepts of **admissible Kähler metrics** [2] on N and the notion of **weighted extremal metrics** as defined in [1]. This method was also employed in [12]. In the current paper this culminates for general S with Proposition 4.8.

In particular, we obtain the following result when S is a compact Riemann surface and N is a regular quotient of a special type of Sasaki manifold; the so-called *Yamazaki fiber join* [11, 13].

Theorem 1.2 *Let M be a Yamazaki fiber join whose regular quotient is a ruled manifold of the form $\mathbb{P}(E_0 \oplus E_\infty) \longrightarrow \Sigma_{\mathfrak{g}}$ where E_0, E_∞ are projectively flat hermitian holomorphic vector bundles on $\Sigma_{\mathfrak{g}}$ of complex ranks (d_0+1) and $(d_\infty+1)$, respectively, and $\Sigma_{\mathfrak{g}}$ is a compact Riemann surface of genus $\mathfrak{g}$. Then the Sasaki cone has at least one cscS ray.*

For $\mathfrak{g} \leq 1$ this is just a special case of Theorem 1.1 of [9], but for $\mathfrak{g} > 1$ a little more work (see Example 5.4 and Theorem 5.5) was needed.

2 Admissible Projective Bundles and Metrics

On admissible projective bundles as defined above we can now construct the *admissible metrics*, [2]. Here we recover the main points of this construction.

We denote the index set $\{1\}$ by $\mathcal{A}$ and let $\hat{\mathcal{A}}$ be the extended index set:

- $\hat{\mathcal{A}} = \mathcal{A}$, if $d_0 = d_\infty = 0$.
- $\hat{\mathcal{A}} = \mathcal{A} \cup \{0\}$, if $d_0 > 0$ and $d_\infty = 0$.
- $\hat{\mathcal{A}} = \mathcal{A} \cup \{\infty\}$, if $d_0 = 0$ and $d_\infty > 0$.
- $\hat{\mathcal{A}} = \mathcal{A} \cup \{0\} \cup \{\infty\}$, if $d_0 > 0$ and $d_\infty > 0$.

Now we set $x_0 = 1$, $x_\infty = -1$, $s_0 = d_0 + 1$ and $s_\infty = -(d_\infty + 1)$. This will correspond to (g_0, ω_0) (or $(g_\infty, \omega_\infty)$) being the induced Fubini-Study structure with scalar curvature $d_0(d_0+1)$ (or $d_\infty(d_\infty+1)$) on each fiber of $P(E_0)$ (or $P(E_\infty)$).

Consider the circle action on $N = \mathbb{P}(E_0 \oplus E_\infty) \longrightarrow S$ induced by the standard circle action on E_0. It extends to a holomorphic $\mathbb{C}^*$ action. The open and dense set N_0 of stable points with respect to the latter action has the structure of a principal circle bundle over the stable quotient. The hermitian norm on the fibers induces via a Legendre transform a function $\mathfrak{z} : N_0 \to (-1, 1)$ whose extension to N consists of the critical manifolds $e_0 := \mathfrak{z}^{-1}(1) = P(E_0 \oplus 0)$ and $e_\infty := \mathfrak{z}^{-1}(-1) = P(0 \oplus E_\infty)$. Letting θ be a connection one form for the Hermitian metric on N_0, with curvature $d\theta = \sum_{a \in \hat{\mathcal{A}}} \omega_a$, an admissible Kähler metric and form are given (up to scale) by the respective formulas

$$g = \sum_{a \in \hat{\mathcal{A}}} \frac{1 + x_a \mathfrak{z}}{x_a} g_a + \frac{d\mathfrak{z}^2}{\Theta(\mathfrak{z})} + \Theta(\mathfrak{z})\theta^2, \quad \omega = \sum_{a \in \hat{\mathcal{A}}} \frac{1 + x_a \mathfrak{z}}{x_a} \omega_a + d\mathfrak{z} \wedge \theta, \tag{1}$$

valid on N_0. Here Θ is a smooth function with domain containing $(-1, 1)$ and $0 < x_1 < 1$. The complex structure yielding this Kähler structure is given by the pullback of the base complex structure along with the requirement $Jd\mathfrak{z} = \Theta\theta$. The function $\mathfrak{z}$ is Hamiltonian with $K = J\, grad\, \mathfrak{z}$ a Killing vector field. In fact, $\mathfrak{z}$ is the moment map on M for the circle action, decomposing N into the free orbits $N_0 = \mathfrak{z}^{-1}((-1, 1))$ and the special orbits $\mathfrak{z}^{-1}(\pm 1)$. Finally, θ satisfies $\theta(K) = 1$.

Now g is a (positive definite) Kähler metric which extends smoothly to all of N if and only if Θ satisfies the following positivity and boundary conditions

$$(i)\ \Theta(\mathfrak{z}) > 0, \quad -1 < \mathfrak{z} < 1, \quad (ii)\ \Theta(\pm 1) = 0, \quad (iii)\ \Theta'(\pm 1) = \mp 2. \tag{2}$$

The Kähler class $\Omega_{\mathbf{x}} = [\omega]$ of an admissible metric as in (1) is also called *admissible* and is uniquely determined by x_1, once the rest of the data associated with N (i.e. $d_1, d_0, d_\infty, s_1, g_1$) are fixed. Indeed, we have

$$\Omega_{\mathbf{x}} = \frac{[\omega_1]}{x_1} + \Xi, \tag{3}$$

where Ξ is a certain fixed cohomology class on N. In the event that $d_0 = d_\infty = 0$, then we have that Ξ is the Poincare dual of $2\pi(e_0 + e_\infty)$. For a more thorough description of Ξ, please consult Section 1.3 of [2].

3 Weighted-Extremal Metrics and Their Application in Sasaki Geometry

In this section, we give a brief introduction of weighted-extremal metrics, with weight $m + 2$, as defined in [1] and summarize how this can be used to find both extremal Sasaki metrics and constant scalar curvature Sasaki (cscS) metrics. For a more in-depth treatment we refer to original source [1] as well the application in [12].

Let (N, g, ω) be a Kähler manifold of complex dimension m and f a positive Killing potential on N. Then the $(f, m+2)$-scalar curvature of g is given by

$$Scal_{f,m+2}(g) = f^2 Scal(g) - 2(m+1) f \Delta_g f - (m+2)(m+1)|df|_g^2, \tag{4}$$

If $Scal_{f,m+2}(g)$ is a Killing potential, g is said to be a *$(f, m+2)$-extremal Kähler metric.*

If the Kähler class $[\omega/2\pi]$ is an integer class, then $\frac{Scal_{f,m+2}(g)}{f}$ is equal to the transverse scalar curvature of a certain Sasaki structure (determined by f) in the Sasaki-Reeb cone of the Boothby-Wang constructed Sasaki manifold M over (N, g, ω). More precisely, if the Reeb vector field χ of the Sasaki structure coming directly from the Boothby-Wang construction M over (N, g, ω) is viewed as a section of the trivial line bundle $TM/\mathcal{D}$ (where $\mathcal{D}$ is the contact bundle) and if f is viewed as a pull-back to M, then $\xi := f\chi$ is a Reeb vector field in the Sasaki-Reeb cone giving a new Sasaki structure. From [1] it now follows that the Sasaki structure determined by f is extremal if and only if g is $(f, m+2)$-extremal and then the corresponding Sasaki structure is CSC if and only if $\frac{Scal_{f,m+2}(g)}{f}$ is constant.

4 Admissible $(c\mathfrak{z}+1, m+2)$-Extremal

In the admissible setting, with $m = (\Sigma_{a\in\hat{\mathcal{A}}} d_a) + 1 \geq 2$, we are considering the $(c\mathfrak{z}+1, m+2)$-extremal condition for $-1 < c < 1$. From the work in Section 2.2 of [4] and in the notation of [12] we have the following set-up and result. Let A_1 and A_2 be the unique solutions to

$$\begin{aligned} \alpha_{1,-(3+m)}A_1 + \alpha_{0,-(3+m)}A_2 &= 2\beta_{0,-(1+m)} \\ \alpha_{2,-(3+m)}A_1 + \alpha_{1,-(3+m)}A_2 &= 2\beta_{1,-(1+m)}, \end{aligned} \tag{5}$$

with

$$\begin{aligned} \alpha_{r,k} &= \int_{-1}^{1} (ct+1)^k t^r p_c(t)dt, \qquad p_c(t) = \prod_{a\in\hat{\mathcal{A}}} (1+x_a t)^{d_a}, \\ \beta_{r,k} &= \int_{-1}^{1} \Big(\sum_{a\in\hat{\mathcal{A}}} \frac{x_a d_a s_a}{1+x_a t}\Big) p_c(t) t^r (ct+1)^k dt \\ &\quad + \big((-1)^r (1-c)^k p_c(-1) + (1+c)^k p_c(1)\big). \end{aligned} \tag{6}$$

Let

$$Q(\mathfrak{z}) = \frac{p_c(\mathfrak{z})}{(c\mathfrak{z}+1)^{m+1}} \Big(\sum_{a\in\hat{\mathcal{A}}} \frac{2d_a s_a x_a}{1+x_a\mathfrak{z}}\Big) - \frac{(A_1 z + A_2)p_c(\mathfrak{z})}{(c\mathfrak{z}+1)^{m+3}} \tag{7}$$

and let

$$G(\mathfrak{z}) = \frac{2p_c(-1)}{(1-c)^{m+1}}(\mathfrak{z}+1) + \int_{-1}^{\mathfrak{z}} Q(t)(\mathfrak{z}-t)\,dt. \tag{8}$$

Then

$$G(\pm 1) = 0, \qquad G'(-1) = \frac{2p_c(-1)}{(1-c)^{m+1}}, \qquad G'(1) = -\frac{2p_c(1)}{(1+c)^{m+1}}. \tag{9}$$

We now have the following lemma which is a special case of Proposition 2.2 in [4].

Lemma 4.1 ([4]) *If*

$$G(\mathfrak{z}) > 0, \quad \textit{for all } \mathfrak{z} \in (-1, 1), \tag{10}$$

then the admissible metric given by $\Theta(\mathfrak{z}) = \frac{(c\mathfrak{z}+1)^{m+1}G(\mathfrak{z})}{p_c(\mathfrak{z})}$ *and (1) is indeed* $(c\mathfrak{z}+1, m+2)$*-extremal with weighted scalar curvature* $Scal_{c\mathfrak{z}+1,m+2} = A_1\mathfrak{z} + A_2$.

Remark 4.2 In the case where E_0 and E_∞ are both line bundles, (10) is a necessary condition for the existence of weighted $(c\mathfrak{z}+1, m+2)$-extremal metrics. This follows from Theorem 7.2 of [3]. Related to this remark, let us take the opportunity of point out that the "only if" part of Proposition 3.10 in [12] (which uses a corollary to Theorem 7.2 of [3]) only pertains to the case where E_0 and E_∞ are both line bundles and not to the more general admissible setting which allows E_0 and E_∞ to be projectively flat Hermitian holomorphic vector bundles of arbitrary rank. As is stands Proposition 3.10 in [12], is misleading and is a result of not reading the pre-amble to Theorem 7.2 of [3] sufficiently carefully.

Remark 4.3 The following observations will be useful below: It follows from (8) that $G''(\mathfrak{z}) = Q(\mathfrak{z})$. Further, $G(\mathfrak{z})$, which vanishes at $\mathfrak{z} = \pm 1$, is positive to the immediate right of $\mathfrak{z} = -1$ and to the immediate left of $\mathfrak{z} = 1$. This follows from the fact that $\Theta'(\pm 1) = \mp 2$ (see discussion above Proposition 2.2 of [4]).

From the proof of Theorem 3.1 in [12], we know that $(A_1\mathfrak{z} + A_2)/(c\mathfrak{z} + 1)$ is a constant (giving rise to a cscS metric when (10) is satisfied and the admissible Kähler class is rational) if and only if

$$\alpha_{0,-(m+2)}\beta_{1,-(m+1)} - \alpha_{1,-(m+2)}\beta_{0,-(m+1)} = 0. \tag{11}$$

Assuming that (11) is solved and hence $A_1\mathfrak{z}+A_2 = k(c\mathfrak{z}+1)$ for some real constant k, we can simplify (7) as follows:

$$\begin{aligned} Q(\mathfrak{z}) \underbrace{=}_{(11)} & \frac{p_c(\mathfrak{z})}{(c\mathfrak{z}+1)^{m+1}}\left(\sum_{a\in\hat{\mathcal{A}}} \frac{2d_a s_a x_a}{1+x_a\mathfrak{z}}\right) - \frac{k p_c(\mathfrak{z})}{(c\mathfrak{z}+1)^{m+2}} \\ = & \frac{p_c(\mathfrak{z})}{(c\mathfrak{z}+1)^{m+2}}\left((c\mathfrak{z}+1)\sum_{a\in\hat{\mathcal{A}}} \frac{2d_a s_a x_a}{1+x_a\mathfrak{z}} - k\right) \\ = & \frac{\prod_{a\in\hat{\mathcal{A}}}(1+x_a\mathfrak{z})^{d_a-1}}{(c\mathfrak{z}+1)^{m+2}}\Big((c\mathfrak{z}+1)\sum_{a\in\hat{\mathcal{A}}} 2d_a s_a x_a \prod_{b\in\hat{\mathcal{A}}\setminus\{a\}}(1+x_b\mathfrak{z}) \\ & -k\prod_{a\in\hat{\mathcal{A}}}(1+x_a\mathfrak{z})\Big) \\ = & \frac{\prod_{a\in\hat{\mathcal{A}}}(1+x_a\mathfrak{z})^{d_a-1}}{(c\mathfrak{z}+1)^{m+2}} P(\mathfrak{z}), \end{aligned} \tag{12}$$

where $P(\mathfrak{z}) = (c\mathfrak{z}+1)\sum_{a\in\hat{\mathcal{A}}} 2d_a s_a x_a \prod_{b\in\hat{\mathcal{A}}\setminus\{a\}}(1+x_b\mathfrak{z}) - k\prod_{a\in\hat{\mathcal{A}}}(1+x_a\mathfrak{z})$ is a polynomial of degree at most $\#\hat{\mathcal{A}}$. Note that for $-1 < \mathfrak{z} < 1$, the sign of $Q(\mathfrak{z})$ equals the sign of $P(\mathfrak{z})$ and since, by Remark 4.3 above, $G''(\mathfrak{z}) = Q(\mathfrak{z})$, this means that for $-1 < \mathfrak{z} < 1$, the concavity of G is determined by the sign of $P(\mathfrak{z})$.

Lemma 4.4 *If* (11) *is true, then the corresponding* $G(\mathfrak{z})$ *satisfies the positivity condition* (10).

Proof For simplicity in notation we denote s_1 by s, x_1 by x and d_1 by d. Recall also that $x_0 = 1$, $s_0 = (d_0 + 1)$, $x_\infty = -1$, and $s_\infty = -(d_\infty + 1)$. Let us now assume by contradiction that (10) fails for $G(\mathfrak{z})$. From Remark 4.3, this must mean that there exist values $-1 < \mathfrak{z}_{max_1} < \mathfrak{z}_{min} < \mathfrak{z}_{max_2} < 1$ such that G has a local maximum at $\mathfrak{z}_{max_1}$ and $\mathfrak{z}_{max_2}$ and a local minimum at $\mathfrak{z}_{min}$. The required concavity of G then demands that there exist a sub-interval of $(-1, 1)$, where $P(\mathfrak{z})$ displays the sign variation *negative, positive, negative* going from left to right. In particular, $P(\mathfrak{z})$ has to have a root between $\mathfrak{z}_{max_1}$ and $\mathfrak{z}_{min}$ as well as a root between $\mathfrak{z}_{min}$ and $\mathfrak{z}_{max_2}$.

- If $d_0 = d_\infty = 0$, $P(\mathfrak{z}) = (c\mathfrak{z}+1)2dsx - k(1+x\mathfrak{z})$ and hence the above scenario is clearly not possible.
- If $d_0 = 0$ and $d_\infty > 0$, $P(\mathfrak{z}) = (c\mathfrak{z} + 1)\Big(2dsx(1 - \mathfrak{z}) + 2d_\infty(d_\infty + 1)(1 + x\mathfrak{z})\Big) - k(1 + x\mathfrak{z})(1 - \mathfrak{z})$ is of degree at most two. Since further $P(1) = (c + 1)2d_\infty(d_\infty + 1)(1 + x) > 0$, the above scenario would require a third—impossible—root between $\mathfrak{z}_{max_2}$ and 1.
- Likewise, if $d_0 > 0$ and $d_\infty = 0$, $P(\mathfrak{z}) = (c\mathfrak{z} + 1)\Big(2dsx(1 + \mathfrak{z}) + 2d_0(d_0 + 1)(1 + x\mathfrak{z})\Big) - k(1 + x\mathfrak{z})(1 + \mathfrak{z})$ is of degree at most two. Since further $P(-1) = (1 - c)2d_0(d_0 + 1)(1 - x) > 0$, the above scenario would require a third—impossible—root between -1 and $\mathfrak{z}_{max_1}$.
- Finally, if $d_0, d_\infty > 0$,

$$\begin{aligned}P(\mathfrak{z}) = (c\mathfrak{z} + 1)\Big(&2dsx(1 - \mathfrak{z}^2) + 2d_0(d_0 + 1)(1 + x\mathfrak{z})(1 - \mathfrak{z})\\ &+2d_\infty(d_\infty + 1)(1 + x\mathfrak{z})(1 + \mathfrak{z})\Big)\\ &-k(1 + x\mathfrak{z})(1 - \mathfrak{z}^2),\end{aligned}$$

  and has degree at most three. Now, $P(-1) = (1 - c)4d_0(d_0 + 1)(1 - x) > 0$ and $P(1) = (c + 1)4d_\infty(d_\infty + 1)(1 + x) > 0$, so the above scenario would require an impossible total of four roots; one between -1 and $\mathfrak{z}_{max_1}$, one between $\mathfrak{z}_{max_1}$ and $\mathfrak{z}_{min}$, one between $\mathfrak{z}_{min}$ and $\mathfrak{z}_{max_2}$, and one between $\mathfrak{z}_{max_2}$ and 1.

Since all cases lead to a contradiction, we can conclude that (10) holds for G. □

Lemma 4.4 together with the discussion at the beginning of the section now gives the following lemma which simplifies the hunt for cscS rays into solving (11) for some $-1 < c < 1$.

Lemma 4.5 *Suppose $x_1 \in (0, 1)$ is rational. If there exists a solution $c \in (-1, 1)$ of (11), then we have a corresponding $(c\mathfrak{z} + 1, m + 2)$-extremal admissible metric which satisfies that $Scal_{c\mathfrak{z}+1,m+2}/(c\mathfrak{z} + 1)$ is a constant. This in turn corresponds to a cscS ray in the Sasaki cone of the Boothby-Wang constructed Sasaki manifold, M, over an appropriate rescale of the admissible Kähler form.*

Lemma 4.6 *For all values of $d_1 \in \mathbb{Z}^+$, $s_1 \in \mathbb{R}$, $d_0 \in \mathbb{Z}^{\geq 0}$, and $d_\infty \in \mathbb{Z}^{\geq 0}$, $\exists a \in (0, 1]$ such that when $0 < x_1 < a$, (11) has a solution $c \in (-1, 1)$.*

Proof We already know from the proof of Theorem 3.1 of [12] that a solution $c \in (-1, 1)$ of (11) always exist when $s_1 \geq 0$. Hence, in that case the lemma holds with $a = 1$.

When $s_1 < 0$ we need to be more careful and we will have to adapt the proof of Theorem 3.1 of [12]. Again, for simplicity in notation we denote s_1 by s, x_1 by x and d_1 by d. We will start with the case of $d_0 = d_\infty = 0$. In this case we are able to inspect the left hand side of (11) more carefully. First we note that $m = d + 1$ and that the degree of $p_c(t)$ is equal to d. Using integration by parts on $\alpha_{0,-(m+2)}$, $\alpha_{1,-(m+2)}$, $\beta_{0,-(m+1)}$, and $\beta_{1,-(m+1)}$, it is clear that the left hand side of (11) is some rational function of c which is smooth for $c \in (-1, 1)$ and has vertical asymptotes at $c = \pm 1$. Let us call this function $f(c)$. Recall that $p_c(t) = (1 + xt)^d$. Similarly, we will denote $(1 + xt)^{d-1}$ by $\hat{p}_c(t)$. Note that in the case when $d_0 = d_\infty = 0$, $p_c(\pm 1) > 0$. Now we do some strategic and partial integration by parts:

$$
\begin{aligned}
\alpha_{0,-(m+2)} &= \int_{-1}^{1} p_c(t)(ct + 1)^{-(d+3)} dt \\
&= \frac{-1}{c(d+2)} \left(p_c(1)(c+1)^{-(d+2)} - p_c(-1)(1-c)^{-(d+2)} \right) \\
&+ \frac{1}{c(d+2)} \int_{-1}^{1} p_c'(t)(ct + 1)^{-(d+2)} dt \\
&= \frac{-1}{c(d+2)} \left(p_c(1)(c+1)^{-(d+2)} - p_c(-1)(1-c)^{-(d+2)} \right) \\
&- \frac{1}{c^2(d+2)(d+1)} \left(p_c'(1)(c+1)^{-(d+1)} - p_c'(-1)(1-c)^{-(d+1)} \right) \\
&+ \frac{1}{c^2(d+2)(d+1)} \int_{-1}^{1} p_c''(t)(ct + 1)^{-(d+1)} dt
\end{aligned}
\tag{13}
$$

$$
\begin{aligned}
\alpha_{1,-(m+2)} =& \int_{-1}^{1} t p_c(t)(c\,t+1)^{-(d+3)}dt \\
=& \frac{-1}{c(d+2)}\left(p_c(1)(c+1)^{-(d+2)} + p_c(-1)(1-c)^{-(d+2)}\right) \\
&+ \frac{1}{c(d+2)} \int_{-1}^{1} p_c(t)(c\,t+1)^{-(d+2)}dt \\
&+ \frac{1}{c(d+2)} \int_{-1}^{1} t p_c'(t)(c\,t+1)^{-(d+2)}dt \\
=& \frac{-1}{c(d+2)}\left(p_c(1)(c+1)^{-(d+2)} + p_c(-1)(1-c)^{-(d+2)}\right) \\
&- \frac{1}{c^2(d+2)(d+1)}\left(p_c(1)(c+1)^{-(d+1)} - p_c(-1)(1-c)^{-(d+1)}\right) \\
&+ \frac{1}{c^2(d+2)(d+1)} \int_{-1}^{1} p_c'(t)(c\,t+1)^{-(d+1)}dt \\
&- \frac{1}{c^2(d+2)(d+1)}\left(p_c'(1)(c+1)^{-(d+1)} + p_c'(-1)(1-c)^{-(d+1)}\right) \\
&+ \frac{1}{c^2(d+2)(d+1)} \int_{-1}^{1} p_c'(t)(c\,t+1)^{-(d+1)}dt \\
&+ \frac{1}{c^2(d+2)(d+1)} \int_{-1}^{1} t p_c''(t)(c\,t+1)^{-(d+1)}dt
\end{aligned}
\tag{14}
$$

$$
\begin{aligned}
\beta_{0,-(m+1)} =& xs \int_{-1}^{1} \hat{p}_c(t)(c\,t+1)^{-(d+2)}dt \\
&+ \left((1-c)^{-(d+2)} p_c(-1) + (1+c)^{-(d+2)} p_c(1)\right) \\
=& \frac{-xs}{c(d+1)}\left(\hat{p}_c(1)(c+1)^{-(d+1)} - \hat{p}_c(-1)(1-c)^{-(d+1)}\right) \qquad (15) \\
&+ \frac{xs}{c(d+1)} \int_{-1}^{1} \hat{p}_c{}'(t)(c\,t+1)^{-(d+1)}dt \\
&+ \left((1-c)^{-(d+2)} p_c(-1) + (1+c)^{-(d+2)} p_c(1)\right)
\end{aligned}
$$

$$
\begin{aligned}
\beta_{1,-(m+1)} =& xs\int_{-1}^{1} t\hat{p}_c(t)(ct+1)^{-(d+2)}dt \\
&+\big(-(1-c)^{-(d+2)}p_c(-1)+(1+c)^{-(d+2)}p_c(1)\big) \\
=& \frac{-xs}{c(d+1)}\big(\hat{p}_c(1)(c+1)^{-(d+1)}+\hat{p}_c(-1)(1-c)^{-(d+1)}\big) \\
&+\frac{xs}{c(d+1)}\int_{-1}^{1}\hat{p}_c(t)(ct+1)^{-(d+1)}dt \\
&+\frac{xs}{c(d+1)}\int_{-1}^{1} t\hat{p}_c{}'(t)(ct+1)^{-(d+1)}dt \\
&+\big(-(1-c)^{-(d+2)}p_c(-1)+(1+c)^{-(d+2)}p_c(1)\big).
\end{aligned}
\tag{16}
$$

In the final expressions of each of (13)–(16) the definite integrals remaining yield rational functions in c with powers of $(1 \pm c)$ less negative than $-(d+1)$. With that in mind, it is not hard to see that $f(c) = \alpha_{0,-(m+2)}\beta_{1,-(m+1)} - \alpha_{1,-(m+2)}\beta_{0,-(m+1)}$ has the following one-sided limits near $c = \pm 1$:

$$\lim_{c\to -1^+} f(c) = \lim_{c\to -1^+} \frac{(p_c(1))^2}{c^2(d+2)(d+1)}(c+1)^{-(2d+3)} = +\infty$$

and

$$\lim_{c\to 1^-} f(c) = \lim_{c\to 1^-} \frac{-(p_c(-1))^2}{c^2(d+2)(d+1)}(1-c)^{-(2d+3)} = -\infty.$$

Since $f(c)$ is continuous for $-1 < c < 1$, we know by the intermediate value theorem that there exists some $\hat{c} \in (-1, 1)$ such that $f(\hat{c}) = 0$. This $\hat{c}$ solves equation (11). Therefore the lemma holds for $d_0 = d_\infty = 0$ and here we can pick $a = 1$.

Following the notation of Section 3.1 of [12], we know from Theorem 5.4 in [10] that the Hilbert-Einstein functional

$$H_S(c) := \frac{(\beta_{0,-m})^{m+1}}{(\alpha_{0,-(m+1)})^m}$$

is smooth for $c \in (-1, 1)$, goes to $+\infty$ when $c \to \pm 1^{\mp}$ (i.e. when we get close to a wall of the Sasaki cone), and therefore has a minimum value at some $\tilde{c} \in (-1, 1)$. Obviously, $H_S'(\tilde{c}) = 0$. From the calculations in the proof of Theorem 3.1 in [12] we also know that $H_S'(\tilde{c}) = 0$ if and only if (11) holds or $\beta_{0,-m} = 0$ (the latter being equivalent to $H_S(\tilde{c}) = 0$). As in the proof of Theorem 3.1 in [12], if $s \geq 0$, then $\beta_{0,-m} > 0$ for all $c \in (-1, 1)$ and thus $\tilde{c}$ will be a solution to (11). Thus the lemma holds with $a = 1$ when d_0, d_∞ are any values in $\mathbb{Z}^{\geq 0}$ and $s \geq 0$.

It remains to prove the lemma when $s < 0$ and at least one of the values of d_0 and d_∞ is non-zero. The strategy will be to show that there exists $a \in (0, 1)$ such that if $0 < x < a$, then $\beta_{0,-m}$ is positive for all $c \in (-1, 1)$. This will then show that, assuming $0 < x < a$, the $\tilde{c}$ from the above discussion solves (11). From the definition we know that

$$\beta_{0,-(m)} = \int_{-1}^{1} \Big(\frac{xsd}{1+xt}+\frac{d_0(d_0+1)}{1+t}+\frac{d_\infty(d_\infty+1)}{1-t}\Big)p_c(t)(ct+1)^{-(d+d_0+d_\infty+1)}$$
$$+(1-c)^{-(d+d_0+d_\infty+1)}p_c(-1)+(1+c)^{-(d+d_0+d_\infty+1)}p_c(1). \tag{17}$$

Now (bearing in mind that $s < 0$) we note that for $-1 \le t \le 1$,

$$\frac{xsd}{1+xt}+\frac{d_0(d_0+1)}{1+t}+\frac{d_\infty(d_\infty+1)}{1-t} \ge \frac{xsd}{1-x}+\frac{d_0(d_0+1)}{2}+\frac{d_\infty(d_\infty+1)}{2}.$$

Since $\frac{x}{1-x}$ is monotonically increasing for $0 < x < 1$ and vanishes for $x = 0$ and $(\frac{d_0(d_0+1)}{2} + \frac{d_\infty(d_\infty+1)}{2})$ is positive, we may conclude that there exists $a \in (0, 1)$ such that when $0 < x < a$, $\frac{xsd}{1-x} + \frac{d_0(d_0+1)}{2} + \frac{d_\infty(d_\infty+1)}{2} > 0$. Since also $p_c(t) > 0$ for $-1 < t < 1$ and $p_c(\pm 1) \ge 0$, this ensures that for $0 < x < a$, $\beta_{0,-(m)} > 0$ for all $c \in (-1, 1)$.

□

Remark 4.7 Based on the evidence from Example 5.4 below it might be that we can choose $a = 1$ in all cases (meaning (11) always has a solution), but a rigorous proof of this would probably require an elaborate integration by parts argument similar in spirit to the special $d_0 = d_\infty = 0$ case. This however seems like a rather daunting task and further exploration is left to the interested reader.

We may now conclude with the following result.

Proposition 4.8 *Suppose the admissible data d_1, s_1, d_0 and d_∞ are fixed. Then there exists some $a \in (0, 1]$ such that when $x_1 \in (0, a) \cap \mathbb{Q}$ the Sasaki cone of the Boothby-Wang constructed Sasaki manifold M over an appropriate rescale of the admissible Kähler form ω_1 (determined by x_1) contains a cscS ray.*

Remark 4.9 In the case when $d_0 = d_\infty = 0$ is should be noted that M can also be realized as the regular Sasaki structure in a $S^3_{\mathbf{w}}$-join as defined in [6]. This follows from (the $m_1 = m_2 = 1$ case of) Proposition 4.22 of [11]. Then the existence of a cscS ray follows from Theorem 1.1 of [6]. Thus, together with Remark 4 we see that the only truly novel part of Proposition 4.8 is the case where $s_1 < 0$ and at least one of d_0, d_∞ are non-zero.

5 Yamazaki's Fiber Join in the Sasaki Setting

In this section we give a brief review of Yamazaki's fiber join construction [13] for K-contact manifolds as adapted to Sasaki manifolds in [11]. This generalizes the standard construction of Sasaki manifolds as S^1-bundles over a smooth projective algebraic variety S to arbitrary odd dimensional sphere bundles over S. We begin with a smooth projective algebraic variety S with $d+1$ Kähler forms ω_j, not necessarily independent. For each integral Kähler form ω_j we construct a principal S^1 bundle M_j with its standard Sasakian structure together with its associated complex line bundle L_j. Then the Yamazaki fiber join $M_1 \star_f \cdots \star_f M_{d+1}$ is the unit sphere bundle $S(\oplus_j L_j^*)$. The main theorem is:

Theorem 5.1 *Let S be a smooth projective algebraic variety with integral Kähler forms ω_j for $j = 1, \ldots, d+1$. Let L_j be holomorphic line bundles on S defined by $c_1(L_j) = [\omega_j]$. Then the unit sphere bundle M in $\oplus_j L_j^*$ has a natural Sasakian structure denoted by $\mathcal{S}_{\mathbf{a}} = (\xi_{\mathbf{a}}, \eta_{\mathbf{a}}, \Phi_{\mathbf{a}}, g_{\mathbf{a}})$ for each $\mathbf{a} \in (\mathbb{R}^+)^{d+1}$ such that the Sasaki automorphism group $\mathfrak{Aut}(\mathcal{S}_{\mathbf{a}})$ contains the torus T^{d+1}.*

The Sasakian geometry of such structures was further studied in [7] and [9]. It is important to realize that the Kähler forms ω_j are not necessarily distinct and we get Sasakian structures for each $(d+1)$-tuple $(\omega_1, \ldots, \omega_{d+1})$ of Kähler forms on S such that $[\omega_j] \in \mathcal{K}(S) \cap H^2(S, \mathbb{Z})$ for all $j = 1, \ldots, d+1$ where $\mathcal{K}(S)$ denotes the Kähler cone of S. Note that $\mathbf{a} \in (\mathbb{Q}^+)^{d+1}$ correspond to quasi-regular Sasaki structures on M and in particular $\mathbf{a} = (1, \ldots, 1)$ corresponds to a regular Sasaki structure with regular quotient equal to projective bundle over S with fiber $\mathbb{CP}^d$:

$$
\begin{array}{ccccc}
S^1 & \longrightarrow & S^{2d+1} & \longrightarrow & \mathbb{CP}^d \\
\downarrow{\scriptstyle id} & & \downarrow & & \downarrow \\
S^1 & \longrightarrow & M & \longrightarrow & \mathbb{P}(\oplus_{j=1}^{d+1} L_j^*) \\
 & & \downarrow & & \downarrow \\
 & & S & \xrightarrow{id} & S.
\end{array}
\tag{18}
$$

Remark 5.2 In Proposition 3.4 of [9] we determined the induced transverse Kähler class of quasi-regular quotients in the special case of $d = 1$. At this point in time we do not have an explicit description of the induced transverse Kähler class of quasi-regular quotients (or even regular quotients) when $d > 1$.

A Yamazaki fiber join of particular interest is the monotone case given by $\omega_j = b_j \omega$ with $b_j \in \mathbb{Z}^+$. We now describe many such examples when S is a Riemann surface.

Example 5.3 We take $S = \Sigma_{\mathfrak{g}}$ a Riemann surface of genus $\mathfrak{g}$ with its standard (CSC) Kähler form $\omega_{\mathfrak{g}}$. For all $j = 1, d+1$ we let M_j be the S^1 bundle over $\Sigma_{\mathfrak{g}}$ with Euler class $[\omega_{\mathfrak{g}}]$. In this case the line bundles L_j are all isomorphic as complex line bundles, and we have $c_1(L_j) = [\omega_{\mathfrak{g}}]$. Thus, $c_1(E) = (d+1)[\omega_{\mathfrak{g}}]$. If we choose the regular Reeb vector field $\xi_{\mathbf{a}}$ with $\mathbf{a} = (1, \ldots, 1)$, we have $\mathbb{P}(\oplus_j L_j^*) = \mathbb{P}(\mathbb{1} \oplus \cdots \oplus \mathbb{1}) = \Sigma_{\mathfrak{g}} \times \mathbb{CP}^d$. So in this case the fiber join M is the S^1 bundle over $\Sigma_{\mathfrak{g}} \times \mathbb{CP}^d$ corresponding to the Kähler class $[\omega_{\mathfrak{g}}] + [\omega_{FS}]$ where ω_{FS} is the standard Fubini-Study Kähler form on $\mathbb{CP}^d$. Since we have a CSC (product) Kähler metric in $[\omega_{\mathfrak{g}}] + [\omega_{FS}]$, $\xi_{\mathbf{a}}$ itself admits a CSC Sasaki structure and thus the Sasaki cone certainly have a cscS ray in this case.

Example 5.4 We will now consider the special case of Definition 1.1 where S is a compact Riemann surface $\Sigma_{\mathfrak{g}}$ of genus $\mathfrak{g} \geq 2$ with Euler class $[\omega_{\mathfrak{g}}]$. Note that in this case, every Kähler class on $N = \mathbb{P}(E_0 \oplus E_\infty) \longrightarrow \Sigma_{\mathfrak{g}}$ is admissible, that is, up to scale, it is of the form (3). (See Remark 2 in Section 1.3 of [2].) We define E_0 and E_∞ as follows: Let $b_0, b_\infty \in \mathbb{Z}^+$ and assume for simplicity that $b_0 > b_\infty$. For $j = 1, 2, ..., d_0+1$ let L_j be a holomorphic line bundle over $\Sigma_{\mathfrak{g}}$ satisfying that $c_1(L_j) = b_0[\omega_{\mathfrak{g}}]$, and for $j = d_0+2, ..., d_0+d_\infty+2$ let L_j be a holomorphic line bundle over $\Sigma_{\mathfrak{g}}$ satisfying that $c_1(L_j) = b_\infty[\omega_{\mathfrak{g}}]$. As in the example above, We then define

$$E_0 = \bigoplus_{1 \leq j \leq d_0+1} L_j^*, \quad E_\infty = \bigoplus_{d_0+2 \leq j \leq d_0+d_\infty+2} L_j^*.$$

Now E_0 and E_∞ admit the structure of holomorphic projectively flat Hermitian vector bundles over $\Sigma_{\mathfrak{g}}$ of ranks (d_0+1) and $(d_\infty+1)$, respectively, and we have

$$\frac{c_1(E_\infty)}{(d_\infty+1)} - \frac{c_1(E_0)}{(d_0+1)} = (b_0 - b_\infty)[\omega_{\mathfrak{g}}]. \tag{19}$$

Note that an even more specialized case (described in Section 5.2 of [7]) of this case would be if $L_j = L_0^*$ for $j = 1, 2, ..., d_0+1$ and $L_j = L_\infty^*$ for $j = d_0 + 2, ..., d_0+d_\infty+2$, where L_0 is a holomorphic line bundle over $\Sigma_{\mathfrak{g}}$ satisfying that $c_1(L_0) = -b_0[\omega_{\mathfrak{g}}]$ and L_∞ is a holomorphic line bundle over $\Sigma_{\mathfrak{g}}$ satisfying that $c_1(L_\infty) = -b_\infty[\omega_{\mathfrak{g}}]$. In that case $E_0 = L_0 \otimes \mathbb{C}^{d_0+1}$ and $E_\infty = L_\infty \otimes \mathbb{C}^{d_\infty+1}$.

With M_j denoting the S^1 bundle over $\Sigma_{\mathfrak{g}}$ with Euler class $c_1(L_j)$, N will be the regular quotient of the corresponding Yamazaki fiber join M from Theorem 5.1. Indeed, M is in fact, *super admissible* according to Definition 4.1 in [7] and hence *strongly admissible* according to Definition 3.6 in [9].

Comparing (19) with Definition 1.1 we see that the admissible bundle $N = \mathbb{P}(E_0 \oplus E_\infty)$ has admissible data consisting of $s_1 = \frac{2(1-\mathfrak{g})}{(b_0-b_\infty)}$. As mentioned in Remark 5.2, unless $d_0 = d_\infty = 0$, we have not determined exactly which Kähler class on N corresponds to the transverse Kähler structure of the regular Sasaki structure. However, since every Kähler class on N is admissible, we do know that there is some specific (unknown) value $x_1 \in (0, 1)$ which determines this Kähler class up to scale.

Since we do not know the value of x_1 we cannot use Proposition 4.8 to prove the existence of a cscS ray in the Sasaki cone of M. On the other hand, the fact that S has dimension $d_1 = 1$ does simplify (6) (and hence (11)) into expressions that can be handled efficiently using e.g. Mathematica. From the proof of Lemma 4.6 we know that (11) always has a solution if $d_0 = d_\infty = 0$. Thus, we only need to consider the cases where at least one of d_0 and d_∞ is non-zero. For convenience we are setting $s_1 = s$ and $x_1 = x$.

Case 1: $d_0 = 0$ and $d_\infty > 1$ (the Case $d_\infty = 0$ and $d_0 > 0$ is Similar) Here $m = (\Sigma_{a \in \hat{\mathcal{A}}} d_a) + 1 = d_\infty + 2$ in (11). We have

$$\begin{aligned}
\alpha_{0,-(m+2)} &= \int_{-1}^{1} (ct+1)^{-(d_\infty+4)}(1+xt)(1-t)^{d_\infty}\, dt \\
\alpha_{1,-(m+2)} &= \int_{-1}^{1} t(ct+1)^{-(d_\infty+4)}(1+xt)(1-t)^{d_\infty}\, dt \\
\beta_{0,-(m+1)} &= \int_{-1}^{1} \Big(xs(1-t) + d_\infty(d_\infty+1)(1+xt)\Big) \\
&\quad (1-t)^{d_\infty-1}(ct+1)^{-(d_\infty+3)}\, dt \\
&\quad + 2^{d_\infty}(1-c)^{-(d_\infty+3)}(1-x) \\
\beta_{1,-(m+1)} &= \int_{-1}^{1} t\Big(xs(1-t) + d_\infty(d_\infty+1)(1+xt)\Big) \\
&\quad (1-t)^{d_\infty-1}(ct+1)^{-(d_\infty+3)}\, dt \\
&\quad - 2^{d_\infty}(1-c)^{-(d_\infty+3)}(1-x).
\end{aligned} \tag{20}$$

Using Mathematica, each of the expressions in (20) can be evaluated and then in turn the left hand side of (11) becomes $\frac{-2^{2d_\infty+3}H(c)}{(d_\infty+1)(d_\infty+2)^2(d_\infty+3)(1+c)^5(1-c)^{2d_\infty+5}}$, where

$$\begin{aligned}
H(c) =& 2s(1-c^2)x(x-c) \\
& -2((2+d_\infty)x - d_\infty x^2) \\
& +((2+d_\infty)(5+d_\infty) - 2d_\infty(4+d_\infty)x + (2+d_\infty+d_\infty^2)x^2)c \\
& +2(d_\infty(3+d_\infty) - (6+5d_\infty+2d_\infty^2)x + d_\infty(2+d_\infty)x^2)c^2 \\
& +(d_\infty^2 - d_\infty - 2 - 2d_\infty(2+d_\infty)x + (2+d_\infty)(3+d_\infty)x^2)c^3.
\end{aligned} \tag{21}$$

It is now clear that (11) has a solution $-1 < c < 1$ iff $H(c) = 0$ for some $-1 < c < 1$. This follows immediately from the fact that $H(c)$ is a cubic in c and $H(-1) = -8(1+x)^2 < 0$ while $H(1) = 4(1+d_\infty)(2+d_\infty)(1-x)^2 > 0$.

Case 2: d_0 and d_∞ Are Both Non-zero Here $m = (\Sigma_{a\in\hat{\mathcal{A}}} d_a) + 1 = d_0 + d_\infty + 2$ in (11). We have

$$\begin{aligned}
\alpha_{0,-(m+2)} &= \int_{-1}^{1} (ct+1)^{-(d_0+d_\infty+4)}(1+xt)(1+t)^{d_0}(1-t)^{d_\infty}\,dt \\
\alpha_{1,-(m+2)} &= \int_{-1}^{1} (ct+1)^{-(d_0+d_\infty+4)}t(1+xt)(1+t)^{d_0}(1-t)^{d_\infty}\,dt \\
\beta_{0,-(m+1)} &= xs\int_{-1}^{1} (1+t)^{d_0}(1-t)^{d_\infty}(ct+1)^{-(d_0+d_\infty+3)}\,dt \\
&\quad + d_0(d_0+1)\int_{-1}^{1} (1+xt)(1+t)^{d_0-1}(1-t)^{d_\infty}(ct+1)^{-(d_0+d_\infty+3)}\,dt \\
&\quad + d_\infty(d_\infty+1)\int_{-1}^{1} (1+xt)(1+t)^{d_0}(1-t)^{d_\infty-1}(ct+1)^{-(d_0+d_\infty+3)}\,dt \\
\beta_{1,-(m+1)} &= xs\int_{-1}^{1} t(1+t)^{d_0}(1-t)^{d_\infty}(ct+1)^{-(d_0+d_\infty+3)}\,dt \\
&\quad + d_0(d_0+1)\int_{-1}^{1} t(1+xt)(1+t)^{d_0-1}(1-t)^{d_\infty}(ct+1)^{-(d_0+d_\infty+3)}\,dt \\
&\quad + d_\infty(d_\infty+1)\int_{-1}^{1} t(1+xt)(1+t)^{d_0}(1-t)^{d_\infty-1} \\
&\qquad (ct+1)^{-(d_0+d_\infty+3)}\,dt.
\end{aligned} \tag{22}$$

Using Mathematica, each of the expressions in (22) can be simplified and then in turn the left hand side of (11) simplifies to

$$\frac{-(d_0+1)(d_\infty+1)2^{2d_0+2d_\infty+3}\left(B_{\frac{1-c}{2}}(d_\infty+1,d_0+1)+B_{\frac{1+c}{2}}(d_0+1,d_\infty+1)\right)^2 H(c)}{(d_0+d_\infty+2)^2(d_0+d_\infty+3)(1+c)^{2d_0+5}(1-c)^{2d_\infty+5}},$$

where $B_z(a,b) := \int_0^z u^{a-1}(1-u)^{b-1}\,du$ is the so-called *Incomplete Beta Function* and $H(c)$ is a third degree polynomial in c;

$$
\begin{aligned}
H(c) = {} & 2s(1-c^2)x(x-c) \\
& -2((2+d_0+d_\infty)x+(d_0-d_\infty)x^2) \\
& +((2+d_0+d_\infty)(5+d_0+d_\infty)+2(d_0-d_\infty)(4+d_0+d_\infty)x \\
& \quad +(2+d_0+d_0^2+d_\infty-2d_0d_\infty+d_\infty^2)x^2)c \\
& -2((d_0-d_\infty)(3+d_0+d_\infty)+(6+5d_0+2d_0^2+5d_\infty+2d_\infty^2)x \\
& \quad +(d_0-d_\infty)(2+d_0+d_\infty)x^2)c^2 \\
& +((d_0-d_\infty)^2-(d_0+d_\infty)-2+2(d_0-d_\infty)(2+d_0+d_\infty)x \\
& \quad +(2+d_0+d_\infty)(3+d_0+d_\infty)x^2)c^3.
\end{aligned}
\tag{23}
$$

Note that, since $0 < \frac{1\pm c}{2} < 1$ when $-1 < c < 1$, $B_{\frac{1\pm c}{2}}(d_\infty+1, d_0+1) > 0$ for $-1 < c < 1$. Therefore, (11) has a solution $-1 < c < 1$ iff $H(c) = 0$ for some $-1 < c < 1$. This follows immediately from the fact that $H(c)$ is a cubic in c and $H(-1) = -4(1+d_0)(2+d_0)(1+x)^2 < 0$ while $H(1) = 4(1+d_\infty)(2+d_\infty)(1-x)^2 > 0$.

Note that (21) is just a special case (with $d_0 = 0$) of (23). This is as expected, but in particular $\beta_{0,-(m+1)}$ and $\beta_{1,-(m+1)}$ needed to be handled differently in the two cases.

We may now conclude Examples 5.3 and 5.4 with the following companion result to Theorem 1.1 of [9].

Theorem 5.5 *Let M be a Yamazaki fiber join whose regular quotient is a ruled surface of the form $\mathbb{P}(E_0 \oplus E_\infty) \longrightarrow \Sigma_{\mathfrak{g}}$ where E_0, E_∞ are projectively flat hermitian holomorphic vector bundles on N of complex dimensions (d_0+1) and $(d_\infty+1)$, respectively, and $\Sigma_{\mathfrak{g}}$ is a compact Riemann surface of genus $\mathfrak{g} \geq 2$. Then the Sasaki cone has at least one cscS ray.*

Remark 5.6 Disclaimer: We used Mathematica to simplify our expressions in Example 5.4. As such that proof of Theorem 5.5 does hinge on technology.

Remark 5.7 In Example 5.4 we can deduce the following information about the second Stiefel-Whitney class which is an obstruction to spin. From Proposition 3.12 of [7] we have that

$$
\begin{aligned}
c_1(\mathcal{D}) &= \pi_M^*\left(\sum_{j=1}^{j=d_0+d_\infty+2} c_1(L_j^*) + c_1(\Sigma_{\mathfrak{g}})\right) \\
&= (-(d_0+1)b_0 - (d_\infty+1)b_\infty + 2(1-\mathfrak{g}))\pi_M^*[\omega_{\mathfrak{g}}].
\end{aligned}
$$

Therefore the second Stiefel-Whitney class $w_2(M)$ (which is the mod 2 reduction of $c_1(\mathcal{D})$) vanishes iff and only if $(d_0+1)b_0+(d_\infty+1)b_\infty$ is even, i.e., if and only if the parity of $(d_0+1)b_0$ and $(d_\infty+1)b_\infty$ are equal. We have

- If d_0 and d_∞ are both odd, then $w_2(M)=0$.
- If d_0 and d_∞ are both even, then
  - if b_0 and b_∞ are both even or both odd, then $w_2(M)=0$
  - if one of b_0 and b_∞ is odd while the other is even, then $w_2(M)\neq 0$
- If d_0 is even and d_∞ is odd, then
  - if b_0 is odd, then $w_2(M)\neq 0$
  - if b_0 is even, then $w_2(M)=0$
- If d_0 is odd and d_∞ is even, then
  - if b_∞ is odd, then $w_2(M)\neq 0$
  - if b_∞ is even, then $w_2(M)=0$.

References

1. V. Apostolov, D.M.J. Calderbank, The CR geometry of weighted extremal Kähler and Sasaki metrics. Math. Ann. **379**(3–4), 1047–1088 (2021). MR 4238260
2. V. Apostolov, D.M.J. Calderbank, P. Gauduchon, C.W. Tønnesen-Friedman, Hamiltonian 2-forms in Kähler geometry. III. Extremal metrics and stability. Invent. Math. **173**(3), 547–601 (2008). MR 2425136 (2009m:32043)
3. V. Apostolov, D.M.J. Calderbank, E. Legendre, Weighted K-stability of polarized varieties and extremality of Sasaki manifolds. Adv. Math. **391**, 63 (2021). Paper No. 107969. MR 4300921
4. V. Apostolov, G. Maschler, C.W. Tønnesen-Friedman, Weighted extremal Kähler metrics and the Einstein-Maxwell geometry of projective bundles. Commun. Anal. Geom. **30**(4), 689–744 (2022). MR 4545849
5. C.P. Boyer, C.W. Tønnesen-Friedman, The Sasaki join and admissible Kähler constructions. J. Geom. Phys. **91**, 29–39 (2015). MR 3327046
6. C.P. Boyer, C.W. Tønnesen-Friedman, The Sasaki join, Hamiltonian 2-forms, and constant scalar curvature. J. Geom. Anal. **26**(2), 1023–1060 (2016). MR 3472828
7. C.P. Boyer, C.W. Tønnesen-Friedman, Sasakian geometry on sphere bundles. Differ. Geom. Appl. **77**, 37 (2021). Paper No. 101765. MR 4253896
8. C.P. Boyer, C.W. Tønnesen-Friedman, Constant scalar curvature Sasaki metrics and projective bundles, in *Birational Geometry, Kähler-Einstein Metrics and Degenerations, Springer Proc. Math. Stat.*, vol. 409 (Springer, Cham, 2023), pp. 95–128. MR 4606632
9. C.P. Boyer, C.W. Tønnesen-Friedman, Sasakian geometry on sphere bundles II. Constant scalar curvature. Differ. Geom. Appl. **96** (2024). Paper No. 102166
10. C. Boyer, H. Huang, E. Legendre, An application of the Duistermaat–Heckman theorem and its extensions in Sasaki geometry. Geom. Topol. **22**(7), 4205–4234 (2018). MR 3890775
11. C.P. Boyer, H. Huang, E. Legendre, C.W. Tønnesen-Friedman, Reducibility in Sasakian geometry. Trans. Am. Math. Soc. **370**(10), 6825–6869 (2018). MR 3841834

12. C.P. Boyer, H. Huang, E. Legendre, C.W. Tønnesen-Friedman, Existence and non-existence of constant scalar curvature and extremal Sasaki metrics. Math. Z. **304**(4) (2023), 29. Paper No. 61. MR 4617164
13. T. Yamazaki, A construction of K-contact manifolds by a fiber join. Tohoku Math. J. (2) **51**(4), 433–446 (1999). MR 2001e:53094

A Mapping Tori Construction of Strong HKT and Generalized Hyperkähler Manifolds

Beatrice Brienza, Anna Fino, and Gueo Grantcharov

Dedicated to Paul Gauduchon for his 80th birthday

Abstract In the present paper we provide a construction via mapping tori of (non Bismut flat) strong HKT and generalized hyperkähler structures on compact manifolds. The skew-symmetric torsion is parallel, but the manifolds are not a product of a hyperkähler manifold and a compact Lie group.

Keywords Mapping tori · HKT structure · Generalized hyperkähler structure

2020 Mathematics Subject Classification 53C26

1 Introduction

A HKT structure on a manifold is a quadruple of three anti-commuting complex structures and a compatible metric, with all structures parallel with respect to a (necessarily unique) connection with skew-symmetric torsion. When the torsion 3-form is closed the structure is called strong HKT. The HKT and strong HKT structures appeared first in String theory as the structures on the target space of

B. Brienza
Dipartimento di Matematica "G. Peano", Università degli studi di Torino, Torino, Italy
e-mail: beatrice.brienza@unito.it

A. Fino (✉)
Dipartimento di Matematica "G. Peano", Università degli studi di Torino, Torino, Italy

Department of Mathematics and Statistics, Florida International University, Miami, FL, USA
e-mail: annamaria.fino@unito.it; afino@fiu.edu

G. Grantcharov
Department of Mathematics and Statistics, Florida International University, Miami, FL, USA
e-mail: grantchg@fiu.edu

L. Ornea (ed.), *Real and Complex Geometry*,
https://doi.org/10.1007/978-3-031-92297-8_4

a (0,4)-supersymmetric sigma model with Wess-Zumino term (see [28, 33]). They also appear in superconformal quantum mechanics [39] and the geometry of black hole moduli spaces [32].

In mathematics, the HKT structures were considered first in [30] and are regarded as the closest quaternionic analogs of Kähler structures. They admit a local potentials [6]. When the canonical bundle of one of the complex structures is trivial, HKT structures lead to a form of Hodge Theory [49] and a characterization in real 8-dimensional case [31] similar to the characterization of compact complex surfaces admitting a Kähler metric. They provide a natural venue for a quaternionic Monge-Ampere equation [2, 3, 10]. The torsion of the HKT connection coincides with the torsion of a connection considered by Bismut [11] and Strominger [47] and is often called Bismut or Bismut-Strominger connection. In particular the HKT structures on $4n$-manifolds are characterized as connection with skew-symmetric torsion and local holonomy in $Sp(n)$.

The strong HKT structures also appear in physics as part of the structure on the target space of (4,4) SUSY sigma model with B-field (or Wess-Zumino term). Such symmetry leads to a pair of strong HKT structures with an opposite torsion, which in mathematics are known as generalized hyperkähler structures [13]. In a recent preprint [50] the strong HKT and generalized hyperkahler structures are studied in relation to AdS/CFT duality.

The strong HKT structures are also pluriclosed and Bismut Hermitian-Einstein (or BHE), and an open question is whether there exist such a structure on a compact manifold which is not a local product of a Kähler space and a Bismut flat space [15, 25]. Note that a remarkable result of Gauduchon and Ivanov [27] states that in complex dimension 2 the only BHE manifolds are Bismut flat, and in fact quotients of the standard Bismut flat Hopf surface and recently a detailed study of compact BHE manifolds in complex dimension 3 is given in [5]. The similar question for strong HKT structures is considered in [50].

In this note we provide, using the previous results in [15], a construction of strong HKT and generalized Kähler structures on compact manifolds which are not a product of hyperkähler manifold and a compact Lie group. According to our knowledge these are the first examples of non Bismut flat compact HKT and generalized Kähler manifolds, but the Bismut torsion is still parallel, so they do not provide an answer to the questions above.

The paper is organized as follows. In Sect. 2 we briefly review some relevant facts about the HKT manifolds. In Sect. 3 we note that strong HKT manifolds which are not hyperkähler do not admit a balanced hyperhermitian metric and consider the known constructions of strong HKT structures. In Sect. 4 we present the construction of strong HKT and generalized hyperkähler compact manifolds via mapping tori.

2 HKT Structures

We start recalling the following

Definition 2.1 A *hypercomplex structure* on a $4n$-dimensional manifold M is a triple (I, J, K) of complex structures on M satisfying the imaginary quaternionic relations.

Furthermore, if there exists a Riemannian metric g with respect to which I, J, K are skew, the structure (I, J, K, g) is said hyperhermitian.

Remark 2.1 For quaternionic dimension $n = 1$, compact manifolds admitting a hyperhermitian structure have been classified in [12]. Any compact 4-dimensional hyperhermitian manifold is either a torus with its flat metric, a K3 surface with its hyperkähler metric, or the Hopf Surface with its standard locally conformally flat metric.

Let us consider an Hermitian manifold (M, I, g). The Levi-Civita connection is Hermitian, i.e., $\nabla^{LC} g = 0$, $\nabla^{LC} J = 0$, if and only if $d\omega = 0$, where $\omega = gI$ is the fundamental form of (I, g). Therefore, when $d\omega \neq 0$, one needs to modify the Levi-Civita connection to obtain a connection which is compatible with both g and I.

In [26], the affine line of Hermitian connections on the tangent bundle TM, known as Gauduchon or canonical connections, with expression

$$g(\nabla^t_X Y, Z) = g(\nabla^{LC}_X Y, Z) + \frac{t-1}{4}(d^c\omega)(X, Y, Z) + \frac{t+1}{4}(d^c\omega)(X, IY, IZ), \tag{1}$$

is introduced, where $d^c\omega = -Id\omega$. From now on, we will adopt the convention $Id\omega(X, Y, Z) := d\omega(IX, IY, IZ)$. When (M, I, g) is Kähler, $d^c\omega$ vanishes, so that the line collapses to a single point, namely, the Levi-Civita connection. Nonetheless, when (M, I, g) is not Kähler, the line is not trivial and the connections ∇^t have non vanishing torsion. For particular values of $t \in \mathbb{R}$, the Chern and the Bismut connections are recovered, i.e., $\nabla^1 = \nabla^{Ch}_I$ and $\nabla^{-1} = \nabla^B_I$ [11, 16]. Despite ∇^{LC}, ∇^B_I, and ∇^{Ch}_I being mutually different connections, any one of them completely determines the other two.

In this work, we will focus mainly on the Bismut connection ∇^B, also known as the Strominger connection [11, 47]. The Bismut connection can be characterized as the only Hermitian connection with totally skew-symmetric torsion and it follows from (1) that its expression is given by

$$g(\nabla^B_X Y, Z) = g(\nabla^{LC}_X Y, Z) - \frac{1}{2}d^c\omega(X, Y, Z),$$

with its torsion 3-form H given by

$$H(X, Y, Z) = g(T^B(X, Y), Z) = d\omega(IX, IY, IZ) = -d^c\omega(X, Y, Z),$$
$$X, Y, Z \in \Gamma(TM).$$

Definition 2.2 A $4n$-dimensional hyperhermitian manifold (M, I, J, K, g) is said *hypekähler with Torsion* (HKT, in short) if $\nabla^B_I = \nabla^B_J = \nabla^B_K =: \nabla^B$.

The study of HKT structures is motivated by the fact that these structures pop up in string theory, in the context of certain supersymmetric sigma models. In fact, HKT manifolds first appeared in [33] as target spaces of (4, 0)-SUSY sigma model with Wess-Zumino term. HKT structures also appear in supergravity. For instance, it has been shown that the geometry of the moduli space of a class of black holes in five dimensions is HKT [43].

Remark 2.2 Recall that an Hermitian manifold (M, J, g) is called Calabi-Yau with torsion (CYT in short) if the restricted holonomy of the Bismut connection is contained in $SU(n)$. This is equivalent to ask that the (first) Bismut Ricci curvature vanishes.

For an HKT manifold, $\mathrm{Hol}(\nabla^B) \subseteq Sp(n)$, since $\nabla^B g = 0$ and $\nabla^B_I I = \nabla^B_J J = \nabla^B_K K = 0$ [33]. In particular, any of the three Hermitian structures is CYT.

Proposition 2.3 ([30]) *Given a hyperhermitian manifold (M, I, J, K, g) the following conditions are equivalent*

(1) $\nabla^B_I = \nabla^B_J = \nabla^B_K$,
(2) $I d\omega_I = J d\omega_J = K d\omega_K$,
(3) $\partial_I(\omega_J + i\omega_K) = 0$,
(4) $\overline{\partial}_I(\omega_J - i\omega_K) = 0$.

where ∂_I and $\overline{\partial}_I$ denote the ∂ and $\overline{\partial}$ operators induced by the complex structure I.

Remark 2.3 It has been shown in [38] that an almost hyperhermitian structure satisfying the condition (2.3) in the previous proposition is, in fact, hyperhermitian.

Given an hyperhermitian structure (I, J, K, g) with fundamental forms

$$\omega_I := gI, \quad \omega_J := gJ, \quad \omega_K := gK,$$

the form $\Omega := \omega_J + i\omega_K$ is a $(2, 0)$-form with respect to I and one can easily observe that $J\Omega = \overline{\Omega}$ and that $\Omega(X, JX) > 0$, for every non-zero vector field X. Conversely, any $(2, 0)$-form with respect to I satisfying the above conditions induces an hyperhermitian metric defined by $g(X, Y) := \Omega(X, JY)$ (with a slight abuse of notation, we will sometimes refer to Ω as the hyperhermitian metric).

By Proposition 2.3, it is immediate to observe that in quaternionic dimension 1, any hyperhermitian structure is HKT, since $\partial_I\Omega = 0$. In higher dimension, given a hypercomplex structure, in general it is not true that there exists a compatible HKT metric. Some counterexamples arise on nilmanifolds of quaternionic dimension $n \geq 2$ [21]. Here we briefly recall their construction.

Definition 2.4 Let $\mathfrak{g}$ be a Lie algebra endowed with a hypercomplex structure (I, J, K). The hypercomplex structure is said to be *abelian* if

$$[IX, IY] = [X, Y], \ [JX, JY] = [X, Y], \ [KX, KY] = [X, Y],$$

for every $X, Y \in \mathfrak{g}$.

It turns out that any inner product compatible with an abelian hypercomplex structure is HKT. It has been shown in [19] that if G is a 2-step nilpotent Lie group, the hypercomplex structure arising from an invariant HKT one is abelian. As a consequence of the symmetrization process, one has the following.

Theorem 2.5 ([21]) *Any compact quotient $M = \Gamma\backslash G$ of a 2-step nilpotent Lie group with a non-abelian left invariant hyper-complex structure (I, J, K) admits no HKT metric compatible with such a hypercomplex structure.*

Example 2.6 We exhibit an explicit counterexample. Let $G = \mathbb{R} \times H_7$ be the 2-step nilpotent Lie group whose Lie algebra is defined by the structure equations

$$[e_1, e_2] = [e_3, e_4] = -e_6, \ [e_1, e_3] = -[e_2, e_4] = -e_7, \ [e_1, e_4] = [e_2, e_3] = -e_8.$$

It is straightforward to note that G admits compact quotients by [37], since the structure equations of its Lie algebra are rational. The non-abelian hypercomplex structure is given by

$$I(e_1) = e_2, \ I(e_3) = e_4, \ I(e_5) = e_6, \ I(e_7) = e_8,$$
$$J(e_1) = e_3, \ I(e_2) = -e_4, \ J(e_5) = e_7, \ J(e_6) = -e_8.$$

Furthermore, it was also shown in [21] that G admits a one parameter family of non-abelian invariant hypercomplex structures except for a unique value of the parameter, to which corresponds an abelian one. Hence, the next Theorem is proved.

Theorem 2.7 ([21]) *The HKT structure is not stable under deformations, i.e., there exists a small hypercomplex deformation of the HKT structure which is not HKT.*

In a suitable sense, HKT metrics play in hypercomplex geometry the same role as Kähler metrics play in complex geometry. In fact, in analogy to the Kähler case, one has the existence of a local potential.

Theorem 2.8 ([6]) *Let (M, I, J, K) be a hypercomplex manifold. A form $\Omega \in \Omega_I^{2,0}(M)$ satisfying $J\Omega = \overline{\Omega}$ can be written locally as $\Omega = \partial_I \partial_J u$ for some smooth real-valued local function u, if and only if it is HKT, i.e., $\partial_I \Omega = 0$.*

Twistor theory for hypercomplex manifolds have been investigated in literature [44, 45]. It is not surprising that when a hypercomplex manifold has a HKT-structure, the geometry of the twistor space is much more rich. In order to enunciate the main result, we need to fix some preliminary notations.

Let (M, I, J, K) be a $4n$-dimensional hypercomplex manifold. The smooth manifold $Z = M \times S^2$ admits an integrable complex structure defined as follows. For a unit vector $v = (v_1, v_2, v_3) \in \mathbb{R}^3$ let I_v be the complex structure $v_1 I + v_2 J + v_3 K$ and let J_v be the complex structure on S^2 defined by the cross product in $\mathbb{R}^3$: $J_v w = v \times w$.

The complex structure on $Z = M \times S^2$ at the point (p, v) is $\mathcal{J}_{(p,v)} = I_v \oplus J_v$. We shall have to consider another almost complex structure, $\mathcal{J}_{2_{(p,v)}} = I_v \oplus -J_v$, which always fails to be integrable. In the next theorem, we assume Z endowed with the integrable complex structure $\mathcal{J}$.

Theorem 2.9 ([30, 33]) *Let (M, I, J, K) be a $4n$-dimensional HKT manifold. Then the twistor space Z is a complex manifold such that*

(1) the fibers of the projection $\pi : Z \to M$ are rational curves with holomorphic normal bundle $\oplus^{2n} \mathcal{O}(1)$,
(2) there is a holomorphic projection $p : Z \to \mathbb{CP}^1$ such that the fibers are the manifold M equipped with complex structures of the hypercomplex structure generated by $\{I, J, K\}$,
(3) denoted by $\mathcal{D}$ the sheaf of kernel of the differential dp, we have the exact sequence

$$0 \to \mathcal{D} \to \Theta_Z \to p^*\overline{\Theta}_{\mathbb{CP}^1} \xrightarrow{dp} 0,$$

and there is a $\mathcal{J}_2$-holomorphic section of $\bigwedge^{(0,2)} \mathcal{D} \otimes p^*\overline{\Theta}_{\mathbb{CP}^1}$ defining a positive definite $(0, 2)$-form on each fiber,
(4) there is an anti-holomorphic map τ compatible with (1), (2) and (3) and inducing the antipodal map on $\mathbb{CP}^1$.

Conversely, if Z is a complex manifold with a non-integrable almost complex structure $\mathcal{J}_2$ with the above four properties, then the parameter space of real sections of the projection p is a $4n$-dimensional manifold M with a natural HKT-structure for which Z is the twistor space.

Another construction of HKT manifold is given by HKT reductions. This construction, originally developed by Joyce for hypercomplex manifolds [34], has been improved for HKT manifolds in [29]. We recall it here.

Let (M, I, J, K, g) be an HKT manifold and let G be a compact group of hypercomplex isometries of M. Denote the algebra of hyper-holomorphic vector fields by $\mathfrak{g}$. Suppose that $\mu = (\mu_1, \mu_2, \mu_3) : M \to \mathbb{R}^3 \otimes \mathfrak{g}$ is a G-equivariant map satisfying: the Cauchy-Riemann condition, i.e., $Id\mu_1 = Jd\mu_2 = Kd\mu_3$, and the transversality condition, i.e., $Id\mu_1 \neq 0$, $Jd\mu_2 \neq 0$, $Kd\mu_3 \neq 0$ for any $X \in \mathfrak{g}$. Any map satisfying these conditions is called a G-moment map.

Theorem 2.10 ([29, 34]) *Let (M, I, J, K, g) be a HKT manifold. Suppose that G is a compact group of hypercomplex isometries admitting a G-moment map μ. Then hypercomplex reduced space $N = M//G$ inherits a HKT structure.*

3 Strong HKT Structures

Let (I, J, K, g) be a HKT structure on a 4n-dimensional manifold M. Setting

$$H := Id\omega_I = Jd\omega_J = Kd\omega_K,$$

we have that the HKT structure is said strong if $dH = 0$ and weak otherwise. This is equivalent to require that any of the three Hermitian structures (g, I), (g, J), (g, K) is SKT (or pluriclosed). Observe that when H is identically zero, then $\nabla^B = \nabla^{LC}$ and the metric is hyperkähler.

Definition 3.1 Let (M, I, J, K, g) be a hyperhermitian manifold. (I, J, K, g) is said *balanced hyperhermitian* if any of the three Hermitian structures is balanced.

It turns out that the balanced hyperhermitian condition is orthogonal (in a suitable sense) to the strong HKT one. In fact, a natural generalization of the Fino-Vezzoni conjecture holds true in the hypercomplex setting.

Theorem 3.2 ([24]) *Let (M, I, J, K, g) be a compact non-hyperKähler strong HKT manifold. Then there is no balanced hyperhermitian metric on (M, I, J, K).*

The original definition of HKT structures assumed the strong HKT condition; however in literature it was dropped because of the lack of examples. In fact, up to know, the only known examples of compact (non-hyperkähler) strong HKT manifolds are due to Joyce [35] and Barberis-Fino [8]. Moreover, all of them are homogeneous.

In the non-compact case, a non-homogeneous example has been constructed in [40] in the following way

Theorem 3.3 *Let (M, I, J, K, g) be a compact strong HKT-manifold of real dimension 4, and E be a smooth complex vector bundle on M. De-note by $\mathcal{M}$ the moduli space of gauge-equivalence classes of anti-self-dual connections (instantons) on E. Then $\mathcal{M}$ is equipped with a natural strong HKT-structure.*

Except for the 4-dimensional case, where each HKT manifold is strong HKT, very few is known about the geometry of strong HKT manifolds.

In the locally homogeneous setting, HKT structures on manifolds have been studied in [4, 9, 18, 20]. In particular, it is known that a nilmanifold and a almost abelian solvmanifold admit strong HKT structures if and only if they are hyper-Kähler [4, 9, 19], but a general result on solvmanifolds is still unknown.

In the following sub-sections we describe the constructions of Joyce [35] and Barberis-Fino [8] in details. We recall that for both the examples the Bismut connection is flat. In the last section of this work, we will construct an example of a (non-homogeneous) compact manifold admitting a non Bismut flat strong HKT structure.

3.1 Joyce's Construction

Let G be a compact Lie group, which after a covering argument, we may assume to be semisimple.

Let then G be a compact semisimple Lie group and let $H \subseteq G$ be a maximal torus in it.

Theorem 3.4 ([35]) *The Lie algebra* $\mathfrak{g}$ *of* G *decomposes as*

$$\mathfrak{g} = \mathfrak{b} \oplus \bigoplus_{j=1}^{m} \mathfrak{d}_j \oplus \bigoplus_{j=1}^{m} \mathfrak{f}_j$$

where $\mathfrak{b}$ *is abelian and of dimension* $\operatorname{rank}(G) - m$, $\mathfrak{d}_j \cong \mathfrak{su}(2)$ *and* $\mathfrak{f}_j$ *satisfy the following conditions*

(1) $[\mathfrak{d}_j, \mathfrak{b}] = 0$ *and* $\mathfrak{b} \oplus \bigoplus_{j=1}^{m} \mathfrak{d}_j$ *contains the Lie algebra of* H
(2) $[\mathfrak{d}_j, \mathfrak{d}_i] = 0$ *for* $i \neq j$
(3) $[\mathfrak{d}_j, \mathfrak{f}_i] = 0$ *for* $j < i$
(4) $[\mathfrak{d}_j, \mathfrak{f}_j] \subseteq \mathfrak{f}_j$ *where the Lie bracket action of* $\mathfrak{d}_j$ *on* $\mathfrak{f}_i$ *is isomorphic to the sum of a finite amount of copies of the action of* $\mathfrak{su}(2)$ *on* $\mathbb{C}^2$ *by left multiplication.*

Let $\mathbb{T}^{2m-r}$, with $r = \operatorname{rank}(G)$, be the flat torus, so that the Lie algebra of $\mathbb{T}^{2m-r} \times G$ decomposes as

$$\mathbb{R}^{2m-r} \oplus \mathfrak{g} = \mathbb{R}^m \oplus \bigoplus_{j=1}^{m} \mathfrak{d}_j \oplus \bigoplus_{j=1}^{m} \mathfrak{f}_j.$$

Let $(e_1, \ldots, e_m)$ be a basis of $\mathbb{R}^m$ and ket φ_j be the Lie algebra isomorphism $\varphi : \mathfrak{su}(2) \to \mathfrak{d}_j$, where we assume that $\mathfrak{su}(2)$ is endowed with the basis (i_1, i_2, i_3) satisfying

$$[i_1, i_2] = 2i_3, \ [i_2, i_3] = 2i_4, \ [i_3, i_4] = 2i_1.$$

The definition of the hypercomplex structure is the following

(1) I_1, I_2, I_3 act on $\mathbb{R}^m \oplus \bigoplus_{j=1}^{m} \mathfrak{d}_j$ as

$$I_a(e_j) = \varphi_j(i_a), \ I_a(\varphi_j(i_a)) = -e_j, \ I_a(\varphi_j(i_b)) = \varphi_j(i_c), \ I_a(\varphi_j(i_c)) = -\varphi_j(i_b),$$

where (a, b, c) is an even permutation of $(1, 2, 3)$.
(2) I_1, I_2, I_3 acts on $\mathfrak{f}_j$ as $I_a(v) = [\varphi_j(i_a), v]$, for any $v \in \mathfrak{f}_j$.

By definition is clear that (I_1, I_2, I_3) is an hypercomplex structure on $\mathbb{T}^{2m-r} \times G$, which by [46] is integrable. Actually, all the invariant hypercomplex structures on compact Lie groups are obtained by this construction.

Furthermore, the (opposite) of the Killing Cartan form of $\mathfrak{g}$ extends to a hyperhermitian bi-invariant metric b on $\mathbb{T}^{2m-r} \times G$ [30], which is HKT. In fact, using that I_a are integrable complex structures and b is bi-invariant, one gets that

$$d_a^c\, \omega_a(X, Y, Z) = -b([X, Y], Z),$$

for any X, Y, Z left invariant vector fields. In addition, since $d_a^c \omega_a$ are bi-invariant 3-forms on $\mathbb{T}^{2m-r} \times G$, they must be closed, implying that the HKT structure (I_a, b) is strong HKT. The remark that the metric is HKT is originally due to Opfermann and Papadopoulos [42], who also generalized the construction to certain homogeneous spaces, showing that they carry HKT structures.

3.2 *Barberis-Fino Construction*

Let $\mathfrak{g}$ be a $4n$-dimensional Lie algebra endowed with an hypercomplex structure (I, J, K) and let $\rho : \mathfrak{g} \to \mathfrak{gl}(k, \mathbb{H})$ be a Lie algebra homomorphism. $T_\rho \mathfrak{g} := \mathfrak{g} \ltimes_\rho \mathbb{H}$ has the structure of Lie algebra with Lie bracket

$$[(X, U), (Y, V)] = [([X, Y]), \rho_X(V) - \rho_Y(U)],$$

where $X, Y \in \mathfrak{g}$ and $U, V \in \mathbb{H}^k$.

The hypercomplex structure on $\mathfrak{g}$ induces an hypercomplex structure on $T_\rho \mathfrak{g}$

$$\tilde{I}(X, U) := (IX, iU), \quad \tilde{J}(X, U) := (JX, jU), \quad \tilde{K}(X, U) := (KX, kU),$$

where i, j, k are the quaternionic units of $\mathbb{H}$. Furthermore, given an hyperhermitian metric g on $\mathfrak{g}$, g and the standard metric on $\mathbb{H}^k$ induce a hyperhermitian metric $\tilde{g}$ on $T_\rho \mathfrak{g}$ in such a way $\mathfrak{g}$ and $\mathbb{H}^k$ are orthogonal.

Theorem 3.5 ([8]) *When $\rho : \mathfrak{g} \to \mathfrak{sp}(k)$, the Lie algebra $(T_\rho \mathfrak{g}, \tilde{I}, \tilde{J}, \tilde{K}, \tilde{g})$ is strong HKT if and only if $(\mathfrak{g}, I, J, K, g)$ is strong HKT.*

We exhibit an explicit example. By Joyce's construction the Lie algebra $\mathfrak{sp}(1) \oplus \mathbb{R}$ admits a strong HKT structure. Let $\rho : \mathbb{R} \oplus \mathfrak{sp}(1) \to \mathfrak{sp}(1)$ be a Lie algebra homomorphism defined as follows: fixed the standard basis $\{e_1, e_2, e_3\}$ basis of

$\mathfrak{sp}(1)$ and a generator e_4 of $\mathbb{R}$,

$$\rho(e_1) := \frac{1}{2}\begin{pmatrix} 0 & 0 & 0 & -1 \\ 0 & 0 & -1 & 0 \\ 0 & 1 & 0 & 0 \\ 1 & 0 & 0 & 0 \end{pmatrix}, \quad \rho(e_2) := \frac{1}{2}\begin{pmatrix} 0 & 0 & 1 & 0 \\ 0 & 0 & 0 & -1 \\ -1 & 0 & 0 & 0 \\ 0 & 1 & 0 & 0 \end{pmatrix},$$

$$\rho(e_3) := \frac{1}{2}\begin{pmatrix} 0 & -1 & 0 & 0 \\ 1 & 0 & 0 & 0 \\ 0 & 0 & 0 & -1 \\ 0 & 0 & 1 & 0 \end{pmatrix},$$

and $\rho(e_4) := 0$.

As a consequence of the Theorem, the 8-dimensional Lie algebra $T_\rho\mathfrak{g}$ with structure equations

$$[e_1, e_2] = e_3, \ [e_2, e_3] = e_1, \ [e_3, e_1] = e_2,$$
$$[e_1, e_8] = -[e_2, e_7] = [e_3, e_6] = \frac{1}{2}e_5,$$
$$[e_1, e_7] = [e_2, e_8] = -[e_3, e_5] = -\frac{1}{2}e_6,$$
$$[e_1, e_6] = -[e_2, e_5] = -[e_3, e_8] = \frac{1}{2}e_7,$$
$$[e_1, e_5] = [e_2, e_6] = -[e_3, e_7] = \frac{1}{2}e_8,$$

admits a strong HKT structure. The corresponding connected and simply connected Lie group is $SU(2) \ltimes \mathbb{R}^4 \times \mathbb{R}$. It has been shown in [22] that the 7-dimensional Lie group $SU(2) \ltimes \mathbb{R}^4$ admits a compact quotient M^7, giving an example of strong HKT 8-dimensional manifold $M^7 \times S^1$. We recall here the construction of the lattice.

Let β be a p-th rooth of the unity with p prime and let us consider the discrete subgroup of $SU(2)$ generated by the matrix

$$A_\beta := \begin{pmatrix} \exp(i\beta) & 0 \\ 0 & \exp(-i\beta) \end{pmatrix}.$$

The set $\Gamma_\beta := \langle A_\beta \rangle \ltimes \mathbb{Z}^4$ is then a closed and discrete subgroup of $SU(2) \ltimes \mathbb{R}^4$. In particular, fixed any point (A', q') of $SU(2) \ltimes \mathbb{R}^4$, then $[(A', q')] = [(A', q' + r)]$ where $r \in \mathbb{Z}^4$, and so the restriction of $\pi : SU(2) \ltimes \mathbb{R}^4 \to (SU(2) \ltimes \mathbb{R}^4)/\Gamma_\beta$ to $SU(2) \times [0, 1]^4$ is surjective. This proves that $(SU(2) \ltimes \mathbb{R}^4)/\Gamma_\beta$ is compact.

We point out that this example, and each example constructed in this way, is Bismut flat. Indeed, the strong HKT metric on the universal cover is given by $\sum_{i=1}^8 (e^i)^2$, which is not bi-invariant on $SU(2) \ltimes \mathbb{R}^4 \times \mathbb{R}$, but it is bi-invariant on $SU(2) \times \mathbb{R} \times \mathbb{R}^4$ with the standard Lie group structure.

4 Construction of Strong HKT Manifolds via Mapping Tori

We start by recall the following

Definition 4.1 Let M be a smooth manifold and let $f \in \text{Diff}(M)$. The *mapping torus* M_f is the smooth manifold

$$M \times \mathbb{R} / \mathbb{Z},$$

where the action of $\mathbb{Z}$ on $M \times \mathbb{R}$ is given by $n \cdot (p, x) = (f^n(p), x + n)$, for any $n \in \mathbb{Z}$, $p \in M$ and $x \in \mathbb{R}$.

By [48], any compact manifold endowed with a non vanishing closed 1-form has the structure of a mapping torus. In particular, any solvmanifold is a mapping torus.

Definition 4.2 ([36]) A *co-Kähler* manifold is a mapping torus M_ψ, where (M, J, g) be a Kähler manifold, and ψ is a Kähler isometry of (J, g).

Co-Kähler manifolds are the odd counterpart of Kähler manifolds, as they share some topological properties, e.g. for instance the formality in the sense of Sullivan [17].

When a co-Kähler manifold is obtained by a mapping torus of an hyperkähler manifold (M, I, J, K, g) via an hyperkähler isometry, then it is said a co-hyperkähler manifold.

It is well known that when the hyperkähler manifold is a K3 surface, the hyperkähler metric is in general not explicit. Henceforth, determining an explicit non-trivial hyperkähler isometry is challenging. Here we exhibit an explicit example.

Example 4.3 The Fermat quartic is the complex surface in $\mathbb{CP}^3$ defined by the equation

$$\mathcal{F} = \{z_0^4 + z_1^4 + z_2^4 + z_3^4 = 0\},$$

where z_0, z_1, z_2, z_3 are the standard homogeneous coordinates on $\mathbb{CP}^3$. It is well known, that it is a K3 surface. We will always assume that $\mathcal{F}$ is endowed with the standard complex structure I, i.e., I is such that $\iota : \mathcal{F} \to \mathbb{CP}^3$ is holomorphic, where ι is the natural embedding. Let $\iota^*\omega_{FS}$ be the Kähler metric on $\mathcal{F}$ induced by the Fubini-Study metric on $\mathbb{CP}^3$ and define g to be the unique Kähler Ricci flat metric on $\mathcal{F}$ whose Kähler form in cohomologous to $\iota^*\omega_{FS}$, which is hyperkähler. We will denote by (I, J, K, g) the hyperkähler structure.

The explicit form of g is not known, nevertheless, in [1] the group of (not necessarily holomorphic) isometries of g is characterized. In particular, $\text{Isom}(g)$ is identified with the group of all holomorphic and antiholomorphic isometries of $\mathbb{CP}^3$ which preserve $\mathcal{F}$.

Let $\sigma : \mathbb{CP}^3 \to \mathbb{CP}^3$ be defined by $[z_0, z_1, z_2, z_3] \mapsto [-z_0, -z_1, z_2, z_3]$. By the identification above $\psi := \sigma_{|\mathcal{F}}$ is clearly an isometry of $(\mathcal{F}, g)$, and it is straightforward to prove that ψ is holomorphic with respect to I. Hence, $\psi \in \mathrm{Isom}_{hol}(g)$. It remains to prove that ψ is an hyperkähler isometry, i.e., that ψ preserves the fundamental forms $\omega_J = gJ$ and $\omega_K = gK$.

Since $\mathcal{F}$ is a $K3$ surface, up to complex scalar multiples, there exists a unique $(2, 0)$ holomorphic form Ω on $\mathcal{F}$. The fixed points of σ in $\mathbb{CP}^3$ are the lines $z_0 = z_1 = 0$ and $z_2 = z_3 = 0$, and any of them intersects the Fermat quartic in four isolated fixed points. Since the fixed loci of ψ is given by eight isolated and distinct points, ψ is symplectic by [41]. In fact, since ψ is a holomorphic involution of $\mathcal{F}$ then either it preserves Ω or takes Ω in $-\Omega$. So, it suffices to observe that the second case leads to a contradiction.

Let p be any of the eight fixed points of ψ. Then the eigenvalues of $d\psi$ at $T_p\mathcal{F}$ are ± 1, since it is a involution. However, if there is an eigenspace with eigenvalue 1, then the geodesics tangent to this eigenspace would be also fixed by ψ, since it is an isometry. But this contradicts the fact that the fixed points are isolated. Therefore, $d\psi = -Id$ on $T_p\mathcal{F}$ and $\psi^*\Omega = \Omega$.

Let $k \in \mathbb{C}^*$ be such that $|k|^2\Omega \wedge \overline{\Omega} = 2\omega_I^2$. Then, $\omega_J = \mathrm{Re}(k\Omega)$ and $\omega_K = \mathrm{Im}(k\Omega)$. Since $\psi^*\Omega = \Omega$, $\psi^*(k\Omega) = k\Omega$.

To conclude the proof, we observe that since ψ is holomorphic

$$
\begin{aligned}
\mathrm{Re}(k\Omega) + i\,\mathrm{Im}(k\Omega) = k\Omega = \psi^*(k\Omega) &= \psi^*(\mathrm{Re}(k\Omega) + i\,\mathrm{Im}(k\Omega)) \\
&= \psi^*\,\mathrm{Re}(k\Omega) + i\psi^*\,\mathrm{Im}(k\Omega),
\end{aligned}
$$

from which follows that $\psi^*\omega_J = \omega_J$ and $\psi^*\omega_K = \omega_K$.

The mapping torus construction turns out to be particularly adaptable to provide examples of complex manifolds endowed with special Hermitian metrics [14, 15, 23]. In the next Theorem we show that the product of a co-hyperkähler manifold with the 3-sphere S^3 admits a strong HKT metric.

Theorem 4.4 *Let (N, I, J, K, g) be a compact hyperkähler manifold let ψ be a hyperkähler isometry. Then, the product $N_\psi \times S^3$ admits a strong HKT structure.*

Proof We first observe that $M_\psi \times S^3$ is diffeomorphic to the mapping torus $M_f = (N \times S^3)_{(\psi, Id)}$. On the other hand, $M_f \cong N \times \mathbb{C}^2 \setminus \{(0,0)\} / \mathbb{Z}$, where the action of $\mathbb{Z}$ on $N \times \mathbb{C}^2 \setminus \{(0,0)\}$ is given by

$$
n \cdot (p, \underline{x}) \mapsto (\psi^n(p), 2^n \underline{x}). \tag{2}
$$

The identification above is provided by the map

$$
\alpha : N \times \mathbb{S}^3 \times \mathbb{R} \to N \times \mathbb{C}^2 \setminus \{(0,0)\}, \quad (p, q, t) \mapsto (p, 2^t \cdot q)
$$

with inverse

$$\alpha^{-1} : N \times \mathbb{C}^2 \setminus \{(0,0)\} \to N \times \mathbb{S}^3 \times \mathbb{R}, \quad (p, \underline{x}) \mapsto (p, \frac{\underline{x}}{\|\underline{x}\|}, \log_2 \|\underline{x}\|)$$

induced on the quotients.

Hence, it suffices to prove that $N \times \mathbb{C}^2 \setminus \{(0,0)\} /_{\mathbb{Z}}$ admits a strong HKT structure. Let us fix on $N \times \mathbb{C}^2 \setminus \{(0,0)\}$ the product hypercomplex structure $(I \times I_-, J \times J_-, K \times K_-)$, where I_- is the standard complex structure on $\mathbb{C}^2 \setminus \{(0,0)\}$ with associated holomorphic coordinates (z^1, z^2), J_- is defined by $J_-(dz^1) = -d\bar{z}^2$ and $K_- = I_- J_-$.

We consider the product metric $g+b$, where $b := \frac{1}{|z|}(dz_1 d\bar{z}_1 + dz_2 d\bar{z}_2)$. It is well known that (I_-, J_-, K_-, b) is a strong HKT structure on $\mathbb{C}^2 \setminus \{(0,0)\}$, and therefore, $(I \times I_-, J \times J_-, K \times K_-, g+b)$ is a strong HKT structure on $N \times \mathbb{C}^2 \setminus \{(0,0)\}$.

Finally, since the automorphisms ϕ_n associated to the action (2) are manifestly holomorphic with respect to each complex structure and they preserve the product metric $g + g_-$, the strong HKT structure descend to the quotient, concluding the proof of the Theorem.

We recall the following definition

Definition 4.5 Let M be a $4n$-dimensional manifold endowed with a pair of HKT structures (I, J, K, g) and (I', J', K', g). The structure (I, J, K, I', J', K', g) is said to be *generalized hyperkähler* if the Bismut torsions H and H' of (I, J, K, g) and (I', J', K', g) satisfy $H = -H'$ and $dH = dH' = 0$.

Proposition 4.6 *The manifold $N_\psi \times S^3$ admits a generalized Hyperkähler structure.*

Proof The proof proceeds as the one of Theorem 4.4. In fact, we observe that the hypercomplex structure (I_-, J_-, K_-) on $\mathbb{C}^2 \setminus \{(0,0)\}$ is precisely the left multiplication by i, j, and k on $\mathbb{H} \setminus \{0\}$. If we consider the other hypercomplex structure (I_+, J_+, K_+) on $\mathbb{H} \setminus \{0\}$ corresponding to the right multiplication by i, j, and k, then $(I_-, J_-, K_-, I_+, J_+, K_+, b)$ is a generalized hyperkähler structure on $\mathbb{C}^2 \setminus \{(0,0)\} \cong \mathbb{H} \setminus \{0\}$ (see [40] for the details). Therefore the structure $(I \times I_-, J \times J_-, K \times K_-, I \times I_+, J \times J_+, K \times K_+, g + b)$ is a generalized hyperkähler on $K \times \mathbb{C}^2 \setminus \{(0,0)\}$, which descends to the quotient $N_\psi \times S^3$ by the definition of the action (2). In fact, on the hyperKähler factor, $\mathbb{Z}$ acts by hyperholomorphic isometries, and on $\mathbb{C}^2 \setminus \{(0,0)\} \cong \mathbb{H} \setminus \{0\}$ it acts by multiplication by 2, which clearly commute with the action of right multiplication by i, j, and k, concluding the proof.

In the next Proposition we give some topological properties of $N_\psi \times S^3$.

Proposition 4.7 *Let $N_\psi \times S^3$ be a manifold constructed as in Theorem 4.4. Then*

(1) $N_\psi \times S^3$ is formal;
(2) $N_\psi \times S^3$ is non-Kähler.

Proof As N_ψ is a co-Kähler manifold, it is formal. The results follows by the fact that the formality is preserved under the cartesian product.

The second statement is a consequence of the Künneth formula. Indeed, by the Künneth formula, $b_1(N_\psi \times S^3) = b_1(N_\psi)$. Since the first Betti number of a co-Kähler manifold is odd [17], the result easily follows.

Example 4.8 Let $N = \mathcal{F}$ be the Fermat quartic, and let ψ be the hyperkähler isometry constructed in the Example 4.3. By Theorem 4.4, the manifold $N_\psi \times S^3$ admits a strong HKT structure. We show that the Example is not trivial, i.e., it does not split as a product of $N \times S^3 \times S^1$. Since $\psi \neq Id$, the result follows by [15]. We also observe that it is a non-trivial example of a non-Kähler Bismut Hermitian Einstein manifold.

Remark 4.1 We note the difference of the geometry induced on a mapping tori in the case when ψ is a holomorphic symplectic automorphism of infinite order. According to [23], the space $N_\psi \times S^1$ admits a balanced metric and no SKT metric. We point out that the example constructed in Example 4.8 admits a finite cover isomorphic to $\mathcal{F} \times S^3 \times S^1$, according to [7]. In fact, since any hyperKähler isometry is of finite order, there must exists a $k \in \mathbb{Z}$ such that $\psi^k = Id$, then $\tilde{X} = \mathcal{F} \times S^3 \times \mathbb{R}/k\mathbb{Z}$ where $k\mathbb{Z} \subset \mathbb{Z}$ acts by $kn \cdot (q, p, t) = (\psi^{kn} q, p, t + kn)$ is a finite cover of $N_\psi \times S^3$ and it is isomorphic to $\mathcal{F} \times S^3 \times S^1$.

In particular, our example $N_\psi \times S^3$ is a local product of two strong HKT manifolds. Indeed, it admits a $\mathbb{Z}_2$ quotient which is a product of a Hopf surface and a K3 orbifold $N/\langle\psi\rangle$.

Acknowledgments The authors would like to thank the anonymous referee for their suggestions and valuable comments which helped to improve this paper. Beatrice Brienza and Anna Fino are partially supported by Project PRIN 2022 "Geometry and Holomorphic Dynamics" and by GNSAGA (Indam). Anna Fino is also supported by a grant from the Simons Foundation (#944448). Gueo Grantcharov is partially supported by a grant from the Simons Foundation (#853269).

References

1. D.V. Alekseevsky, M.M. Graev, Calabi-Yau metric on the Fermat surface. Isometries and totally geodesic submanifolds. J. Geom. Phys. **7**(1), 21–43 (1980)
2. S. Alesker, Solvability of the quaternionic Monge-Ampère equation on compact manifolds with a flat hyperKähler metric. Adv. Math. **241**, 192–219 (2013)
3. S. Alesker, M. Verbitsky, Quaternionic Monge-Ampère equations and Calabi problem for HKT-manifolds. Isr. J. Math. **176**, 109–138 (2010)
4. A. Andrada, M.L. Barberis, Hypercomplex almost abelian solvmanifolds. J. Geom. Anal. **33** (2023). https://doi.org/10.1007/s12220-023-01277-y
5. V. Apostolov, G. Barbaro, K.-H. Lee, J. Streets, The classification of non-Kähler Calabi-Yau geometries of threefolds. Preprint arXiv: 2408.09648
6. B. Banos, A. Swann, Potentials for hyper-Kähler metrics with torsion. Class. Quantum Gravity **21**, 3127–3135 (2004)

7. G. Barbaro, F. Pediconi, N. Tardini, Pluriclosed manifolds with parallel Bismut torsion. Preprint arXiv:2406.07039
8. M.L. Barberis, A. Fino, New HKT manifolds arising from quaternionic representations. Math. Z. **267**(3), 717–735 (2011)
9. M.L. Barberis, I. Dotti, M. Verbitsky, Canonical bundles of complex nilmanifolds, with applications to hypercomplex geometry. Math. Res. Lett. **16**(2), 331–347 (2009)
10. L. Bedulli, G. Gentili, L. Vezzoni, A parabolic approach to the Calabi-Yau problem in HKT geometry. Math. Z. **302**(2), 917–933 (2022)
11. J.M. Bismut, A local index theorem for non Kähler manifolds. Math. Ann. **284**(4), 681–699 (1989)
12. C.P. Boyer, A note on hyper-Hermitian four-manifolds. Proc. Am. Math. Soc. **102**(1), 157–164 (1988)
13. A. Bredthauer, Generalized Hyperkähler geometry and supersymmetry. Nucl. Phys. B **773**(3), 172–183 (2007)
14. B. Brienza, A. Fino, Generalized Kähler manifolds via mapping tori. J. Sympl. Geom. Preprint arXiv: 2305.11075 (in press)
15. B. Brienza, A. Fino, G. Grantcharov, CYT and SKT manifolds with parallel Bismut torsion. arXiv:2401.07800. Published online 2024:1-26. https://doi.org/10.1017/prm.2024.115
16. S.S. Chern, *Complex Manifolds Without Potential Theory*. Universitext (Springer, New York-Heidelberg, 1979)
17. D. Chinea, M.D. de León, J.C. Marrero, Topology of cosymplectic manifolds. J. Math. Pures Appl. (9) **72**(6), 567–591 (1993)
18. I. Dotti, A. Fino, Abelian hypercomplex 8-dimensional nilmanifolds. Ann. Glob. Anal. Geom. **18**, 47–59 (2000)
19. I.G. Dotti, F. Fino, HyperKähler torsion structures invariant by nilpotent Lie groups. Class. Quantum Gravity **19**(3), 551–562 (2002)
20. I. Dotti, A. Fino, Hypercomplex eight-dimensional nilpotent Lie groups. J. Pure Appl. Algebra **184**(1), 41–57 (2003)
21. A. Fino, G. Grantcharov, Properties of manifolds with skew-symmetric torsion and special holonomy. Adv. Math. **189**(2), 439–450 (2004)
22. A. Fino, A. Tomassini, Generalized G_2-manifolds and $SU(3)$-structures. Int. J. Math. **19**(10), 1147–1165 (2008)
23. A. Fino, G. Grantcharov, M. Verbitsky, Special Hermitian structures on suspensions. Preprint arXiv:2208.12168
24. E. Fusi, G. Gentili, Special metrics in hypercomplex geometry. Preprint arXiv:2401.13056
25. M. Garcia-Fernandez, J. Jordan, J. Streets, Non-Kähler Calabi-Yau geometry and pluriclosed flow. J. Math. Pures Appl. **177**, 329–367 (2023)
26. P. Gauduchon, Hermitian connections and Dirac operators. Boll. Un. Mat. Ital. B (7) **11**(2), 257–288 (1997)
27. P. Gauduchon, S. Ivanov, Einstein-Hermitian surfaces and Hermitian Einstein-Weyl structures in dimension 4. Math. Z. **226**, 317–326 (1997)
28. G. Gibbons, G. Papadopoulos, K. Stelle, HKT and OKT geometries on soliton Black Hole moduli spaces. Nucl. Phys. **B508**, 623–658 (1997)
29. G. Grantcharov, G. Papadopoulos, Y.S. Poon, Reduction of HKT-structures. J. Math. Phys. **43**(7), 3766–3782 (2002)
30. G. Grantcharov, Y.S. Poon, Geometry of hyper-Kähler connections with torsion. Commun. Math. Phys. **213**, 19–37 (2000)
31. G. Grantcharov, M. Lejmi, M. Verbitsky, Existence of HKT metrics on hypercomplex manifolds of real dimension 8. Adv. Math. **320**, 1135–1157 (2017)
32. J. Gutowski, G. Papadopoulos, The dynamics of very special black holes. Phys. Lett. B **472**(1–2), 45–53 (2000)
33. P.S. Howe, G. Papadopoulos, Twistor spaces for hyper-Kähler manifolds with torsion. Phys. Lett. B **379**(1–4), 80–86 (1996)

34. D. Joyce, The hypercomplex quotient and the quaternionic quotient. Math. Ann. **290**, 323–340 (1991)
35. D. Joyce, Compact hyper-complex and quaternionic manifolds. J. Differ. Geom. **35**(3), 743–761 (1992)
36. H. Li, Topology of co-symplectic/Co-Kähler manifolds. Asian J. Math. **12**(4), 527–543 (2008)
37. A.I. Malcev, On a class of homogeneous spaces. Reprinted in Amer. Math. Soc. Translations Ser. 1, vol. 9 (1962), pp. 276–307
38. F. Martín Cabrera, A. Swann, The intrinsic torsion of almost quaternion-Hermitian manifolds. Ann. Inst. Fourier. **58**(5), 1455–1497 (2008)
39. J. Michelson, A. Strominger, Superconformal multi-Black Hole quantum mechanics. J. High Energy Phys. 9 (1999), 16 pp. Paper 5
40. R. Moraru, M. Verbitsky, Stable bundles on hypercomplex surfaces. Cent. Eur. J. Math. **8**(2), 327–337 (2010)
41. V.V. Nikulin, Finite groups of automorphisms of Kählerian K3 surfaces (Russian). Trudy Moskov. Mat. Obshch. **38**, 75–137 (1979). English translation: Trans. Moscow Math. Soc. **38**, 71–135 (1980)
42. A. Opfermann, G. Papadopoulos, Homogeneous HKT and QKT manifolds. Preprint mathph/9807026
43. G. Papadopoulos, KT and HKT geometries in strings and in black hole moduli spaces. Preprint hep-th/0201111
44. H. Pedersen, Y.S. Poon, A. Swann, Hypercomplex structures associated to quaternionic manifolds. Differ. Geom. Appl. **9**(3), 273–292 (1998)
45. S. Salamon, Differential geometry of quaternionic manifolds. Ann. Sci. Ecole Norm. Sup. (4) **19**(1), 31–55 (1986)
46. H. Samelson, A class of complex-analytic manifolds. Port. Math. **12**(4), 129–132 (1953)
47. A. Strominger, Superstrings with torsion. Nucl. Phys. B. **274**, 254–284 (1986)
48. D. Tischler, On fibering certain foliated manifolds over S^1. Topology **9**(2), 153–154 (1970)
49. M. Verbitsky, Hyper-Kähler manifolds with torsion, supersymmetry and Hodge theory. Asian J. Math. **6**(4), 679–712 (2002)
50. E. Witten, Instantons and the large N = 4 algebra. Preprint arXiv:2407.20964

Classification of Odd Generalized Einstein Metrics on 3-Dimensional Non-unimodular Lie Groups

Vicente Cortés and Liana David

Dedicated to Paul Gauduchon on the occasion of his 80th birthday

Abstract An odd generalized metric E_- on a Lie group G of dimension n with Lie algebra $\mathfrak{g}$ is a left-invariant generalized metric defined on a Courant algebroid $E_{H,F}$ of type B_n over G, with left-invariant twisting forms H and F. We say that E_- is generalized Einstein with left-invariant divergence $\delta \in (\mathfrak{g} \oplus \mathfrak{g}^* \oplus \mathbb{R})^*$ if the Ricci tensor of the pair (E_-, δ) vanishes. In this paper we describe all odd generalized Einstein metrics (together with their divergences and Courant algebroids $E_{H,F}$ on which they are defined) on 3-dimensional non-unimodular Lie groups.

Keywords Courant algebroids · Lie groups · Invariant generalized metrics · Generalized Einstein equation

Primary MSC Classification Numbers 53D18 (Generalized geometry a la Hitchin), 35Q76 (Einstein's equations)

Secondary MSC Classification Numbers 53C30 (Homogeneous manifolds)

V. Cortés
Department of Mathematics and Center for Mathematical Physics, University of Hamburg, Hamburg, Germany
e-mail: vicente.cortes@math.uni-hamburg.de

L. David (✉)
Institute of Mathematics 'Simion Stoilow' of the Romanian Academy, Bucharest, Romania
e-mail: liana.david@imar.ro

L. Ornea (ed.), *Real and Complex Geometry*,
https://doi.org/10.1007/978-3-031-92297-8_5

1 Introduction

Generalized geometry is an active research area in mathematics with applications in mathematical physics. It studies classical geometric structures (metrics, complex structures, Kähler structures, etc.) on Courant algebroids. In this paper we are interested in a class of generalized metrics (odd generalized metrics), which satisfy the so-called generalized Einstein equation.

It was shown in [2] that the bosonic equations of motion of type II supergravity (with the Ramond-Ramond fields switched off) can be formulated as a generalized Einstein equation, that is, vanishing of the generalized Ricci curvature, supplemented by a scalar equation of motion. The underlying exact Courant algebroid is a generalized tangent bundle twisted by a closed 3-form H. More general transitive Courant algebroids for which H is no longer closed but satisfies an equation of the form $dH + \langle F \wedge F \rangle = 0$, where F is the curvature of a gauge bundle, allow to formulate also the bosonic equations of motion of heterotic supergravity (including α'-terms) as a generalized Einstein equation [5]. These Courant algebroids are known as string algebroids [5] or heterotic Courant algebroids [1], depending on the authors. Unlike in classical geometry, where the Ricci tensor is canonically associated to a metric, in generalized geometry there is a further piece of information (the divergence) which is involved in the Einstein equation for a generalized metric.

The present paper is a continuation of [3], where a class of generalized metrics (odd generalized metrics) on arbitrary Lie groups G is studied systematically. Such metrics are left-invariant and live on a simple class of Courant algebroids, introduced in [9] as odd exact or Courant algebroids of type B_n. Their underlying bundle is $TG \oplus T^*G \oplus \mathbb{R}$ and their Dorfman bracket is twisted by a pair (H, F), where $H \in \Omega^3(G)$, $F \in \Omega^2(G)$ are left-invariant and satisfy $dH + F \wedge F = 0$, $dF = 0$. A main result of [3] is a description of odd generalized metrics on G which satisfy the generalized Einstein equation, when G is 3-dimensional unimodular. In this paper we consider the case when G is 3-dimensional non-unimodular and we develop a similar description. When G is non-unimodular (and 3-dimensional), the Lie algebra $\mathfrak{g} = \mathrm{Lie}(G)$ is a semi-direct product $\mathbb{R} \ltimes_{\mathbf{A}} \mathbb{R}^2$ of $\mathbb{R}$ and an abelian Lie algebra $\mathfrak{u} = \mathbb{R}^2$ (the unimodular kernel of $\mathfrak{g}$), where $\mathbb{R}$ acts on $\mathbb{R}^2$ by an endomorphism $\mathbf{A} \in \mathrm{End}\,\mathbb{R}^2$ with non-zero trace. The isomorphism classes of 3-dimensional non-unimodular Lie algebras correspond to a list of canonical forms for the endomorphism $\mathbf{A}$, see (8), which, in the notation of [8], define the Lie algebras $\tau_2(\mathbb{R}) \oplus \mathbb{R}$, $\tau_3(\mathbb{R})$, $\tau_{3,\lambda}(\mathbb{R})$ ($\lambda \in \mathbb{R} \setminus \{-1\}$, $0 < |\lambda| \leq 1$) and $\tau'_{3,\lambda}(\mathbb{R})$ ($\lambda \in \mathbb{R} \setminus \{0\}$). Our main results are summarized as follows (as the case $F = 0$ can be directly recovered from [4], we assume in the statement below that the generalized Einstein metrics are defined on Courant algebroids $E_{H,F}$ with $F \neq 0$):

Theorem 1 *Let G be a 3-dimensional non-unimodular Lie group with Lie algebra $\mathfrak{g}$ and let $\mathfrak{u}$ be the unimodular kernel of $\mathfrak{g}$.*

(i) There are odd generalized Einstein metrics $E_- = \{X - g(X) \mid X \in TG\}$ on G with $g|_{\mathfrak{u} \times \mathfrak{u}}$ degenerate. If $\mathfrak{g}$ is isomorphic to $\tau_2(\mathbb{R}) \oplus \mathbb{R}$, $\tau_3(\mathbb{R})$ or $\tau_{3,\lambda}(\mathbb{R})$,

then E_- can be chosen to be divergence-free or with non-zero divergence. If $\mathfrak{g}$ is isomorphic to $\tau'_{3,\lambda}(\mathbb{R})$ then E_- has necessarily non-zero divergence.

(ii) There are odd generalized Einstein metrics $E_- = \{X - g(X) \mid X \in TG\}$ on G with $g|_{\mathfrak{u}\times\mathfrak{u}}$ non-degenerate, if and only if $\mathfrak{g}$ is isomorphic to $\tau_2(\mathbb{R})\oplus\mathbb{R}$, $\tau_{3,\frac{1}{2}}(\mathbb{R})$ or $\tau_{3,-\frac{1}{2}}(\mathbb{R})$. All such metrics have non-zero divergence.

The results presented here and in [3] may be seen as the odd analogue of [4], where all (ordinary) generalized Einstein metrics on 3-dimensional Lie groups were described. We recall that a generalized metric on a Lie group G lives on an exact (rather than odd exact) Courant algebroid, with left-invariant twisting 3-form $H \in \Omega^3(G)$, and is left-invariant. It is worth to make some comments concerning the main differences between the methods and results we employ here and those of [4]. The algebraic system expressing the generalized Einstein equation in the left-invariant odd exact setting, which leads to our main results, involves the additional unknowns $F \in \Lambda^2\mathfrak{g}^*$ and $\delta_0 \in \mathbb{R}$. Here F is the closed twisting 2-form and δ_0 is the value of the divergence on the canonical basis of $\mathbb{R} \subset \mathfrak{g}\oplus\mathfrak{g}^*\oplus\mathbb{R}$. These additional variables occur in a coupled algebraic system involving all the data, which makes the system considerably more difficult to solve. Once the system is solved we identify the isomorphism type of the Lie algebra $\mathfrak{g}$ on which any given solution lives. This is done in a systematic way (rather than ad-hoc as in [4]) by an analysis of the characteristic polynomial of the endomorphism $\mathbf{A}$.

We end the paper with a brief summary on the existence of ordinary and odd generalized Einstein metrics on 3-dimensional Lie groups, which collects the main results from [3, 4] and the present paper.

2 Preliminary Material

We preserve the setting and notation from Sections 2, 3 and 4 of [3]. For the theory of generalized metrics on an arbitrary Courant algebroids, see e.g. [6].

Let G be a 3-dimensional Lie group with Lie algebra $\mathfrak{g}$ and E_- an odd generalized metric on G. Assume that E_- is defined on a Courant algebroid $E_{H,F}$ of type B_3 over G, with $H \in \Omega^3(G)$ and $F \in \Omega^2(G)$ left-invariant (and F closed). Let $\delta \in (\mathfrak{g}\oplus\mathfrak{g}^*\oplus\mathbb{R})^*$ be a left-invariant divergence operator. Without loss of generality, we assume that

$$E_- = \{X - g(X) \mid X \in TG\} \tag{1}$$

where g is a left-invariant pseudo-Riemannian on G. Let $E_H := \mathbb{T}G = TG\oplus T^*G$ be the generalized tangent bundle with Dorfman bracket

$$[X+\xi, Y+\eta]_H := \mathcal{L}_X(Y+\eta) - i_Y d\xi + i_X i_Y H, \tag{2}$$

twisted by H. Then E_- can be considered as a generalized metric on E_H. Let $\mathrm{Ric}^{\tilde{\delta}}$ be the Ricci tensor of the pair $(E_-, \tilde{\delta} := \delta|_{\mathfrak{g}\oplus\mathfrak{g}^*})$ defined on E_H. Let $(v_a)_{1\leq a\leq 3}$ be a g-orthonormal basis of $\mathfrak{g}$ and $(e_A)_{0\leq A\leq 6}$ the induced basis of $\mathfrak{g} \oplus \mathfrak{g}^* \oplus \mathbb{R}$, defined by

$$e_0 := 1,\ e_a := v_a + g(v_a),\ a \in \{1, \dots, 3\}$$
$$e_i := v_{i-3} - g(v_{i-3}),\ i \in \{4, \dots, 6\}$$

Unless otherwise stated, indices a, b, c will run from 1 to 3, indices i, j, k will run from 4 to 6 and capital indices A, B, C will run from 1 to 6. Let

$$F_{ab} := F(v_a, v_b),\ h := H(v_1, v_2, v_3),\ \delta_D = \delta(e_D)$$

(where $0 \leq D \leq 6$) and define the Dorfman coefficients

$$B_{ABC} := \langle [e_A, e_B]_H, e_C \rangle, \tag{3}$$

where $\langle \cdot, \cdot \rangle$ is the scalar product of E_H. The next result has been proved in Corollary 19 of [3].

Proposition 2 *The metric E_- is odd generalized Einstein with divergence δ if and only if the components $R^{\tilde{\delta},+}_{ia} := \mathrm{Ric}^{\tilde{\delta}}(e_i, e_a)$ of the Ricci tensor $\mathrm{Ric}^{\tilde{\delta},+}$ of the pair $(E_-, \tilde{\delta})$ defined on E_H satisfy*

$$\begin{aligned}
R^{\tilde{\delta},+}_{41} &= \epsilon_2 (F_{12})^2 + \epsilon_3 (F_{13})^2,\ R^{\tilde{\delta},+}_{51} = \delta_0 F_{12} + \epsilon_3 F_{23} F_{13},\\
R^{\tilde{\delta},+}_{61} &= \delta_0 F_{13} + \epsilon_2 F_{32} F_{12}\\
R^{\tilde{\delta},+}_{42} &= -\delta_0 F_{12} + \epsilon_3 F_{13} F_{23},\ R^{\tilde{\delta},+}_{52} = \epsilon_1 (F_{12})^2 + \epsilon_3 (F_{23})^2,\\
R^{\tilde{\delta},+}_{62} &= \delta_0 F_{23} + \epsilon_1 F_{12} F_{13}\\
R^{\tilde{\delta},+}_{43} &= -\delta_0 F_{13} + \epsilon_2 F_{12} F_{32},\ R^{\tilde{\delta},+}_{53} = -\delta_0 F_{23} + \epsilon_1 F_{21} F_{31},\\
R^{\tilde{\delta},+}_{63} &= \epsilon_1 (F_{13})^2 + \epsilon_2 (F_{23})^2,
\end{aligned} \tag{4}$$

the components of the 2-form F and divergence δ satisfy

$$\begin{aligned}
&F_{12}\epsilon_2(\delta_2 - \epsilon_1 B_{145}) + F_{13}\epsilon_3(\delta_3 - \epsilon_1 B_{146}) + F_{23}\epsilon_2\epsilon_3(B_{345} - B_{246}) = 0\\
&F_{12}\epsilon_1(\epsilon_2 B_{254} - \delta_1) + \epsilon_1\epsilon_3 F_{13}(B_{354} - B_{156}) + F_{23}\epsilon_3(\delta_3 - \epsilon_2 B_{256}) = 0\\
&F_{12}\epsilon_1\epsilon_2(B_{264} - B_{165}) + F_{13}\epsilon_1(\epsilon_3 B_{364} - \delta_1) + F_{23}\epsilon_2(\epsilon_3 B_{365} - \delta_2) = 0\\
&F_{12}\epsilon_2(\delta_2 - \delta_5) + F_{13}\epsilon_3(\delta_3 - \delta_6) = 0
\end{aligned}$$

$$F_{21}\epsilon_1(\delta_1-\delta_4)+F_{23}\epsilon_3(\delta_3-\delta_6)=0$$
$$F_{31}\epsilon_1(\delta_1-\delta_4)+F_{32}\epsilon_2(\delta_2-\delta_5)=0, \tag{5}$$

the products $\delta_0 F_{ab}$ are given by

$$\begin{aligned}\delta_0 F_{12} &= \epsilon_2 B_{512}\delta_2+\epsilon_3 B_{513}\delta_3-\epsilon_1 B_{514}\delta_4-\epsilon_3 B_{516}\delta_6\\ \delta_0 F_{13} &= \epsilon_2 B_{612}\delta_2+\epsilon_3 B_{613}\delta_3-\epsilon_1 B_{614}\delta_4-\epsilon_2 B_{615}\delta_5\\ \delta_0 F_{23} &= \epsilon_1 B_{621}\delta_1+\epsilon_3 B_{623}\delta_3-\epsilon_1 B_{624}\delta_4-\epsilon_2 B_{625}\delta_5\end{aligned} \tag{6}$$

and the following additional relations hold:

$$\begin{aligned}&\epsilon_2 B_{412}\delta_2+\epsilon_3 B_{413}\delta_3-\epsilon_2 B_{415}\delta_5-\epsilon_3 B_{416}\delta_6=0\\ &\epsilon_1 B_{521}\delta_1+\epsilon_3 B_{523}\delta_3-\epsilon_1 B_{524}\delta_4-\epsilon_3 B_{526}\delta_6=0\\ &\epsilon_1 B_{631}\delta_1+\epsilon_2 B_{632}\delta_2-\epsilon_1 B_{634}\delta_4-\epsilon_2 B_{635}\delta_5=0\\ &\epsilon_1 B_{421}\delta_1+\epsilon_2 B_{512}\delta_2+\epsilon_3(B_{513}+B_{423})\delta_3-\epsilon_1 B_{514}\delta_4-\epsilon_2 B_{425}\delta_5\\ &-\epsilon_3(B_{516}+B_{426})\delta_6=0\\ &\epsilon_1 B_{431}\delta_1+\epsilon_2(B_{612}+B_{432})\delta_2+\epsilon_3 B_{613}\delta_3-\epsilon_1 B_{614}\delta_4-\epsilon_2(B_{615}+B_{435})\delta_5\\ &-\epsilon_3 B_{436}\delta_6=0\\ &\epsilon_1(B_{621}+B_{531})\delta_1+\epsilon_2 B_{532}\delta_2+\epsilon_3 B_{623}\delta_3-\epsilon_1(B_{624}+B_{534})\delta_4-\epsilon_2 B_{625}\delta_5\\ &-\epsilon_3 B_{536}\delta_6=0.\end{aligned} \tag{7}$$

Remark 3

(i) The Ricci tensor $\mathrm{Ric}^{\tilde{\delta}}$ and Dorfman coefficients B_{ABC} from the previous corollary were computed in [4] when $\mathfrak{g}$ is unimodular (and 3-dimensional). As part of our arguments, in the next section we compute them when $\mathfrak{g}$ is non-unimodular (some partial computations can be found in [4]).
(ii) There is a conventional difference of sign (in the 3-form H), between the Dorfman bracket (2) and the Dorfman bracket considered in [4]. Owing to this, in various expressions from [4] we will replace H with $-H$.

3 Our Main Result

Let $\mathfrak{g}$ be a 3-dimensional non-unimodular Lie algebra with unimodular kernel $\mathfrak{u} := \{x \in \mathfrak{g} \mid \mathrm{tr}\,\mathrm{ad}_x = 0\}$. Since $\mathfrak{u}$ is a 2-dimensional abelian ideal of $\mathfrak{g}$ (see e.g. [7]), $\mathfrak{g}$ is a semi-direct product $\mathbb{R} \ltimes_{\mathbf{A}} \mathbb{R}^2$ of $\mathbb{R}$ and $\mathbb{R}^2$, where $\mathbb{R}$ acts on $\mathbb{R}^2$ by an endomorphism $\mathbf{A} \in \mathrm{End}\,\mathbb{R}^2$ of non-zero trace. The endomorphism $\mathbf{A}$ is uniquely determined by $\mathfrak{g}$ up to conjugation and multiplication by a non-zero scalar. It follows that there are four classes of non-unimodular 3-dimensional Lie algebras (see [8], p. 212), according to the following four classes of canonical forms for the endomorphism $\mathbf{A}$:

$$\mathbf{A} = \begin{pmatrix} 1 & 0 \\ 0 & 0 \end{pmatrix}, \ \mathbf{A} = \begin{pmatrix} 1 & 1 \\ 0 & 1 \end{pmatrix}, \ \mathbf{A} = \begin{pmatrix} 1 & 0 \\ 0 & \lambda \end{pmatrix}, \ \mathbf{A} = \begin{pmatrix} \lambda & -1 \\ 1 & \lambda \end{pmatrix}, \tag{8}$$

where in the third matrix $\lambda \in \mathbb{R} \setminus \{-1\}$, $0 < |\lambda| \leq 1$ and in the fourth matrix $\lambda \in \mathbb{R} \setminus \{0\}$. Following [8] we will denote the corresponding Lie algebras by $\tau_2(\mathbb{R}) \oplus \mathbb{R}$, $\tau_3(\mathbb{R})$, $\tau_{3,\lambda}(\mathbb{R})$ and $\tau'_{3,\lambda}(\mathbb{R})$ respectively.

Let G be a Lie group with Lie algebra $\mathfrak{g}$ and E_- an odd generalized metric on G, defined on a Courant algebroid $E_{H,F}$ of type B_3 over G. We assume that E_- is in the form (1). With respect to a g-orthonormal basis, we write

$$H = h v_1^* \wedge v_2^* \wedge v_3^*, \ F = \frac{1}{2} F_{ab} v_a^* \wedge v_b^*,$$

where (v_a^*) is the dual basis of (v_a) and $h, F_{ab} \in \mathbb{R}$. Let $\delta \in (\mathfrak{g} \oplus \mathfrak{g}^* \oplus \mathbb{R})^*$ be a left-invariant divergence operator and $\delta_A := \delta(e_A)$, where $(e_A)_{0 \leq A \leq 6}$; is the basis of $\mathfrak{g} \oplus \mathfrak{g}^* \oplus \mathbb{R}$ induced by (v_a).

Assumption 4 From now we assume, unless otherwise stated, that $F \neq 0$ (see Sect. 4 for more explanations on this assumption).

We consider separately the cases when $g|_{\mathfrak{u}\times\mathfrak{u}}$ is degenerate and, respectively, non-degenerate.

3.1 The case $g|_{\mathfrak{u}\times\mathfrak{u}}$-degenerate

As proved in Proposition 3.7 of [4], when $g|_{\mathfrak{u}\times\mathfrak{u}}$ is degenerate there is a g-orthonormal basis (v_a) of $\mathfrak{g}$ with $\epsilon_1 = \epsilon_2 = -\epsilon_3$ and

$$\begin{aligned} \mathcal{L}_{v_1} v_2 &= \epsilon_1(\lambda v_1 + \mu v_2 + \mu v_3) \\ \mathcal{L}_{v_2} v_3 &= -\epsilon_1(\nu v_1 + \rho v_2 + \rho v_3) \\ \mathcal{L}_{v_3} v_1 &= \epsilon_1(\lambda v_1 + \mu v_2 + \mu v_3), \end{aligned} \tag{9}$$

where $\lambda, \mu, \rho \in \mathbb{R}$, $\lambda + \rho \neq 0$ and $\epsilon_a := g(v_a, v_a)$. The Lie algebra $\mathfrak{g}$ is the semi-direct product $\mathbb{R} \ltimes_{\mathbf{A}} \mathbb{R}^2$ where

$$\mathbf{A} = -\epsilon_1 \begin{pmatrix} \lambda & \frac{\nu}{2} \\ 2\mu & \rho \end{pmatrix}. \tag{10}$$

Remark 5 In [4], the minus sign in the expression of $\mathcal{L}_{v_2} v_3$ from (9) was forgotten. As a consequence, the Lie algebra coefficients k_{23a} $(1 \leq a \leq 3)$ from the proof of

Proposition 3.7 of [4] need to be replaced with $-k_{23a}$. The correct expressions of the Dorfman coefficients and Ricci tensor components from the proof of Propositions 3.7 and 3.16 of [4] are given in the proof of Theorem 6 below.

From (9) and $dF = 0$ we obtain that

$$F_{12} + F_{13} = 0. \tag{11}$$

Theorem 6 *There are two families (1) and (2) (see below) of generalized Einstein metrics* $E_- = \{X - g(X) \mid X \in TG\}$ *with divergence* δ *defined on* $E_{H,F}$. *In the basis* (v_a) *above, the corresponding constants* $\lambda, \nu, \mu, \rho, h, F_{ab}, \delta_A$ *are described as follows:*

(1) $\lambda + \rho \neq 0$, $\nu = F_{23} = h = 0$, $F_{13} = -F_{12}$,

$$(F_{12})^2 = -\epsilon_1(\lambda^2 + \rho^2) + \rho\delta_2 > 0 \tag{12}$$

and

$$\rho(\delta_2 - \delta_5) = 0,\ \delta_3 = -\delta_2,\ \delta_6 = -\delta_5,\ \delta_0 = \delta_1 = \delta_4 = 0. \tag{13}$$

(2) $\nu(\lambda + \rho) \neq 0$, $F_{23} = \epsilon\nu$ *(where* $\epsilon = \pm 1$*),* $F_{12} = -F_{13} = -\epsilon\lambda$, $\epsilon_1 = -1$, $h = \nu$, $\delta_0 = -\epsilon\delta_1$ *and*

$$(\lambda\rho - \mu\nu)(\delta_1 + \nu) = (\lambda\rho - \mu\nu)(\delta_1 - \delta_4) = 0$$

$$\delta_3 = -\delta_2 = \lambda + \rho + \frac{\lambda}{\nu}\delta_1,\ \delta_6 = -\delta_5 = \lambda + \rho + \frac{\lambda}{\nu}\delta_4. \tag{14}$$

Proof From Remarks 3 and 5, we obtain the Dorfman coefficients B_{ABC} $(1 \leq A, B, C \leq 6)$ by replacing $h \mapsto -h$, $\nu \mapsto -\nu$, $\rho \mapsto -\rho$ and leaving λ, μ unchanged in the formulae from the proof of Proposition 3.7 of [4]:

$$B_{145} = \lambda,\ B_{146} = -\lambda,\ B_{156} = -\frac{1}{2}(\nu + h),\ B_{245} = \mu,\ B_{613} = -\mu,\ B_{623} = -\rho,$$

$$B_{246} = \frac{1}{2}(h - 2\mu - \nu),\ B_{256} = -\rho,\ B_{345} = \frac{1}{2}(-h - 2\mu + \nu),\ B_{346} = \mu$$

$$B_{356} = \rho,\ B_{412} = -\lambda,\ B_{413} = \lambda,\ B_{423} = \frac{1}{2}(\nu - h),\ B_{512} = -\mu,$$

$$B_{513} = \frac{1}{2}(h + 2\mu + \nu),\ B_{523} = \rho,\ B_{612} = -\frac{1}{2}(h - 2\mu + \nu). \tag{15}$$

The components $R^{\tilde{\delta},+}_{ia}$ (with $1 \leq a \leq 3$ and $4 \leq j \leq 6$) of the Ricci tensor $\mathrm{Ric}^{\tilde{\delta},+}$ from Proposition 2 are computed from relations (15), together with

$$R^{\tilde{\delta},+}_{ia} = \sum_{b=1}^{3}(B^{j}_{bi}B^{b}_{aj} + B^{b}_{ia}\delta_b),\ B^{C}_{AB} := B_{ABC}\eta^{CC}, \tag{16}$$

where $\eta^{AB} := \delta_{AB}\langle v_A, v_B\rangle$ (see Theorem 2.25 of [4]). Note that

$$B^{j}_{bi} = -\epsilon_{j-3}B_{bij},\ B^{b}_{aj} = \epsilon_b B_{ajb},\ B^{c}_{ia} = \epsilon_c B_{iac}. \tag{17}$$

Since $\epsilon_1 = \epsilon_2 = -\epsilon_3$ we obtain

$$\begin{aligned}
R^{\tilde{\delta},+}_{41} &= \frac{1}{2}(h^2 - \nu^2) - \epsilon_1\lambda(\delta_2 + \delta_3)\\
R^{\tilde{\delta},+}_{42} &= \frac{\nu}{2}(\lambda - \rho) + \frac{h}{2}(\lambda + \rho) + \epsilon_1\lambda\delta_1 + \frac{\epsilon_1}{2}(h - \nu)\delta_3\\
R^{\tilde{\delta},+}_{43} &= -\frac{h}{2}(\lambda + \rho) - \frac{\nu}{2}(\lambda - \rho) - \epsilon_1\lambda\delta_1 + \frac{\epsilon_1}{2}(h - \nu)\delta_2\\
R^{\tilde{\delta},+}_{51} &= \frac{\nu}{2}(\lambda - \rho) - \frac{h}{2}(\lambda + \rho) - \epsilon_1\mu\delta_2 - \frac{\epsilon_1}{2}(h + 2\mu + \nu)\delta_3\\
R^{\tilde{\delta},+}_{52} &= -\lambda^2 - \rho^2 + \frac{1}{2}(h^2 + \nu^2) - \mu\nu + \epsilon_1(\mu\delta_1 - \rho\delta_3)\\
R^{\tilde{\delta},+}_{53} &= \lambda^2 + \rho^2 + \mu\nu - \frac{\epsilon_1}{2}(h + 2\mu + \nu)\delta_1 - \epsilon_1\rho\delta_2\\
R^{\tilde{\delta},+}_{61} &= \frac{h}{2}(\lambda + \rho) - \frac{\nu}{2}(\lambda - \rho) - \frac{\epsilon_1}{2}(h - 2\mu + \nu)\delta_2 + \epsilon_1\mu\delta_3\\
R^{\tilde{\delta},+}_{62} &= \lambda^2 + \rho^2 + \mu\nu + \frac{\epsilon_1}{2}(h - 2\mu + \nu)\delta_1 + \epsilon_1\rho\delta_3\\
R^{\tilde{\delta},+}_{63} &= -\lambda^2 - \rho^2 - \frac{1}{2}(h^2 + \nu^2) - \mu\nu + \epsilon_1(\mu\delta_1 + \rho\delta_2).
\end{aligned} \tag{18}$$

Using now $F_{13} = -F_{12}$, $\epsilon_1 = \epsilon_2 = -\epsilon_3$ and relations (18), it is easy to see that relations (4) from Proposition 2 are equivalent to

$$\begin{aligned}
&h^2 - \nu^2 = 2\epsilon_1\lambda(\delta_2 + \delta_3)\\
&h(\lambda + \rho) + \epsilon_1\mu(\delta_2 + \delta_3) + \epsilon_1(\lambda\delta_1 + h\delta_3) = -2\delta_0 F_{12}\\
&(h + \nu)(\delta_2 + \delta_3) = 0\\
&(\lambda - \rho)\nu - \epsilon_1\mu(\delta_2 + \delta_3) + \epsilon_1(\lambda\delta_1 - \nu\delta_3) = 2\epsilon_1 F_{12}F_{23}\\
&\lambda^2 + \rho^2 + \mu\nu - \epsilon_1\mu\delta_1 - \frac{\epsilon_1\rho}{2}(\delta_2 - \delta_3) = -\epsilon_1(F_{12})^2
\end{aligned}$$

$$
\begin{aligned}
&(h-\nu)(\delta_2+\delta_3)=0\\
&(h+\nu)\delta_1+\rho(\delta_2+\delta_3)=2\epsilon_1\delta_0 F_{23}\\
&(h^2+\nu^2)-\epsilon_1\rho(\delta_2+\delta_3)=-2\epsilon_1(F_{23})^2.
\end{aligned}
\tag{19}
$$

We now write the remaining relations (5), (6) and (7) from Proposition 2. For this, we first claim that they imply

$$
\delta_3-\delta_6=-(\delta_2-\delta_5). \tag{20}
$$

In order to prove this claim we consider the first three relations in (7). Using the Dorfman coefficients given in (15), they become:

$$
\begin{aligned}
&\lambda(\delta_2+\delta_3-\delta_5-\delta_6)=0\\
&\mu(\delta_1-\delta_4)-\rho(\delta_3-\delta_6)=0\\
&\mu(\delta_1-\delta_4)+\rho(\delta_2-\delta_5)=0.
\end{aligned}
\tag{21}
$$

The last two relations in (21) imply that

$$
\rho(\delta_2+\delta_3-\delta_5-\delta_6)=0,
$$

and, since $\lambda+\rho\neq 0$, we obtain relation (20), as required. Using $F_{13}=-F_{12}$, $\epsilon_1=\epsilon_2=-\epsilon_3$ and relations (15), (18) and (20), we obtain that relations (5), (6) and (7) from Proposition 2 are equivalent to

$$
\begin{aligned}
&\epsilon_1 F_{12}(\delta_2+\delta_3)=F_{23}(\nu-h)\\
&F_{12}(h-\epsilon_1\delta_1)=F_{23}(\rho+\epsilon_1\delta_3)\\
&F_{23}(\delta_2+\delta_3)=0\\
&\delta_3-\delta_6=-(\delta_2-\delta_5)\\
&F_{12}(\delta_1-\delta_4)=F_{23}(\delta_2-\delta_5)\\
&(h+\nu)(\delta_2-\delta_5)-2\mu(\delta_2+\delta_3)-2\lambda\delta_4=2\epsilon_1\delta_0 F_{12}\\
&(h-2\mu)(\delta_1-\delta_4)+\nu(\delta_1+\delta_4)+2\rho(\delta_3+\delta_5)=2\epsilon_1\delta_0 F_{23}\\
&\mu(\delta_1-\delta_4)+\rho(\delta_2-\delta_5)=0\\
&\lambda(\delta_1-\delta_4)+\nu(\delta_2-\delta_5)=0.
\end{aligned}
\tag{22}
$$

We proved that E_- is generalized Einstein with divergence δ if and only if relations (19) and (22) hold. Moreover, notice that in the above argument we used nowhere Assumption 4. Therefore, the generalized Einstein condition on E_- is equivalent to relations (19) and (22) also when $F = 0$, a fact which will be used in the proof of Proposition 8 below.

Remark that

$$\delta_2 + \delta_3 = 0, \tag{23}$$

which follows from the 1st and 3rd relations in (22), combined with $F \neq 0$ and $F_{13} = -F_{12}$. Using relation (23), the 2nd relation in (19) and the last relation in (22), we can replace the 6th relation in (22) with

$$h(\lambda + \rho) = \epsilon_1(\lambda\delta_4 + h\delta_5). \tag{24}$$

Similarly, comparing the 7th relation in (19) with the 7th relation in (22) and using the 8th relation in (22) together with relation (23), we can replace the 7th relation in (22) with

$$(h - \nu)\delta_4 = 0. \tag{25}$$

We obtain that relations (19) and (22) are equivalent to relation (23) together with

$$\begin{aligned}
&(h-\nu)(h+\nu) = 0\\
&h(\lambda + \rho - \epsilon_1\delta_2) + \epsilon_1\lambda\delta_1 = -2\delta_0 F_{12}\\
&\epsilon_1\nu(\lambda - \rho) + \lambda\delta_1 + \nu\delta_2 = 2F_{12}F_{23}\\
&\epsilon_1(\lambda^2 + \rho^2 + \mu\nu) - \mu\delta_1 - \rho\delta_2 = -(F_{12})^2\\
&\epsilon_1(h+\nu)\delta_1 = 2\delta_0 F_{23}\\
&h^2 + \nu^2 = -2\epsilon_1(F_{23})^2\\
&F_{23}(h-\nu) = 0\\
&F_{12}(h - \epsilon_1\delta_1) = F_{23}(\rho - \epsilon_1\delta_2)\\
&\delta_6 = -\delta_5\\
&F_{12}(\delta_1 - \delta_4) = F_{23}(\delta_2 - \delta_5)\\
&\lambda\delta_4 + h\delta_5 = \epsilon_1 h(\lambda + \rho)\\
&(h-\nu)\delta_4 = 0\\
&\mu(\delta_1 - \delta_4) + \rho(\delta_2 - \delta_5) = 0\\
&\lambda(\delta_1 - \delta_4) + \nu(\delta_2 - \delta_5) = 0.
\end{aligned} \tag{26}$$

We note for later purpose that the derivation of (26) did not use that $F \neq 0$ but only $\delta_2 + \delta_3 = 0$. (In the proof of Proposition 8 we will actually show that $\delta_2 + \delta_3 = 0$ also when $F = 0$.) The 6th and 7th relations in (26) imply that $h = \nu$. Straightforward computations lead to the first class of solutions listed in the theorem when $F_{23} = 0$ and to the second class of solutions when $F_{23} \neq 0$. □

The next proposition follows from the classification of 3-dimensional non-unimodular Lie algebras recalled at the beginning of Sect. 3. The parameter λ in Proposition 7 refers to the notation explained after Eq. (8) and is not meant to be the parameter λ in Theorem 6.

Proposition 7

(i) There is an odd generalized Einstein metric from the first family stated in Theorem 6 if and only if $\mathfrak{g}$ *is isomorphic to* $\tau_2(\mathbb{R}) \oplus \mathbb{R}$, $\tau_3(\mathbb{R})$ *or* $\tau_{3,\lambda}(\mathbb{R})$. *It can be chosen to be divergence-free or with non-zero divergence.*

(ii) There is an odd generalized Einstein metric from the second family stated in Theorem 6 if and only if $\mathfrak{g}$ *is isomorphic to* $\tau_2(\mathbb{R}) \oplus \mathbb{R}$, $\tau_3(\mathbb{R})$, $\tau_{3,\lambda}(\mathbb{R})$ *with* $\lambda \neq 1$, *or* $\tau'_{3,\lambda}(\mathbb{R})$. *All such generalized Einstein metrics have non-zero divergence.*

Proof Let us explain the argument for the second family. As $\lambda + \rho \neq 0$ we immediately obtain from (14) that any generalized metric E_- from the second family has non-zero divergence. Let $\mathbf{A}$ be the matrix defined by (10), $\mathbf{A}' := -\epsilon_1 \mathbf{A}$ and

$$P_{\mathbf{A}'}(x) = x^2 - (\lambda + \rho)x + \lambda\rho - \mu\nu$$

the characteristic polynomial of $\mathbf{A}'$. When $\lambda\rho = \mu\nu$, $P'_{\mathbf{A}}$ has two real roots, one zero and the other non-zero. We obtain that $\mathfrak{g}$ is isomorphic to $\tau_2(\mathbb{R}) \oplus \mathbb{R}$. When $\lambda\rho \neq \mu\nu$, the obstructions on the constants λ, ν, μ, ρ are

$$\nu \neq 0, \ \lambda + \rho \neq 0, \ \lambda\rho \neq \mu\nu.$$

Letting $\mu := \frac{1}{4\nu}(\Delta - (\lambda - \rho)^2)$, we see that any $\Delta \in \mathbb{R} \backslash \{(\lambda+\rho)^2\}$ can be realized as the discriminant of $P_{\mathbf{A}'}$. When $\Delta \in \mathbb{R}^+ \setminus \{(\lambda+\rho)^2\}$, the roots $x_\pm = \frac{1}{2}(\lambda + \rho \pm \sqrt{\Delta})$ of $P'_{\mathbf{A}}$ are distinct, real, and non-zero. The corresponding Lie algebras $\mathfrak{g}$ cover the entire family $\tau_{3,\lambda}(\mathbb{R})$ except $\tau_{3,1}(\mathbb{R})$ (as $x_+ \neq x_-$). When $\Delta < 0$, $x_\pm$ are non-real and the corresponding Lie algebras $\mathfrak{g}$ cover the family $\tau'_{3,\lambda}(\mathbb{R})$. Finally, when $\Delta = 0$, $x_+ = x_- \in \mathbb{R} \setminus \{0\}$. As $\mathbf{A}'$ is non-diagonal (since $\nu \neq 0$), we obtain that $\mathfrak{g}$ is isomorphic to $\tau_3(\mathbb{R})$. □

We end this section by describing generalized Einstein metrics $E_- = \{X - g(X) \mid X \in TG\}$ on $E_{H,0}$ with $g|_{\mathfrak{u}\times\mathfrak{u}}$-degenerate. As these are ordinary generalized Einstein metrics (see [3] and Sect. 4), the next proposition corrects Proposition 3.16 of [4].

Proposition 8 *Consider the setting from the beginning of this section and assume that* $F = 0$.

(i) The generalized metric $E_- = \{X - g(X) \mid X \in TG\}$ *is generalized Einstein with divergence* δ *if and only if* $h = \nu = 0$, $\lambda + \rho \neq 0$, $\delta_3 = -\delta_2$, $\delta_6 = -\delta_5$,

$\lambda\delta_1 = \lambda\delta_4 = 0$ and

$$\mu\delta_1 + \rho\delta_2 = \mu\delta_4 + \rho\delta_5 = \epsilon_1(\lambda^2 + \rho^2). \tag{27}$$

(ii) A generalized metric like in (i) exists if and only if $\mathfrak{g}$ is isomorphic to $\mathfrak{r}_2(\mathbb{R}) \oplus \mathbb{R}$, $\mathfrak{r}_3(\mathbb{R})$ or $\mathfrak{r}_{3,\lambda}(\mathbb{R})$. Moreover, it has always non-zero divergence.

Proof

(i) As stated in the proof of Theorem 6, E_- is generalized Einstein with divergence δ if and only if relations (19) and (22) hold with $F = 0$. We now prove that

$$\delta_2 + \delta_3 = 0. \tag{28}$$

If, by contradiction, $\delta_2 + \delta_3 \neq 0$, then the 3rd and 6th relations in (19) imply that $h = \nu = 0$. But then the 1st and 7th relations in (19) imply that $\lambda = \rho = 0$, which contradicts $\lambda + \rho \neq 0$. Relation (28) follows. We obtain that E_- is generalized Einstein with divergence δ if and only if relations (28) and (26) hold (the latter with $F = 0$). We arrive at the family of solutions listed in the proposition.

(ii) Relations (27) imply that $\delta \neq 0$ (otherwise $\lambda + \rho = 0$). The identification of the Lie algebra $\mathfrak{g}$ follows from the classification of 3-dimensional non-unimodular Lie algebras recalled at the beginning of Sect. 3.

□

3.2 The case $g|_{\mathfrak{u}\times\mathfrak{u}}$ non-degenerate

Consider the setting from the beginning of Sect. 3 and assume that $g|_{\mathfrak{u}\times\mathfrak{u}}$ is non-degenerate. Let (v_a) be a g-orthonormal basis of $\mathfrak{g}$ such that $v_1, v_3 \in \mathfrak{u}$. The Lie bracket of $\mathfrak{g}$ is given by

$$\begin{aligned} \mathcal{L}_{v_1} v_3 &= 0 \\ \mathcal{L}_{v_2} v_1 &= \epsilon_1 \lambda v_1 + \epsilon_3 \mu v_3 \\ \mathcal{L}_{v_2} v_3 &= \epsilon_1 \nu v_1 + \epsilon_3 \rho v_3, \end{aligned} \tag{29}$$

where $\epsilon_i := g(v_i, v_i)$, $\lambda, \mu, \nu, \rho \in \mathbb{R}$ and $\mathrm{tr}\,\mathrm{ad}_{v_2} = \epsilon_1\lambda + \epsilon_3\rho \neq 0$. The Lie algebra $\mathfrak{g}$ is the semi-direct product $\mathbb{R} \ltimes_{\mathbf{A}} \mathbb{R}^2$ where

$$\mathbf{A} = \begin{pmatrix} \epsilon_1\lambda & \epsilon_1\nu \\ \epsilon_3\mu & \epsilon_3\rho \end{pmatrix}. \tag{30}$$

From (29) and $dF = 0$ we obtain that

$$F_{13} = 0. \tag{31}$$

Case (7) of Theorem 9 below will be treated in the next section.

Theorem 9 *There are seven classes of odd generalized Einstein metrics $E_- = \{X - g(X) \mid X \in TG\}$ with divergence δ defined on $E_{H,F}$. In the basis (v_a) above (with v_1 and v_3 interchanged, if necessary), the corresponding constants $\lambda, \nu, \mu, \rho, h, F_{ab}, \delta_A$ are described as follows (below $\epsilon, \tilde{\epsilon} \in \{\pm 1\}$):*

(1) $\lambda \neq 0$, $h = \mu$, $F_{23} = F_{13} = 0$, $\delta_0 F_{12} = \epsilon_3 \mu \delta_3$, $\delta_1 = \delta_4 = 0$, $\delta_2 = \delta_5 = -\epsilon_1 \lambda$ *and one of the following conditions hold:*

(a) $\rho = \nu = 0$ *and* $(F_{12})^2 = -(\epsilon_1 \lambda^2 + \epsilon_3 \mu^2) > 0$;
(b) $\mu = 0$, $\rho = -2\lambda$, $\nu = 2\epsilon\lambda$, $\epsilon_1 = -1$, $\epsilon_3 = 1$, $F_{12} = \tilde{\epsilon}\lambda\sqrt{3}$, $\delta_0 = \delta_3 = \delta_6 = 0$;

(2) $\lambda = 0$, $\mu \neq 0$, $\nu = -2\mu$, $\rho = \tilde{\epsilon}\mu$, $\epsilon_1 = -1$, $\epsilon_3 = 1$, $h = \mu$, $F_{13} = 0$, $F_{23} = \epsilon\rho$, $F_{12} = \epsilon\mu$, $\delta_0 = \delta_1 = \delta_3 = \delta_4 = \delta_6 = 0$, $\delta_2 = \delta_5 = -\rho$;
(3) $\lambda\nu(\epsilon_1\lambda + \epsilon_3\rho) \neq 0$, $\epsilon_1\lambda^2 + \epsilon_3\mu^2 < 0$, $\lambda\rho = \mu\nu$, $h = \mu - \nu$, $F_{23} = -\frac{\nu}{\lambda} F_{12}$, $(F_{12})^2 = -(\epsilon_1\lambda^2 + \epsilon_3\mu^2)$, $F_{13} = 0$, $\delta_0 F_{12} = \epsilon_1 \frac{\lambda}{\nu} \delta_1 (\nu - \mu)$, $\delta_2 = \delta_5 = -(\epsilon_1\lambda + \epsilon_3\rho)$, $\delta_3 = -\epsilon_1\epsilon_3 \frac{\lambda}{\nu} \delta_1$, $\delta_6 = -\epsilon_1\epsilon_3 \frac{\lambda}{\nu} \delta_4$, *and one of the following conditions holds:*

(a) $\delta_1 = \delta_4 \neq 0$;
(b) $\delta_0 = \delta_1 = \delta_4 = 0$ *and either* $(\rho = \mu = 0, \epsilon_1 = -1)$ *or* $(\mu - \nu)\rho \neq 0$;
(c) $\delta_1 \neq \delta_4$ *and either* $\delta_1 \neq 0$ *or* $\delta_0 = \delta_1 = 0$;

(4) $\lambda = \nu = 0$, $\rho\mu \neq 0$, $\epsilon_3 = -1$, $h = \mu$, $F_{12} = \epsilon\mu$, $F_{23} = -\epsilon\rho$, $F_{13} = 0$, $\delta_2 = \delta_5 = \rho$, $\delta_3 = -\epsilon\delta_0$, $\delta_3 = \epsilon_1 \frac{\mu}{\rho} \delta_1$, $\delta_6 = \epsilon_1 \frac{\mu}{\rho} \delta_4$ *and one of the following conditions hold:*

(a) either $\delta_0 = 0$ *or* $\delta_0(\delta_1 - \delta_4) \neq 0$;
(b) $\delta_4 = \delta_1$ *and* $\delta_1\delta_3 \neq 0$;

(5) $\lambda = \frac{\epsilon}{3}(2\mu + \nu)$, $\rho = \frac{\epsilon}{3}(2\nu + \mu)$, $\epsilon_1(\mu^2 - \nu^2) > 0$, $\mu + 2\nu \neq 0$, $\nu + 2\mu \neq 0$, $\epsilon_3 = -\epsilon_1$, $h = \frac{1}{3}(\mu - \nu)$, $F_{13} = 0$, $F_{23} = -\epsilon F_{12}$, $(F_{12})^2 = \frac{\epsilon_1}{3}(\mu^2 - \nu^2)$, $\delta_0 = \delta_1 = \delta_3 = \delta_4 = \delta_6 = 0$, $\delta_2 = \delta_5 = \frac{\epsilon\epsilon_1}{3}(\nu - \mu)$;
(6) $\rho = -2\lambda \neq 0$, $\mu = 2\epsilon\lambda$, $\nu = 0$, $\epsilon_1 = 1$, $\epsilon_3 = -1$, $h = 0$, $F_{12} = \tilde{\epsilon}\lambda$, $F_{23} = 2\epsilon\tilde{\epsilon}\lambda$, $F_{13} = 0$, $\delta_0 = \delta_1 = \delta_3 = \delta_4 = \delta_6 = 0$, $\delta_2 = \delta_5 = -\lambda$;
(7) all constants μ, ν, F_{12}, F_{23} *and* $\epsilon_1\lambda + \epsilon_3\rho$ *are non-zero,*

$$F_{13} = h = \delta_0 = \delta_1 = \delta_3 = \delta_4 = \delta_6 = 0, \quad \delta_2 = \delta_5 \tag{32}$$

and relations (43) with $h = 0$ *hold.*

We divide the proof of the above theorem into several lemmas. To keep the text reasonably short, we only present the main steps of the computations. The remaining parts of the argument follow easily.

Lemma 10 *The metric E_- is generalized Einstein with divergence δ if and only if relations*

$$\begin{aligned}
&\frac{\epsilon_3}{2}(\nu^2 - h^2 - \mu^2) + \delta_2\lambda = (F_{12})^2\\
&\frac{\epsilon_3\delta_3}{2}(h + \mu - \nu) = \delta_0 F_{12}\\
&\frac{\epsilon_1}{2}\lambda(h - \mu + \nu) + \frac{\epsilon_3}{2}\rho(h + \mu - \nu) + \frac{\delta_2}{2}(h - \mu - \nu) = F_{12}F_{23}\\
&\epsilon_1\delta_1\lambda + \frac{\epsilon_3\delta_3}{2}(h + \mu + \nu) = \delta_0 F_{12}\\
&-\lambda^2 - \frac{\epsilon_1\epsilon_3}{2}h^2 - \frac{\epsilon_1\epsilon_3}{2}(\mu + \nu)^2 - \rho^2 = \epsilon_1(F_{12})^2 + \epsilon_3(F_{23})^2\\
&\frac{\epsilon_1\delta_1}{2}(h - \mu - \nu) - \epsilon_3\delta_3\rho = \delta_0 F_{23}\\
&\frac{\epsilon_1}{2}\lambda(h + \mu - \nu) + \frac{\epsilon_3}{2}\rho(h - \mu + \nu) + \frac{\delta_2}{2}(h + \mu + \nu) = -F_{12}F_{23}\\
&\frac{\epsilon_1\delta_1}{2}(h + \mu - \nu) = \delta_0 F_{23}\\
&\frac{\epsilon_1}{2}(\mu^2 - h^2 - \nu^2) + \delta_2\rho = (F_{23})^2
\end{aligned}\tag{33}$$

together with

$$\delta_2 = \delta_5 \tag{34}$$

and

$$\begin{aligned}
&F_{12}\epsilon_3(\delta_2 + \epsilon_1\lambda) = F_{23}(h + \nu)\\
&F_{12}\epsilon_3\delta_1 = F_{23}\epsilon_1\delta_3\\
&F_{12}(\mu - h) = F_{23}\epsilon_1(\epsilon_3\rho + \delta_2)\\
&F_{12}\epsilon_1(\delta_1 - \delta_4) = F_{23}\epsilon_3(\delta_3 - \delta_6)\\
&\frac{\epsilon_3}{2}(h + \mu - \nu)\delta_3 + \epsilon_1\lambda\delta_4 + \frac{\epsilon_3}{2}(-h + \mu + \nu)\delta_6 = \delta_0 F_{12}\\
&\frac{\epsilon_1}{2}(h - \mu - \nu)\delta_1 - \epsilon_3\rho\delta_3 - \frac{\epsilon_1}{2}(h - \mu + \nu)\delta_4 = \delta_0 F_{23}\\
&\epsilon_1\lambda(\delta_1 - \delta_4) + \epsilon_3\nu(\delta_3 - \delta_6) = 0\\
&\epsilon_1\mu(\delta_1 - \delta_4) + \epsilon_3\rho(\delta_3 - \delta_6) = 0
\end{aligned}\tag{35}$$

hold.

Proof From Remark 3, we obtain the Dorfman coefficients B_{ABC} $(1 \leq A, B, C \leq 6)$ by replacing $h \mapsto -h$ in the Dorfman coefficients computed in the proof of Proposition 3.5 of [4]:

$$
\begin{aligned}
&B_{145} = -\lambda,\ B_{156} = \frac{1}{2}(-h+\mu+\nu),\ B_{246} = \frac{1}{2}(h-\mu+\nu)\\
&B_{345} = -\frac{1}{2}(h+\mu+\nu),\ B_{356} = \rho,\ B_{412} = \lambda\\
&B_{423} = -\frac{1}{2}(h+\mu+\nu),\ B_{513} = \frac{1}{2}(h+\mu-\nu)\\
&B_{612} = \frac{1}{2}(-h+\mu+\nu),\ B_{623} = -\rho\\
&B_{146} = B_{245} = B_{256} = B_{346} = B_{413} = B_{512} = B_{523} = B_{613} = 0.
\end{aligned}
\tag{36}
$$

The components $R^{\tilde{\delta},+}_{ia}$ of the Ricci tensor $\mathrm{Ric}^{\tilde{\delta},+}$ from Proposition 2 are computed from (16), (17) and (36):

$$
\begin{aligned}
R^{\tilde{\delta},+}_{41} &= \frac{\epsilon_2\epsilon_3}{2}(\nu^2-h^2-\mu^2)+\epsilon_2\delta_2\lambda\\
R^{\tilde{\delta},+}_{42} &= -\epsilon_1\delta_1\lambda-\frac{\epsilon_3\delta_3}{2}(h+\mu+\nu)\\
R^{\tilde{\delta},+}_{43} &= \frac{\epsilon_1\epsilon_2}{2}\lambda(h+\mu-\nu)+\frac{\epsilon_2\epsilon_3}{2}\rho(h-\mu+\nu)+\frac{\epsilon_2\delta_2}{2}(h+\mu+\nu)\\
R^{\tilde{\delta},+}_{51} &= \frac{\epsilon_3\delta_3}{2}(h+\mu-\nu)\\
R^{\tilde{\delta},+}_{52} &= -\lambda^2-\rho^2-\frac{\epsilon_1\epsilon_3}{2}((\mu+\nu)^2+h^2)\\
R^{\tilde{\delta},+}_{53} &= -\frac{\epsilon_1\delta_1}{2}(h+\mu-\nu)\\
R^{\tilde{\delta},+}_{61} &= -\frac{\epsilon_1\epsilon_2}{2}\lambda(h-\mu+\nu)-\frac{\epsilon_2\epsilon_3}{2}\rho(h+\mu-\nu)+\frac{\epsilon_2\delta_2}{2}(-h+\mu+\nu)\\
R^{\tilde{\delta},+}_{62} &= \frac{\epsilon_1\delta_1}{2}(h-\mu-\nu)-\epsilon_3\delta_3\rho\\
R^{\tilde{\delta},+}_{63} &= \frac{\epsilon_1\epsilon_2}{2}(\mu^2-h^2-\nu^2)+\epsilon_2\delta_2\rho
\end{aligned}
\tag{37}
$$

(when $\epsilon_1 = \epsilon_2 = -\epsilon_3$, they coincide with the Ricci coefficients from the proof of Proposition 3.15 of [4], with h replaced by $-h$). Our claim follows from Proposition 2. More precisely, relations (33) are equivalent to relations (4), while relations (34) and (35) are equivalent to relations (5), (6) and (7). □

Remark 11

(i) Combining the 2nd with the 4th relation in (33), the 3rd with the 7th relation in (33) and the 6th with the 8th relation in (33), we will often replace in the arguments below the 4th, 6th and 7th relations in (33) with

$$\begin{aligned} \epsilon_1 \lambda \delta_1 + \epsilon_3 \nu \delta_3 &= 0 \\ \epsilon_1 \mu \delta_1 + \epsilon_3 \rho \delta_3 &= 0 \\ h(\delta_2 + \epsilon_1 \lambda + \epsilon_3 \rho) &= 0. \end{aligned} \tag{38}$$

Similarly, using the 1st and last relations in (33), we will often replace the 5th relation in (33) by

$$\delta_2(\epsilon_1 \lambda + \epsilon_3 \rho) = -\lambda^2 - \frac{\epsilon_1 \epsilon_3}{2}(\mu + \nu)^2 - \rho^2 + \frac{\epsilon_1 \epsilon_3}{2} h^2. \tag{39}$$

(ii) Moreover, if $(\delta_1 - \delta_4)(\delta_3 - \delta_6) \neq 0$ or $\delta_1 \delta_3 \neq 0$, then

$$\lambda \rho = \mu \nu. \tag{40}$$

This follows from the last two relations in (35) and the first two relations in (38).

In order to determine the generalized Einstein metrics E_- we need to solve the system formed by the equations (33), (34) and (35). For this, we will consider various cases, namely: (I) $F_{23} = 0$; (II) $F_{23} \neq 0$ and $F_{12} = 0$; (III) $F_{23} F_{12} \neq 0$. Recall that $F_{13} = 0$, as the 2-form F is closed.

Lemma 12 *When $F_{23} = 0$, the system (33), (34) and (35) leads to the classes of solutions (1a) and (1b).*

Proof Since $F_{23} = 0$, we obtain that $F_{12} \neq 0$ (otherwise $F = 0$). From the first four relations in (35) and the 6th relation in (35), we obtain

$$\delta_2 = -\epsilon_1 \lambda, \ \delta_1 = \delta_4 = 0, \ h = \mu, \ \rho \delta_3 = 0. \tag{41}$$

When $\rho = 0$ we obtain the class of solutions (1a). When $\rho \neq 0$ we obtain the class of solutions (1b). □

Lemma 13 *When $F_{23} \neq 0$ and $F_{12} = 0$, the system (33), (34) and (35) leads (after interchanging v_1 and v_3), to the classes of solutions 1a) and 1b).*

Proof The 1st and 3rd relations in (35) imply that $h = -\nu$ and $\delta_2 = -\epsilon_3 \rho$. From the 3rd relation in (33) we obtain that $\mu(\lambda - 2\epsilon_1 \epsilon_3 \rho) = 0$. On the other hand, remark that under the interchange of the vectors v_1 and v_3, the constant λ is interchanged with ρ, μ with ν, δ_1 with δ_3, δ_4 with δ_6, ϵ_1 with ϵ_3 and δ_0, δ_2 and δ_5 remain the same while h is replaced by $-h$. When $\mu = 0$ we obtain (after interchanging v_1 and v_3)

the class of solutions (1a). When $\mu \neq 0$ we obtain (after interchanging v_1 and v_3) the class of solutions (1b). □

The key remark to treat the case $F_{12}F_{23} \neq 0$ is the observation that $(\delta_1 - \delta_4)$ and $(\delta_3 - \delta_6)$ are both zero or both non-zero, and, similarly, for δ_1 and δ_3 (see the 2nd and 4th relation in (35)). Also, from the 3rd relation in (38), either $h = 0$ or $\delta_2 = -(\epsilon_1\lambda + \epsilon_3\rho)$. Using these facts we divide the case $F_{12}F_{23} \neq 0$ into several lemmas, as follows.

Lemma 14 *When $F_{12}F_{23} \neq 0$, $\delta_1 = \delta_4$, $\delta_3 = \delta_6$ and $\delta_1 = \delta_3 = h = 0$, the system (33), (34) and (35) leads (modulo interchanging v_1 and v_3), to the classes of solutions (6) and (7).*

Proof By our assumption, $\delta_i = 0$ for any $i \notin \{0, 2, 5\}$. The 2nd relation in (33) implies that also $\delta_0 = 0$. When $\delta_i = 0$ for any $i \notin \{0, 2, 5\}$, relations (33) (with the 4th, 5th, 6th and 7th relation in (33) replaced by relations (38) and (39)), together with relations (34) and (35) become

$$\delta_0 = 0,\ \delta_2 = \delta_5,\ h(\delta_2 + \epsilon_1\lambda + \epsilon_3\rho) = 0 \tag{42}$$

together with

$$\begin{aligned}
&F_{12}\epsilon_3(\delta_2 + \epsilon_1\lambda) = F_{23}(h + \nu)\\
&F_{12}(\mu - h) = F_{23}\epsilon_1(\epsilon_3\rho + \delta_2)\\
&(F_{12})^2 = \frac{\epsilon_3}{2}(\nu^2 - \mu^2 - h^2) + \delta_2\lambda\\
&\delta_2(\epsilon_1\lambda + \epsilon_3\rho) = -\lambda^2 - \frac{\epsilon_1\epsilon_3}{2}(\mu + \nu)^2 - \rho^2 + \frac{\epsilon_1\epsilon_3}{2}h^2\\
&F_{12}F_{23} = \frac{\epsilon_1}{2}\lambda(h - \mu + \nu) + \frac{\epsilon_3}{2}\rho(h + \mu - \nu) + \frac{\delta_2}{2}(h - \mu - \nu)\\
&(F_{23})^2 = \frac{\epsilon_1}{2}(\mu^2 - \nu^2 - h^2) + \delta_2\rho.
\end{aligned} \tag{43}$$

(Relations (42) and (43) with $h \neq 0$ will be used in the proof of the next lemma.) Recall now our assumption $h = 0$. Since $F_{23}F_{12} \neq 0$ and $h = 0$, μ and ν cannot be both zero (from the 5th relation in (43)). When $\mu \neq 0$ and $\nu = 0$ we obtain the class of solutions (6). In fact, when $\nu = 0$ the 1st equation of (43) with $h = 0$ yields $\delta_2 = -\epsilon_1\lambda$, which can be used to rewrite the 4th and 6th equations as

$$\begin{aligned}
-\epsilon_3\rho^2 &= \epsilon_1(\frac{\mu^2}{2} - \lambda\rho)\\
(F_{23})^2 &= \epsilon_1(\frac{\mu^2}{2} - \lambda\rho).
\end{aligned}$$

This shows that $(F_{23})^2 = -\epsilon_3\rho^2$ and the remaining equations are easily solved building on that. When $\mu = 0$ and $\nu \neq 0$ we obtain again (after interchanging v_1 and v_3) the class of solutions (6). When $\mu\nu \neq 0$ the solutions of the system (43) with $h = 0$ correspond to the class of solutions (7) from the statement of the theorem. □

Lemma 15 *When $F_{12}F_{23} \neq 0$, $\delta_1 = \delta_4$, $\delta_3 = \delta_6$, $\delta_1 = \delta_3 = 0$ and $h \neq 0$, the system (33), (34) and (35) leads (modulo interchanging v_1 and v_3) to the classes of solutions (4a) with $\delta_0 = \delta_4 = 0$, (2), (3b) and (5).*

Proof From the previous proof, what we need to solve is the system (42) and (43) under the assumption $h \neq 0$. From the 3rd relation in (42) and the 1st relation in (43), we obtain

$$\delta_2 = -(\epsilon_1\lambda + \epsilon_3\rho), \quad F_{12}\rho = -F_{23}(h + \nu). \tag{44}$$

Since $F_{23}F_{12} \neq 0$, either $\rho = h + \nu = 0$ or $\rho(h + \nu) \neq 0$. When $\rho = h + \nu = 0$ and $\mu = 0$, we obtain the class of solutions (3b) with $(\rho = \mu = 0, \epsilon_1 = -1)$. When $\rho = h + \nu = 0$ and $\mu \neq 0$, we obtain (after interchanging v_1 and v_3) the class of solutions (2). Assume now that $\rho(h + \nu) \neq 0$. Since $\delta_2 = -(\epsilon_1\lambda + \epsilon_3\rho)$, the 3rd and 6th relations in (43) imply that

$$\begin{aligned}(F_{12})^2 &= \frac{\epsilon_3}{2}(\nu^2 - \mu^2 - h^2) - \epsilon_1\lambda^2 - \epsilon_3\lambda\rho \\ (F_{23})^2 &= \frac{\epsilon_1}{2}(\mu^2 - \nu^2 - h^2) - \epsilon_1\lambda\rho - \epsilon_3\rho^2\end{aligned} \tag{45}$$

and the 4th relation in (43) becomes

$$4\lambda\rho = (\mu + \nu)^2 - h^2. \tag{46}$$

Expressing $\lambda\rho$ in terms of μ, ν and h (using the above relation), we rewrite relations (45) in an equivalent way:

$$\begin{aligned}(F_{12})^2 &= \frac{\epsilon_3}{4}(-h^2 - 3\mu^2 + \nu^2 - 2\mu\nu) - \epsilon_1\lambda^2 \\ (F_{23})^2 &= \frac{\epsilon_1}{4}(-h^2 - 3\nu^2 + \mu^2 - 2\mu\nu) - \epsilon_3\rho^2.\end{aligned} \tag{47}$$

When $\lambda = 0$, $\delta_2 = -\epsilon_3\rho$ and the 2nd relation in (43) implies that $h = \mu$. Relation (46) becomes $\nu(2\mu+\nu) = 0$. We arrive at the class of solutions 4a) with $\delta_0 = \delta_4 = 0$ (when $\nu = 0$) and at the class of solutions (2) (when $\nu = -2\mu \neq 0$). Assume now that $\lambda \neq 0$. From the 2nd relation in (43), and $\delta_2 = -(\epsilon_1\lambda + \epsilon_3\rho)$, we obtain

$$F_{23} = F_{12}(\frac{h - \mu}{\lambda}), \tag{48}$$

which, combined with the 2nd relation in (44), implies

$$\lambda\rho = (\mu - h)(h + \nu). \tag{49}$$

Replacing F_{23} with the right-hand side of (48) we obtain, from relations (47) and the 5th relation in (43), that $(F_{12})^2$ satisfies

$$\begin{aligned}
&(F_{12})^2 = \frac{\epsilon_3}{4}(-h^2 - 3\mu^2 + \nu^2 - 2\mu\nu) - \epsilon_1\lambda^2\\
&(F_{12})^2(\frac{h-\mu}{\lambda}) = \epsilon_1\lambda\nu + \epsilon_3\rho\mu\\
&(F_{12})^2(\frac{h-\mu}{\lambda})^2 = \frac{\epsilon_1}{4}(-h^2 - 3\nu^2 + \mu^2 - 2\mu\nu) - \epsilon_3\rho^2,
\end{aligned} \tag{50}$$

where in the second relation above we used that $\delta_2 = -(\epsilon_1\lambda + \epsilon_3\rho)$. On the other hand, remark that relations (46) and (49) imply that

$$(\mu - \nu - h)(\mu - \nu - 3h) = 0. \tag{51}$$

When $h = \mu - \nu$ straightforward computations lead to the class of solutions (3b) with $\rho(\mu - \nu) \neq 0$. When $h = \frac{1}{3}(\mu - \nu)$, we claim that we arrive at the class of solutions (5). The key point of the argument is to solve the system

$$\begin{aligned}
&9\lambda\rho = (\nu + 2\mu)(\mu + 2\nu)\\
&(F_{12})^2 = \frac{\epsilon_3}{9}(-7\mu^2 + 2\nu^2 - 4\mu\nu) - \epsilon_1\lambda^2\\
&(F_{12})^2 = -\frac{3\lambda}{(2\mu + \nu)}(\epsilon_1\lambda\nu + \epsilon_3\rho\mu)\\
&(F_{12})^2 = \frac{9\lambda^2}{(2\mu + \nu)^2}\left(\frac{\epsilon_1}{9}(-7\nu^2 + 2\mu^2 - 4\mu\nu) - \epsilon_3\rho^2\right),
\end{aligned} \tag{52}$$

which is obtained from relations (46) and (50), using $h = \frac{1}{3}(\mu - \nu)$, under the assumption that $\mu \neq \nu$ (as $h \neq 0$). The compatibility between the 2nd and 3rd relations in (52), respectively the 3rd and 4th relations in (52), reduce to

$$\begin{aligned}
&9\epsilon_1\lambda^2 + \epsilon_3(2\mu + \nu)^2 = 0\\
&\epsilon_1\lambda(\mu + 2\nu) + \epsilon_3\rho(2\mu + \nu) = 0,
\end{aligned} \tag{53}$$

where we have used the 1st relation in (52) to write $\lambda\rho$ in terms of μ and ν and $\mu \neq \nu$. From the 1st relation in (53), we obtain that $\epsilon_3 = -\epsilon_1$ and

$$\lambda = \frac{\epsilon}{3}(2\mu + \nu) \tag{54}$$

(where $\epsilon \in \{\pm 1\}$). Then the 2nd relation in (53) implies

$$\rho = \frac{\epsilon}{3}(2\nu + \mu). \tag{55}$$

Building on these formulae we obtain the class of solutions (5). □

Lemma 16 *When $F_{12}F_{23} \neq 0$, $\delta_1 = \delta_4$, $\delta_3 = \delta_6$ and $\delta_1\delta_3 \neq 0$, the system (33), (34) and (35) leads to the classes of solutions (3a) and (4b).*

Proof Since $\delta_3 \neq 0$, the 2nd relation in (35) expresses F_{23} in terms of F_{12}:

$$F_{23} = \epsilon_1\epsilon_3 F_{12}\frac{\delta_1}{\delta_3}. \tag{56}$$

Since $\delta_1 = \delta_4$ and $\delta_3 = \delta_6$, the 5th relation in (35) can be written as

$$\delta_0 F_{12} = \epsilon_1\lambda\delta_1 + \epsilon_3\mu\delta_3. \tag{57}$$

Combining relation (57) with the 2nd relation in (33) and using the 1st relation in (38) together with $\delta_3 \neq 0$ we obtain that

$$h = \mu - \nu. \tag{58}$$

On the other hand, the 1st relation in (38) and $\delta_1\delta_3 \neq 0$ implies that either $\lambda\nu \neq 0$ or $\lambda = \nu = 0$. Assume first that $\lambda\nu \neq 0$. Then, from the 1st relation in (38) and relation (56), we obtain that

$$\delta_3 = -\epsilon_1\epsilon_3\frac{\lambda}{\nu}\delta_1, \quad F_{23} = -\frac{\nu}{\lambda}F_{12}. \tag{59}$$

We claim that we arrive at the class of solutions (3a). To prove the claim, we replace in relations (33) (without the 4th, 5th, 6th and 7th relations), (35), (38) and (39), h, δ_3 and F_{23} by the right-hand sides of (58) and (59). Recall, from our assumption $\delta_1\delta_3 \neq 0$ and relation (40), that $\lambda\rho = \mu\nu$. From the 1st relation in (35) we obtain that $\delta_2 = -(\epsilon_1\lambda + \epsilon_3\rho)$. Relations (38) and relation (39) give no further restrictions. Relations (35) reduce to relation (57), which, from the 1st relation in (59), is equivalent to $\delta_0 F_{12} = \epsilon_1\frac{\lambda}{\nu}\delta_1(\nu - \mu)$. Relations (33) give in addition $(F_{12})^2 = -(\epsilon_1\lambda^2 + \epsilon_3\mu^2)$. We arrive at the class of solutions (3a), as claimed. By similar computations, when $\lambda = \nu = 0$ we arrive at the class of solutions (4b). □

Lemma 17 *When $F_{12}F_{23} \neq 0$, $\delta_1 \neq \delta_4$, $\delta_3 \neq \delta_6$ and $\delta_1\delta_3 \neq 0$, the system (33), (34) and (35) leads to the class of solutions (3c) with $\delta_1 \neq 0$ and to the class of solutions (4a) with $\delta_0(\delta_1 - \delta_4) \neq 0$.*

Proof From the 7th relation in (35) and $(\delta_1 - \delta_4)(\delta_3 - \delta_6) \neq 0$, either $\lambda = \nu = 0$ or $\lambda\nu \neq 0$. Assume first that $\lambda = \nu = 0$. Then $\rho \neq 0$ (since $\mathrm{tr}\,\mathrm{ad}_{v_2} = \epsilon_1\lambda + \epsilon_3\rho \neq 0$), and, from the last relation (35), also $\mu \neq 0$. The last relation in (35) becomes

$$\frac{\delta_1 - \delta_4}{\delta_3 - \delta_6} = -\epsilon_1 \epsilon_3 \frac{\rho}{\mu}. \tag{60}$$

From relation (60) combined with the 4th relation in (35), we obtain that $F_{23} = -\frac{\rho}{\mu} F_{12}$ and, from the 1st relation in (35) we obtain that

$$\delta_2 = -\frac{\epsilon_3 \rho h}{\mu}. \tag{61}$$

Replacing these expressions of F_{23} and δ_2 into the 3rd relation in (35) we obtain that

$$(\mu - h)(\mu^2 + \epsilon_1 \epsilon_3 \rho^2) = 0. \tag{62}$$

We claim that $h = \mu$. If, by contradiction, $h \neq \mu$, then, from the above relation, $\epsilon_3 = -\epsilon_1$, $\mu = \epsilon \rho$ (where $\epsilon \in \{\pm 1\}$), $F_{12} = -\epsilon F_{23}$ and the 5th relation in (33) implies that $h = -\mu$. From the 1st relation in (33) we conclude that $\epsilon_3 = -1$ and $F_{12} = \tilde{\epsilon} \mu$ (where $\tilde{\epsilon} \in \{\pm 1\}$). From relation (61) we obtain that $\delta_2 = -\epsilon \mu$. But the 3rd relation in (33) implies that $\delta_2 = \epsilon \mu$. We obtain a contradiction. We have proved that $h = \mu$ and we arrive at the class of solutions (4a) with $\delta_0(\delta_1 - \delta_4) \neq 0$. Assume now that $\lambda \nu \neq 0$. Since $\lambda \neq 0$, the 1st relation in (38) implies that $\frac{\delta_1}{\delta_3} = -\epsilon_1 \epsilon_3 \frac{\nu}{\lambda}$. Then $F_{23} = -\frac{\nu}{\lambda} F_{12}$, and, from the 3rd relation in (35), $h = \mu + \epsilon_1 \frac{\nu}{\lambda}(\delta_2 + \epsilon_3 \rho)$. Replacing this expression of h into the 1st relation in (35) and using relation (40) together with $F_{23} = -\frac{\nu}{\lambda} F_{12}$ and $F_{12} \neq 0$, we obtain that

$$(1 + \epsilon_1 \epsilon_3 (\frac{\nu}{\lambda})^2)(\delta_2 + \epsilon_1 \lambda + \epsilon_3 \rho) = 0. \tag{63}$$

If the first factor from the left-hand side of (63) vanishes, then $\epsilon_3 = -\epsilon_1$, $\nu = \epsilon \lambda$ (where $\epsilon \in \{\pm 1\}$), $\rho = \epsilon \mu$ (as $\lambda \rho = \mu \nu$), $F_{23} = -\epsilon F_{12}$ and, from the 1st relation in (35), $\delta_2 = \epsilon \epsilon_1 h$. The 5th relation in (33) implies that $h = \tilde{\epsilon}(\lambda - \rho)$, where $\tilde{\epsilon} \in \{\pm 1\}$. In particular, $h \neq 0$ as $\lambda - \rho = \epsilon_1(\epsilon_1 \lambda + \epsilon_3 \rho) \neq 0$ and $\delta_2 = \epsilon \tilde{\epsilon} \epsilon_1 (\lambda - \rho)$. From the 3rd relation in (38), we obtain that $\tilde{\epsilon} = -\epsilon$. We arrive at the class of solutions (3c) with $\delta_1 \neq 0$, $\nu = \epsilon \lambda$ and $\epsilon_3 = -\epsilon_1$. If the first factor from the left-hand side of (63) is non-zero, we arrive, by similar computations, to the class of solutions (3c) with $\delta_1 \neq 0$ and $1 + \epsilon_1 \epsilon_3 (\frac{\nu}{\lambda})^2 \neq 0$. Therefore, when $\lambda \nu \neq 0$ we arrive at the class of solutions (3c) with $\delta_1 \neq 0$. □

Lemma 18 *When $F_{12} F_{23} \neq 0$, $\delta_1 \neq \delta_4$, $\delta_3 \neq \delta_6$ and $\delta_1 = \delta_3 = 0$, the system (33), (34) and (35) leads to the class of solutions (3c) with $\delta_0 = \delta_1 = 0$ and to the class of solutions (4a) with $\delta_0 = 0$ and $\delta_4 \neq 0$.*

Proof Since $F_{12} \neq 0$ and $\delta_3 = 0$, the 2nd relation (33) implies that $\delta_0 = 0$. The 6th relation in (35) implies that $h = \mu - \nu$. As in the previous proof, either $\lambda \nu \neq 0$

or $\lambda = \nu = 0$. By similar computations, when $\lambda\nu \neq 0$ we arrive at the class of solutions (3c) with $\delta_0 = \delta_1 = 0$. When $\lambda = \nu = 0$ we arrive at the class of solutions (4a) with $\delta_0 = 0$ and $\delta_4 \neq 0$. □

The proof of Theorem 9 is now completed.

Like in Proposition 7, we obtain:

Proposition 19 *Consider the setting from Theorem 9.*

(i) For the class of solutions (1a), (3) and (4), $\mathfrak{g}$ is isomorphic to $\mathfrak{r}_2(\mathbb{R}) \oplus \mathbb{R}$.
(ii) For the class of solutions (1b) and (6), $\mathfrak{g}$ is isomorphic to $\mathfrak{r}_{3,\frac{1}{2}}(\mathbb{R})$.
(iii) For the class of solutions (2) and (5), $\mathfrak{g}$ is isomorphic to $\mathfrak{r}_{3,-\frac{1}{2}}(\mathbb{R})$.

It remains to describe the odd generalized Einstein metrics which correspond to case (7) of Theorem 9. This will be done in the next section.

3.2.1 The Case (7) of Theorem 9

We need to solve the system (43) with $h = 0$ and μ, ν, F_{12}, F_{23}, $\epsilon_1\lambda + \epsilon_3\rho$ all non-zero. It is convenient to replace the variables λ and ρ with

$$\tilde{\lambda} := \delta_2 + \epsilon_1\lambda,\ \tilde{\rho} := \delta_2 + \epsilon_3\rho.$$

Since $\epsilon_1\lambda + \epsilon_3\rho \neq 0$, we obtain that

$$\delta_2 \neq \frac{1}{2}(\tilde{\lambda} + \tilde{\rho}). \tag{64}$$

In terms of the unknowns $(\mu, \nu, F_{12}, F_{23}, \delta_2, \tilde{\lambda}, \tilde{\rho})$, the system (43) takes the form

$$\begin{aligned}
&F_{23} = F_{12}\epsilon_3\frac{\tilde{\lambda}}{\nu}\\
&\mu\nu = \epsilon_1\epsilon_3\tilde{\lambda}\tilde{\rho}\\
&(F_{12})^2 = \frac{\epsilon_3}{2}(\nu^2 - \mu^2) + \epsilon_1\delta_2(\tilde{\lambda} - \delta_2)\\
&\delta_2(\tilde{\lambda} + \tilde{\rho}) = \tilde{\lambda}^2 + \tilde{\rho}^2 + \frac{\epsilon_1\epsilon_3}{2}(\mu + \nu)^2\\
&(F_{12})^2\frac{\tilde{\lambda}}{\nu} = \frac{\epsilon_3}{2}(\nu - \mu)(\tilde{\lambda} - \tilde{\rho}) - \frac{\epsilon_3\delta_2}{2}(\mu + \nu)\\
&(F_{12})^2(\frac{\tilde{\lambda}}{\nu})^2 = \frac{\epsilon_1}{2}(\mu^2 - \nu^2) + \delta_2\rho.
\end{aligned} \tag{65}$$

(The 2nd, 5th and 6th relation in (65) are obtained from the 2nd, 5th and 6th relation in (43), by replacing F_{23} with $F_{12}\epsilon_3\frac{\tilde{\lambda}}{\nu}$, see the 1st relation in (65).) Remark, from the second relation in (65) and $\mu\nu \neq 0$, that $\tilde{\lambda}\tilde{\rho} \neq 0$. The 1st, 2nd and 3rd relations in (65) determine F_{23}, $\tilde{\rho}$ and F_{12} in terms of $\tilde{\lambda}$, μ, ν, δ_2. We are left with the 4th, 5th and 6th relations in (65), which are compatibility relations between the unknowns $\tilde{\lambda}$, μ, ν, δ_2.

Proposition 20

(i) The expression $\tilde{\lambda} + \epsilon_1\epsilon_3\frac{\mu\nu}{\tilde{\lambda}}$ *is non-zero. The 4th relation in (65) determines the unknown* δ_2 *in terms of* $\tilde{\lambda}$, μ, ν, *by*

$$\delta_2(\tilde{\lambda} + \epsilon_1\epsilon_3\frac{\mu\nu}{\tilde{\lambda}}) = \tilde{\lambda}^2 + (\frac{\mu\nu}{\tilde{\lambda}})^2 + \frac{\epsilon_1\epsilon_3}{2}(\mu+\nu)^2. \tag{66}$$

(ii) The 5th and 6th relations in (65) hold if and only if

$$(\mu-\nu)(\tilde{\lambda}^2 + \epsilon_1\epsilon_3\nu^2)\left(\tilde{\lambda}^2 + (\frac{\mu\nu}{\tilde{\lambda}})^2 + \frac{\epsilon_1\epsilon_3}{2}(\mu+\nu)^2 + (\tilde{\lambda} + \epsilon_1\epsilon_3\frac{\mu\nu}{\tilde{\lambda}})^2\right) = 0. \tag{67}$$

Proof Relation (66) follows immediately from the 4th relation in (65), by using $\tilde{\rho} = \epsilon_1\epsilon_3\frac{\mu\nu}{\tilde{\lambda}}$. Using the expression of $(F_{12})^2$ provided by the 3rd relation in (65) and $\tilde{\rho} = \epsilon_1\epsilon_3\frac{\mu\nu}{\tilde{\lambda}}$, we now rewrite the 5th and 6th relation in (65). We obtain that the 5th relation in (65) is equivalent to

$$(\delta_2)^2 - \left(\tilde{\lambda} + \epsilon_1\epsilon_3\frac{(\mu+\nu)\nu}{2\tilde{\lambda}}\right)\delta_2 + \frac{\epsilon_1\epsilon_3}{2}(\mu-\nu)\mu\left(1 + \epsilon_1\epsilon_3\frac{\nu^2}{\tilde{\lambda}^2}\right) = 0. \tag{68}$$

We now prove that the 6th relation in (65) is equivalent to

$$(\delta_2)^2(\epsilon_1\tilde{\lambda}^2 - \epsilon_3\nu^2) + \epsilon_1(\frac{\mu\nu^3}{\tilde{\lambda}} - \tilde{\lambda}^3)\delta_2 + \frac{1}{2}(\epsilon_3\tilde{\lambda}^2 + \epsilon_1\nu^2)(\mu^2 - \nu^2) = 0. \tag{69}$$

In order to prove (69) we remark, from the 3rd relation in (65), that

$$\frac{1}{2}(\mu^2 - \nu^2) = -\epsilon_3(F_{12})^2 + \epsilon_1\epsilon_3\delta_2(\tilde{\lambda} - \delta_2). \tag{70}$$

Replacing in the 6th relation in (65) the term $\frac{\epsilon_1}{2}(\mu^2 - \nu^2)$ by the right-hand side of (70) multiplied with ϵ_1 and using that $\rho = \epsilon_3(\tilde{\rho} - \delta_2)$, we obtain

$$(F_{12})^2((\frac{\tilde{\lambda}}{\nu})^2 + \epsilon_1\epsilon_3) = \epsilon_3\delta_2(\tilde{\lambda} + \tilde{\rho}) - 2\epsilon_3(\delta_2)^2. \tag{71}$$

Using now the expression of $\delta_2(\tilde{\lambda}+\tilde{\rho})$ provided by the 4th relation in (65) and the 2nd relation in (65), we obtain that relation (71) is equivalent to

$$(F_{12})^2((\frac{\tilde{\lambda}}{\nu})^2+\epsilon_1\epsilon_3)=\epsilon_3(\tilde{\lambda}^2+\tilde{\rho}^2+\tilde{\lambda}\tilde{\rho})+\frac{\epsilon_1}{2}(\mu^2+\nu^2)-2\epsilon_3(\delta_2)^2. \tag{72}$$

Replacing in (72) $(F_{12})^2$ by its expression from the 3rd relation in (65) and using (66) together with $\mu\nu=\epsilon_1\epsilon_3\tilde{\lambda}\tilde{\rho}$ we obtain relation (69), as needed.

We now claim that $\tilde{\lambda}+\epsilon_1\epsilon_3\frac{\mu\nu}{\tilde{\lambda}}\neq 0$. Assume, by contradiction, that $\tilde{\lambda}+\epsilon_1\epsilon_3\frac{\mu\nu}{\tilde{\lambda}}=0$. Then the right-hand side of relation (66) vanishes, that is,

$$\epsilon_3=-\epsilon_1,\ \tilde{\lambda}^2+(\frac{\mu\nu}{\tilde{\lambda}})^2=\frac{1}{2}(\mu+\nu)^2. \tag{73}$$

We deduce that $\mu=\nu$, since $\mu\nu=\tilde{\lambda}^2$. Hence $\tilde{\lambda}^2=\nu^2$. But then relation (68) implies that $\delta_2=0$ and the 3rd relation in (65) implies that $F_{12}=0$, which is a contradiction. The claim is proved.

Straightforward computations show that by multiplying relation (68) with the non-zero factor $(\tilde{\lambda}+\epsilon_1\epsilon_3\frac{\mu\nu}{\tilde{\lambda}})^2$ and using relation (66) we obtain relation (67). Similarly, multiplying relation (69) with $(\tilde{\lambda}+\epsilon_1\epsilon_3\frac{\mu\nu}{\tilde{\lambda}})^2$ and using relation (66) again we obtain

$$(\mu^2-\nu^2)(\tilde{\lambda}^2+\epsilon_1\epsilon_3\nu^2)\left(\tilde{\lambda}^2+(\frac{\mu\nu}{\tilde{\lambda}})^2+\frac{\epsilon_1\epsilon_3}{2}(\mu+\nu)^2+(\tilde{\lambda}+\epsilon_1\epsilon_3\frac{\mu\nu}{\tilde{\lambda}})^2\right)=0.$$

Remark that the left-hand side of the above relation coincides with the left-hand side of (67), multiplied by $\mu+\nu$. This concludes our proof. □

We arrive at the description of odd generalized Einstein metrics from case (7) of Theorem 9.

Corollary 21 *There are two classes (i) and (ii) (see below) of odd generalized Einstein metrics generated by case (7) of Theorem 9. The remaining constants $\tilde{\lambda}$, $\tilde{\rho}$, μ, ν, ϵ_1, ϵ_3, F_{12}, F_{23}, δ_2 satisfy the following conditions:*

(i) $\mu=\nu$, $\mu\tilde{\lambda}\neq 0$, $\delta_2=\tilde{\lambda}+\epsilon_1\epsilon_3\frac{\mu^2}{\tilde{\lambda}}$ *and*

$$(F_{12})^2=-(\frac{\mu}{\tilde{\lambda}})^2(\epsilon_3\tilde{\lambda}^2+\epsilon_1\mu^2)>0; \tag{74}$$

(ii) $\tilde{\lambda}^2\neq\nu^2$, $\mu\tilde{\lambda}\neq 0$, $\epsilon_3=-\epsilon_1$, $\delta_2=\frac{\epsilon}{2}(\nu-\mu)$ *(where $\epsilon\in\{\pm 1\}$) and*

$$\tilde{\lambda}^2+\frac{\epsilon}{2}(\nu-\mu)\tilde{\lambda}-\mu\nu=0$$

$$(F_{12})^2=\epsilon_1(\nu-\mu)\nu(\epsilon\frac{\mu}{2\tilde{\lambda}}-1)>0. \tag{75}$$

For both families,

$$F_{23} = F_{12}\epsilon_3\frac{\tilde{\lambda}}{\nu}, \ \tilde{\rho} = \epsilon_1\epsilon_3\frac{\mu\nu}{\tilde{\lambda}}. \tag{76}$$

For the family (i), $\mathfrak{g}$ *is isomorphic to* $\tau_2(\mathbb{R}) \oplus \mathbb{R}$*. For the family (ii),* $\mathfrak{g}$ *is isomorphic to* $\tau_{3,\frac{1}{2}}(\mathbb{R})$*.*

Proof Starting from relation (67) we distinguish various cases. If $\mu = \nu$ then we obtain the family (i). If $\mu \neq \nu$ and $\tilde{\lambda}^2 + \epsilon_1\epsilon_3\nu^2 = 0$ we arrive at $\delta_2 = \frac{\epsilon}{2}(\nu - \mu)$, $\tilde{\lambda} = \epsilon\nu$, $\tilde{\rho} = -\epsilon\mu$ and $\epsilon_3 = -\epsilon_1$. But then relation (64) is not satisfied. Finally, if $\mu \neq \nu$ and $\tilde{\lambda}^2 + \epsilon_1\epsilon_3\nu^2 \neq 0$ then relation (67) becomes

$$\tilde{\lambda}^2 + (\frac{\mu\nu}{\tilde{\lambda}})^2 + \frac{\epsilon_1\epsilon_3}{2}(\mu + \nu)^2 + (\tilde{\lambda} + \epsilon_1\epsilon_3\frac{\mu\nu}{\tilde{\lambda}})^2 = 0. \tag{77}$$

In particular, $\epsilon_3 = -\epsilon_1$, $\tilde{\lambda}^2 \neq \nu^2$ and the above relation reduces to the 1st relation in (75). In fact, relation (77) can be rewritten as

$$\begin{aligned}\frac{1}{2}(\mu + \nu)^2 &= \tilde{\lambda}^2 + (\frac{\mu\nu}{\tilde{\lambda}})^2 + (\tilde{\lambda} - \frac{\mu\nu}{\tilde{\lambda}})^2 \\ &= 2\left(\tilde{\lambda} - \frac{\mu\nu}{\tilde{\lambda}}\right)^2 + 2\mu\nu\end{aligned}$$

and therefore as

$$\frac{1}{2}(\mu - \nu)^2 = 2\left(\tilde{\lambda} - \frac{\mu\nu}{\tilde{\lambda}}\right)^2,$$

which leads to

$$\mu - \nu = 2\epsilon(\tilde{\lambda} - \frac{\mu\nu}{\tilde{\lambda}}), \quad \epsilon \in \{\pm 1\}. \tag{78}$$

That is the 1st relation in (75). From relation (77) we also see that the right-hand side of (66) is precisely $-(\tilde{\lambda} - \frac{\mu\nu}{\tilde{\lambda}})^2$, which proves that

$$\delta_2 = -(\tilde{\lambda} - \frac{\mu\nu}{\tilde{\lambda}}) = \frac{\epsilon}{2}(\nu - \mu),$$

where in the 2nd equality we used relation (78). We now prove the 2nd relation in (75). From the 3rd relation in (65) we obtain

$$(F_{12})^2 = -\frac{\epsilon_1}{2}(\nu^2 - \mu^2) + \frac{\epsilon_1\epsilon}{2}(\nu - \mu)\left(\tilde{\lambda} - \frac{\epsilon}{2}(\nu - \mu)\right)$$
$$= \frac{\epsilon_1}{2}(\nu - \mu)(\epsilon\tilde{\lambda} - \frac{\mu}{2} - \frac{3\nu}{2})$$

where in the 1st relation above we used that $\epsilon_3 = -\epsilon_1$ and $\delta_2 = \frac{\epsilon}{2}(\nu - \mu)$. On the other hand, the 1st relation in (75) implies that

$$\frac{1}{2}(\epsilon\tilde{\lambda} - \frac{\mu}{2} - \frac{3\nu}{2}) = \nu(\epsilon\frac{\mu}{2\tilde{\lambda}} - 1) \tag{79}$$

The 2nd relation in (75) follows and we obtain the family ii).

We remark that relation (64) is satisfied for both families (i) and (ii). For example, for the family (ii),

$$\frac{1}{2}(\tilde{\lambda} + \tilde{\rho}) = \frac{1}{2}(\tilde{\lambda} + \epsilon_1\epsilon_3\frac{\mu\nu}{\tilde{\lambda}}) = \frac{\epsilon}{4}(\mu - \nu) \neq \delta_2$$

as $\mu \neq \nu$.

The statements on the Lie algebra $\mathfrak{g}$ are straightforward, by using that the matrix $\mathbf{A}$ is given by (30) and

$$\lambda = \epsilon_1(\tilde{\lambda} - \delta_2), \;\; \rho = \epsilon_1\frac{\mu\nu}{\tilde{\lambda}} - \epsilon_3\delta_2. \tag{80}$$

For example, for the family (ii) the matrix $\mathbf{A}$ is given by

$$\mathbf{A} = \begin{pmatrix} \tilde{\lambda} - \frac{\epsilon}{2}(\nu - \mu) & \epsilon_1\nu \\ -\epsilon_1\mu & -\frac{\mu\nu}{\tilde{\lambda}} - \frac{\epsilon}{2}(\nu - \mu) \end{pmatrix}. \tag{81}$$

Its the characteristic polynomial is given by

$$P_{\mathbf{A}}(x) = x^2 + \frac{3\epsilon}{2}(\nu - \mu)x + \frac{1}{2}(\nu - \mu)^2. \tag{82}$$

As the roots of $P_{\mathbf{A}}$ are given by $x_{\pm} = \frac{1}{4}(3\epsilon \pm 1)(\mu - \nu)$ and $\mu \neq \nu$, $\epsilon \in \{\pm 1\}$, we obtain that $\mathfrak{g}$ is isomorphic to $\tau_{3,\frac{1}{2}}(\mathbb{R})$, as claimed. □

4 Generalized Versus Odd Generalized Einstein Metrics

We now summarize the results from [3, 4] and this paper, and we comment on the relation between generalized Einstein and odd generalized Einstein metrics. As explained in Remark 18 of [3], an odd generalized metric $E_- = \{X - g(X) \mid X \in TG\}$ defined on a Courant algebroid $E = E_{H,F}$ over a Lie group G with Lie algebra $\mathfrak{g}$, such that $F = 0$, is generalized Einstein with divergence δ, if and only if, considered as an (ordinary) generalized metric on G, it is generalized Einstein with divergence $\tilde{\delta} := \delta|_{\mathfrak{g}\oplus\mathfrak{g}}$. This is a main relation between generalized and odd generalized Einstein metrics and (in view of the material developed in [4]) the reason for our assumption $F \neq 0$ in [3] and in the present paper.

Any 3-dimensional unimodular Lie group supports a generalized Einstein metric, which can be chosen to be divergence-free or with non-zero divergence (see Theorem 1.4 of [4]). Odd generalized Einstein metrics (with $F \neq 0$) exist on a 3-dimensional unimodular Lie group G if and only if (in the notation of [3]),

$$\mathfrak{g} = \mathfrak{so}(3), \mathfrak{so}(2,1), \mathfrak{e}(1,1), \mathfrak{heis} \tag{83}$$

(see Theorem 1 of [3]).

Any 3-dimensional non-unimodular Lie group G with Lie algebra $\mathfrak{r}_2(\mathbb{R}) \oplus \mathbb{R}$, $\mathfrak{r}_3(\mathbb{R})$, $\mathfrak{r}_{3,\lambda}(\mathbb{R})$ or $\mathfrak{r}'_{3,-1}(\mathbb{R})$, supports both ordinary and odd generalized Einstein metrics (with $F \neq 0$), while if $\mathfrak{g} = \mathfrak{r}'_{3,\lambda}(\mathbb{R})$ (with $\lambda \neq -1$), then G supports odd generalized Einstein metrics (all with non-zero divergence) but no ordinary generalized Einstein metrics (see Theorem 1.3 of [4], the first statement in Theorem 1.5 of [4], Theorem 1 and Proposition 8).

Acknowledgments Research of V.C. was supported by the Deutsche Forschungsgemeinschaft (DFG, German Research Foundation) under Germany's Excellence Strategy—EXC 2121 "Quantum Universe"—390833306 and under SFB-Geschäftszeichen 1624—Projektnummer 506632645.

References

1. D. Baraglia, P. Hekmati, Transitive Courant algebroids, string structures and T-duality. Adv. Theor. Math. Phys. **19**(3), 613–672 (2015)
2. A. Coimbra, C. Strickland-Constable, D. Waldram, Supergravity as generalised geometry I: type II theories. J. High Energy Phys. **2011**, 1–35 (2011), 91
3. V. Cortés, L. David, Odd generalized Einstein metrics on Lie groups. Ann. Mat. Pura Appl. https://doi.org/10.1007/s10231-024-01540-1 (see also arXiv:2311.00380 [math.DG])
4. V. Cortés, D. Krusche, Classification of generalized Einstein metrics on 3-dimensional Lie groups. Can. J. Math. **75**(6), 2038–2095 (2023)
5. M. García-Fernández, Torsion-free generalized connections and heterotic supergravity. Commun. Math. Phys. **332**, 89–115 (2014)
6. M. García-Fernández, Ricci flow, Killing spinors and T-duality in generalized geometry. Adv. Math. **350**, 1059–1108 (2019)

7. J. Milnor, Curvatures of left invariant metrics on lie groups. Adv. Math. **21**, 293–329 (1976)
8. A.L. Onishchik, E.B. Vinberg, *Lie Groups and Lie Algebras III; Structure of Lie Groups and Lie Algebras*. Encyclopedia of Mathematical Sciences, vol. 41 (Springer, Berlin, 1994)
9. R. Rubio, Generalized geometry of type B_n. Ph.D thesis, University of Oxford (2014)

Cohomological Lifting of Multi-Toric Graphs

Kael Dixon, Thomas Bruun Madsen, and Andrew Swann

To Paul Gauduchon on the occasion of his 80th birthday

Abstract We study G_2-manifolds obtained from circle bundles over symplectic SU(3)-manifolds with T^2-symmetry. When the geometry is multi-Hamiltonian, we show how the compact part of the resulting multi-moment graph for the G_2-structure may obtained cohomologically from the base. The lifting procedure is illustrated in the context of toric geometry.

Keywords Holonomy G_2 · Circle bundle · Torus symmetry · Multi-Hamiltonian geometry · Tri-valent graph · Toric geometry

2020 Mathematics Subject Classification Primary 53C25; Secondary 14M25, 53C26, 53C29, 53D20

1 Introduction

Symmetry is a powerful tool to understand different classes of geometric structures. For a manifold M with a closed differential form preserved by a Lie group G, the notion of multi-moment map was introduced in [22, 23]. If G is Abelian, this is a smooth invariant map $\nu\colon M \to V$, for a certain finite-dimensional vector space V,

K. Dixon
Montreal, QC, Canada
e-mail: kael.dix@gmail.com

T. B. Madsen
School of Computing and Engineering, University of West London, London, UK
e-mail: thomas.madsen@uwl.ac.uk

A. Swann (✉)
Department of Mathematics and DIGIT, Aarhus University, Aarhus C, Denmark
e-mail: swann@math.au.dk

L. Ornea (ed.), *Real and Complex Geometry*,
https://doi.org/10.1007/978-3-031-92297-8_6

which generalises the usual notion of moment map $\mu\colon M \to \mathfrak{g}^*$ in symplectic geometry. We will study this map for special geometries in dimensions 6 and 7 with torus symmetry. Here, there is a trivalent graph Γ immersed in V given by the image under ν of the points with non-trivial stabiliser. These immersions have edges that are line segments, and satisfy a zero-tension condition at each vertex. For Calabi-Yau three-folds these graphs are often used in the physics literature, see Bouchard [5], where they are planar. In [24], we showed that for 7-manifolds with holonomy G_2, there are also such graphs, but now the graphs live in three-dimensions.

Recently, Foscolo et al. [15] gave a construction of many examples of manifolds of holonomy G_2 as circle bundles over Calabi-Yau three-folds with certain asymptotics; a powerful result, that significantly increases the number of known non-compact G_2-examples, thanks to a wealth of information from the Calabi-Yau literature. This motivates us to study how graphs may be lifted from 6-dimensional geometries to those in dimension 7. One of our main results, see Sect. 3, is that for compact edges of the graph, this information is determined cohomologically by the classes of the symplectic structure on the base and the curvature form of the circle bundle. We derive the statements in a broader setting than that of reduced holonomy, requiring just that the G_2 three-form is closed and by using the symplectic property of the base. Some generality in this direction is needed to apply the results to the construction of [15], since there the base Calabi-Yau structure is deformed to a more general SU(3)-structure before getting a submersion structure, with just the group action and symplectic form surviving the process.

In Sect. 4, we specialise to the case when the base is a toric Calabi-Yau manifold. There we can use data from the associated fans, and show that loops in the SU(3)-graphs lift to closed loops in the G_2-graphs precisely when a cohomological condition in [15] is satisfied. This indicates that the condition is intrinsic to the general set-up and not purely an artefact of the adiabatic limit construction used in [15]. We close the paper with a number of explicit examples, demonstrating the utility of the lifting formulae.

2 Special Multi-Toric Geometries

A G_2-structure on a seven-dimensional manifold M is given by a three-form φ that is pointwise linearly equivalent to

$$e^{123} - e^{147} - e^{167} - e^{246} - e^{275} - e^{347} - e^{356},$$

where e^i is a basis for $(\mathbb{R}^7)^* \cong T_p^*M$ and $e^{abc} = e^a \wedge e^b \wedge e^c$. This basis is orthonormal for a Riemannian metric g and specifies an orientation, so also a Hodge $*$-operator; these objects depend only on φ, see Bryant [6]. The G_2-structure is *closed* if $d\varphi = 0$, [10]. If both $d\varphi = 0 = d*\varphi$, then the (restricted) holonomy of g is contained in G_2, and the structure is called *torsion-free*.

Suppose a three-dimensional torus $T^3 = \mathbb{R}^3/\mathbb{Z}^3$ acts effectively on M preserving φ, and hence g and $*\varphi$. If $\mathsf{U}_1, \mathsf{U}_2, \mathsf{U}_3$ is a basis for $\mathrm{Lie}(T^3)$, then we write U_1, U_2, U_3 for the corresponding vector fields generated on M. When $d\varphi = 0$, repeated use of the Cartan formula gives that the one-form $\varphi(U_i, U_j, \cdot)$ is closed. A closed G_2-structure is *multi-Hamiltonian* if there is an invariant function $\nu = (\nu_1, \nu_2, \nu_3)\colon M \to \mathbb{R}^3 = \Lambda^2 \mathrm{Lie}(T^3)^*$ such that

$$d\nu_i = \varphi(U_j, U_k, \cdot) \qquad \text{for } (ijk) = (123) \text{ as permutations.} \tag{2.1}$$

We call ν a *multi-moment map* and note that ν is guaranteed to exist if $b_1(M) = 0$. In [24, Lemmas 2.5 and 2.6], we showed that $\varphi(U_1, U_2, U_3) = 0$ and that each isotropy is a connected subtorus of dimension at most 2.

If, in addition, the geometry is torsion-free, then there is also a multi-moment map $\hat{\nu}$ associated to $*\varphi$ satisfying $d\hat{\nu} = *\varphi(U_1, U_2, U_3, \cdot)$. It is notable, see [24, Theorem 4.5], that $(\nu, \hat{\nu})$ induces a local homeomorphism of the orbit space M/T^3 to $\mathbb{R}^4$.

Suppose now that U_3 generates a free circle-action on M of fundamental period 1. We set

$$\theta = \frac{g(U_3, \cdot)}{g(U_3, U_3)},$$

so $\theta(U_3) = 1$ and θ is T^3-invariant. Put $B = M/\langle U_3 \rangle$ and write $\pi\colon M \to B$ for the projection. From Apostolov and Salamon [4, cf. Proposition 1], we get the decompositions

$$\begin{aligned} g &= \theta^2 \pi^*(h^{-1}) + \pi^*(h^{1/2} g_B), \\ \varphi &= \theta \wedge \pi^*(\omega) + \pi^*(h^{3/4}\psi_+), \\ *\varphi &= -\theta \wedge \pi^*(h^{1/4}\psi_-) + \pi^*(\tfrac{1}{2}h\omega^2), \end{aligned} \tag{2.2}$$

where $\pi^* h = g(U_3, U_3)^{-1} > 0$ is smooth, and $(g_B, \omega, \psi_c = \psi_+ + i\psi_-)$ defines an SU(3)-structure on B.

Note that $\pi^*(h)$ is also T^3-invariant, so the SU(3)-structure on B is preserved by the two-torus $T^2 = T^3/\langle \exp \mathsf{U}_3 \rangle$. If the G_2-structure is closed, then we have

$$0 = d\varphi = \theta \wedge \pi^*(d\omega) + \pi^*(F \wedge \omega + d(h^{3/2}\psi_+)),$$

where $\pi^* F = d\theta$ is the curvature of the circle bundle $M \to B$. Contracting with U_3, we see that $d\omega = 0$, so the SU(3)-structure is *symplectic*.

We observe that

$$d\nu_1 = \varphi(U_2, U_3, \cdot\,) = -(\pi^*\omega)(U_2, \cdot\,),$$
$$d\nu_2 = \varphi(U_3, U_1, \cdot\,) = (\pi^*\omega)(U_1, \cdot\,).$$

So putting

$$X_1 = -\pi_*(U_2), \quad X_2 = \pi_*(U_1),$$

we have a symplectic moment map $\mu = (\mu_1, \mu_2)\colon B \to \mathbb{R}^2 \cong \mathrm{Lie}(T^2)^*$ given by $\pi^*\mu = (\nu_1, \nu_2)$, that is, μ_i are invariant functions satisfying

$$d\mu_i = \omega(X_i, \cdot\,). \tag{2.3}$$

There is a weak converse: given an SU(3)-structure (g_B, ω, ψ_c) on a six-manifold B, then $M = S^1 \times B$ carries a product G_2-structure built as in (2.2), with $\theta = ds$, for s the standard parameter on $S^1 = \mathbb{R}/\mathbb{Z}$, and $h \equiv 1$. This product structure is closed if and only if B is *symplectic half-flat*, meaning $d\omega = 0 = d\psi_+$, [12], and has holonomy (strictly) contained in G_2 if and only if B is *Calabi-Yau*, meaning $d\omega = 0 = d\psi_c$. Taking $U_1 = X_2$, $U_2 = -X_1$, $U_3 = \partial_s$, we may transfer terminology from above for M to B. In particular, a symplectic (not necessarily half-flat) SU(3)-structure on B is *multi-Hamiltonian* for a two-torus action generated by X_1, X_2 preserving the SU(3)-structure, if there are invariant functions μ_1, μ_2 satisfying (2.3). It follows that all stabilisers are connected sub-tori, and that $\omega(X_1, X_2) = 0$. If B is symplectic half-flat, then at least locally we also have an invariant function $\hat{\mu}_+$ with $d\hat{\mu}_+ = \psi_+(X_1, X_2, \cdot\,) = d\nu_3$. The functions $(\mu_1, \mu_2, \hat{\mu}_+)$ are guaranteed to exist globally if $b_1(B) = 0$. If B is also Calabi-Yau, then there is an additional multi-moment map $\hat{\mu}_-$ satisfying $d\hat{\mu}_- = \psi_-(X_1, X_2, \cdot\,) = d\hat{\nu}$. Again, we have that $(\mu, \hat{\mu}) = (\mu_1, \mu_2, \hat{\mu}_+, \hat{\mu}_-)$ induces a local homeomorphism $B/T^2 \to \mathbb{R}^4$.

A significant source of Calabi-Yau structures is available within the category of toric six-manifolds B. These are connected and have a Hamiltonian action of a three-torus T^3 preserving the symplectic form ω and metric g, and consequently the complex structure given by $\omega = g(J\,\cdot, \cdot\,)$. The complex volume form ψ_c is a parallel section of $\Lambda^{3,0}(B)$ of fixed length, so any other section $\tilde{\psi}_c$ of $\Lambda^{3,0}(B)$ of this length has the form $\tilde{\psi}_c = e^{2\pi i t}\psi_c$ for some local function $t\colon B \to \mathbb{R}$. If $\tilde{\psi}_c$ is parallel then $0 = \nabla\tilde{\psi}_c = 2\pi i\, e^{2\pi i t} dt \otimes \psi_c$, so t is constant and globally defined modulo $\mathbb{Z}$. The action of T^3 on B induces via pull-back an action on the parallel sections of $\Lambda^{3,0}(B)$ of fixed length and hence gives a group homomorphism $T^3 \to S^1$. The kernel of this action is compact of dimension 2, so its component of the identity is a two-torus T^2 that acts on B preserving the SU(3)-structure. (See Li et al. [21, Section 3.1] for an algebraic version of this.) If $b_1(B) = 0$, we see that the T^2-action is multi-Hamiltonian.

A wide class of examples is given by toric asymptotically conical (AC) Calabi-Yau three-folds. These are asymptotic, with polynomial decay rates, to cones over toric Sasaki-Einstein manifolds of dimension five, as constructed from good polytopes by Cho et al. [9]. Taking crepant resolutions of these cones yields smooth AC Calabi-Yau manifolds with T^3-symmetry, classified by Conlon and Hein [11, Theorem 4.3]. These structures are uniquely specified by the cohomology class of their Kähler form. Note that Li [20] provides an example of a non-toric Calabi-Yau manifold, with different asymptotics, that has a multi-Hamiltonian action of T^2.

Suppose B is an AC Calabi-Yau three-fold with Kähler form ω_0, and that $M \to B$ is a circle bundle over B with curvature F such that $[F] \neq 0$ in $H^2(B)$ and $[\omega_0] \cup [F] = 0$ in $H^4(B)$. Then Foscolo et al. [15] construct metrics of holonomy G_2 on M which are asymptotically locally conical: the geometry outside a compact set is approximated by a Riemannian submersion whose fibres are circles of constant length over a base that is a Calabi-Yau cone. These G_2-structures are obtained via a particular type of adiabatic limit, so that in (2.2) we have $\omega = \epsilon\omega_0$ with $h = h_\epsilon$, $\psi_c = \psi_{c,\epsilon}$, for a parameter $\epsilon > 0$, with $\lim_{\epsilon\to 0} h_\epsilon \equiv 1$ and $\lim_{\epsilon\to 0} \psi_{c,\epsilon}$ the Calabi-Yau complex volume form of B. Arguments of Cavalleri [8], show that if the initial Calabi-Yau data is T^2-invariant, then so are the deformed symplectic SU(3)-structures of the solutions.

3 Graph Lifting

Associated to the multi-Hamiltonian geometries above there are certain graphs determined by the points with non-trivial stabilisers. For toric G_2-manifolds these were described in [24, 25], for toric Calabi-Yau structures they are known in the physics literature as pq-webs [2]. However, the existence of such graphs does not need the holonomy reduction.

Proposition 3.1 *Let M be a manifold with a closed G_2-structure and an action of T^3 that is multi-Hamiltonian. Write M^0 for the points of M with non-trivial stabiliser. Then $\Gamma_M = M^0/T^3 \subset M/T^3$ is an embedded trivalent graph.*

Furthermore, the multi-moment map $\nu\colon M \to \Lambda^2 \mathrm{Lie}(T^3)^*$ maps each edge of the graph to a non-trivial part of a straight line with rational tangent in $\mathbb{R}^3 = \Lambda^2 \mathrm{Lie}(T^3)^*$. At the image of each vertex, these lines satisfy a zero-tension condition.

Let us explain some of the terminology from the statement. Graphs are to be understood in the sense of [31, Definiton 12], so there may be both edges that join two vertices and unbounded edges that meet only a single vertex. The graph is *trivalent* if exactly three distinct edges meet at each vertex.

Recall that $T^3 = \mathbb{R}^3/\mathbb{Z}^3$. So we can identify $\mathrm{Lie}(T^3)$ with $\mathbb{R}^3$, and $\mathbb{Z}^3$ is then the integral lattice. To be concrete, given a $\mathbb{Z}$-basis $\mathsf{U}_1, \mathsf{U}_2, \mathsf{U}_3$ of $\mathbb{Z}^3 \subset \mathbb{R}^3 = \mathrm{Lie}(T^3)$, we can consider the dual basis u^1, u^2, u^3. This generates a lattice $\mathbb{Z}^3$ in $\mathrm{Lie}(T^3)^* \cong \mathbb{R}^3$. The target space $\Lambda^2 \mathrm{Lie}(T^3)^* \cong \mathbb{R}^3$ of ν also has a lattice $\mathbb{Z}^3$, generated by

$u^2 \wedge u^3$, $u^3 \wedge u^1$ and $u^1 \wedge u^2$. These identifications are canonical if we fix an orientation of $\mathrm{Lie}(T^3)$, since change of $\mathbb{Z}$-basis is given by an element of $\mathrm{GL}(3, \mathbb{Z})$ and $\det(\mathrm{GL}(3, \mathbb{Z})) = \{\pm 1\}$.

A *straight line* in $\Lambda^2 \mathrm{Lie}(T^3)^*$ with *rational tangent* is an affine line of the form

$$r + \mathbb{R}\mathsf{U} \lrcorner u^{123}, \tag{3.1}$$

where $\mathsf{U} \in \mathbb{Z}^3 \setminus \{0\}$, $r \in \Lambda^2(\mathrm{Lie}(T^3))^*$ and $u^{123} = u^1 \wedge u^2 \wedge u^3$. We can choose U to be *primitive*, meaning that it is not a positive multiple of another vector in $\mathbb{Z}^3$. Note that (3.1) contains the half-line $r + \mathbb{R}_{\geqslant 0}\mathsf{U} \lrcorner u^{123}$. If three half-lines given by primitive vectors $\mathsf{U}, \mathsf{V}, \mathsf{W}$ meet at r, then we say there is *zero-tension* at the vertex r if $\mathsf{U} + \mathsf{V} + \mathsf{W} = 0$. The final claim of the Proposition is that for each vertex x of the graph, the three half-lines of $\nu(M^0)$ meeting at $r = \nu(x)$ have zero-tension at r.

Proof of Proposition 3.1 Consider a point $p \in M^0$. As noted in Sect. 2, the stabiliser of p is a connected subtorus of T^3 of dimension at most 2. So there is some non-zero $\mathsf{U} = \sum_{i=1}^{3} b_i \mathsf{U}_i \in \mathrm{Lie}(T^3)$ such that the generated vector field U fixes p and has fundamental period 1. Let $M_p^U \subset M^0$ be the connected component of the fixed-point set $M^U = \{x \in M \mid U_{|x} = 0\}$ that contains p. We will show that $\nu(M_p^U)$ is a non-trivial part of a straight line in $\mathbb{R}^3 = \Lambda^2 \mathrm{Lie}(T^3)^*$, when p is just stabilised by a circle.

Given $\mathsf{C} \in \Lambda^2 \mathrm{Lie}(T^3)$, we may write

$$\mathsf{C} = \sum_{(jk\ell)=(123)} c_j \mathsf{U}_k \wedge \mathsf{U}_\ell,$$

for some constants c_j. The definition (2.1) of the multi-moment map ν says

$$\langle d\nu, \mathsf{C}\rangle = \sum_{(jk\ell)=(123)} c_j \varphi(U_k, U_\ell, \cdot) = \sum_{(jk\ell)=(123)} c_j U_k \wedge U_\ell \lrcorner \varphi.$$

Since, for $q \in M_p^U$ we have $U_q = 0$, we see that $\langle d\nu_q, \mathsf{C}\rangle = 0$ whenever there is a V such that $\mathsf{C} = \mathsf{U} \wedge \mathsf{V}$. This is the same as $\mathsf{C} \wedge \mathsf{U} = 0$. But

$$\mathsf{C} \wedge \mathsf{U} = \sum_{(ijk)=(123)} c_i b_i \, \mathsf{U}_j \wedge \mathsf{U}_k \wedge \mathsf{U}_i = c^T b \, \mathsf{U}_1 \wedge \mathsf{U}_2 \wedge \mathsf{U}_3.$$

So $\langle d\nu_q, \mathsf{C}\rangle = 0$ whenever $c^T b = 0$. On the other hand,

$$(\mathsf{U} \lrcorner u^{123})(\mathsf{C}) = \sum_{(ijk)=(123)} b_i u^{jk}(\mathsf{C}) = c^T b,$$

so $\ker(\mathsf{U} \lrcorner u^{123}) \leqslant \ker(d\nu(T_q M))$ and is two-dimensional. Thus $\nu(M_p^U)$ is contained in the line $\nu(p) + \mathbb{R}\mathsf{U} \lrcorner u^{123}$.

Suppose the stabilisers of p and q are exactly one-dimensional. We may choose $\mathsf{V}, \mathsf{W} \in \mathrm{Lie}(T^3)$ so that $\mathsf{U} \wedge \mathsf{V} \wedge \mathsf{W} = \mathsf{U}_1 \wedge \mathsf{U}_2 \wedge \mathsf{U}_3$. Then $\langle d\nu_q, \mathsf{V} \wedge \mathsf{W}\rangle = \varphi_q(V, W, \cdot)$ is non-zero and U-invariant. We have $\varphi(V, W, \cdot) = g(V \times W, \cdot)$, by the defining equation for the G_2 cross-product $\times$. The vectors V, W and $V \times W$ are U-invariant and linearly independent at q. But as the T^3-action is effective, U must act non-trivially on their four-dimensional orthogonal complement $Z_q \subset T_qM$. As U preserves the G_2-structure and two linearly independent vectors, we therefore have that U acts on Z_q as a non-zero diagonal matrix $\mathrm{diag}(it, -it)$ in $\mathfrak{su}(2)$ in the standard representation on $\mathbb{C}^2 = \mathbb{R}^4$. We see that the tangent space to M_p^U is spanned by $V, W, V \times W$ at q, cf. [24, Section 4.2.3], so M_p^U is associative. Consequently, $d\nu_q$ has rank one on this submanifold. It follows that ν is injective on the submanifold M_p^U.

At a point p where the stabiliser is of dimension two, this must be a two-torus $T^2 \subset T^3 = \mathbb{R}^3/\mathbb{Z}^3$. The T^3-action stabilises the one-dimensional orbit of $S^1 = T^3/T^2$ through p and preserves the G_2-structure. It follows that T^2 acts on T_pM as $\mathbb{R} \oplus \mathbb{C}^3$, with the action on $\mathbb{C}^3$ as the maximal torus in SU(3). We get a $\mathbb{Z}$-basis $\mathsf{U}_1, \mathsf{U}_2, \mathsf{U}_3$ of $\mathbb{Z}^3$ such that $\mathsf{U}_1, \mathsf{U}_2$ span $\mathrm{Lie}(T^2)$ and U_3 generates the quotient S^1. The point p has a tubular neighbourhood that is equivariantly diffeomorphic to $S^1 \times \mathbb{C}^3$. Viewing this as a circle bundle over $\mathbb{C}^3$, the discussion of Sect. 2 shows that $\mathbb{C}^3$ inherits a symplectic form. By the equivariant Darboux Theorem for the T^2-action, this is symplectomorphic to the standard symplectic form on $\mathbb{C}^3$. We may choose a basis so that T^2 acts via the diagonal matrices $D_{\theta,\phi} = \mathrm{diag}(e^{2\pi i\theta}, e^{2\pi i\phi}, e^{-2\pi i(\theta+\phi)})$ in SU(3). In this neighbourhood, $S^1 \times \{0\}$ has stabiliser T^2 and each $S^1 \times L$, for L a complex coordinate axis, has one-dimensional stabiliser. Thus in M/T^3, points with two-dimensional stabilisers give a discrete set, and at each such point three lines, corresponding to points with one-dimensional stabilisers, meet. We thus have a trivalent graph $\Gamma_M = M^0/T^3$ in M/T^3.

To get the zero-tension condition, write (z_1, z_2, z_3) for the complex coordinates on the $\mathbb{C}^3$-factor in the tubular neighbourhood. The stabiliser acts via $D_{\theta,\phi}$, so the Kähler moment map for the T^2-action is $\mu = \frac{1}{2}(|z_3|^2 - |z_1|^2, |z_3|^2 - |z_2|^2)$, see [24, Section 4.1.1]. Along each coordinate axis L we have $U_1 \wedge U_2 = 0$, so $d\nu_3 = 0$ and ν_3 is constant on L. The derivatives of μ are $(-1, 0)$, $(0, -1)$ and $(1, 1)$ along the z_i-axis, $i = 1, 2, 3$, respectively. As these are primitive vectors that sum to zero, we get the zero-tension condition under μ. Since $(\nu_1, \nu_2) = \pi^*(\mu_1, \mu_2)$, we also get the zero-tension under ν. □

As the above proof shows, by considering $S^1 \times B$, we get a corresponding result for symplectic SU(3)-structures with multi-Hamiltonian T^2-action. There is then a trivalent graph Γ_B and its image under μ satisfies the zero-tension condition for each vertex.

Remark 3.2 Note that if we have a torsion-free G_2-structure, then the fourth multi-moment map $\hat{\nu}$ has zero derivative on M^0, since $U_1 \wedge U_2 \wedge U_3 = 0$ at each $p \in M^0$. Thus for each component γ of Γ_M, its image $(\nu, \hat{\nu})(\gamma) \subset \mathbb{R}^4$ lies in an affine hyperplane given by $\hat{\nu} \in \mathbb{R}$ being constant. Similarly, for Calabi-Yau structures on B, we have $X_1 \wedge X_2 = 0$ on B^0 and each component γ of Γ_B has image $(\mu, \hat{\mu})(\gamma)$ in an affine plane on which $\hat{\mu} \in \mathbb{R}^2$ is constant.

In particular, if the graphs are connected, then when determining their images, it is sufficient to work in the corresponding affine subspace, and we only need to focus on the effect of μ and ν.

Taking B to be a product $N \times (S^1 \times \mathbb{R})$ of a hyperKähler four-manifold N obtained from the Gibbons-Hawking Ansatz [18] and a cylinder $S^1 \times \mathbb{R} = \mathbb{C}/\mathbb{Z}$, gives a Calabi-Yau structure with multi-Hamiltonian T^2-symmetry [25]. In this case, Γ_B embeds in $\mathbb{R}^4$ as a collection of parallel straight lines, corresponding to the fixed points of the circle action on N. Thus Γ_B can have an arbitrary number of connected components.

The problem we now wish to study is how graphs on a base B for an SU(3)-structure lift to graphs for a G_2-structure on a principal circle bundle M over B. We assume that B has a T^2-symmetry preserving the SU(3)-structure generated by $\mathsf{X}_1, \mathsf{X}_2 \in \mathrm{Lie}(T^2)$. If the action lifts to a T^3-action on a circle bundle M over B, then the curvature two-form $F \in \Omega^2(B)$ is T^2-invariant and has integral periods. Conversely, any such two-form specifies a principal circle bundle M and by Lashof et al. [19], the T^2-action on B lifts to a T^2-action on M that commutes with the principal action. Hence we get an effective action of T^3, generated by $\mathsf{U}_1, \mathsf{U}_2$ with $\pi_*(U_1) = X_2$ and $\pi_*(U_2) = -X_1$, and by the principal action of U_3.

As explained in [27, Propositions 2.1 and 2.3], a necessary and sufficient condition for there to be a T^2-invariant connection one-form θ on M with curvature F is that there are functions λ_i such that

$$d\lambda_i = -F(X_i, \cdot), \tag{3.2}$$

for $i = 1, 2$. These are then T^2-invariant, and we get $F(X_1, X_2) = 0$. In other words, the T^2-action is Hamiltonian with respect to the (possibly degenerate) two-form F. The generators U_1, U_2 are then given by

$$\tilde{X}_1 + \lambda_1 U_3 = -U_2, \quad \tilde{X}_2 + \lambda_2 U_3 = U_1,$$

where $\tilde{X} \in \ker\theta$ denotes the horizontal lift of X from B, and we have written λ_i for $\pi^*\lambda_i$.

Remark 3.3 Note that (3.2) only specifies λ_i up to an additive constant. By Lashof et al. [19], there is some choice so that the lifted action is effective. On the other hand, for any choice of λ_i, we get the same action of T^3, just with respect to a different real basis of $\mathrm{Lie}(T^3)$.

Let $\gamma : I = [0, 1] \to M$ be any path. Suppose X is a constant linear combination of X_1 and X_2, so X lifts to $U = \tilde{X} + \lambda U_3$ on M with $d\lambda = -X \lrcorner F$. Then we have

$$\begin{aligned}\lambda(\gamma(1)) - \lambda(\gamma(0)) &= \int_{t=0}^{1} \frac{d}{dt}\lambda(\gamma(t))\,dt = \int_{t=0}^{1} d\lambda(\dot{\gamma}(t))\,dt \\ &= -\int_{t=0}^{1} F(X, \dot{\gamma}(t))\,dt = -\int_{t=0}^{1} F(X, \dot{\tau}(t))\,dt,\end{aligned} \tag{3.3}$$

for $\tau = \pi \circ \gamma$, the projection of γ to B.

If $\mathsf{X} = r_1\mathsf{X}_1 + r_2\mathsf{X}_2$ is an integral linear combination of X_1 and X_2, then X generates a circle action on B, via a one-parameter group φ_s, with $\varphi_1 = \varphi_0 = \mathrm{Id}_B$. The combination of φ_s and τ defines a, perhaps singular, parameterised surface $S = \sigma_X([0, 1] \times [0, 1]) \subset B$ via $\sigma_{\mathsf{X}}(s, t) = \varphi_s(\tau(t))$. We may rewrite (3.3) as

$$\begin{aligned}\lambda(\tau(1)) - \lambda(\tau(0)) &= -\int_{t=0}^{1}\int_{s=0}^{1} F(X, \dot{\tau}(t))\,ds\,dt \\ &= -\int_{t=0}^{1}\int_{s=0}^{1} F(\partial_s, \partial_t)\,ds\,dt = -\int_S F.\end{aligned} \tag{3.4}$$

If the endpoints $\tau(0)$ and $\tau(1)$ are fixed points for X, then, as a singular 2-chain, σ_{X} is a cycle. Equation (3.4) now reads

$$\lambda(\tau(1)) - \lambda(\tau(0)) = -[F] \cap [\sigma_{\mathsf{X}}], \tag{3.5}$$

where $[F] \in H^2(B)$ and $[\sigma_{\mathsf{X}}] \in H_2(B, \mathbb{Z})$. As F has integral periods, (3.5) is an integer.

Similar considerations for the Kähler moment map give

$$\langle \mu(\tau(1)) - \mu(\tau(0)), \mathsf{X} \rangle = [\omega] \cap [\sigma_{\mathsf{X}}], \tag{3.6}$$

when X fixes $\tau(0)$ and $\tau(1)$.

Suppose now the SU(3)-structure on B is symplectic. Let $e = [\mu(b), \mu(c)]$ in $\mu(\Gamma_B)$ be the image of a compact edge in Γ_B. The corresponding points in B have a one-dimensional stabiliser generated by a primitive vector $\mathsf{X} = r_1\mathsf{X}_1 + r_2\mathsf{X}_2$.

Let $\mathsf{Z} = z_1\mathsf{X}_1 + z_2\mathsf{X}_2$ be a choice of primitive vector so that X, Z is a $\mathbb{Z}$-basis for $\mathrm{Lie}(T^2)$ with the same orientation as $\mathsf{X}_1, \mathsf{X}_2$, so

$$r_1 z_2 - r_2 z_1 = 1.$$

Then Z is a primitive generator for the action of $T^2/\langle \exp \mathsf{X} \rangle$.

Let $\tau\colon [0,1] \to B$ be a path such that $\mu \circ \tau$ parameterises the line segment e affinely: $(\mu \circ \tau)(a) = (1-a)\mu(b) + a\mu(c)$. Then the surface $\mathcal{E} \subset B$ corresponding to the edge e is given by $T^2\tau$ and is generated effectively by Z and τ. As $X = 0$ on τ, we have $[\sigma_{\mathsf{X}}] = 0$ in $H_2(B, \mathbb{Z})$, so

$$r_1[\sigma_{\mathsf{X}_1}] + r_2[\sigma_{\mathsf{X}_2}] = 0.$$

We find now that

$$\begin{aligned} r_1[\mathcal{E}] &= r_1 z_1[\sigma_{\mathsf{X}_1}] + r_1 z_2[\sigma_{\mathsf{X}_2}] = r_1 z_1[\sigma_{\mathsf{X}_1}] + (1 + r_2 z_1)[\sigma_{\mathsf{X}_2}] \\ &= [\sigma_{\mathsf{X}_2}] + z_1(r_1[\sigma_{\mathsf{X}_1}] + r_2[\sigma_{\mathsf{X}_2}]) = [\sigma_{\mathsf{X}_2}] \end{aligned}$$

and similarly

$$\begin{aligned} r_2[\mathcal{E}] &= r_2 z_1[\sigma_{\mathsf{X}_1}] + r_2 z_2[\sigma_{\mathsf{X}_2}] = (-1 + r_1 z_2)[\sigma_{\mathsf{X}_1}] + r_2 z_2)[\sigma_{\mathsf{X}_2}] \\ &= -[\sigma_{\mathsf{X}_1}] + z_2(r_1[\sigma_{\mathsf{X}_1}] + r_2[\sigma_{\mathsf{X}_2}]) = -[\sigma_{\mathsf{X}_1}]. \end{aligned}$$

Using (3.6), we get in coordinates that

$$\mu(c) - \mu(b) = [\omega] \cap ([\sigma_{\mathsf{X}_1}], [\sigma_{\mathsf{X}_2}]) = [\omega] \cap [\mathcal{E}](-r_2, r_1) \tag{3.7}$$

and similarly, for $\lambda = (\lambda_1, \lambda_2)$,

$$\lambda(c) - \lambda(b) = -[F] \cap ([\sigma_{\mathsf{X}_1}], [\sigma_{\mathsf{X}_2}]) = -[F] \cap [\mathcal{E}](-r_2, r_1). \tag{3.8}$$

Given a smooth positive function h on B, we define a T^3-invariant G_2-structure on the circle bundle M via (2.2). Assume that this G_2-geometry has φ closed. Then the multi-moment map ν for M has $(\nu_1, \nu_2) = \pi^*(\mu_1, \mu_2)$, up to an additive constant. We also have

$$\begin{aligned} d\nu_3 &= U_1 \wedge U_2 \lrcorner \varphi \\ &= \theta(U_1)(\pi^*\omega)(U_2, \cdot) - \theta(U_2)(\pi^*\omega)(U_1, \cdot) \\ &\quad + (\pi^*\omega)(U_1, U_2)\theta + \pi^*(h^{3/4}\psi_+)(U_1, U_2, \cdot) \\ &= -\lambda_2 d\mu_1 + \lambda_1 d\mu_2 + \pi^*(\omega(X_1, X_2))\theta + \pi^*(h^{3/4}\psi_+(X_1, X_2, \cdot)). \end{aligned}$$

Above e we have $X_1 \wedge X_2 = 0$, and using (3.7) gives $d\nu = (d\nu_1, d\nu_2, d\nu_3)$ is proportional to

$$(-r_2, r_1, r_1\lambda_1 + r_2\lambda_2).$$

Note that $d(r_1\lambda_1 + r_2\lambda_2) = -(r_1 X_1 + r_2 X_2) \lrcorner F = 0$ along e, so we get a straight line in $\Lambda^2 \operatorname{Lie}(T^3)^*$, as expected, with tangent vector $(-r_2, r_1, r_1\lambda_1 + r_2\lambda_2)$. Thus, for some $t \in \mathbb{R}$, we have

$$\begin{aligned}\nu(c) - \nu(b) &= t(-r_2, r_1, r_1\lambda_1(b) + r_2\lambda_2(b)) \\ &= (\mu_1(c) - \mu_1(b), \mu_2(c) - \mu_2(b), \nu_3(c) - \nu_3(b)),\end{aligned}$$

for any $b, c \in M$ projecting to $b, c \in B$. Using (3.7) and the first two components, we get

$$t = [\omega] \cap [\mathcal{E}], \tag{3.9}$$

and hence

$$\begin{aligned}\nu_3(c) &= \nu_3(b) + t(r_1\lambda_1(b) + r_2\lambda_2(b)) \\ &= \nu_3(b) + t(r_1\lambda_1(c) + r_2\lambda_2(c)).\end{aligned} \tag{3.10}$$

Remark 3.4 Note that

$$\nu(c) - \nu(b) = (x, y, \lambda_1(b)y - \lambda_2(b)x),$$

so if λ is replaced by $\lambda + \lambda^0$, then ν is changed by the linear transformation

$$\nu \mapsto \begin{pmatrix} 1 & 0 & 0 \\ 0 & 1 & 0 \\ -\lambda_2^0 & \lambda_1^0 & 1 \end{pmatrix} \nu.$$

When determining the lifted graphs, we may therefore choose to have $\lambda = 0$ at some basepoint.

4 Toric Calabi-Yau Manifolds

Let us now consider a situation where we can be more concrete. We take B to be a toric Calabi-Yau manifold, so there is an action of T^3 preserving the complex and symplectic structures, but not the complex volume form. Singular examples may be obtained by considering cones $C(S) = \mathbb{R}_{\geqslant 0} \times S$ on toric Sasaki-Einstein manifolds S of dimension 5; these give smooth examples, asymptotic to the cone, by taking a crepant resolution of the singularity [9, 17, 26, 28, 29]. Note that there is one such Calabi-Yau solution for each Kähler class on the resolution.

For $C(S)$, the image of the symplectic moment map is a cone on a plane convex polygon. Dually this is described by the fan, which is the convex set generated by rays $\mathbb{R}_{\geqslant 0}u$ of normals u to each face of the moment cone. Bases may be chosen so that each u is integral with last coordinate 1. The fan is then a cone on its cross-section Δ at height 1, and this cross-section is a convex polygon that is good, meaning that no edge of Δ contains primitive vectors that are not endpoints of the edge.

Smooth toric resolutions B are obtained by triangulating Δ, so that each vertex u_i is integral, so lies in $\mathbb{Z}^3$ with last coordinate 1, and the vertices of each triangle form an integral basis for $\mathbb{Z}^3$.

Each ray corresponds to a divisor D_i in B. These satisfy the relation

$$\sum_{i=1}^{r} D_i u_i = 0. \tag{4.1}$$

Indeed, the Picard group of B is generated by the D_i subject to the relations (4.1), see Cox et al. [13, Theorems 4.1.3 and 4.2.1]. By [13, Theorem 12.3.2], this Picard group is isomorphic to the second cohomology group $H^2(B, \mathbb{Z})$.

4.1 Quadrilateral Relations

Suppose we have a quadrilateral in the fan picture with vertices u_1, u_2, v_1, v_2 such that $v_1 v_2$ is an internal edge. Denote the corresponding divisors by D_1, D_2, A_1, A_2. Then $C = A_1 \cap A_2$ is a compact curve. The quadrilateral is a union of two triangles u_i, v_1, v_2, for $i = 1, 2$. From smoothness, the vectors u_i, v_1, v_2 form a $\mathbb{Z}$-basis for $\mathbb{Z}^3$, so the determinants satisfy $\det(u_i\ v_1\ v_2) = -\varepsilon_i$ with $\varepsilon_i \in \{\pm 1\}$. We may change the basis of $\mathbb{Z}^2 \leqslant \mathbb{Z}^3$, so that $v_1 = (0, 0, 1)$ and $v_2 = (0, 1, 1)$. Then the values of the determinants give $u_i = (\varepsilon_i, a_i, 1)$, for some $a_i \in \mathbb{Z}$. As u_1 and u_2 lie on different sides of $v_1 v_2$, we have $\varepsilon_1 = -\varepsilon_2$. Swapping u_1 and u_2, we may assume $\varepsilon_1 = 1$. The four integral vectors u_i, v_j are linearly dependent, so there exist integers $x_i, y_j \in \mathbb{Z}$ such that

$$0 = x_1 u_1 + x_2 u_2 + y_1 v_1 + y_2 v_2 = \begin{pmatrix} x_1 - x_2 \\ x_1 a_1 + x_2 a_2 + y_2 \\ x_1 + x_2 + y_1 + y_2 \end{pmatrix}.$$

We get $x_2 = x_1$ and this common value divides y_2, by the second equation, and hence also y_1, by the final component. We may thus divide through by x_1 and get the quadrilateral relations

$$u_1 + u_2 + y_1 v_1 + y_2 v_2 = 0, \qquad y_1 + y_2 = -2. \tag{4.2}$$

See Cox et al. [13, Equation 6.4.4].

Divisors D outside of the quadrilateral do not meet C, so the corresponding triple intersections DA_1A_2 are zero. By [13, Corollary 6.4.3 and Proposition 6.4.4], the relations (4.2) give

$$D_1 A_1 A_2 = 1 = D_2 A_1 A_2, \quad A_1^2 A_2 = y_1, \quad A_1 A_2^2 = y_2. \tag{4.3}$$

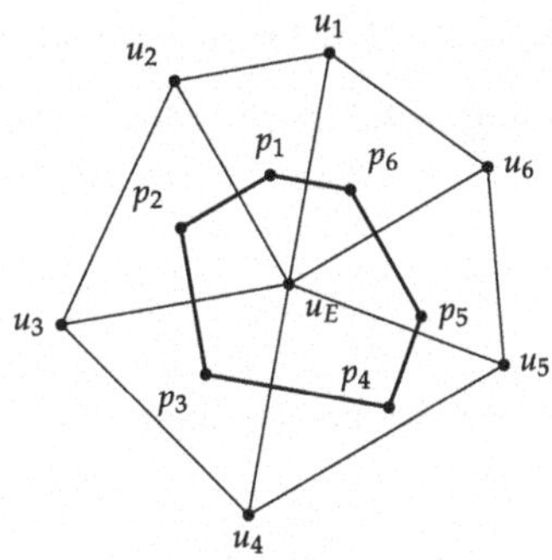

Fig. 1 A polygon $p_1 \dots p_6$ in toric picture from a fan $u_1 \dots u_6 u_E$ at level 1

4.2 Closed Polygons

Suppose $R_E = \mathbb{R}_{\geqslant 0} u_E$ is an interior ray, with corresponding (compact) divisor E. The cones containing R_E are spanned by u_E together with vectors $u_1, \dots, u_m$ that we may order anti-clockwise as in Fig. 1. Write A_j for the divisor corresponding to u_j, for $j = 1, \dots, m$. See Example 5.2, below, for an explicit example.

The edges in the corresponding image of the moment map are orthogonal to differences $u_j - u_E$, $u_j - u_k$. As all u_i have third coordinate 1, we may regard these differences as vectors in $\mathbb{Z}^2 \subset \mathbb{R}^2$. We may thus write the directions of the first set of edges as $J_2(u_i - u_E)$, where $J_2 = \begin{pmatrix} 0 & -1 \\ 1 & 0 \end{pmatrix}$. Let $p_j = \mu(q_j)$ be the point in the image of the moment map corresponding to the cone generated by u_E, u_j, u_{j+1}, reading indices modulo m where necessary. By (3.7), we have

$$p_j - p_{j-1} = t_j J_2(u_j - u_E), \qquad t_j = [\omega] \cap [\mathcal{E}_j],$$

where $\mathcal{E}_j = A_j \cap E$.

Proposition 4.1 *$p_1, \dots, p_m$ are the vertices of a closed polygon with $p_{m+1} = p_1$.*

For the quadrilateral $u_{j-1}, u_{j+1}, u_j, u_E$, Eq. (4.2) gives

$$u_{j-1} + u_{j+1} + a_j u_j + b_j u_E = 0, \qquad a_j + b_j = -2. \tag{4.4}$$

Writing $[\omega] = \sum_{i=1}^{r} w_{D_i} D_i$ and $w_j = w_{A_j}$, we have

$$t_j = \sum_{i=1}^{r} w_{D_i} D_i A_j E = w_{j-1} + w_{j+1} + a_j w_j + b_j w_E,$$

by (4.3). Now Proposition 4.1 will follow directly from

Lemma 4.2 *In the above notation, we have*

$$\sum_{j=1}^{m} t_j(u_j - u_E) = 0.$$

Proof We first compute

$$\sum_{j=1}^{m} t_j(u_j - u_E) = \sum_{j=1}^{m} (w_{j-1} + w_{j+1} + a_j w_j + b_j w_E) u_j$$
$$- \sum_{j=1}^{m} ((2 + a_j) w_j + b_j w_E) u_E.$$

The coefficient of w_j is $u_{j+1} + u_{j-1} + a_j u_j - (2 + a_j) u_E$, which is zero, since $b_j = -(2 + a_j)$. The coefficient of w_E is

$$\sum_{j=1}^{m} b_j u_j - b_j u_E = \sum_{j=1}^{m} -(2 + a_j) u_j - b_j u_E$$
$$= - \sum_{j=1}^{m} u_{j-1} + u_{j+1} + a_j u_j + b_j u_E = 0,$$

and the assertion follows. □

In this context, we now have the following interpretation of the condition $[\omega] \cup [F] = 0$ of Foscolo et al. [15].

Theorem 4.3 *The lifted polygon closes, so $\nu_3(q_{m+1}) = \nu_3(q_1)$, if and only $[\omega] \cup [F] = 0$.*

The change in ν_3 is given by (3.10), so

$$\nu_3(q_j) - \nu_3(q_{j-1}) = t_j \langle u_j - u_E, \lambda(q_j) \rangle. \tag{4.5}$$

So we need to compute $\lambda(q_j)$. As for $[\omega]$, we write $[F] = \sum_{i=1}^{r} f_{D_i} D_i$ and $f_j = f_{A_j}$. From (3.8), we have

$$\lambda(q_j) - \lambda(q_{j-1}) = -s_j J_2(u_j - u_E), \tag{4.6}$$

where $s_j = f_{j-1} + f_{j+1} + a_j f_j + b_j f_E$.

Lemma 4.4 *Putting* $d_j = f_E(u_j - u_{j+1}) + f_j(u_{j+1} - u_E) - f_{j+1}(u_j - u_E)$, *we have*

$$\lambda(q_j) - \lambda(q_1) = J_2(d_j - d_1).$$

Proof It is enough to show $\lambda(q_j) - \lambda(q_{j-1}) = J_2(d_j - d_{j-1})$. We have

$$J_2(\lambda(q_j) - \lambda(q_{j-1})) = s_j(u_j - u_E) = (f_{j-1} + f_{j+1} + a_j f_j + b_j f_E)(u_j - u_E).$$

On the other hand

$$\begin{aligned} d_j - d_{j-1} &= f_E(-u_{j-1} + 2u_j - u_{j+1}) + f_j(u_{j+1} + u_{j-1} - 2u_E) \\ &\quad - (f_{j+1} + f_{j-1})(u_j - u_E) \\ &= f_E((2 + a_j)u_j + b_j u_E) + f_j(-a_j u_j - (2 + b_j)u_E) \\ &\quad - (f_{j+1} + f_{j-1})(u_j - u_E) \\ &= (b_j f_E + f_{j-1} + f_{j+1} + a_j f_j)(u_E - u_j) = -J_2(\lambda(q_j) - \lambda(q_{j-1})), \end{aligned}$$

and the result follows. □

Now (4.5) gives

$$\nu_3(q_{m+1}) - \nu_3(q_1) = \sum_{j=1}^{m} t_j \langle u_j - u_E, J_2 d_j + J_2(d_1) + \lambda(q_1)\rangle.$$

As the last two terms are independent of j, Lemma 4.2 implies that, after summing, their contribution is zero. So

$$\begin{aligned} \nu_3(q_{m+1}) - \nu_3(q_1) &= \sum_{j=1}^{m} t_j \langle u_j - u_E, J_2 d_j\rangle \\ &= \sum_{j=1}^{m} t_j \big\langle u_j - u_E, J_2\big((f_E - f_{j+1})(u_j - u_E) + (f_j - f_E)(u_{j-1} - u_E)\big)\big\rangle \\ &= \sum_{j=1}^{m} t_j (f_j - f_E)\langle u_j - u_E, J_2(u_{j-1} - u_E)\rangle = \sum_{j=1}^{m} t_j (f_j - f_E), \end{aligned} \tag{4.7}$$

since u_E, u_{j-1}, u_j is a positively oriented $\mathbb{Z}$-basis.

On the other hand, $[\omega] \cup [F] = 0$ if and only if it pairs to zero on each compact divisor. We compute

$$([\omega] \cup [F]).E = (w_E E + \sum_{j=1}^{m} w_j A_j)(f_E E + \sum_{j=1}^{m} f_j A_j)E$$
$$= w_E f_E E^3 + \sum_{j=1}^{m} (w_j f_E + w_E f_j) A_j E^2 + \sum_{j,k=1}^{m} w_j f_k A_j A_k E.$$

We know the values of the triple intersections apart from E^3. But this may be computed by noting that the last component of (4.1) gives $E = -\sum_{D_i \neq E} D_i$, so $E^3 = -\sum_{j=1}^{m} A_j E^2 = -\sum_{j=1}^{m} b_j$. Now

$$([\omega] \cup [F]).E$$
$$= -w_E f_E \sum_{j=1}^{m} b_j + \sum_{j=1}^{m} b_j (w_j f_E + w_E f_j) + \sum_{j=1}^{m} w_j f_{j-1} + w_j f_{j+1} + a_j w_j f_j$$
$$= f_E \sum_{j=1}^{m} b_j (w_j - w_E) + \sum_{j=1}^{m} t_j f_j = -f_E \sum_{j=1}^{m} (2 + a_j) w_j + b_j w_E) + \sum_{j=1}^{m} t_j f_j$$
$$= \sum_{j=1}^{m} t_j (f_j - f_E).$$

Thus $([\omega] \cup [F]).E$ agrees with (4.7), completing the proof of Theorem 4.3.

5 Examples

We illustrate the above theory with three examples. The first does not require the machinery of Sect. 4 and is tackled more directly.

Example 5.1 (The Spaces $M_{m,n}$) Let $B = K$ be the canonical bundle of $\mathbb{C}P(1)_+ \times \mathbb{C}P(1)_-$. Following Foscolo et al. [16], $M = M_{m,n}$ denotes the total space over K of the circle bundle with first Chern class

$$[F] = m[\omega_+] - n[\omega_-],$$

where $[\omega_\pm]$ is the class of the Fubini-Study metric on $\mathbb{C}P(1)_\pm$, normalised so that $\int_{\mathbb{C}P(1)_\pm} \omega_\pm = 1$. The condition $[\omega] \cup [F] = 0$ implies

$$[\omega] = k(m[\omega_+] + n[\omega_-])$$

for some $k \in \mathbb{R}$ so that $km > 0$ and $kn > 0$.

The T^2-action on $\mathbb{C}P(1)_+ \times \mathbb{C}P(1)_-$ is given by circle actions on each factor. On $\mathbb{C}P(1)$ we have the action $[z_0, z_1] \mapsto [e^{i\theta} z_0, z_1]$ with fixed points $p = [1, 0]$, $q = [0, 1] \in \mathbb{C}P(1)$. Let X_1 be the generator of the action of $\mathbb{C}P(1)_+$ and X_2 that on $\mathbb{C}P(1)_-$. The T^2-action has fixed points

$$\mathrm{A} = (p_+, p_-), \quad \mathrm{B} = (p_+, q_-), \quad \mathrm{C} = (q_+, p_-), \quad \mathrm{D} = (q_+, p_+).$$

The curves

$$\mathcal{E}_{\mathrm{AB}} = \{p_+\} \times \mathbb{C}P(1)_-, \quad \mathcal{E}_{\mathrm{CD}} = \{q_+\} \times \mathbb{C}P(1)_-$$

are fixed by X_1, and the element X_2 fixes the curves

$$\mathcal{E}_{\mathrm{AC}} = \mathbb{C}P(1)_+ \times \{p_-\}, \quad \mathcal{E}_{\mathrm{BD}} = \mathbb{C}P(1)_+ \times \{q_-\}.$$

Each of these sets maps to an edge under μ, and there are no other points with non-trivial stabiliser.

As μ, λ, ν are only defined up to additive constants, we choose them so that

$$\mu(\mathrm{A}) = (0, 0), \quad \lambda(\mathrm{A}) = (0, 0), \quad \nu(\mathrm{A}) = (0, 0, 0).$$

Let us now compute $\mu(\mathrm{B})$ and $\lambda(\mathrm{B})$. The points A and B are joined by $\mathcal{E}_{\mathrm{AB}}$ with stabiliser $\mathsf{X}_1 = 1\mathsf{X}_1 + 0\mathsf{X}_2$, giving $(r_1, r_2) = (1, 0)$, up to sign. Now $t_{\mathrm{AB}} = [\omega] \cap [\mathcal{E}_{\mathrm{AB}}] = kn$ and $s_{\mathrm{AB}} = [F] \cap [\mathcal{E}_{\mathrm{AB}}] = n$, so by (3.7), (3.8), and (3.10)

$$\mu(\mathrm{B}) = kn(0, 1), \quad \lambda(\mathrm{B}) = n(0, 1), \quad \nu_3(\mathrm{B}) = kn(1 \times 0 + 0 \times n) = 0.$$

Similarly, for $\mathcal{E}_{\mathrm{BD}}$ the stabiliser is generated by $\mathsf{X}_2 = 0\mathsf{X}_1 + 1\mathsf{X}_2$ up to sign. We choose $(r_1, r_2) = (0, -1)$, so that $t_{\mathrm{BD}} = km$, $s_{\mathrm{BD}} = m$ and

$$\mu(\mathrm{D}) = \mu(\mathrm{B}) + km(1, 0) = (km, kn), \quad \lambda(\mathrm{D}) = \lambda(\mathrm{B}) - (m, 0) = (-m, n),$$
$$\nu_3(\mathrm{D}) = \nu_3(\mathrm{B}) + km(0 \times -m + (-1) \times n) = -kmn.$$

Similarly, $\nu(\mathrm{C}) = (km, 0, 0)$. So the lifted graph is as in Fig. 2, cf. [14].

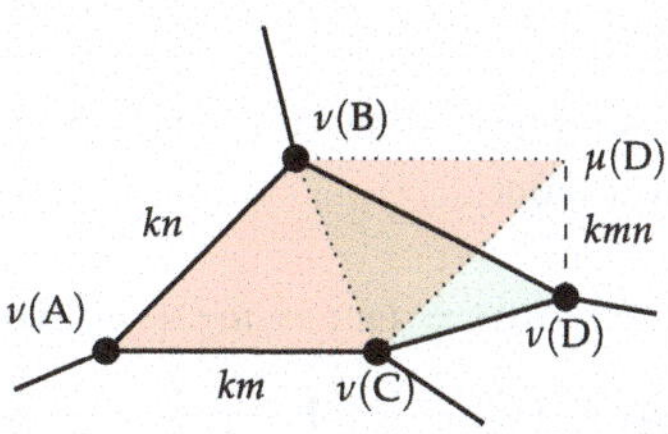

Fig. 2 The G_2-trivalent graph for $M_{m,n}$, with the red region giving the compact part of underlying planar SU(3)-graph

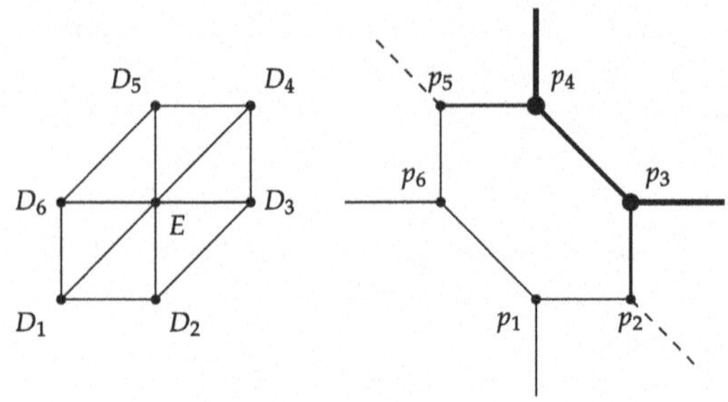

Fig. 3 Fan diagram for the star triangulation of the hexagon, and the corresponding toric diagram

Example 5.2 (Hexagon, Star Triangulation) The fan for this example is given on the left in Fig. 3, with the vectors $u_1, u_2, \ldots, u_6, u_E \in \mathbb{Z}^2$ chosen so that

$$u_1 - u_E = (-1, -1), \quad u_2 - u_E = (0, -1), \quad u_3 - u_E = (1, 0),$$
$$u_4 - u_E = (1, 1), \quad u_5 = (0, 1) - u_E, \quad u_6 - u_E = (-1, 0).$$

The second cohomology is generated by the divisors modulo (4.1), so each class may be specified as linear combination of four divisors. Following the approach of Acharya et al. [1], we choose D_1, D_2, D_3, D_4 as our generating set and put

$$[F] = f_1 D_1 + f_2 D_2 + f_3 D_3 + f_4 D_4, \quad [\omega] = w_1 D_1 + w_2 D_2 + w_3 D_3 + w_4 D_4.$$

Triple product computations, using the relations (4.2) and (4.3), cf. [1, 30], give the volumes of the six compact curves $\mathcal{E}_j = A_j \cap E$:

$$t_1 = w_2 - w_1, \quad t_2 = w_1 - w_2 + w_3, \quad t_3 = w_2 - w_3 + w_4,$$
$$t_4 = w_3 - w_4, \quad t_5 = w_4, \quad t_6 = w_1.$$

From these we see that $[\omega]$ is a Kähler class if

$$w_1 > 0, \qquad w_4 > 0, \qquad w_3 > w_2 - w_1 > 0, \qquad w_2 > w_3 - w_4 > 0.$$

We choose our additive constants so that

$$p_1 = \mu(q_1) = (0, 0), \quad \lambda(q_1) = (0, 0), \quad \nu(q_1) = (0, 0, 0).$$

Using the values of t_j, we get the remaining five vertices, $p_i = \mu(q_i)$, of the planar hexagon:

$$p_2 = (w_1 - w_2 + w_3, 0), \quad p_3 = (w_1 - w_2 + w_3, w_2 - w_3 + w_4),$$
$$p_4 = (w_1 - w_2 + w_4, w_2), \quad p_5 = (w_1 - w_2, w_2), \quad p_6 = (w_1 - w_2, w_2 - w_1).$$

Next, we compute the lifts, using (4.5). First, we need to compute the numbers $\lambda(q_i)$. In order to express these in terms of the coefficients $f_1, \ldots, f_4$, we use the formula (3.8) in the form (4.6). Here $s_j = [F] \cap [\mathcal{E}_j]$ and this is computed in the same way as t_j, just with f_i replacing each w_i. We therefore find

$$\lambda(q_2) = (-f_1 + f_2 - f_3, 0), \quad \lambda(q_3) = (-f_1 + f_2 - f_3, -f_2 + f_3 - f_4),$$
$$\lambda(q_4) = (-f_1 + f_2 - f_4, -f_2), \quad \lambda(q_5) = (-f_1 + f_2, -f_2),$$
$$\lambda(q_6) = (-f_1 + f_2, -f_1 - f_2).$$

Based on this, we compute the lifts $\nu_3(q_j)$ using (4.5)

$$\nu_3(q_2) = 0, \quad \nu_3(q_3) = (-f_1 + f_2 - f_3)(w_2 - w_3 + w_4),$$
$$\nu_3(q_4) = -f_1 w_2 + (f_2 - f_3)(w_2 - w_3 + w_4) - f_4(w_3 - w_4),$$
$$\nu_3(q_5) = -f_1 w_2 + f_2(w_2 - w_3) - f_3(w_2 - w_3 + w_4) - f_4(w_3 - w_4),$$
$$\nu_3(q_6) = f_1(w_1 - w_2) - f_2(w_1 - w_2 + w_3) - f_3(w_2 - w_3 + w_4)$$
$$- f_4(w_3 - w_4).$$

In this case, the explicit expression for $[\omega] \cup [F] = 0$ reads

$$f_1(w_2 - w_1) + f_2(w_1 - w_2 + w_3) + f_3(w_2 - w_3 + w_4) + f_4(w_3 - w_4) = 0.$$

Note that this is a linear condition in the f_i. However each f_i is integral, so this is also a discrete constraint on the w_i, and certain choices of $[\omega]$ will give no non-zero solution. Also note that each coefficient of f_i is positive, so at least two of the f_i have opposite signs.

With $w_1 = w_4 = 1$, $w_2 = w_3 = 2$, $f_1 = f_2 = 1 = -f_3 = -f_4$ we obtain the toric diagram on the right in Fig. 3. The thickness of lines and points indicate the relative values of ν_3. The half-lines out of p_1 and p_6 lie in the plane of $p_5 p_6 p_1 p_2$, those out of p_2 and p_6 descend from this plane. The points p_3, p_4 are higher than this plane, and the half lines out of p_3 and p_4 rise out of a parallel plane through $p_3 p_4$. Other values of w_i give more irregular figures, with different lengths for $p_1 p_2$, $p_2 p_3$ and $p_3 p_4$. Varying the f_j also enables moving p_5 out of the $p_6 p_1 p_2$-plane, making $p_3 p_4$ skew relative to $p_1 p_6$.

Remark 5.3 According to Altmann [3], the hexagon singularity admits two types of versal deformation: one into three skew lines and one into two triangles. We expect that in such cases the graph Γ_B from such a deformation has two separate components lying in distinct parallel planes. A further example in [3], based on an octahedron, has versal deformations that admit non-trivial circle bundles, and again we expect the G_2-graphs Γ_M to have components lying in distinct parallel affine subspaces.

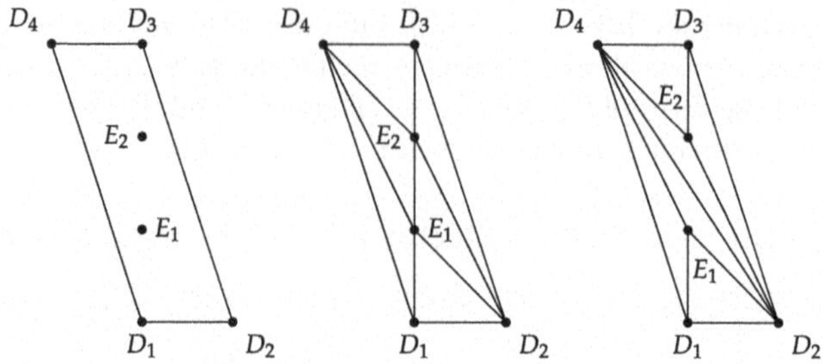

Fig. 4 The fan diagrams of $\mathbb{Z}/3\mathbb{Z}$-quotient of the conifold (left) and its two resolutions

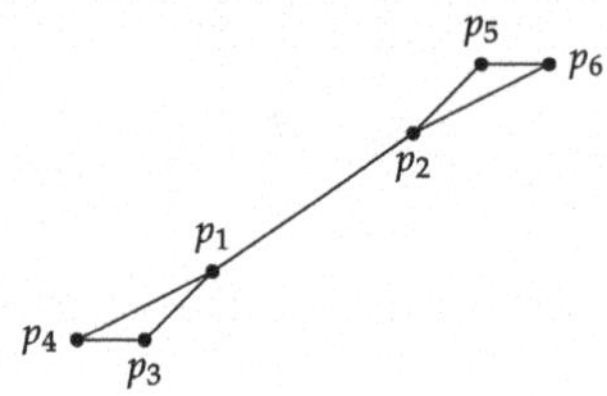

Fig. 5 Compact part of the moment graph, before lifting, corresponding to the second resolution of the $\mathbb{Z}/\mathbb{Z}3$-quotient of the conifold

Example 5.4 (Second Resolution of the Three-Fold Quotient of the Conifold) As in Acharya et al. [1], let us consider the $\mathbb{Z}/3\mathbb{Z}$-quotient of the conifold. The rays of the fan are given by

$$u_2 - u_1 = (1, 0), \quad u_3 - u_1 = (0, 3), \quad u_4 - u_1 = (-1, 3),$$
$$v_1 - u_1 = (0, 1), \quad v_2 - u_1 = (0, 2),$$

with u_i corresponding to divisors D_i, and v_j to compact divisors E_j. There are two resolutions as shown in Fig. 4. We will work with the second resolution, whose corresponding toric diagram is given in Fig. 5.

The space of divisors is spanned by D_1, D_2, D_3, so we write $[\omega] = w_1 D_1 + w_2 D_2 + w_3 D_3$. This defines a Kähler structure if $w_1 > -w_2 > 0$ and $w_3 > -w_2 > 0$. Writing $[F] = f_1 D_1 + f_2 D_2 + f_3 D_3$, the condition $[\omega] \cup [F] = 0$ is

$$(f_1 + f_2)(w_1 + w_2) = 0 = (f_2 + f_3)(w_2 + w_3),$$

so $f_1 = -f_2 = f_3$.

Putting $p_1 = 0$, we have

$$p_2 = -w_2(3, 2), \quad p_3 = -(w_1 + w_2)(1, 1), \quad p_6 = (2w_3 - w_2, w_3 - w_2).$$

Similarly, we compute

$$\lambda(p_2) - \lambda(p_1) = f_2(3, 2), \quad \lambda(p_3) - \lambda(p_1) = (f_1 + f_2)(1, 1) = (0, 0),$$
$$\lambda(p_6) - \lambda(p_2) = -(f_2 + f_3)(2, 1) = (0, 0),$$

and so, if $\lambda(p_1) = 0$, we get

$$\nu_3(p_2) - \nu_3(p_1) = 0, \quad \nu_3(p_3) - \nu_3(p_1) = 0,$$
$$\nu_3(p_6) - \nu_3(p_2) = f_2(w_2 + w_3) \neq 0.$$

So the lifted graph is not planar, with the two triangles rotated relative to each other along the central axis. This behaviour may be compared with the graph described in [24] of the Bryant-Salamon metric from [7].

Acknowledgments We are grateful for partial support from the Aarhus University Research Foundation. We thank Lorenzo Foscolo for useful discussions of results in this area.

References

1. B.S. Acharya, L. Foscolo, M. Najjar, E.E. Svanes, New G_2-conifolds in M-theory and their field theory interpretation. J. High Energy Phys. **2021**(5), 32 pp. (2021)
2. O. Aharony, A. Hanany, B. Kol, Webs of (p, q) 5-branes, five dimensional field theories and grid diagrams. J. High Energy Phys. **1998**(1), 51 pp. (1998)
3. K. Altmann, The versal deformation of an isolated toric Gorenstein singularity. Invent. Math. **128**(3), 443–479 (1997)
4. V. Apostolov, S. Salamon, Kähler reduction of metrics with holonomy G_2. Comm. Math. Phys. **246**(1), 43–61 (2004)
5. V. Bouchard, Lectures on complex geometry, Calabi-Yau manifolds and toric geometry, preprint arXiv: hep-th/0702063 (2007)
6. R.L. Bryant, Metrics with exceptional holonomy. Ann. Math. **126**, 525–576 (1987)
7. R.L. Bryant, S.M. Salamon, On the construction of some complete metrics with exceptional holonomy. Duke Math. J. **58**, 829–850 (1989)
8. N. Cavalleri, Complete non-compact Spin(7)-manifolds from T^2-bundles over AC Calabi-Yau manifolds, preprint arXiv: 2407.19486 [math.DG] (2024)
9. K. Cho, A. Futaki, H. Ono, Uniqueness and examples of compact toric Sasaki-Einstein metrics. Comm. Math. Phys. **277**(2), 439–458 (2008)
10. R. Cleyton, A.F. Swann, Cohomogeneity-one G_2-structures. J. Geom. Phys. **44**(2–3), 202–220 (2002)
11. R.J. Conlon, H.-J. Hein, Classification of asymptotically conical Calabi-Yau manifolds. Duke Math. J. **173**(5), 947–1015 (2024)
12. D. Conti, A. Tomassini, Special symplectic six-manifolds. Q. J. Math. **58**(3), 297–311 (2007)
13. D.A. Cox, J.B. Little, H.K. Schenck, *Toric varieties, Graduate Studies in Mathematics*, vol. 124 (American Mathematical Society, Providence, RI, 2011)
14. K. Dixon, Toric ALC G_2 manifolds, in online talk *Simons Collaboration on Special Holonomy in Geometry, Analysis, and Physics*, https://sites.duke.edu/scshgap/kael-dixon-lectures/ (2020)
15. L. Foscolo, M. Haskins, J. Nordström, Complete non-compact G_2-manifolds from asymptotically conical Calabi-Yau 3-folds. Duke Math. J. **170**(15), 3323–3416 (2021)
16. L. Foscolo, M. Haskins, J. Nordström, Infinitely many new families of complete cohomogeneity one G_2-manifolds: G_2 analogues of the Taub-NUT and Eguchi-Hanson spaces. J. Eur. Math. Soc. (JEMS) **23**(7), 2153–2220 (2021)
17. A. Futaki, H. Ono, G. Wang, Transverse Kähler geometry of Sasaki manifolds and toric Sasaki-Einstein manifolds. J. Differential Geom. **83**(3), 585–635 (2009)

18. G.W. Gibbons, S.W. Hawking, Gravitational multi-instantons. Phys. Lett. B **78**, 430–432 (1978)
19. R.K. Lashof, J.P. May, G.B. Segal, Equivariant bundles with Abelian structural group, in *Proceedings of the Northwestern Homotopy Theory Conference (Evanston, Ill., 1982)*, Contemp. Math., vol. 19 (Amer. Math. Soc., 1983), pp. 167–176
20. Y. Li, SYZ geometry for Calabi-Yau 3-folds: Taub-NUT and Ooguri-Vafa type metrics. Mem. Amer. Math. Soc. **292**(1453), 126 pp. (2023)
21. J. Li, C.-C.M. Liu, K. Liu, J. Zhou, A mathematical theory of the topological vertex. Geom. Topol. **13**(1), 527–621 (2009)
22. T.B. Madsen, A.F. Swann, Multi-moment maps. Adv. Math. **229**, 2287–2309 (2012)
23. T.B. Madsen, A.F. Swann, Closed forms and multi-moment maps. Geom. Dedicata **165**(1), 25–52 (2013)
24. T.B. Madsen, A.F. Swann, Toric geometry of G_2-manifolds. Geom. Topol. **23**(7), 3459–3500 (2019)
25. T.B. Madsen, A.F. Swann, Multi-toric geometries with larger compact symmetry, Q. J. Math. **76**(1), 349–365 (2025)
26. D. Martelli, J. Sparks, Toric geometry, Sasaki-Einstein manifolds and a new infinite class of AdS/CFT duals. Comm. Math. Phys. **262**(1), 51–89 (2006)
27. A.F. Swann, Twisting Hermitian and hypercomplex geometries. Duke Math. J. **155**(2), 403–431 (2010)
28. C. van Coevering, Ricci-flat Kähler metrics on crepant resolutions of Kähler cones. Math. Ann. **347**(3), 581–611 (2010)
29. C. van Coevering, Examples of asymptotically conical Ricci-flat Kähler manifolds. Math. Z. **267**(1–2), 465–496 (2011)
30. D. Xie, S.-T. Yau, Three dimensional canonical singularity and five dimensional $\mathcal{N} = 1$ SCFT. J. High Energy Phys. **2017**(6), 33 pp. (2017)
31. K. Yeats, *A Combinatorial Perspective on Quantum Field Theory*, Springer Briefs in Mathematical Physics, vol. 15 (Springer, Cham, 2017)

On Classification of Compact Complex Surfaces of Class VII

Georges Dloussky

Dedicated to Paul Gauduchon whose metric is most useful in non-Kähler geometry

Abstract Let S be a minimal compact complex surface with Betti numbers $b_1(S) = 1$ and $b_2(S) \geq 1$ i.e. a compact surface in class VII_0^+. We show that if there exists a twisted logarithmic 1-form $\theta \in H^0(S, \Omega^1(\log D) \otimes \mathcal{L}_\lambda)$, where D is a non zero divisor and $\mathcal{L} \in H^1(S, \mathbb{C}^\star)$, then S is a Kato surface. It is known that λ is in fact real and we show that $\lambda \geq 1$ and unique if S is not a Inoue-Hirzebruch surface. Moreover $\lambda = 1$ if and only if S is a Enoki surface. When $\lambda > 1$ these conditions are equivalent to the existence of a negative PSH function $\hat{\tau}$ on the cyclic covering $p : \hat{S} \to S$ of S which is PH outside $\hat{D} := p^{-1}(D)$ with automorphic constant being the **same** automorphy constant λ for a suitable automorphism of $\hat{S}$. With previous results obtained with V.Apostolov it suggests a strategy to prove the GSS conjecture.

Keywords Non Kähler · Classification · Compact complex surfaces · Complex foliation

---

G.D. is grateful to CIRGET of UQAM in Montréal and I2M, UMR 7373 of Aix-Marseille University for the financial support. The author thanks Vestislav Apostolov for insightful discussions and invitations all along the preparation of this article. The author is also grateful to Karl Oeljeklaus for discussions on his work [18] in order to obtain the adapted result needed in the main theorem.

G. Dloussky (✉)
Georges Dloussky Aix-Marseille Université, Institut de Mathématiques de Marseille (I2M), UMR 7373, Marseille Cedex 3, France
e-mail: georges.dloussky@univ-amu.fr

L. Ornea (ed.), *Real and Complex Geometry*,
https://doi.org/10.1007/978-3-031-92297-8_7

1 Introduction

The classification of minimal compact complex surfaces is incomplete for non Kähler surfaces S in VII_0^+ class, that is minimal compact complex surfaces which have Betti numbers $b_1(S) = 1$ and $b_2(S) \geq 1$. All known examples of such surfaces contain Global Spherical Shells (GSS) and are called Kato surfaces [7, 17]. The GSS conjecture asserts that there is no other surfaces. Given a surface S in class VII_0^+ several criteria allow us to assert that S is a Kato surface

- Existence on S of $b_2(S)$ rational curves [12],
- Existence of a non trivial section of a power of the anticanonical bundle twisted by a topologically trivial line bundle $K_S^{-m} \otimes \mathcal{L}$, $m \geq 1$, [8],
- Existence of a non trivial global vector field [11],
- $b_2(S) = 1$ by Teleman theorem [19].

In [5] M. Brunella proved that if the cyclic covering $\pi : \hat{S} \to S$ of S admits a negative plurisubharmonic (PSH) function $\hat{u} : \hat{S} \to [-\infty, 0[$ which is pluriharmonic (PH) outside a non empty analytic subset and for a generator $\varphi : \hat{S} \to \hat{S}$, $\hat{\tau} \circ \varphi = \lambda \hat{\tau}$ for some positive λ, then S is a (possibly) blown-up Kato surface.

It can be noticed that λ cannot be equal to 1 and replacing if necessary φ by φ^{-1}, we may suppose that $\lambda > 1$. Therefore, by [10], S is an intermediate Kato surface (i.e. the maximal divisor D in S is a cycle of rational curves with branches) or a Inoue-Hirzebruch surface.

Given a compact complex surface in class VII_0^+, the main results (see Theorem 4.2) of the article are

(1) The existence of a non trivial twisted logarithmic 1-form

$$\theta \in H^0(S, \Omega^1(\log D) \otimes \mathcal{L}_\lambda),$$

where D is a non trivial divisor and $\mathcal{L}_\lambda \in H^1(S, \mathbb{C}^\star) \simeq \mathbb{C}^\star$ implies that S is a Kato surface,

(2) The existence of such a twisted logarithmic 1-form θ is equivalent to the existence of a negative PSH function $\hat{u} : \hat{S} \to [-\infty, 0[$ on the cyclic covering $\pi : \hat{S} \to S$ of S which is PH outside $\hat{D} := \pi^{-1}(D)$ such that for a suitable generator $\gamma : \hat{S} \to \hat{S}$ of the deck group of automorphism of $\hat{S}$,

$$\hat{u} \circ \gamma = \lambda \hat{u}.$$

(3) If $\lambda \geq 1$ λ is unique and it is the same constant for both twisted logarithmic 1-form θ and PSH function $\hat{u}$.

A locally conformally symplectic structure (LCS) tamed by the complex structure is a 2-form ω on S such that $d_\alpha \omega = 0$, for a closed 1-form α and whose (1, 1) part $\omega^{1,1}$ is positive. In a sequence of articles [1–3] it is proved that the set of LCS structures tamed by the complex structure of $S \in \mathrm{VII}_0^+$, denoted by $\mathcal{T}(S)$ can be represented by a non empty open interval in $]-\infty, 0[$.

In this article there are two ways to represent a holomorphically trivial line bundle: by closed one forms or by a real number. More precisely, let $\pi : \hat{S} \to S$ be the cyclic covering of S. For α a closed real 1-form on S, let $\hat{\alpha} = \pi^\star\alpha$ which is exact. If $\gamma : \hat{S} \to \hat{S}$ is a generator of the deck group of automorphisms there is a constant $c \in \mathbb{R}$ such that for $\hat{\alpha} = d\hat{f}$, $\hat{f} \circ \gamma - \hat{f} = c$. Replacing γ by γ^{-1} if necessary we may suppose that $c < 0$. Consider the mapping

$$]-\infty, 0[\to]1, +\infty[, \quad c \mapsto C = e^{-c}.$$

This mapping gives the correspondence between a line bundle $\mathcal{L}_\alpha$, $\alpha \in H^1_{dR}(S, \mathbb{R})$, and $\mathcal{L}_C$, $C > 1$ setting $C = \varphi(c)$. Then we have

Conjecture 1.1 If $\mathcal{T}(S) =]a, b[$, $b \leq 0$, we have $\lambda = \varphi(b) = e^{-b} \geq 1$.

For hyperbolic Kato surfaces, it is shown that $\varphi(b) \geq \lambda$ (φ is decreasing) [3, section 4.3]. Moreover we have for Enoki surfaces [13] $0 = \lambda = \varphi(0)$ because in this case $\mathcal{T}(S) =]-\infty, 0[$.

This conjecture gives the expected unique value for which we have to look for a PSH function, PH outside D with automorphy constant λ. Therefore a strategy to prove the GSS conjecture would be to show the existence of a negative PSH function, as in theorem 2.17, with twisting constant being defined by the upper bound of the open interval $\mathcal{T}(S)$.

A sketch of proof is the following: Given a twisted logarithmic 1-form we prove first that the Camacho-Sad indices [6] are negative, therefore this assumption in Lemma 6.18 of [3] becomes superfluous. We obtain

Proposition 1.2 *Let S be a minimal surface in the class VII_0^+ which is not a Inoue-Hirzebruch surface. Suppose that S admits a non-trivial logarithmic 1-form*

$$0 \neq \theta \in H^0(S, \Omega^1(log D) \otimes \mathcal{L}_\alpha), \quad [\alpha] \in H^1_{dR}(S, \mathbb{R})$$

with pole along a non-trivial connected divisor D. Then we can multiply θ by a non zero complex number such that the residues $Res_\alpha(\theta)$ are real valued and positive, and the real part $\mathrm{Re}(P_\theta)$ *of the current P_θ associated to θ is exact, i.e. satisfy* $\mathrm{Re}(P_\theta) = d_\alpha\tau$ *for a degree zero current τ on S. Furthermore,*

$$d_\alpha d^c_\alpha \tau = 2\pi\, Res_\alpha(\theta) T_D > 0$$

and $[\alpha] \notin \mathcal{T}(S)$.

Notice that the support of $d_\alpha d^c_\alpha \tau$ is D. With this proposition we find a PSH function $\hat{u} = e^{-\hat{f}}\hat{\tau}$, PH outside $\hat{D}$. Conversely, $i\partial\hat{u}$ is holomorphic outside $\hat{D}$ since $\hat{u}$ is there PH. The end of the proof consists to extend $i\partial\hat{u}$ on whole $\hat{S}$ into a logarithmic 1-form.

2 Negativity of Camacho-Sad Coefficients

We start with an explicit example and after we use a family of polynomials, already used in [11] to solve a system of equations which determine the Camacho-Sad indices of the foliation defined by a global twisted logarithmic 1-form. We want to show that this system of equations has positive solutions. When $p = 2$ and $\delta = (1, 1)$ the solution is not real, therefore we have to use the geometric properties of the foliation.

2.1 An Explicit Example

Example 2.1 We use notations and explicit construction of a Kato surface used in [7]. Consider the sequence $(4223) = (s_2 r_1 s_1)$ of opposite self-intersections. The associate Kato surface has four rational curves of self-intersections

$$(D_0^2, D_1^2, D_2^2, D_3^2) = (-4, -2, -2, -3).$$

Its opposite intersection matrix is

$$M(S) = \begin{pmatrix} 4 & 0 & 0 & -1 \\ 0 & 2 & -1 & -1 \\ 0 & -1 & 2 & -1 \\ -1 & -1 & -1 & 3 \end{pmatrix}$$

therefore we have a cycle with three curves D_1, D_2, D_3 and a tree D_0 which intersects D_3. The normal form of the contraction $F(z_1, z_2) = (\star, z_2^k)$, associated to the corresponding sequence of blow-ups, has a second member z_2^k where

$$k = k(S) = \sqrt{\det M(S)} + 1 = \sqrt{9} + 1 = 4$$

by a result of C. Favre (see [10] or [14] for the normal form of the contraction).

We compute now Camacho-Sad indices setting

$$\alpha_1 = CS(\mathcal{F}, D_1, D_1 \cap D_2), \quad \alpha_2 = CS(\mathcal{F}, D_2, D_2 \cap D_3), \quad \alpha_3 = CS(\mathcal{F}, D_3, D_3 \cap D_1)$$

hence we have chosen an orientation of the cycle. We have, by Camacho-Sad formula

$$\begin{aligned}
-4 = D_0^2 &= CS(\mathcal{F}, D_0, D_0 \cap D_3) \\
-2 = D_1^2 &= CS(\mathcal{F}, D_1, D_1 \cap D_2) + CS(\mathcal{F}, D_1, D_1 \cap D_3) = \alpha_1 + \frac{1}{\alpha_3} \\
-2 = D_2^2 &= CS(\mathcal{F}, D_2, D_2 \cap D_3) + CS(\mathcal{F}, D_2, D_2 \cap D_1) = \alpha_2 + \frac{1}{\alpha_1} \\
-3 = D_3^2 &= CS(\mathcal{F}, D_3, D_3 \cap D_1) + CS(\mathcal{F}, D_3, D_3 \cap D_2) \\
&\quad + CS(\mathcal{F}, D_3, D_3 \cap D_0) \\
&= \alpha_3 + \frac{1}{\alpha_2} - \frac{1}{4}
\end{aligned}$$

We obtain a system of equations for the cycle

$$\begin{cases} \alpha_1 + \dfrac{1}{\alpha_3} = -2 \\ \alpha_2 + \dfrac{1}{\alpha_1} = -2 \\ \alpha_3 + \dfrac{1}{\alpha_2} = -\dfrac{11}{4} \end{cases}$$

We shall show that Camacho-Sad indices are negative. For conveniency we change all the signs (α_i is changed into $-\alpha_i$) and we solve

$$\begin{cases} \alpha_1 + \dfrac{1}{\alpha_3} = 2 \\ \alpha_2 + \dfrac{1}{\alpha_1} = 2 \\ \alpha_3 + \dfrac{1}{\alpha_2} = \dfrac{11}{4} \end{cases}$$

and we shall show that all the solutions are positive. For example look at α_1: it is equal to a periodic continued fraction or is a solution of an equation of degree two:

$$\alpha_1 = 2 - \frac{1}{\alpha_3} = 2 - 1/\frac{11}{4} - 1/2 - \frac{1}{\alpha_1}$$

$$6\alpha_1^2 - 13\alpha_1 + 6 = 0, \quad \Delta = 25$$

- If $\alpha_1 = \frac{3}{2}$ then $\alpha_2 = 2$, $\alpha_3 = \frac{4}{3}$.
- If $\alpha_1 = \frac{2}{3}$, then $\alpha_2 = \frac{3}{4}$, $\alpha_3 = \frac{1}{2}$.

The two sets of solutions are inverse of each other, corresponding of the same foliation but changing the orientation of the cycle. By [10] there is only one foliation on S.

We notice that

$$\frac{3}{2}.2.\frac{4}{3} = 4 = k(S), \quad \frac{2}{3}.\frac{3}{4}.\frac{1}{2} = \frac{1}{4} = \frac{1}{k(S)}.$$

It gives the multiplicative factor, hence the coefficient of the torsion line bundle of the logarithmic 1-form. In these relations, only Camacho-Sad indices of intersection points of the cycle come into play and no other singularities. However the values of these indices depend also on the tree.

These relation reveal that, in general, for a root of one of the degree two equations, its inverse is a root of another degree two equation. Therefore the roots exist by pairs of the same sign.

2.2 Algebraic Part of the Proof

Lemma 2.2 *We consider the family of polynomials*

$$\begin{aligned} P_0 &= 1, \\ P_1(X_1) &= X_1, \\ P_i(X_1, \ldots, X_i) &= X_1 P_{i-1}(X_2, \ldots, X_i) - P_{i-2}(X_3, \ldots, X_i),\ i \geq 2 \end{aligned}$$

Then for any $n \geq 1$, $P_n(X_1, \ldots, X_n) = P_n(X_n, \ldots, X_1)$.

Proof Remark that the definition gives

$$P_i(X_i, \ldots, X_1) = X_i P_{i-1}(X_{i-1}, \ldots, X_1) - P_{i-2}(X_{i-2}, \ldots, X_1).$$

Therefore we have to show by induction on $n \geq 2$ that

$$P_i(X_1, \ldots, X_i) = X_i P_{i-1}(X_{i-1}, \ldots, X_1) - P_{i-2}(X_{i-2}, \ldots, X_1),$$

It is easy to check that for $n = 2, 3$ the two expressions give the same polynomials. Suppose that it is true for i and $i - 1$; then for $i + 1$ we have by the induction hypothesis

$$\begin{aligned} P_{i+1}(X_1, \ldots, X_{i+1}) &= X_1 P_i(X_2, \ldots, X_{i+1}) - P_{i-1}(X_3, \ldots, X_{i+1}) \\ &= X_1\big[X_{i+1} P_{i-1}(X_i, \ldots, X_2) - P_{i-2}(X_{i-1}, \ldots, X_2)\big] \\ &\quad -\big[X_{i+1} P_{i-2}(X_i, \ldots, X_3) - P_{i-3}(X_{i-1}, \ldots, X_3)\big] \\ &= X_{i+1}\big[X_1 P_{i-1}(X_i, \ldots, X_2) - P_{i-2}(X_i, \ldots, X_3)\big] \\ &\quad -\big[X_1 P_{i-2}(X_{i-1}, \ldots, X_2) - P_{i-3}(X_{i-1}, \ldots, X_3)\big] \\ &= X_{i+1} P_i(X_i, \ldots, X_1) - P_{i-1}(X_{i-1}, \ldots, X_1) \end{aligned}$$

□

Lemma 2.3 *For $p \geq 2$, the polynomials*

$$Q_p(X_0, \dots, X_{p-1}) := P_p(X_0, \dots, X_{p-1}) - P_{p-2}(X_1, \dots, X_{p-2})$$

are invariant under cyclic permutation.

Proof It is sufficient to prove that

$$D := Q_p(X_0, \dots, X_{p-1}) - Q_p(X_1, \dots, X_{p-1}, X_0) = 0.$$

In fact, by Lemma 2.2

$$\begin{aligned} D &:= \big(P_p(X_0, \dots, X_{p-1}) - P_{p-2}(X_1, \dots, X_{p-2})\big) \\ &\quad -\big(P_p(X_1, \dots, X_{p-1}, X_0) - P_{p-2}(X_2, \dots, X_{p-1})\big) \\ &= \big(X_0 P_{p-1}(X_1, \dots, X_{p-1}) - P_{p-2}(X_2, \dots, X_{p-1}) - P_{p-2}(X_1, \dots, X_{p-2})\big) \\ &\quad -\big(X_0 P_{p-1}(X_1, \dots, X_{p-1}) - P_{p-2}(X_1, \dots, X_{p-2}) \\ &\qquad -P_{p-2}(X_2, \dots, X_{p-1})\big) \\ &= 0. \end{aligned}$$

□

Example 2.4 $P_0 = 1$, $P_1(X_0) = X_0$, $P_2(X_0, X_1) = X_0 X_1 - 1$,

$$P_3(X_0, X_1, X_2) = X_0 X_1 X_2 - X_0 - X_2,$$

$$P_4(X_0, X_1, X_2, X_3) = X_0 X_1 X_2 X_3 - X_0 X_1 - X_0 X_3 - X_2 X_3 + 1,$$

$$P_5 = X_0 X_1 X_2 X_3 X_4 - X_0 X_1 X_2 - X_2 X_3 X_4 - X_3 X_4 X_0 - X_4 X_0 X_1 + X_0 + X_2 + X_4$$

$$\begin{aligned} P_6 = {} & X_0 X_1 X_2 X_3 X_4 X_5 \\ & -X_0 X_1 X_2 X_3 - X_2 X_3 X_4 X_5 - X_3 X_4 X_5 X_0 - X_4 X_5 X_0 X_1 - X_5 X_0 X_1 X_2 \\ & +X_0 X_1 + X_2 X_3 + X_4 X_5 + X_5 X_0 + X_0 X_3 + X_2 X_5 - 1 \end{aligned}$$

$$Q_2(X_0, X_1) = X_0 X_1 - 2$$

$$Q_3(X_0, X_1, X_2) = X_0 X_1 X_2 - X_0 - X_1 - X_2$$

$$Q_4(X_0, X_1, X_2, X_3) = X_0 X_1 X_2 X_3 - X_0 X_1 - X_1 X_2 - X_2 X_3 - X_3 X_0 + 2$$

$$\begin{aligned}Q_5 = &X_0X_1X_2X_3X_4\\ &-X_0X_1X_2 - X_1X_2X_3 - X_2X_3X_4 - X_3X_4X_0 - X_4X_0X_1\\ &+X_0 + X_1 + X_2 + X_3 + X_4\end{aligned}$$

Notice that new monomials appear in Q_6:

$$\begin{aligned}Q_6 = &X_0X_1X_2X_3X_4X_5\\ &-X_0X_1X_2X_3 - X_1X_2X_3X_4 - X_2X_3X_4X_5 - X_3X_4X_5X_0\\ &-X_4X_5X_0X_1 - X_5X_0X_1X_2\\ &+X_0X_1 + X_1X_2 + X_2X_3 + X_3X_4 + X_4X_5 + X_5X_0 + \mathbf{X_0X_3}\\ &+\mathbf{X_1X_4} + \mathbf{X_2X_5} - 2\end{aligned}$$

Lemma 2.5 *If δ_i are real numbers such that $\delta_i \geq 2$ for $i \geq 0$, then for any $j \geq 0$,*

$$\frac{P_{j+1}(\delta_i, \dots, \delta_{i+j})}{P_j(\delta_i, \dots, \delta_{i+j-1})} \geq \frac{j+2}{j+1} > 1,$$

$$P_{j+1}(\delta_i, \dots, \delta_{i+j}) \geq j+2,$$

and for $p \geq 2$,

$$P_p(\delta_i, \dots, \delta_{i+p-1}) - P_{p-2}(\delta_{i+1}, \dots, \delta_{i+p-2}) \geq 2.$$

Moreover, if $\delta = (2, \dots, 2)$ then the previous inequalities are equalities.

Proof By induction on $j \geq 0$. If $j = 0$, $\dfrac{P_1(\delta_i)}{P_0} \geq 2$. Suppose the inequality fulfilled for $j \geq 0$, then

$$\begin{aligned}\frac{P_{j+2}(\delta_i, \dots, \delta_{i+j+1})}{P_{j+1}(\delta_i, \dots, \delta_{i+j})} &= \frac{\delta_{i+j+1}P_{j+1}(\delta_i, \dots, \delta_{i+j}) - P_j(\delta_i, \dots, \delta_{i+j-1})}{P_{j+1}(\delta_i, \dots, \delta_{i+j})}\\ &= \delta_{i+j+1} - \frac{P_j(\delta_i, \dots, \delta_{i+j-1})}{P_{j+1}(\delta_i, \dots, \delta_{i+j})}\\ &\geq \delta_{i+j+1} - \frac{j+1}{j+2} \geq 2 - \frac{j+1}{j+2} = \frac{j+3}{j+2}.\end{aligned}$$

Therefore

$$P_{j+1}(\delta_i, \dots, \delta_{i+j}) = \frac{P_{j+1}(\delta_i, \dots, \delta_{i+j})}{P_j(\delta_i, \dots, \delta_{i+j-1})} \cdots \frac{P_1(\delta_i)}{P_0} \geq \frac{j+2}{j+1} \cdots \frac{2}{1} = j+2.$$

Using previous inequalities,

$$\begin{aligned}
&P_p(\delta_i,\dots,\delta_{i+p-1}) - P_{p-2}(\delta_{i+1},\dots,\delta_{i+p-2})\\
&\quad = P_{p-2}(\delta_{i+1},\dots,\delta_{i+p-2})\\
&\qquad \left\{\frac{P_p(\delta_i,\dots,\delta_{i+p-1})}{P_{p-1}(\delta_i,\dots,\delta_{i+p-2})}\frac{P_{p-1}(\delta_i,\dots,\delta_{i+p-2})}{P_{p-2}(\delta_{i+1},\dots,\delta_{i+p-2})} - 1\right\}\\
&\quad \geq (p-1)\left\{\frac{p+1}{p}.\frac{p}{p-1} - 1\right\} = 2.
\end{aligned}$$

It can be easily checked by induction on $p \geq 0$ that $P_p(2,\dots,2) = p+1$. □

Lemma 2.6 *Let $\alpha = (\alpha_0,\dots,\alpha_{p-1})$ be a solution (real or complex) of the system*

$$(E_p)\qquad \begin{cases}
\alpha_0 + \dfrac{1}{\alpha_1} & = \delta_0\\
\quad\vdots & \ \vdots\\
\alpha_i + \dfrac{1}{\alpha_{i+1}} & = \delta_i\\
\quad\vdots & \ \vdots\\
\alpha_{p-1} + \dfrac{1}{\alpha_0} & = \delta_{p-1}
\end{cases}$$

where $\alpha_i \neq 0$ and δ_i are real, for $i = 0,\dots,p-1$. Then, we have (with $P_{-1} = 0$),

$$\alpha_{i+1} = \frac{\alpha_0 P_{i-1}(\delta_1,\dots,\delta_{i-1}) - P_i(\delta_0,\dots,\delta_{i-1})}{\alpha_0 P_i(\delta_1,\dots,\delta_i) - P_{i+1}(\delta_0,\dots,\delta_i)} \neq 0 \quad i = 0,\dots,p-1,$$

where

$$\alpha_0 \neq \frac{P_{i+1}(\delta_0,\dots,\delta_i)}{P_i(\delta_1,\dots,\delta_i)}, \qquad i = 0,\dots,p-1.$$

In particular, α_0 satisfies the degree two equation

$$(\mathcal{E}_0)\qquad a_0(\delta)X^2 + b_0(\delta)X + c_0(\delta) = 0$$

where

$$a_0(\delta) = P_{p-1}(\delta_1,\dots,\delta_{p-1}) \neq 0, \quad c_0(\delta) = P_{p-1}(\delta_0,\dots,\delta_{p-2}) \neq 0$$

$$b_0(\delta) = -\Big(P_p(\delta_0,\dots,\delta_{p-1}) + P_{p-2}(\delta_1,\dots,\delta_{p-2})\Big)$$

By circular permutation we have also similar relations with α_j, $0 \leq j \leq p-1$:

For $i = 0, \ldots, p-1$,

$$\alpha_{j+i+1} = \frac{\alpha_j P_{i-1}(\delta_{j+1}, \ldots, \delta_{p-1}, \delta_0, \ldots, \delta_{j+i-1}) - P_i(\delta_j, \ldots, \delta_{p-1}, \delta_0, \ldots, \delta_{j+i-1})}{\alpha_j P_i(\delta_{j+1}, \ldots, \delta_{p-1}, \delta_0, \ldots, \delta_{j+i}) - P_{i+1}(\delta_j, \ldots, \delta_{p-1}, \delta_0, \ldots, \delta_{j+i})},$$

where

$$\alpha_j \neq \frac{P_{i+1}(\delta_j, \ldots, \delta_{p-1}, \delta_0, \ldots, \delta_{j+i})}{P_i(\delta_{j+1}, \ldots, \delta_{p-1}, \delta_0, \ldots, \delta_{j+i})} \neq 0, \quad i = 0, \ldots, p-1.$$

In particular, α_j *satisfies the degree two equation*

$$(\mathcal{E}_j) \qquad a_j(\delta)X^2 + b_j(\delta)X + c_j(\delta) = 0$$

where

$$a_j(\delta) = P_{p-1}(\delta_{j+1}, \ldots, \delta_{p-1}, \delta_0, \ldots, \delta_{j-1}) \neq 0,$$
$$c_j(\delta) = P_{p-1}(\delta_j, \ldots, \delta_{p-1}, \delta_0, \ldots, \delta_{j-2}) \neq 0$$

$$b_j(\delta) = -\Big(P_p(\delta_j, \ldots, \delta_{p-1}, \delta_0, \ldots, \ldots, \delta_{j-1}) + P_{p-2}(\delta_{j+1}, \ldots, \ldots, \delta_{p-1}, \delta_0, \ldots, \delta_{j-2})\Big)$$

We have $a_j(\delta) = c_{j+1}(\delta) \neq 0$ *and*

$$\prod_{j=0}^{p-1} P_{p-1}(\delta_j, \ldots, \delta_{j-2}) = \prod_{j=0}^{p-1} a_j(\delta) = \prod_{j=0}^{p-1} c_j(\delta) \neq 0.$$

Moreover if one root of $(\mathcal{E}_j)$ *is real then all the roots of all the equations* $(\mathcal{E}_i)$, $i = 0, \ldots, p-1$ *are real.*

Remark 2.7 For some values of δ the system of equations (E_p) has no solution. For example, for $p = 2$ and $\delta = (0, 1)$.

Proof For $i = 0$, $\alpha_1 = \dfrac{P_0}{P_1(\delta_0) - \alpha_0 P_0}$ and $\alpha_0 \neq \dfrac{P_1(\delta_0)}{P_0}$.

As

$$\alpha_2 = \frac{P_1(\delta_0) - \alpha_0 P_0}{P_2(\delta_0, \delta_1) - \alpha_0 P_1(\delta_1)}$$

it suggests that for an integer $i \geq 0$,

$$\alpha_{i+1} = \frac{P_i(\delta_0, \ldots, \delta_{i-1}) - \alpha_0 P_{i-1}(\delta_1, \ldots, \delta_{i-1})}{P_{i+1}(\delta_0, \ldots, \delta_i) - \alpha_0 P_i(\delta_1, \ldots, \delta_i)}$$

with

$$\alpha_0 \neq \frac{P_{j+1}(\delta_0, \ldots, \delta_j)}{P_j(\delta_1, \ldots, \delta_j)}, \qquad j = 0, \ldots, i.$$

and this will be proved by induction.

From $\dfrac{1}{\alpha_{i+2}} = \delta_{i+1} - \alpha_{i+1}$ we deduce

$$\begin{aligned}\alpha_{i+2} &= \cfrac{1}{\delta_{i+1} - \cfrac{P_i(\delta_0, \ldots, \delta_{i-1}) - \alpha_0 P_{i-1}(\delta_1, \ldots, \delta_{i-1})}{P_{i+1}(\delta_0, \ldots, \delta_i) - \alpha_0 P_i(\delta_1, \ldots, \delta_i)}} \\ &= \frac{P_{i+1}(\delta_0, \ldots, \delta_i) - \alpha_0 P_i(\delta_1, \ldots, \delta_i)}{P_{i+2}(\delta_0, \ldots, \delta_{i+1}) - \alpha_0 P_{i+1}(\delta_1, \ldots, \delta_{i+1})}\end{aligned}$$

and $\alpha_0 \neq \dfrac{P_{i+2}(\delta_0, \ldots, \delta_{i+1})}{P_{i+1}(\delta_1, \ldots, \delta_{i+1})}$.

For $i = p - 1$ we have

$$\alpha_0 = \frac{\alpha_0 P_{p-2}(\delta_1, \ldots, \delta_{p-2} - P_{p-1}(\delta_0, \ldots, \delta_{p-2})}{\alpha_0 P_{p-1}(\delta_1, \ldots, \delta_{p-1}) - P_p(\delta_0, \ldots, \delta_{p-1})}$$

which gives the announced degree two equation satisfied by α_0.

The similar formulas with α_j are clear. Since $a_j(\delta) = c_{j+1}(\delta)$, j mod p,

$$\prod_{j=0}^{p-1} a_j(\delta) = \prod_{j=0}^{p-1} c_j(\delta)$$

and solution of the system (E_p) exist if and only if $c_j(\delta) \neq 0$ for $j = 0, \ldots, p-1$. □

Lemma 2.8 *The polynomial discriminant*

$$\begin{aligned}\Delta_p(X_0, \ldots, X_{p-1}) := \big(P_p(X_0, \ldots, X_{p-1}) + P_{p-2}(X_1, \ldots, X_{p-2})\big)^2 \\ -4P_{p-1}(X_0, \ldots, X_{p-2})P_{p-1}(X_1, \ldots, X_{p-1})\end{aligned}$$

satisfies the equality

$$\Delta_p(X_0, \ldots, X_{p-1}) = \big(P_p(X_0, \ldots, X_{p-1}) - P_{p-2}(X_1, \ldots, X_{p-2})\big)^2 - 4$$

and is invariant by circular permutation.

Proof

(1) Let $0 \leq i \leq p-1$ and

$$\begin{aligned}A_p(X_i, \ldots, X_{p-1}, X_0, \ldots, X_{i-1}) := {} & P_p(X_i, \ldots, X_{i-1})P_{p-2}(X_{i+1}, \ldots, X_{i-2})\\ & -P_{p-1}(X_i, \ldots, X_{i-2})\\ & \times P_{p-1}(X_{i+1}, \ldots, X_{i-1}).\end{aligned}$$

Developing from both sides, thanks to the definition of these polynomials and Lemma 2.2, we have

$$\begin{aligned}
& A_p(X_i, \ldots, X_{i-1})\\
&\quad = \Big\{X_{i-1}X_i P_{p-2}(X_{i+1}, \ldots, X_{i-2})^2 - X_i P_{p-3}(X_{i+1}, \ldots, X_{i-3})\\
&\qquad \times P_{p-2}(X_{i+1}, \ldots, X_{i-2}) - P_{p-2}(X_{i+2}, \ldots, X_{i-1})\\
&\qquad \times P_{p-2}(X_{i+1}, \ldots, X_{i-2})\Big\}\\
&\quad - \Big\{X_{i-1}X_i P_{p-2}(X_{i+1} \ldots, X_{i-2})^2 - X_i P_{p-3}(X_{i+1}, \ldots, X_{i-3})\\
&\qquad \times P_{p-2}(X_{i+1}, \ldots, X_{i-2}) - X_{i-1} P_{p-2}(X_{i+1}, \ldots, X_{i-2})\\
&\qquad \times P_{p-3}(X_{i+2}, \ldots, X_{i-2})\\
&\qquad + P_{p-3}(X_{i+2}, \ldots, X_{i-2})P_{p-3}(X_{i+1}, \ldots, X_{i-3})\Big\}\\
&\quad = \Big[X_{i-1}P_{p-2}(X_{i+1}, \ldots, X_{i-2}) - P_{p-3}(X_{i+1}, \ldots, X_{i-3})\Big]\\
&\qquad \times P_{p-3}(X_{i+2}, \ldots, X_{i-2})\\
&\qquad - P_{p-2}(X_{i+1}, \ldots, X_{i-2})P_{p-2}(X_{i+2}, \ldots, X_{i-1})\\
&\quad = P_{p-1}(X_{i+1}, \ldots, X_{i-1})P_{p-3}(X_{i+2}, \ldots, X_{i-2})\\
&\qquad - P_{p-2}(X_{i+1}, \ldots, X_{i-2})P_{p-2}(X_{i+2}, \ldots, X_{i-1})\\
&\quad = A_{p-1}(X_{i+1}, \ldots, X_{i-1})
\end{aligned}$$

Therefore,

$$\begin{aligned}A_p(X_i, \ldots, X_{i-1}) &= A_{p-1}(X_{i+1}, \ldots, X_{i-1}) = \cdots = A_2(X_{i-2}, X_{i-1})\\ &= P_2(X_{i-2}, X_{i-1})P_0 - P_1(X_{i-2})P_1(X_{i-1}) = -1.\end{aligned}$$

(2) We have by (1),

$$\Delta_p(X_0,\dots,X_{p-1}) = \big(P_p(X_0,\dots,X_{p-1}) - P_{p-2}(X_1,\dots,X_{p-2})\big)^2 - 4$$

By Lemma 2.3, Δ_p is invariant by cyclic permutation.

□

Lemma 2.9 *For $q=(q_0,\dots,q_{j-1})$ where $q_i=0$ or $q_i=1$, $i=0,\dots,j-1$, the partial derivatives of the polynomials P_j satisfy the relation*

$$\frac{\partial^{|q|}P_j(X_0,\dots,X_{j-1})}{\partial^{q_0}X_0\cdots\partial^{q_{p-1}}X_{j-1}} = P_{j-|q|}(X_0^{\epsilon_0},\dots,X_{j-1}^{\epsilon_{j-1}})$$

where

$$\epsilon_i = \begin{cases} 0 \text{ if } q_i = 1 \\ 1 \text{ if } q_i = 0 \end{cases}.$$

The condition $\epsilon_i = 0$ means that X_i is omitted.

In particular, by Lemma 2.5, for $q^1=(q_0^1,\dots,q_{p-1}^1)$ and $q^2=(q_0^2,\dots,q_{p-1}^2)$ such that $|q^1|=|q^2|$,

$$\frac{\partial^{|q^1|}P_j(X_0,\dots,X_{p-1})}{\partial^{q_0^1}X_0\cdots\partial^{q_{p-1}^1}X_{p-1}}(2,\dots,2) = \frac{\partial^{|q^2|}P_j(X_0,\dots,X_{p-1})}{\partial^{q_0^2}X_0\cdots\partial^{q_{p-1}^2}X_{p-1}}(2,\dots,2)$$
$$= P_{j-|q_1|}(2,\dots,2) = j - |q_1| + 1.$$

Proof By induction on $j\geq 1$ for $q=(q_0,\dots,q_{j-1})$ such that $|q|\leq j$.

The relation is trivial for $j=1$. Suppose it is true for an integer $j\geq 1$, then by induction hypothesis,

$$\frac{\partial^{|q|}P_{j+1}(X_0,\dots,X_j)}{\partial^{q_0}X_0\cdots\partial^{q_j}X_j}$$
$$= \frac{\partial^{|q|}}{\partial^{q_0}X_0\cdots\partial^{q_j}X_j}\big(X_0P_j(X_1,\dots,X_j) - P_{j-1}(X_2,\dots,X_j)\big)$$
$$= \begin{cases} \dfrac{\partial^{|q|-1}P_j(X_1,\dots,X_j)}{\partial^{q_1}X_1\cdots\partial^{q_j}X_j} = P_{j-|q|+1}(X_1^{\epsilon_1},\dots,X_j^{\epsilon_j}) \text{ if } q_0 = 1 \\[2ex] X_0\dfrac{\partial^{|q|}P_j(X_1,\dots,X_j)}{\partial^{q_1}X_1\cdots\partial^{q_j}X_j} - \dfrac{\partial^{|q|}P_{j-1}(X_1,\dots,X_j)}{\partial^{q_1}X_1\cdots\partial^{q_j}X_j} \quad \text{if } q_0 = 0 \end{cases}$$
$$= P_{j-|q|+1}(X_0^{\epsilon_0},X_1^{\epsilon_1},\dots,X_j^{\epsilon_j}).$$

□

2.3 Geometric Part of the Proof

If the surface S admits a twisted logarithmic 1-form it is known [3, lemma 6.12] that S contains a cycle of rational curves. Moreover by [9] extra curves are composed of branches (called also trees), each curve of the cycle meeting at most one branch.

Lemma 2.10 *The foliation $\mathcal{F}$ associated to a non trivial logarithmic 1-form*

$$\theta \in H^0(S, \Omega^1(log D) \otimes \mathcal{L}_\lambda)$$

has no other singularities on D than the intersection points of rational curves.

Proof Let p be a regular point of D. Let $\Pi : \hat{S} \to S$ be the cyclic covering and we set $\hat{D} = \Pi^{-1}(D), \hat{\alpha} := \Pi^\star\alpha, \hat{\alpha} = d\hat{h}, \hat{\theta} = \Pi^\star\theta$. In a local chart in a neighbourhood of $p \in D_j = \{z_2 = 0\}$,

$$e^{\hat{h}}\hat{\theta} = \theta_0 + g_2(z)\frac{dz_2}{z_2}$$

where θ_0 and g_2 are holomorphic. Let γ_1 be a closed curve in $\{z_1 = c\}$ around $(c, 0)$. Since $e^{\hat{h}}\hat{\theta}$ is closed in $\hat{S} \setminus \hat{D}$,

$$\int_{\gamma_1} e^{\hat{h}}\hat{\theta} = \int_{\gamma_1} g_2(z)\frac{dz_2}{z_2} = 2\pi i Res_{z_2=0}(e^{\hat{h}}\hat{\theta}_{|z_1=c}) = 2\pi i g_2(c, 0)$$

does not depend on c, therefore g_2 depends only on z_2. Hence

$$e^{\hat{h}}\hat{\theta} = \theta_0' + g_2(0)\frac{dz_2}{z_2}$$

for θ_0' holomorphic and $g_2(0) \neq 0$. As $\mathcal{F}$ is also defined by the holomorphic form

$$\omega = z_2\theta_0' + g_2(0)dz_2$$

which has no singularity at $p = (0, 0)$, p is a regular point of $\mathcal{F}$. □

Lemma 2.11 *Let S be in class VII_0^+ endowed with a twisted logarithmic 1-form $\theta \in H^0(S, \Omega^1(log(D) \otimes \mathcal{L}_\lambda)$ with a cycle of rational curves $\Gamma = D_0 + \cdots + D_{p-1}$, and denote by $\mathcal{F}$ the associated foliation. Let $\alpha = (\alpha_0, \ldots, \alpha_{p-1})$ be a real solution of the system*

$$(E_p) \begin{cases} \alpha_0 + \dfrac{1}{\alpha_1} & = \delta_0 \\ \vdots & \vdots \\ \alpha_i + \dfrac{1}{\alpha_{i+1}} & = \delta_i \\ \vdots & \vdots \\ \alpha_{p-1} + \dfrac{1}{\alpha_0} & = \delta_{p-1} \end{cases}$$

where $\alpha_i = -CS(\mathcal{F}, D_i, D_i \cap D_{i-1})$, $i \bmod p$. *If* $A_j = C_0 + \cdots + C_{k-1}$ *is a chain meeting the cycle* Γ *at a point* $A_j \cap D_j = C_{k-1} \cap D_j$ *then*

$$-\frac{k}{k+1} \leq CS(\mathcal{F}, D_j, D_j \cap C_{k-1}) \leq \frac{1}{C_{k-1}^2}.$$

For indices i corresponding of curves of self-intersection -2 *and with a tree, we have* $1 < \delta_i < 2$. *For all other indices* $\delta_i \geq 2$.

Proof Let $A_j = C_0 + \cdots + C_{k-1}$, a chain, where C_0 is its top, which meets the curve D_j of the cycle, By Lemma 2.10, the intersection points of the rational curves are the only singularities of $\mathcal{F}$, therefore Camacho-Sad formula for C_0 is

$$CS(\mathcal{F}, C_0, C_0 \cap C_1) = C_0^2 \leq -2,$$

with notations of [3, lemma 6.14]. By induction we shall show that $CS(\mathcal{F}, C_i, C_i \cap C_{i+1})$ is a negative rational which satisfies

$$C_i^2 \leq CS(\mathcal{F}, C_i, C_i \cap C_{i+1}) \leq -\frac{i+2}{i+1}.$$

In fact, it is true for $i = 0$. Camacho-Sad formula for C_{i+1} is

$$C_{i+1}^2 = CS(\mathcal{F}, C_{i+1}, C_{i+1} \cap C_i) + CS(\mathcal{F}, C_{i+1}, C_{i+1} \cap C_{i+2}).$$

We recall that $CS(\mathcal{F}, C_{i+1}, C_i \cap C_{i+1}) = CS(\mathcal{F}, C_i, C_i \cap C_{i+1})^{-1}$. By induction hypothesis,

$$C_{i+1}^2 \leq CS(\mathcal{F}, C_{i+1}, C_{i+1} \cap C_{i+2}) = C_{i+1}^2 - CS(\mathcal{F}, C_{i+1}, C_{i+1} \cap C_i)$$

$$\leq -2 + \frac{i+1}{i+2} = -\frac{i+3}{i+2}$$

For $i = k-1$, we obtain

$$C_{k-1}^2 \leq CS(\mathcal{F}, C_{k-1}, C_{k-1} \cap D_j) \leq -\frac{k+1}{k}$$

hence

$$-\frac{k}{k+1} \leq CS(\mathcal{F}, D_j, C_{k-1} \cap D_j) \leq \frac{1}{C_{k-1}^2} < 0.$$

□

Corollary 2.12 *Let S be a surface in class VII_0^+ endowed with a twisted logarithmic 1-form $\theta \in H^0(S, \Omega^1(\log D)\otimes \mathcal{L}_\lambda)$. If each tree meets a curve with self-intersection at most -3 all Camacho-Sad indices are negative.*

Proof There exists a non-trivial twisted logarithmic 1-form, therefore S contains a cycle of rational curves, hence the maximal divisor of S is a cycle with chains of rational curves (called trees), each chain meeting the cycle at exactly one point and each curve of the cycle meets at most one chain (see [9]). By Lemma 2.11 and Camacho-Sad formula,

$$D_j^2 = CS(\mathcal{F}, D_j, D_j\cap D_{j-1}) + CS(\mathcal{F}, D_j, D_j\cap D_{j+1}) + CS(\mathcal{F}, D_j, D_j\cap C_{k-1})$$

with $0 < -CS(\mathcal{F}, D_j, D_j \cap C_{k-1}) < 1$, therefore, for $\alpha_j = -CS(\mathcal{F}, D_j, D_j \cap D_{j-1})$

$$\alpha_j + \frac{1}{\alpha_{j+1}} = \delta_j = -D_j^2 + CS(\mathcal{F}, D_j, D_j\cap C_{k-1}) > 2.$$

as self-intersections of the curves are less than -2. We conclude by Lemma 2.5 that coefficients $a_j(\delta) > 0$, $-b_j(\delta) > 0$, $c_j(\delta) > 0$. Since product and sum of roots are positive, roots are positive. □

Lemma 2.13 *Let $p \geq 1$ and $x_0, \dots, x_{p-1} \in \mathbb{R}$. Then we have the identity*

$$\sum_{j=0}^{p}(p+1-j)\sum_{q\mid |q|=j} x_0^{q_0}\cdots x_{p-1}^{q_{p-1}} = \prod_{i=0}^{p-1}(1+x_i) + \sum_{k=0}^{p-1}\prod_{i\neq k}(1+x_i).$$

Proof The relation is clear for $p = 1$. If it is true for an integer $p \geq 1$ we have

$$\begin{aligned}
&\sum_{j=0}^{p+1}(p+2-j)\sum_{q\mid |q|=j} x_0^{q_0}\cdots x_p^{q_p}\\
&= \sum_{j=0}^{p+1}\sum_{q\mid |q|=j} x_0^{q_0}\cdots x_p^{q_p} + \sum_{j=0}^{p}(p+1-j)\sum_{q\mid |q|=j} x_0^{q_0}\cdots x_p^{q_p}\\
&= \prod_{j=0}^{p}(1+x_j) + \sum_{j=0}^{p}(p+1-j)\Big(\sum_{\substack{|q|=j\\ q_p=1}} x_0^{q_0}\cdots x_p^{q_p} + \sum_{\substack{|q|=j\\ q_p=0}} x_0^{q_0}\cdots x_{p-1}^{q_{p-1}}\Big)\\
&= \prod_{j=0}^{p}(1+x_j) + x_p\sum_{j=1}^{p}(p+1-j)\sum_{|q'|=j-1} x_0^{q_0}\cdots x_{p-1}^{q_{p-1}}
\end{aligned}$$

$$
\begin{aligned}
&+\sum_{j=0}^{p}(p+1-j)\sum_{|q'|=j} x_0^{q_0}\cdots x_{p-1}^{q_{p-1}}\\
&=\prod_{j=0}^{p}(1+x_j)+x_p\sum_{j'=0}^{p-1}(p-j')\sum_{|q'|=j'} x_0^{q_0}\cdots x_{p-1}^{q_{p-1}}\\
&+\prod_{j=0}^{p-1}(1+x_j)+\sum_{k=0}^{p-1}\prod_{i\neq k}(1+x_i)\\
&=\prod_{j=0}^{p}(1+x_j)+x_p\sum_{j'=0}^{p}(p+1-j')\sum_{|q'|=j'} x_0^{q_0}\cdots x_{p-1}^{q_{p-1}}-x_p\sum_{|q'|\leq p} x_0^{q_0}\cdots x_{p-1}^{q_{p-1}}\\
&+\prod_{j=0}^{p-1}(1+x_j)+\sum_{k=0}^{p-1}\prod_{i\neq k}(1+x_i)\\
&=\prod_{j=0}^{p}(1+x_j)+x_p\left\{\prod_{j=0}^{p-1}(1+x_j)+\sum_{k=0}^{p-1}\prod_{i\neq k}(1+x_i)\right\}+\sum_{k=0}^{p-1}\prod_{i\neq k}(1+x_i)\\
&=\prod_{j=0}^{p}(1+x_j)+(1+x_p)\sum_{k=0}^{p-1}\prod_{i\neq k}(1+x_i)+\prod_{j=0}^{p-1}(1+x_j)\\
&=\prod_{j=0}^{p}(1+x_j)+\sum_{k=0}^{p}\prod_{i\neq k}(1+x_i).
\end{aligned}
$$

□

Lemma 2.14 *Let $\delta=(\delta_0,\ldots,\delta_{p-1})$ be the second member of a system (E_p) which corresponds to a surface $S\in VII_0$ with a cycle of rational curves, in particular admits real solutions.*

Then for any index $0\leq i\leq p-1$,

$$
P_p(\delta_i,\ldots,\delta_{i-1})-P_{p-2}(\delta_{i+1},\ldots,\delta_{i-2})>0,
$$

$$
P_{p-1}(\delta_i,\ldots,\delta_{i-2})>0
$$

Proof By Lemma 2.11, $\delta_i>1$ for $i=0,\ldots,p-1$. The highest possible power of an indeterminacy in a monomial is one. Therefore the only non vanishing partial derivatives

$$
\frac{\partial^{|q|}\big(P_p(X_0,\ldots,X_{p-1})-P_{p-2}(X_1,\ldots,X_{p-2})\big)}{\partial^{q_0}X_0\cdots\partial^{q_{p-1}}X_{p-1}}
$$

are those for which $q_i = 0$, or $q_i = 1$, $i = 0, \dots, p-1$, where $|q| = \sum_{i=0}^{p-1} q_i$. Besides, for

$$Q_p(X_0, \dots, X_{p-1}) := P_p(X_0, \dots, X_{p-1}) - P_{p-2}(X_1, \dots, X_{p-2})$$

and fixed $|q|$, $0 \le |q| \le p$, we have by Lemma 2.5,

$$\frac{\partial^{|q|} Q_p(X_0, \dots, X_{p-1})}{\partial^{q_0} X_0 \cdots \partial^{q_{p-1}} X_{p-1}}(2, \dots, 2) = P_{p-|q|}(2, \dots, 2) - P_{p-2-|q|}(2, \dots, 2) = 2.$$

By Taylor formula for polynomials,

$$\begin{aligned}
&Q_p(\delta_0, \dots, \delta_{p-1}) \\
&\quad = \sum_{0 \le j \le p} \sum_{q, |q|=j} (\delta_0 - 2)^{q_0} \cdots (\delta_{p-1} - 2)^{q_{p-1}} \frac{\partial^{|q|} Q_p(X_0, \dots, X_{p-1})}{\partial^{q_0} X_0 \cdots \partial^{q_{p-1}} X_{p-1}}(2, \dots, 2) \\
&\quad = \sum_{0 \le j \le p} \sum_{q, |q|=j} (\delta_0 - 2)^{q_0} \cdots (\delta_{p-1} - 2)^{q_{p-1}} Q_{p-|q|}(2, \dots, 2) \\
&\quad = 2 \sum_{0 \le j \le p} \sum_{q, |q|=j} (\delta_0 - 2)^{q_0} \cdots (\delta_{p-1} - 2)^{q_{p-1}} \\
&\quad = 2 \prod_{j=0}^{p-1} \big(1 + (\delta_j - 2)\big) = 2 \prod_{j=0}^{p-1} (\delta_j - 1) > 0.
\end{aligned}$$

Hence

$$P_p(\delta_i, \dots, \delta_{i-1}) - P_{p-2}(\delta_{i+1}, \dots, \delta_{i-2}) > 0.$$

In a similar way, for $\delta = (\delta_0, \dots, \delta_{p-1})$, $\delta_i > 1$, $i = 0, \dots, p-1$,

$$\begin{aligned}
P_{p-1}(\delta_0, \dots, \delta_{p-2}) &= \sum_{0 \le j \le p-1} \sum_{q, |q|=j} (\delta_0 - 2)^{q_0} \cdots (\delta_{p-1} - 2)^{q_{p-1}} \\
&\quad \times \frac{\partial^{|q|} P_{p-1}(X_0, \dots, X_{p-2})}{\partial^{q_0} X_0 \cdots \partial^{q_{p-1}} X_{p-1}}(2, \dots, 2) \\
&= \sum_{0 \le j \le p-1} (p-j) \sum_{|q|=j} (\delta_0 - 2)^{q_0} \cdots (\delta_{p-2} - 2)^{q_{p-2}} \\
&= \prod_{i=0}^{p-2} (\delta_i - 1) + \sum_{k=0}^{p-2} \prod_{i \ne k} (\delta_i - 1) > 0
\end{aligned}$$

We have the result by circular permutation.

□

Proposition 2.15 *Let* $\alpha = (\alpha_0, \ldots, \alpha_{p-1})$ *be a real solution of the system*

$$(E_p)\begin{cases} \alpha_0 + \dfrac{1}{\alpha_1} & = \delta_0 \\ \quad\vdots & \quad\vdots \\ \alpha_i + \dfrac{1}{\alpha_{i+1}} & = \delta_i \\ \quad\vdots & \quad\vdots \\ \alpha_{p-1} + \dfrac{1}{\alpha_0} & = \delta_{p-1} \end{cases}$$

where $\alpha_i \neq 0$ *and* $\delta_i > 1$ *are real, for* $i = 0, \ldots, p-1$. *If the system* (E_p) *is associated to a surface S in class* VII_0^+ *endowed with a twisted logarithmic* 1*-form then for* $i = 0, \ldots, p-1$, α_i *satisfies a degree two equation*

$$a_i(\delta)\alpha_i^2 + b_i(\delta)\alpha_i + c_i(\delta) = 0$$

where

$$\frac{c_i(\delta)}{a_i(\delta)} > 0 \quad \text{and} \quad -\frac{b_i(\delta)}{a_i(\delta)} > 0.$$

In particular α_j *are positive and Camacho-Sad indices are negative.*

Proof The discriminant of all the equations is by 2.8

$$\Delta_p(X_0, \ldots, X_{p-1}) = \big(P_p(X_0, \ldots, X_{p-1}) - P_{p-2}(X_1, \ldots, X_{p-2})\big)^2 - 4$$

Camacho-Sad indices are real by lemma 6.17 of [3] therefore $\Delta_p(\delta) \geq 0$, therefore

$$|P_p(\delta) - P_{p-2}(\delta)| \geq 2.$$

As by Lemma 2.14, $P_p(\delta) - P_{p-2}(\delta) > 0$, we have

$$P_p(\delta) - P_{p-2}(\delta) \geq 2$$

From the expressions of a_j, b_j and c_j, $a_j(\delta) > 0$, $-b_j(\delta) > 0$ and $c_j(\delta) > 0$, the roots are real, their sum and product are positive hence α_i are positive and Camacho-Sad indices are negative. □

Lemma 2.16 *Let* $\theta \in H^0(S, \Omega^1(\log D) \otimes \mathcal{L}_\mu)$. *Denote by* $\Gamma = D_0 + \cdots + D_{p-1}$ *the cycle in S. If* $\alpha_i = -CS(\mathcal{F}, D_i, D_i \cap D_{i+1})$, *i mod p, then*

$$\mu = \prod_{i=0}^{p-1} \alpha_i > 0.$$

Proof Denote $a_i := \int_{\gamma_i} e^{\hat{h}}\hat{\theta}$ where γ_i is a closed curve around a lift of D_i in $\hat{S}$. Let $\hat{g} : \hat{S} \to \hat{S}$ be a generator of the automorphism group of $Aut_S(\hat{S})$ and $\gamma_q = \hat{g} \circ \gamma_0$, we have

$$\int_{\gamma_q} e^{\hat{h}}\hat{\theta} = \int_{\gamma_0} \hat{g}^\star e^{\hat{h}}\hat{\theta} = \mu \int_{\gamma_0} e^{\hat{h}}\hat{\theta}$$

hence

$$\begin{aligned}\mu &= \frac{\int_{\gamma_q} e^{\hat{h}}\hat{\theta}}{\int_{\gamma_0} e^{\hat{h}}\hat{\theta}} = \frac{\int_{\gamma_q} e^{\hat{h}}\hat{\theta}}{\int_{\gamma_{q-1}} e^{\hat{h}}\hat{\theta}} \cdots \frac{\int_{\gamma_i} e^{\hat{h}}\hat{\theta}}{\int_{\gamma_{i-1}} e^{\hat{h}}\hat{\theta}} \cdots \frac{\int_{\gamma_1} e^{\hat{h}}\hat{\theta}}{\int_{\gamma_0} e^{\hat{h}}\hat{\theta}} = \frac{a_q}{a_{q-1}} \cdots \frac{a_i}{a_{i-1}} \cdots \frac{a_1}{a_0} \\ &= \prod_{i=0}^{q-1} \alpha_i > 0\end{aligned}$$

By Proposition 2.15, $\alpha_i > 0$ for $i = 0, \ldots, q-1$. □

The following theorem summarises the previous results:

Theorem 2.17 *Let S be a surface in the class* VII_0^+, *endowed with a non-trivial logarithmic* 1*-form* $0 \neq \theta \in H^0(S, \Omega^1(logD) \otimes \mathcal{L}_\mu)$ *with poles along a non-trivial divisor D. Then*

- *All Camacho-Sad indices of the associated foliation* $\mathcal{F}$ *are negative,*
- *The coefficient* μ *is a positive real number,*
- *We have* $\mu \geq 1$ *for a suitable choice of the generator of* $Aut_S(\hat{S})$ *and* $\mu = 1$ *if and only if S is a Enoki surface.*

Proof Proposition 2.15 and Lemma 2.16. □

Remark 2.18 If $\delta = (2, \ldots 2)$ discriminants vanish therefore $\prod \alpha_i = \prod \frac{1}{\alpha_i}$ hence $\mu = \mu^{-1} = 1$ we have a non-trivial logarithmic 1-form $\theta \in H^0(S, \Omega^1(\log D))$. We derive that S is a Enoki surface.

By Lemma 2.16, Lemma 6.18 of [3] is improved: the hypothesis "characteristic numbers at the intersection of curves are negative" is superfluous. More precisely

Proposition 2.19 *Let S be a minimal surface in class* VII_0^+, *which is neither Enoki nor Inoue-Hirzebruch, endowed with a non trivial logarithmic* 1*-form*

$$0 \neq \theta \in H^0(S, \Omega^1(logD) \otimes \mathcal{L}_\alpha), [\alpha] \in H^1_{dR}(S, \mathbb{R}),$$

then there exists a degree zero current τ *such that*

$$d_\alpha d^c_\alpha \tau = 2\pi \, Res_\alpha(\theta) T_D > 0.$$

We have therefore on the cyclic covering $\hat{S}$ *of S a PSH function* $\hat{\tau}$, *PH outside* $\hat{D}$.

3 Torsion of Intermediate Kato Surfaces

Intermediate Kato surfaces are surfaces containing a global spherical shell (GSS) whose maximal divisor D is a cycle of rational curves with at least one tree (or branch). See [7, 10] for their properties. A contraction associated to these surfaces has the normal form

$$F(z_1, z_2) = (*, z_2^k),$$

where $k = k(S) \geq 2$ is an integer. This integer gives the twisting line bundle of a logarithmic 1-form

$$\theta \in H^0(S, \Omega^1(\log D) \otimes \mathcal{L}^k)$$

and the automorphy constant of a PSH function $\tilde{\tau}(z_1, z_2) = \log|z_2|$.

Besides, there is a unique smaller integer $m \geq 1$ such that there exists a numerically m-anticanonical divisor [8] i.e. a divisor D_m such that

$$mK_S + F + [D_m] = 0$$

for a flat line F. Equivalently $m \geq 1$ is the least integer for which exists $\kappa \in \mathbb{C}^\star$ such that $H^0(S, K_S^{-m} \otimes \mathcal{L}_\kappa) \neq 0$.

In the following proposition we give a construction which associates to any intermediate Kato surface S another Kato surface S' of index one. The aim is to show that they have the same torsion

$$k(S) = k(S').$$

In the following proposition "special" means that the surface S has $b_2(S)$ rational curves. In [12] it is proved that a surface is special if and only if it is a Kato surface.

Proposition 3.1 ([12, p. 286]) *Let S be a special surface of intermediate type with index $m := m(S) > 1$. Then there exists a diagram*

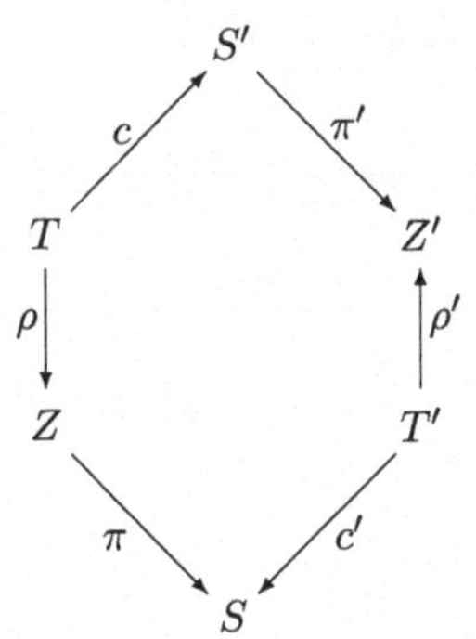

where

(1) (Z, π, S) *is a* m*-fold cyclic ramified covering space of* S*, branched over* D*, endowed with an automorphism group isomorphic to* $\mathbb{U}_m$ *which acts transitively on the fibers,*
(2) (T, ρ, Z) *is the minimal desingularization of* Z*,*
(3) (T, c, S') *is the contraction of the (possible) exceptional curves of the first kind,*
(4) S' *is a special surface with* $b_2(S') > 0$ *with action of the group of* m*-th roots of unity* $\mathbb{U}_m$*, with index* $m(S') = 1$*,*
(5) (S', π', Z') *is the quotient space of* S' *by* $\mathbb{U}_m$*,*
(6) (T', ρ', Z') *is the minimal desingularization of* Z'*,*
(7) (T', c', S) *is the contraction of the (possible) exceptional curves of the first kind,*

such that the restriction over $S \setminus D$ *is commutative, i.e.*

$$\theta := \pi \circ \rho \circ c^{-1} = c' \circ \rho'^{-1} \circ \pi' : S' \setminus D' \to S \setminus D$$

and $(S' \setminus D', \theta, S \setminus D)$ *is a* m*-fold non ramified covering. Moreover*

- S' *has a GSS if and only if* S *has a GSS,*
- *The maximal divisors* D *and* D' *of* S *and* S' *respectively have the same number of cycles and branches.*

We have the following proposition of Oeljeklaus-Toma in [18, pp. 330–331]:

Proposition 3.2 ([18]) *The two intermediate Kato surfaces* S *and* S' *of the diagram of Proposition 3.1 have the same torsion of logarithmic* 1*-forms i.e.*

$$k(S) = k(S').$$

Proof Following the diagram of Proposition 3.1

$$c' \circ \rho'^{-1} \circ \pi' : S' \setminus D' \to S \setminus D$$

is a Galois covering with fiber $\mathbb{Z}/m\mathbb{Z}$. By Remark 4.4. in [18] every surface of index $m \geq 2$ is obtained in this way. Recall that any intermediate Kato surface S is defined by a contracting germ [14] in normal form

$$F(z, \zeta) = (\lambda \zeta^s z + P(\zeta) + c\, \zeta^{\frac{sk}{k-1}}, \zeta^k)$$

where

$$k, s \in \mathbb{Z}, \quad k > 1, \quad s > 0, \quad \lambda \in \mathbb{C}^\star,$$

P is a complex polynomial,

$$P(\zeta) = c_j \zeta^j + c_{j+1}\zeta^{j+1} + \cdots + c_s \zeta^s,$$
$$0 < j < k, \quad j \le s, \quad c_j = 1, \quad c = c_{\frac{sk}{k-1}} \in \mathbb{C}$$

with $c = 0$ whenever $\frac{sk}{k-1} \notin \mathbb{Z}$ or $\lambda \neq 1$ and $gcd\{k, p \mid c_p \neq 0\} = 1$.

Moreover (Remark 4.3 [18]) intermediate surfaces of index one are exactly those for which $(k-1) \mid s$. Therefore S of index $m \geq 2$ is associated to a germ

$$F(z, \zeta) = (\lambda \zeta^s z + \sum_{p=j}^{s} c_p \zeta^p, \zeta^k)$$

with $(k-1) \nmid s$. For a positive divisor q of $k-1$ such that $k-1$ divides qs, the authors define a new contracting germ in the following way: set

$$r' := \lfloor \frac{qj}{k} \rfloor, \quad s' := qs - r'(k-1), \quad j' = qj - r'k, \quad P'(\zeta) := \sum_{p=j'}^{s'} c_p \zeta^{qp - r'k}$$

then

$$F'(z, \zeta) := (\lambda \zeta^{s'} + P'(\zeta), \zeta^k)$$

is associated to a surface of index one with action of $\mathbb{Z}/q\mathbb{Z}$ given by $(z, \zeta) \mapsto (\epsilon^{-r} z, \epsilon \zeta)$, where ϵ is a primitive root of unity of order q.

As the index of an intermediate surface is given by [18, remark 4.5]

$$m = \frac{k-1}{gcd\{k-1, s\}}$$

we may take $q = m$ and obtain the expression of a germ associated to S'. We notice that the second members of contracting germ F and F' are equal, therefore $k(S) = k(S')$. □

4 Surfaces in Class VII_0^+ with Twisted Logarithmic 1-Form

Example 4.1 Let S be a Inoue-Hirzebruch surface with maximal divisor $D = \Gamma_+ + \Gamma_-$ composed of two cycles of rational curves. The surface S is defined by a contracting germ $F(z_1, z_2) = (z_1^p z_2^q, z_1^r z_2^s)$ (see [10] for details). The matrix $\begin{pmatrix} p & q \\ r & s \end{pmatrix}$ has two eigenvalues λ_1 and λ_2 which satisfy $\lambda_1 > 1$ and $0 < |\lambda_2| < 1$ with eigenvectors $x_1 = \begin{pmatrix} a_1 \\ b_1 \end{pmatrix}$ and $x_2 = \begin{pmatrix} a_2 \\ b_2 \end{pmatrix}$ respectively such that, a_1, b_1 are of the same sign and a_2, b_2 of opposite sign, say $a_1 > 0, b_1 > 0, a_2 > 0, b_2 < 0$. We have then for each curve C in the universal covering space $\tilde{S} = \hat{S}$ two functions

$$G_{C,i}(z) = a_i \log|z_1| + b_i \log|z_2|, \quad i = 1, 2.$$

Recall that the inverse image of D in $\tilde{S}$ gives two infinite chains of rational curves $\tilde{D} = \tilde{\Gamma}_+ + \tilde{\Gamma}_-$. The function $G_{C,1}$ yields a PSH function $\hat{u}_1$ on $\tilde{S}$ with polar set $\tilde{D}$, PH outside $\tilde{D}$; the function $G_{C,2}$ yields a PSH function $\hat{u}_2$ outside $\tilde{D}$ with value $-\infty$ on one chain, say $\tilde{\Gamma}_-$, and $+\infty$ of the other $\tilde{\Gamma}_+$. In the first situation $\hat{u}_1 < 0$ is negative, however $\hat{u}_2$ takes all values of $[-\infty, +\infty]$, in particular there is a Levi flat real hypersurface $\hat{H} = \{\hat{u}_2 = 0\}$ which splits $\tilde{S}$ into two parts and gives a decomposition of $\tilde{S}$ into

$$\tilde{S} = \{\hat{u}_2 = -\infty\} \cup \{\hat{u}_2 < 0\} \cup \tilde{H} \cup \{\hat{u}_2 > 0\} \cup \{\hat{u}_2 = +\infty\}.$$

As all these parts of the decomposition are invariant by the automorphism of the covering we derive a decomposition

$$S = \Gamma_- \cup S_- \cup H \cup S_+ \cup \Gamma_+.$$

The current $dd^c\hat{u}_1 \geq 0$ on $\tilde{S}$ and $dd^c\hat{u}_2 \geq 0$ on $\tilde{S} \setminus \tilde{\Gamma}_+$.

Suppose there exists a locally integrable function τ such that $d_\beta d^c_\beta \tau \geq 0$, $\hat{u} = e^{-\hat{f}}\hat{\tau}$ is a multiplicatively automorphic PSH function $\hat{u} : \hat{S} \to \mathbb{R} \cup \{-\infty\}$ with $\hat{D}$ as polar set, PH outside $\hat{D}$ and multiplicative constant $\mu > 1$. We want to prove that $\hat{u}$ is negative.

We denote by $\mathcal{T}(S)$ the subset of $H^1_{dR}(S, \mathbb{R})$ of classes of Lee forms of locally conformally symplectic forms taming the complex structure J

$$\mathcal{T}(S) = \{a \in H^1_{dR}(S, \mathbb{R}) \mid \exists \omega \in \Omega^2(S), \omega^{1,1} > 0, d_\alpha \omega = 0\}.$$

By [3], $\mathcal{T}(S)$ is represented by an open interval contained in $(-\infty, 0)$. Recall that given $\alpha \in a$ and $\hat{f}$ on $\hat{S}$ such that $\hat{\alpha} = d\hat{f}$ we have $\hat{f} \circ \gamma - \hat{f} = c < 0$ for the suitable choice of generator of $Aut_S(\hat{S})$. The constant c depends only on the class a. Denote by $\varphi :]-\infty, 0[\to \mathbb{R}_+$ the decreasing mapping $a \mapsto C = e^{-c}$ which sends $]-\infty, 0]$ onto $[1, +\infty[$.

Recall that by lemma 6.11 of [3] (see also the remark 5.9 in [2]) S in the class VII_0 admits a non-trivial twisted holomorphic 1-form $\theta \in H^0(S, \Omega^1 \otimes \mathcal{L})$ only if S is a Hopf surface, an Inoue-Bombieri surface, or a Enoki surface. In the following theorem we shall suppose that $D \neq 0$.

Theorem 4.2 *Let S in class VII_0^+. The following two conditions are equivalent:*

- *There exists* $0 \neq \theta \in H^0(S, \Omega^1(logD) \otimes \mathcal{L}_\mu)$, $\mu > 1$, $D \neq 0$,
- *There exists a negative locally integrable function* τ *such that* $d_\beta d^c_\beta \tau \geq 0$, *and for* $\hat{\beta} = d\hat{f}$, $\hat{u} = e^{-\hat{f}}\hat{\tau}$ *is a multiplicatively automorphic negative PSH function* $\hat{u} : \hat{S} \to \mathbb{R}_- \cup \{-\infty\}$ *with polar set* $\hat{D} \neq 0$, *PH outside* $\hat{D}$ *and multiplicative constant* $\mu > 1$.

Moreover if one of these conditions is satisfied,

- *S is a Kato surface, in particular the torsion* μ *is an integer if S is intermediate and* $\mu > 1$ *is a quadratic number if S is a Inoue-Hirzebruch surface,*
- *The PSH function* $\hat{u}$ *on* $\hat{S}$ *is unique up to multiplication by a positive constant.*

Proof

(1) If there exists a non trivial logarithmic form $\theta \in H^0(S, \Omega^1(\log D) \otimes \mathcal{L}_\mu)$, with $\mu > 1$, D contains (at least) a cycle of rational curves [3, lemma 6.10]. As $\mu > 1$, $\mu = \varphi(\beta)$ with $\beta < 0$. Besides, with notations of [3], we have the relation

$$CS(\mathcal{F}, D_i, D_i \cap D_{i-1}) = -\frac{Res_\beta(\theta)_{\hat{D}_{i-1}}}{Res_\beta(\theta)_{\hat{D}_i}}$$

Since Camacho-Sad index are negative by Theorem 2.17, quotients of residues are positive and up to multiplication of θ by a suitable complex number, residues are all positive. By Proposition 2.19, there exists τ such that $T := d_\beta d^c_\beta \tau \geq 0$. Recall that for $\hat{\beta} = d\hat{f}$,

$$dd^c(e^{-\hat{f}}\hat{\tau}) = e^{-\hat{f}} d_{\hat{\beta}} d^c_{\hat{\beta}} \hat{\tau}$$

therefore $\hat{u} = e^{-\hat{f}}\hat{\tau}$ is PSH multiplicatively automorphic with positive constant $\mu > 1$ and PH outside $\hat{D}$. The set $\mathcal{T}(S)$ of LCS structures is an interval which satisfies $\mathcal{T}(S) \subset]-\infty, 0[$ with $\beta \notin \mathcal{T}(S)$ therefore $\mathcal{T}(S) \subset]-\infty, b[$ or $\mathcal{T}(S) \subset]b, 0[$. In the first case $\tau < 0$ by [3, Prop 3.8]. In the second case, by [3, Prop 3.7] there exists for any $\beta \in b$, a weakly positive current $\tau' > 0$ of degree zero such that $T = d_\beta d^c_\beta \tau' \geq 0$ and $T \neq 0$. Hence $\hat{u} = e^{-\hat{f}}\hat{\tau}'$ is a non constant PSH multiplicatively automorphic function with positive constant $\mu > 1$. As in $\hat{S}$, $\hat{D}$ contains an infinite chain of rational curves, $\hat{u}$ is constant on this chain, and $\mu = 1$...contradiction and this case does not exist.

(2) Conversely suppose that τ exists such that $\hat{u} = e^{-\hat{f}}\hat{\tau}$ is negative, PSH, PH outside $\hat{D}$, with constant $\mu > 1$ of automorphy. By Brunella's theorem S is a hyperbolic Kato surface. By lemma 5.4 in [10],

$$l := \text{Card}\{\lambda \in \mathbb{C}^\star \mid h^0(S, \Omega^1(\log D) \otimes \mathcal{L}_\lambda) > 0\} \leq 2$$

and equality holds if and only if S is a Inoue-Hirzebruch surface. Therefore in all cases there is a unique real, $\lambda > 1$ for which $h^0(S, \Omega^1(\log D) \otimes \mathcal{L}_\lambda) \neq 0$.

Moreover any holomorphic foliation is defined by a twisted logarithmic 1-form θ [10, thm 5.5], therefore if multiplicative parameter λ is supposed $\lambda > 1$ it is unique.

We consider first the case of intermediate Kato surfaces.

- Suppose that there exists a global twisted vector field

$$X \in H^0(S, \Theta(-\log D) \otimes \mathcal{L}_{\lambda(S)}).$$

This situation happens if and only if $H^0(S, K_S^{-1} \otimes \mathcal{L}_{\kappa(S)}) \neq 0$, where $\lambda(S) = k(S)\kappa(S)$, i.e. the index of S is 1 (see [12]). By unicity of the holomorphic foliation, X is in the kernel of a twisted logarithmic 1-form

$$\theta \in H^0(S, \Omega^1(\log D) \otimes \mathcal{L}_k),$$

(unique up to multiplication by a scalar) where $k = k(S)$ is uniquely defined by the intersection matrix $M(S)$ of the rational curves. The PSH function $\hat{u}$ defines a PSH section u of $\mathcal{L}_\mu$ on S, PH outside D. As well $\omega = \partial u$ is a twisted holomorphic $(1, 0)$ form outside D in $H^0(S \setminus D, \Omega^1 \otimes \mathcal{L}_\mu)$, with the same twisting factor μ. Therefore $f = \langle X, \omega\rangle$ is a holomorphic section of $\mathcal{L}_{\lambda(S)\mu}$ on $S \setminus D$. As the intersection matrix $M(S)$ is negative definite, D has a strictly pseudoconvex neighbourhood U and f extends to S by Ma.Kato theorem [16, Prop 1].

If f does not vanish on an open set, we would have two different holomorphic foliations which is impossible [10, Thm 5.5]. Therefore $f \equiv 0$, ω and θ define the same foliation. We can consider ω/θ which is a holomorphic section of $\mathcal{L}_{\mu/k}$ on $S \setminus D$. It extends to S by the previous Kato theorem and cannot vanish because S is not Enoki. Hence $\mathcal{L}_{\mu/k}$ is holomorphically trivial, $\mu = k$, ω/θ is a constant, and $\omega \in H^0(S, \Omega^1(\log D) \otimes \mathcal{L}_k)$.
- If the index of S is $m \geq 2$, i.e. m is the least integer for which there exists $\kappa \in \mathbb{C}^\star$ such that $H^0(S, K_S^{-m} \otimes \mathcal{L}_\kappa) \neq 0$, there is a commutative diagram given by Proposition 3.1. The left-hand side of this diagram can be completed by fiber products

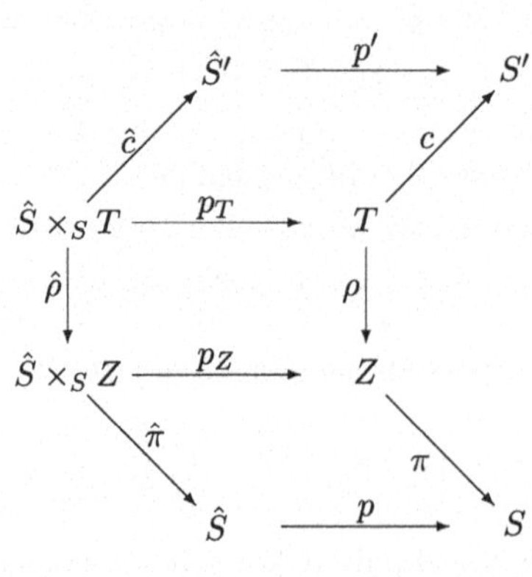

All the mappings of this diagram exist by construction of the fiber products, except $\hat{c}$. The covering $(\hat{S} \times_S T, p_T, T)$ is cyclic, therefore $\hat{c}$ contracts the exceptional curves of the first kind. Let $\hat{u}$ be a negative PSH function on $\hat{S}$, PH outside $\hat{D}$ with polar set $\hat{D}$ and constant of automorphy μ. The ramified covering (Z, π, S) has possible singularities only over singular points of D. The PSH function $\hat{u}_T := \hat{u} \circ \hat{\pi} \circ \hat{\rho}$ is PH outside $\hat{D}_T := (\hat{\pi} \circ \hat{\rho})^{-1}(\hat{D})$ with polar set $\hat{D}_T$ which is the maximal divisor in $\hat{S} \times_S T$. The cyclic covering $(\hat{S} \times_S T, p_T, T)$ is smooth however perhaps non minimal. An exceptional curve of the first kind belongs to $\hat{D}_T$ therefore is an irreducible component of the polar set. The PSH function $\hat{u}' := \hat{u}_T \circ (\hat{c})^{-1}$ on the complement of the points images of these exceptional curves extends continuously to $\hat{S}'$ as a PSH function.

By the case $m = 1$, $\hat{\theta}' = \partial \hat{u}'$ induces on S' a twisted logarithmic 1-form $\theta' \in H^0(S', \Omega^1(\log D') \otimes \mathcal{L}_{k'})$ invariant by the action of $\mathbb{Z}/m\mathbb{Z}$, since it is the case for $\hat{u}_T$ and $\hat{u}'$, with polar set the maximal divisor D' of S'. Besides, by Proposition 3.2, $\mu = k' = k(S') = k(S)$.

Let $\gamma' : \hat{S}' \to \hat{S}'$ be the generator of $Aut_{S'}(\hat{S}') \simeq \mathbb{Z}$ such that $\gamma^{\star}\hat{\theta}' = k(S)\hat{\theta}'$. The quotient $Z' \simeq S'/\mathbb{U}_m$ is a normal surface with has at most isolated quotient singularities image by π' of intersection points of the rational curves of S', therefore $\hat{\theta}'$ induces on the regular part Z'_{reg} of Z' a twisted logarithmic 1-form θ'. The singular points of Z' are Jung-Hirzebruch singularities, hence their desingularization is a chain of rational curves (see [4]). Let P such a point, U a Stein neighbourhood of P and $\rho' : V \to U$ a minimal desingularization, $E = f^{-1}(P)$. On V, simply connected, $\mathcal{L}_{k(S)}$ is trivial hence $\rho'^{\star}\theta'$ extends on V as a logarithmic 1-form by [15, Thm 1.1 and Rem. 1.2] (recall that a reflexive coherent sheaf is normal). Finally θ' extends globally on T' and gives a twisted logarithmic 1-form $\theta \in H^0(S, \Omega^1(\log D) \otimes \mathcal{L}_{k(S)})$.

For Inoue-Hirzebruch surfaces the Example 4.5. in [3] shows that there exist a PSH function $\hat{v}$ on $\hat{S}$ with multiplicative automorphic constant $\lambda > 1$. Moreover there are twisted logarithmic 1-form θ and twisted vector field X with same quadratic real number $\lambda > 1$ by [10, p. 1526].

We show as for intermediate case of index $m = 1$ that ∂v extends to S with same torsion.

By the previous arguments, if $\hat{u}$ and $\hat{v}$ are negative PSH function, PH outside the rational curves, $\hat{\theta} = \partial \hat{u}$ and $\hat{\omega} = \partial \hat{v}$ have the same coefficient of torsion and induce on S two twisted logarithmic 1-form θ and ω respectively, in the same space $H^0(S, \Omega^1(\log D) \otimes \mathcal{L}_{k(S)})$ of dimension one. Therefore there exists a complex number t such that $\theta = t\omega$. We have then $\partial(\hat{u} - t\hat{v}) = 0$, i.e. $\hat{g} := \hat{u} - t\hat{v}$ is antiholomorphic on $\hat{S}$ and gives an antiholomorphic section g of $\mathcal{L}_{k(S)}$. The holomorphic section $\bar{g}$ of $\mathcal{L}_{k(S)}$ vanishes identically because S is not a Enoki surface, therefore $\hat{u} = t\hat{v}$ and since both $\hat{u}$ and $\hat{v}$ are negative t is a positive constant.

□

Definition 4.3 *Let S in class VII_0^+. A negative locally integrable function $\hat{u} : \hat{S} \to \mathbb{R}_-$ multiplicatively automorphic PSH function with $\hat{D}$ as polar set, PH outside $\hat{D}$ and multiplicative constant $\mu > 1$ will be called a Green function.*

Here we give a geometric conjectural meaning of the upper bound b of $\mathcal{T}(S)$:

Conjecture 4.4 Let S in class VII_0 and $\mathcal{T}(S) =]a, b[, \ -\infty \leq a, b < 0$ be the moduli space of LCS structures taming the complex structure of S. Let φ be the decreasing mapping

$$\varphi :]-\infty, 0] \to [1, +\infty[, \quad a \mapsto C = e^{-c}$$

If there is a Green function $\hat{u}$ on $\hat{S}$ with multiplicity constant $\lambda > 1$ or if there is a twisted logarithmic 1-form $\theta \in H^0(S, \Omega^1(\log D) \otimes \mathcal{L}_\lambda)$ with twisting constant $\lambda > 1$, then $\lambda = \varphi(b)$.

5 Appendix

We give here explicit examples for Proposition 3.2:

Example 5.1

(1) For $b_2(S) = 3$ there is only one configuration of curves for which the index is different from one, that is $a(S) = (42\,2) = (s_1 r_1) = -(D_0^2, D_1^2, D_2^2)$. The opposite intersection matrix is

$$M(S) = \begin{pmatrix} 2 & 0 & -1 \\ 0 & 2 & -1 \\ -1 & -1 & 2 \end{pmatrix}.$$

Then $\det M(S) = 4$, $k(S) = \sqrt{\det M(S)} + 1 = 3$ and the index is $m(S) = 2$ because we have $D_{-K} = \frac{3}{2}D_0 + \frac{1}{2}D_1 + D_2$. The associated contracting germ is of simple type because there is only $t = 1$ tree. By the procedure which allows to determine the sequence of blow-ups and gluing in the construction of the Kato surface S, given in [18, p. 337] we have with the notations there, $q = 0$,

$$[\alpha_1 + 1, \underbrace{2, \ldots, 2}_{\alpha_1 - 2}, \underbrace{2, \ldots, 2}_{s-j+1}] = [42\,2]$$

hence $\alpha_1 = 3$ and $s = j$.

Besides, the index is given by the formula [18, p. 331]

$$m(S) = \frac{k-1}{gcd\{k-1, s\}}$$

therefore s is odd. Since $c_j = 1$ and $j < k = 3$ for all normal forms, $j = 1$ and the germ associated to S is of the form

$$F(z,\zeta) = (\lambda z\zeta + \zeta, \zeta^3), \quad \lambda \in \mathbb{C}^\star.$$

We determine now the germ associated to S'. We follow the computations of [18, p. 330]. Set

$$r' := \lfloor \frac{mj}{k} \rfloor = 0, \quad s' := 2s = 2, \quad j' = 2j = 2, \quad P'(\zeta) := c_1\zeta^2 = \zeta^2$$

hence

$$F'(z,\zeta) = (\lambda z\zeta^{s'} + P'(\zeta), \zeta^k) = (\lambda z\zeta^2 + \zeta^2, \zeta^3) = \big(\zeta^2(\lambda z + 1), \zeta^3)\big)$$

Following procedure given in [18, 337] we obtain $F' = \Pi_0 \circ \Pi_1 \circ \Pi_2 \circ \sigma$ with

$$\sigma(z,\zeta) = \left((\lambda z + 1)^3 - 1, \frac{\zeta}{\lambda z + 1}\right),$$

$$\Pi_2(u,v) = \big((u+1)v, v\big), \quad \Pi_1(u',v') = (v', u'v'), \quad \Pi_0(v',u') = v', u'v').$$

We have $a(S') = [332]$. We check that $det M(S') = 4$ and $D_{-K} = 2D_0 + D_1 + 2D_2$ in particular $m(S') = 1$.

(2) For $b_2(S) = 4$, there are several configurations of index $m \geq 2$. For instance, suppose that $a(S) = (3\,3\,22) = (s_1 s_1 r_2) = -(D_0^2, D_1^2, D_2^2, D_3^2)$. We have

$$M(S) = \begin{pmatrix} 3 & 0 & -1 & -1 \\ 0 & 3 & 0 & -1 \\ -1 & 0 & 2 & -1 \\ -1 & -1 & -1 & 2 \end{pmatrix}$$

$$\det M(S) = 4, \quad k(S) = 3, \quad D_{-K} = \frac{5}{2}D_0 + \frac{3}{2}D_1 + 3D_2 + \frac{7}{2}D_3, \quad m(S) = 2.$$

$$[\alpha_1 + 2, \underbrace{2,\ldots,2}_{\alpha_1 - 1}, \alpha_2 + 1, \underbrace{2,\ldots,2}_{\alpha_2 - 2}, \underbrace{2,\ldots,2}_{s-j+1}] = [3322]$$

therefore $\alpha_1 = 1$, $\alpha_2 = 2$, $s - j = 1$, and we recover that $b_2(S) = \alpha_1 + \alpha_2 + (s - j) = 4$. Besides $t = 1$ therefore the germ associated to S is simple,

$$m(S) = 2 = \frac{k - 1}{gcd\{k - 1, s\}}$$

hence s is odd and j is even. We have $j < k = 3$ hence $j = 2$ and $s = 3$.

$$F(z, \zeta) = (\lambda z\zeta^3 + \zeta^2 + c_3\zeta^3, \zeta^3)$$

$$r = \lfloor \frac{2j}{k} \rfloor = 1, \quad s' = ms - r(k-1) = 4, \quad j' = mj - rk = 1,$$

$$P'(\zeta) = \sum_{p=j}^{s} c_p \zeta^{mp-rk} = \zeta + c_3\zeta^3$$

$$\begin{aligned} F'(z, \zeta) &= (\lambda z\zeta^4 + \zeta + c_3\zeta^3, \zeta^3) \\ &= (\zeta A, \zeta^3), \quad \text{with } A = A(z, \zeta) = \lambda z\zeta^3 + 1 + c_3\zeta^2 \end{aligned}$$

Set

$$A^{-3} - 1 = \zeta^2 B, \quad B = -3c_3 + \cdots, \quad \text{and} \quad BA^{-2} = -3c_3 + \zeta C,$$

then a sequence of six blow-ups of the unit ball $B^2 \subset \mathbb{C}^2$

$$\Pi_0(u', v') = (v', u'v'), \quad \Pi_1(u, v) = (uv, v), \quad \Pi_2(u, v) = (uv, v),$$
$$\Pi_3(u, v) = (uv + 1, v)$$

$$\Pi_4(u, v) = (uv, v), \quad \Pi_5(u, v) = (uv - 3c_3, v), \quad \sigma(z, \zeta) = (CA^{-1}, \zeta A)$$

σ is a germ of biholomorphism and

$$F'(z, \zeta) = \Pi_0 \circ \Pi_1 \circ \Pi_2 \circ \Pi_3 \circ \Pi_4 \circ \Pi_5 \circ \sigma(z, \zeta).$$

Setting $C_i = \Pi^{-1}(p_{i-1})$ with $p_{i-1} \in C_{i-1}$, $i \geq 0$, p_{-1} the origin of the ball and $C_{-1} = \sigma^{-1}(C_5)$, we obtain the sequence of self intersection (in S), $(C_0^2, C_1^2, C_2^2, C_3^2, C_4^2, C_5^2) = (-2, -2, -2, -2, -2, -4)$, i.e. setting by a circular permutation C_5 at the first place and denoting $D_0 = C_5$, a.s.o. we obtain $a(S) = (42\ 2222) = (s_2, r_4)$. We check that $D_{-K} = 3D_0 + 2D_1 + 4D_2 + 6D_3 + 5D_4 + 4D_5$, therefore the index is $m(S') = 1$ as wanted.

6 Declarations

My manuscript has no associated data.

Conflicts of Interest The author states that there is no conflict of interests.

References

1. V. Apostolov, G. Dloussky, Locally conformally symplectic structures on compact non-Kähler complex surfaces. Int. Math. Res. Not. (IMRN) **9**, 2717–2747 (2016)
2. V. Apostolov, G. Dloussky, On the Lee classes of locally conformally symplectic complex surfaces. J. Sympl. Geom. **16**, 931–958 (2018)
3. V. Apostolov, G. Dloussky, Twisted differentials and Lee classes of locally symplectic complex surfaces. Math. Z. **303**, 76. https://doi.org/10.1007/s00209-023-03242-5
4. W. Barth, K. Hulek, C. Peters, A. Van de Ven, *Compact Complex Surfaces*, 2nd edn. (Springer, Heidelberg, 2004)
5. M. Brunella, A characterization of hyperbolic Kato surfaces. Publ. Mat. **58**, 251–261 (2014)
6. C. Camacho, P. Sad, Invariant varieties through singularities of holomorphic vector fields. Ann. Math. **115**, 579–595 (1982)
7. G. Dloussky, Structure des surfaces de Kato. Mém. S.M.F $2^{\text{ème}}$ Sér. **14**, I1–I120 (1984)
8. G. Dloussky, On surfaces of class VII_0^+ with numerically anticanonical divisor. Amer. J. Math. **128**, 639–670 (2006)
9. G. Dloussky, Non-Kählerian surfaces with a cycle of rational curves. Complex Manifolds **8**, 208–222 (De Gruyter, 2021). https://doi.org/10.1515/coma-2020-0114
10. G. Dloussky, K. Oeljeklaus, Vector fields and foliations on compact surfaces of class VII_0. Ann. Inst. Fourier **49**, 1503–1545 (1999)
11. G. Dloussky, K. Oeljeklaus, M. Toma, Surfaces de la classe VII_0 admettant un champ de vecteurs, II. Comment. Math. Helv. **76**, 640–664 (2001)
12. G. Dloussky, K. Oeljeklaus, M. Toma, Class VII_0 surfaces with b_2 curves. Tohoku Math. J. **55**, 283–309 (2003)
13. I. Enoki, Surfaces of class VII_o with curves. Tôhoku Math. J. (2) **33**, 543–492 (1981)
14. C. Favre, Classification of 2-dimensional contracting rigid germs. J. Math. Pures Appl. **79**, 475–514 (2000)
15. D. Greb, S. Kebekus, S.J. Kovács, Extension theorems for differential forms and Bogomolov-Sommese vanishing on log canonical varieties. Compos. Math. **146**, 193–219 (2010). https://doi.org/10.1112/S0010437X09004321
16. M. Kato, Riemann-Roch theorem for strongly pseudoconvex manifolds of dimension 2. Math. Ann. **222**, 243–250 (1976)
17. M. Kato, Compact complex surfaces containing "global spherical shells", in *"Proceedings of the Int. Symp. Alg. Geometry" (Kyoto Univ., Kyoto, 1977)* (Kinokuniya Book Store, Tokyo, 1978), pp. 45–84
18. K. Oeljeklaus, M. Toma, Logarithmic moduli spaces for surfaces of class VII. Math. Ann. **341**, 323–345 (2008). https://doi.org/10.1007/s00208-007-0192-1
19. A. Teleman, Donaldson theory on non-Kählerian surfaces and class VII surfaces with $b_2 = 1$. Invent. Math. **162**, 493–521 (2005)

Conformal Vector Fields on LCP Manifolds

Brice Flamencourt and Andrei Moroianu

Dedicated to Paul Gauduchon on the occasion of his 80th birthday.

Abstract We show that conformal vector fields on compact locally conformally product manifolds are orthogonal to the flat distribution and Killing with respect to the Gauduchon metric.

Keywords Conformal geometry · Weyl connections · LCP manifolds · Gauduchon metric

2020 Mathematics Subject Classification 53C05, 53C18, 53C29

1 Introduction

Locally conformally product (LCP) structures arise naturally in the theory of Weyl connections. They consist in a closed non-exact Weyl connection D on a compact conformal manifold (M, c) with (non-zero) reducible holonomy. LCP structures are closely related to LCK structures, which can be defined in a similar way, just by replacing the reducibility condition for the holonomy of D with the existence of a D-parallel complex structure.

The study of LCP structures started with the Belgun-Moroianu conjecture [3], which stated that they should not exist. However, the works of Matveev-

B. Flamencourt
UMPA - ENS Lyon, CNRS, Lyon, France
e-mail: brice.flamencourt@ens-lyon.fr

A. Moroianu (✉)
Laboratoire de mathématiques d'Orsay, Université Paris-Saclay, CNRS, Orsay, France

Institute of Mathematics "Simion Stoilow" of the Romanian Academy, Bucharest, Romania
e-mail: andrei.moroianu@math.cnrs.fr

L. Ornea (ed.), *Real and Complex Geometry*,
https://doi.org/10.1007/978-3-031-92297-8_8

Nikolayevsky [9, 10] and Kourganoff [8] after them exhibited examples of such manifolds, and also showed that their structure was very special. Indeed, when lifted to the universal cover $\tilde{M}$ of M, D becomes the Levi-Civita connection of a Riemannian metric h and $(\tilde{M}, h)$ is isometric to $(\mathbb{R}^q, h_0) \times (N, h_N)$, where (N, h_N) is an irreducible Riemannian manifold.

After the discovery of this result, LCP manifolds have been analyzed from various points of view by several authors. LCP structures on solvmanifolds [1] and more generally on compact quotients of Lie groups [4] have been studied by Andrada, del Barco and the second author, and their description under some additional assumptions concerning their characteristic group has been given by the first author in [6]. Obstructions to the existence of LCP structures on conformal manifolds are also discussed in [2], where it is shown that the conformal class of an LCP manifold cannot contain Einstein of Kähler metrics for example.

An important tool in conformal geometry is the Gauduchon metric [7], which on a compact conformal manifold (M, c) endowed with a Weyl connection D is the unique (up to a multiplicative constant) Riemannian metric $g \in c$ with the property that the Lee form of D with respect to g is coclosed.

Recently, the second author together with Pilca [12] studied the behavior of the Gauduchon metric of Weyl connections defining an LCP structure, and proved that it is adapted. Recall that if (M, c, D) is an LCP manifold, a Riemannian metric g in the conformal class c is called adapted if the lift to the universal cover of the Lee form of D with respect to g vanishes on the flat distribution $T\mathbb{R}^q$.

In another recent work [11], the same authors generalized to the LCK setting the well-known fact that all conformal vector fields on a compact Kähler manifold are Killing, by proving that every conformal vector field on a compact LCK manifold is Killing with respect to the Gauduchon metric.

It is thus natural to investigate the corresponding question in the LCP setting, all the more that a classical result of Tashiro and Miyashita [13] states that on complete non-flat Riemannian products, all complete conformal vector fields are Killing.

This is precisely what we achieve in this paper, by proving the following:

Theorem 1.1 *Let (M, c, D) be an LCP manifold. Then the conformal vector fields of (M, c) are exactly the Killing vector fields of the Gauduchon metric of the Weyl structure D.*

The strategy of the proof is roughly as follows. Every conformal vector field $\bar{\xi}$ on (M, c) can be lifted to a (complete) conformal vector field ξ of the Riemannian product structure $(\mathbb{R}^q, h_0) \times (N, h_N)$ on the universal cover of M. However, the Tashiro-Miyashita result does not apply, since this product metric is not complete. Nonetheless, we can show that the covariant derivative of ξ with respect to tangent vectors to N, restricted to each flat leaf $\mathbb{R}^q \times \{y\}$, is a gradient conformal vector field. Then using the explicit description of these vector fields and the fact that ξ has bounded geometry, we obtain that ξ has to be the pull-back of a Killing vector field on (N, h_N). This implies in particular that $\bar{\xi}$ is affine with respect to the Weyl connection D, and we conclude by the uniqueness property of the Gauduchon metric.

2 Preliminaries on LCP Structures

We start by introducing the basic concepts of conformal geometry.

Definition 2.1 A conformal structure on a smooth manifold M is an equivalence class of the equivalence relation defined on the space of Riemannian metrics on M by $g \sim g'$ if and only if there exists a smooth function φ such that $g' = e^{2\varphi} g$.

Conformal structures are usually denoted by c. There is no natural connection on a conformal manifold as for the Riemannian case. However, one can consider a class of connections, called Weyl connections, which extends in a certain way the concept of Levi-Civita connection to the conformal case.

Definition 2.2 Let (M, c) be a conformal manifold. A Weyl connection on (M, c) is a torsion-free connection D on TM which preserves the conformal structure, in the sense that for any metric $g \in c$, there is a 1-form θ_g, called the *Lee form* of D with respect to g, such that $Dg = -2\theta_g \otimes g$.

Clearly if $g' = e^{2\varphi} g$ then $\theta_{g'} = \theta_g - d\varphi$, so the cohomological nature of the Lee form is independent of the metric. This motivates the following definition:

Definition 2.3 A Weyl connection D on a conformal manifold (M, c) is called *closed* if the Lee form of D with respect to one metric—and then to all metrics—in c is closed. Similarly, D is called *exact* if the Lee form of D with respect to one metric—and then to all metrics—in c is exact.

Note that every closed Weyl connection on a simply connected conformal manifold is exact.

Let now D be a closed Weyl connection on a conformal manifold (M, c). Its pull-back $\tilde{D}$ to the universal covering $\tilde{M}$ of M is a Weyl connection for the conformal structure $\tilde{c}$ obtained by pulling-back c. This Weyl connection is exact since $\tilde{M}$ is simply connected, thus there exists a metric $h \in \tilde{c}$, unique up to a multiplication by a constant, such that $\nabla^h = \tilde{D}$, where ∇^h is the Levi-Civita connection of h.

More precisely, if g is any Riemannian metric on M in the conformal class c, and θ_g is the Lee form of D with respect to g, then its pull-back $\tilde{\theta}_g$ to the universal cover $\tilde{M}$ is exact, so there exists a function $\varphi \in C^\infty(\tilde{M})$ such that $d\varphi = \tilde{\theta}_g$. Moreover, $\tilde{\theta}_g$ is also the Lee form of the pull-back Weyl connection $\tilde{D}$ on $\tilde{M}$, so by the above formula, the Lee form of $\tilde{D}$ with respect to the metric $h := e^{2\varphi}\tilde{g}$ vanishes, i.e. $\tilde{D}$ is the Levi-Civita connection of h. Note that the fundamental group $\pi_1(M)$ acts on $\tilde{M}$ by h-homotheties, and this action preserves h if and only if D is exact.

LCP structures arise when one considers closed, non-exact Weyl connections on a compact conformal manifold. In this situation, one has a remarkable result proved by Kourganoff [8]:

Theorem 2.4 ([8, Thm. 1.5]) *Let (M, c) be a conformal manifold endowed with a closed, non-exact Weyl connection D. Let h be a metric on $\tilde{M}$, the universal cover* *of M, such that $\nabla^h = \tilde{D}$ where $\tilde{D}$ is the pull-back of D to $\tilde{M}$. Then, one of the three following cases occurs:*

- *$(\tilde{M}, h)$ is flat;*
- *$(\tilde{M}, h)$ is irreducible;*
- *$(\tilde{M}, h)$ is a Riemannian product $(\mathbb{R}^q, h_0) \times (N, h_N)$ where $q \geq 1$, $(\mathbb{R}^q, h_0)$ is the usual Euclidean space and (N, h_N) is a non-flat, incomplete Riemannian manifold.*

The third case in Theorem 2.4 corresponds to LCP structures. More precisely, we have the following:

Definition 2.5 An LCP manifold is a triple (M, c, D) where M is a compact manifold, c is a conformal structure on M and D is a closed, non-exact Weyl connection, which is non-flat and reducible (i.e. the representation of its restricted holonomy group $\mathrm{Hol}_0(D)$ is reducible).

With the notations above, $(\mathbb{R}^q, h_0)$ is called the *flat part* of the LCP manifold, while (N, h_N) is called the non-flat part. The distributions $T\mathbb{R}^q$ and TN descend to D-parallel distributions on M, respectively called the *flat distribution* and the *non-flat distribution* of the LCP manifold.

Definition 2.6 Let (M, c, D) be an LCP manifold. A Riemannian metric $g \in c$ is called *adapted* if the Lee form θ_g vanishes on the flat distribution (or equivalently, if the primitive φ of the pull-back to the universal cover of θ_g is the pull-back of a function defined on the non-flat factor N.

Remark 2.7 Adapted metrics always exist (see [5, 12]). The above considerations show that their lift to $\tilde{M}$ can be written $\tilde{g} = e^{-2\alpha}h$, where $\alpha \in C^\infty(N)$.

3 Conformal Vector Fields on LCP Manifolds

Let (M, c, D) be an n-dimensional LCP manifold. Let h be the Riemannian metric on its universal cover whose Levi-Civita connection is the lift of D, so that $(\tilde{M}, h) = (\mathbb{R}^q, h_0) \times (N, h_N)$, where $(\mathbb{R}^q, h_0)$ is the flat part and (N, h_N) is the non-flat part. We will denote the Levi-Civita connection of h by ∇.

Denoting by p_1, p_2 the projections from $\tilde{M}$ onto $\mathbb{R}^q$ and N respectively, the tangent bundle of $\mathbb{R}^q \times N$ can be identified with the direct sum $\pi_1^* T\mathbb{R}^q \oplus \pi_2^* TN$. Correspondingly, one can write every vector field $\xi \in \Gamma(T\tilde{M})$ as a sum

$$\xi = \xi_1 + \xi_2, \tag{1}$$

where $\xi_1(\cdot, y) \in \Gamma(T\mathbb{R}^q)$ for every $y \in N$, and $\xi_2(x, \cdot) \in \Gamma(TN)$ for every $x \in \mathbb{R}^q$.

Let $\bar{\xi} \in \Gamma(TM)$ be a conformal vector field, and let $\xi \in \Gamma(T\tilde{M})$ be its lift to $\tilde{M}$. In particular $\gamma^*\xi = \xi$ for any $\gamma \in \pi_1(M)$. We decompose $\xi = \xi_1 + \xi_2$ as in (1).

The fact that ξ is a conformal vector field on $\tilde{M}$ is equivalent to the identity

$$h(\nabla_X\xi, Y) + h(X, \nabla_Y\xi) = fh(X, Y), \qquad (\forall)X, Y \in T\tilde{M}, \tag{2}$$

where f is a real-valued smooth function on $\tilde{M}$. Applying (2) with $X, Y \in T\mathbb{R}^q$ shows that for any $y \in N$, $\xi_1(\cdot, y)$ is a conformal vector field on $\mathbb{R}^q$. In the same way, we show that for any $x \in \mathbb{R}^q$, $\xi_2(x, \cdot)$ is a conformal vector field on N.

Let $(e_1, \ldots, e_q)$ be the canonical basis of $\mathbb{R}^q$ and denote by the same letters the induced vector fields on $\tilde{M}$ constant along N. We also fix an arbitrary vector field Z on N, identified with the induced vector field on $\tilde{M}$ constant along $\mathbb{R}^q$. Then $\nabla_{e_i} Z = \nabla_Z e_i = 0$ for every $1 \leq i \leq q$.

Using (2) for $X = e_i$ and $Y = Z$ we obtain

$$h(e_i, \nabla_Z\xi_1) = -h(\nabla_{e_i}\xi_2, Z) = -\partial_{e_i}(h(\xi_2, Z)), \tag{3}$$

which implies that for every fixed $y \in N$, the vector field $\nabla_Z\xi_1(\cdot, y)$ on $\mathbb{R}^q$ is the gradient in $\mathbb{R}^q$ of the function $-h_{(\cdot,y)}(\xi_2, Z)$.

Moreover, taking $X = e_i$, $Y = e_j$ in (2), differentiating with respect to Z, and using the commutation of ∇_Z with ∇_{e_i} for $1 \leq i \leq q$, shows that $\nabla_Z\xi_1(\cdot, y)$ is a conformal vector field on $(\mathbb{R}^q, h_0)$ for every $y \in N$.

Synthesizing the previous analysis, for any $y \in N$ and $Z \in T_yN$, $\nabla_Z\xi_1(\cdot, y)$ is a gradient conformal vector field on $(\mathbb{R}^q, h_0)$. These vector fields are well understood:

Lemma 3.1 *Let X be a gradient conformal vector field on $(\mathbb{R}^q, h_0)$. Then, there exist real numbers $b, b_1, \ldots, b_q$ such that $X = b\sum_{i=1}^q x_i e_i + \sum_{i=1}^q b_i e_i$, where $(e_1, \ldots, e_q)$ is the canonical basis of $\mathbb{R}$ and x_i is the i-th coordinate function.*

Proof In this proof, ∇ stands for the gradient in $\mathbb{R}^q$ and the usual scalar product on $\mathbb{R}^q$ is denoted by $\langle\cdot,\cdot\rangle$.

By assumption, there is a function $\psi : \mathbb{R}^q \to \mathbb{R}$ such that $X = \nabla\psi = \sum_{i=1}^q (\partial_{e_i}\psi)e_i$. Since X is a conformal vector field one has

$$\langle\partial_{e_i}X, e_j\rangle + \langle e_i, \partial_{e_j}X\rangle = \chi\langle e_i, e_j\rangle, \qquad (\forall)1 \leq i, j \leq q,$$

for some smooth function $\chi : \mathbb{R}^q \to \mathbb{R}$. This is equivalent to

$$\partial_{e_i}\partial_{e_j}\psi = \delta_i^j\chi, \qquad (\forall)1 \leq i, j \leq q.$$

In particular, for any $1 \leq i \leq q$, $\partial_{e_i}\psi$ is a function depending only on the i-th coordinate, and thus χ has the same property. Consequently, χ is constant and ψ belongs to the vector space

$$E := \{\phi \in C^\infty(\mathbb{R}^q) \mid \exists\mu \in \mathbb{R},\ \forall i, j,\ \partial_{e_i}\partial_{e_j}\phi = \delta_i^j\mu\}. \tag{4}$$

Now, we remark that

$$E_0 := \{\phi \in C^\infty(\mathbb{R}^q) \mid \forall i, j,\ \partial_{e_i}\partial_{e_j}\phi = 0\} \tag{5}$$

is a hyperplane of E since it is the kernel of the linear form $E \ni \phi \mapsto \partial^2_{e_1}\phi$. Moreover, denoting by x_i the i-th coordinate function, $\mathrm{Span}(\sum_{i=1}^q (x_i)^2)$ is a supplementary of E_0 in E, whence

$$E = \mathrm{Span}(\sum_{i=1}^q (x_i)^2) \oplus E_0.$$

Since $E_0 = \mathrm{Span}(x_1, \dots, x_q)$, we obtain $\psi = \frac{b}{2}\sum_{i=1}^q (x_i)^2 + \sum_{i=1}^q b_i x_i$ for some $(b, b_1, \dots, b_q) \in \mathbb{R}^{q+1}$, and $X = \nabla\psi = b\sum_{i=1}^q x_i e_i + \sum_{i=1}^q b_i e_i$.

Conversely, any vector field of this form is a conformal gradient vector field. □

From Lemma 3.1 we conclude that there are functions $\tilde{b}, \tilde{b}_1, \dots, \tilde{b}_q$ from TN to $\mathbb{R}$ such that

$$\nabla_Z \xi_1(x, y) = \tilde{b}(y, Z)\sum_{i=1}^q x_i e_i + \sum_{i=1}^q \tilde{b}_i(y, Z)e_i, \tag{6}$$

for all $(x, y) \in \mathbb{R}^q \times N$, and $Z \in T_yN$. Applying (6) to $x = 0$ and $x = e_1$ shows that

$$\tilde{b}_i(y, Z) = h_{(0,y)}(\nabla_Z\xi_1, e_i), \quad \tilde{b}(y, Z) = h_{(e_1,y)}(\nabla_Z\xi_1, e_1) - \tilde{b}_1(y, Z) \tag{7}$$

for any $y \in N$, $Z \in T_yN$ and $1 \le i \le q$, which implies that the functions $\tilde{b}, \tilde{b}_1, \dots, \tilde{b}_q$ are smooth, and linear in the variable Z.

We fix $y_0 \in N$. Let $y \in N$ and $c : [0, 1] \to N$ be a smooth path from y_0 to y, which exists by connectedness. One has, for any $x \in \mathbb{R}^q$,

$$\begin{aligned}
\xi_1(x, y) - \xi_1(x, y_0) &= \int_0^1 \nabla_{\dot{c}(t)}\xi_1(x, c(t))\mathrm{d}t \\
&= \int_0^1 \tilde{b}(c(t), \dot{c}(t))\sum_{i=1}^q x_i e_i + \sum_{i=1}^q \tilde{b}_i(c(t), \dot{c}(t))e_i \mathrm{d}t \\
&= \left(\int_0^1 \tilde{b}(c(t), \dot{c}(t))\mathrm{d}t\right)\sum_{i=1}^q x_i e_i \\
&\quad + \sum_{i=1}^q \left(\int_0^1 \tilde{b}_i(c(t), \dot{c}(t))\mathrm{d}t\right) e_i.
\end{aligned}$$

We define the smooth functions

$$b(y) := \int_0^1 \tilde{b}(c(t), \dot{c}(t))\mathrm{d}t, \qquad b_i(y) := \int_0^1 \tilde{b}_i(c(t), \dot{c}(t))\mathrm{d}t, \tag{8}$$

which by the above computation do not depend on the chosen path. Then, writing $\xi_1(\cdot, y_0) =: \sum_{i=1}^q \beta_i e_i$ with $\beta_i \in C^\infty(\mathbb{R}^q)$, we have

$$\xi_1(x, y) = \sum_{i=1}^q (\beta_i(x)e_i + b(y)x_i e_i + b_i(y)e_i), \quad (\forall)(x, y) \in \mathbb{R}^q \times N. \tag{9}$$

Using (3), one obtains for every $Z \in \Gamma(TN)$ and $1 \leq i \leq q$:

$$\partial_{e_i}(h(\xi_2, Z)) = -(Z(b)x_i + Z(b_i)),$$

which implies that there exists some vector field $V \in \Gamma(TN)$ such that

$$\xi_2(x, y) = V(y) - \sum_{i=1}^q \left(\frac{(x_i)^2}{2} \nabla^N b(y) + x_i \nabla^N b_i(y) \right) \tag{10}$$

for all $(x, y) \in \mathbb{R}^q \times N$, where ∇^N denotes the gradient defined by the Levi-Civita connection of h_N.

Let g be an adapted metric on N (Remark 2.7) whose pull-back to $\tilde{M}$ can be written as $\tilde{g} = e^{-2\alpha} h$ for some $\alpha \in C^\infty(N)$. Note that the function α is not bounded from above or from below. Indeed, there exists an element $\gamma \in \pi_1(M)$ which is a contraction with respect to h (i.e. $\rho^* h = \lambda h$ with $\lambda \in (0, 1)$), and since $\pi_1(M)$ acts isometrically with respect to $\tilde{g}$, we get $\gamma^*\alpha = \alpha + \frac{1}{2}\ln\lambda$.

Since M is compact, there exists a constant $C > 0$ such that

$$\sup_{\tilde{M}} \|\xi\|_{\tilde{g}} = \sup_{M} \|\bar{\xi}\|_g \leq C.$$

Consequently,

$$\|e^{-\alpha}\xi_2\|_h^2 = \|\xi_2\|_{\tilde{g}}^2 \leq \|\xi\|_{\tilde{g}}^2 \leq C^2,$$

whence for any $y \in N$, the estimate

$$\|\xi_2(x, y)\|_h \leq e^{\alpha(y)} C \tag{11}$$

(independent on x) holds. Now, Eq. (10) shows that for every fixed $y_0 \in N$, the square norm $\|\xi_2(x, y_0)\|_h$ is polynomial in $(x_1, \ldots, x_q)$. Clearly this norm is bounded on $\mathbb{R}^q$ if and only if $\nabla^N b(y_0) = \nabla^N b_i(y_0) = 0$ for every $1 \leq i \leq q$ and for every $y_0 \in N$, showing that the functions $b, b_1, \ldots, b_q$ are constant on N.

Consequently, $\xi_2(x, y) = V(y)$ is induced on $\tilde{M}$ by a conformal vector field of (N, g_N).

On the other hand, one also has $\|e^{-\alpha}\xi_1\|_h^2 = \|\xi_1\|_{\tilde{g}}^2 = \|\xi_1\|_{\tilde{g}}^2 \leq \|\xi\|_{\tilde{g}}^2 \leq C^2$. Since $b, b_1, \ldots, b_q$ are constant, (9) shows that ξ_1 depends only on the variable x of $\mathbb{R}^q$. Therefore, for every $x \in \mathbb{R}^q$ and $y \in N$ one has $\|\xi_1(x)\|_h \leq e^{-\alpha(y)}C$. However, we have seen that the function α is unbounded from above. We conclude that $\|\xi_1(x)\|_h = 0$, so finally $\xi_1 = 0$.

The conformal vector field $\xi = \xi_2 = V$ is thus tangent to N and constant in the direction of $\mathbb{R}^q$. Taking non-zero vectors $X = Y \in T\mathbb{R}^q$ in Eq. (3), one gets

$$fh(X, X) = 0,$$

which in turn implies $f = 0$, so ξ is a Killing vector field for h.

Altogether, we have proved the following result:

Theorem 3.2 *Let (M, c, D) be an LCP manifold. Then the lift of any conformal vector field of (M, c) to the universal cover of M endowed with its canonical Riemannian decomposition $(\mathbb{R}^q, h_0) \times (N, h_N)$ is a Killing vector field of (N, h_N).*

The last step in our analysis of conformal vector fields on LCP manifolds is to link them to the Killing fields of the Gauduchon metric [7] of the Weyl structure (M, c, D). We first recall the definition of this particular metric:

Definition 3.3 If (M, c) is a compact conformal manifold of dimension larger than 2, then for any Weyl connection D on (M, c) there exists a unique (up to a positive multiplicative constant) Riemannian metric g_G such that the Lee form of D with respect to g_G is coclosed. This metric is called the *Gauduchon metric* of the Weyl structure.

We now return to our particular setting, with $\bar{\xi} \in TM$ a conformal vector field of (M, c). By Theorem 3.2, the lift ξ of $\bar{\xi}$ to $\tilde{M}$ is a Killing vector field on N. In particular, it is a Killing vector field for $(\tilde{M}, h)$, so it preserves the Levi-Civita connection ∇ of h. But ∇ is the lift of the Weyl connection D, and we deduce that $\bar{\xi}$ is an affine vector field for D, i.e. $\mathcal{L}_{\bar{\xi}}(D) = 0$.

Let g_G be the Gauduchon metric of the Weyl structure (M, c, D). We denote by θ_G the Lee form of D with respect to g_G, by ∇^{g_G} the Levi-Civita connection of g_G, and by $(\varphi_t)_{t \in \mathbb{R}}$ the flow of $\bar{\xi}$. Since $\bar{\xi}$ is affine, we obtain for any $t \in \mathbb{R}$:

$$D(\varphi_t^* g_G) = \varphi_t^*(D g_G) = -2\varphi_t^*(\theta_G \otimes g_G) = -2(\varphi_t^* \theta_G) \otimes (\varphi_t^* g_G), \tag{12}$$

and we get that $\varphi_t^* \theta_G$ is the Lee form of D with respect to $\varphi_t^* g_G$. The Levi-Civita connection of g_G is $\varphi_t^* \nabla^{g_G}$, hence $\delta^{\varphi_t^* g_G} = \varphi_t^* \delta^{g_G}$. We thus have:

$$\delta^{\varphi_t^* g_G} \theta_{\varphi_t^* g_G} = (\varphi_t^* \delta^{g_G})(\varphi_t^* \theta_{g_G}) = \varphi_t^*(\delta^{g_G} \theta_{g_G}) = 0. \tag{13}$$

Consequently, $\varphi_t^* g_G$ still is a Gauduchon metric of (M, c, D), so by the uniqueness property there exists a constant $\lambda_t > 0$ such that $(\varphi_t)^* g_G = \lambda_t g_G$. It follows that there exists $\lambda > 0$ such that

$$\mathcal{L}_{\bar{\xi}} g_G = \lambda g_G. \tag{14}$$

Taking the trace in (14) yields $2\delta\bar{\xi} = -n\lambda$. Integrating this equality over M and using the divergence theorem, one obtains $\lambda = 0$, i.e.

$$\mathcal{L}_{\bar{\xi}} g_G = 0, \tag{15}$$

thus showing that $\bar{\xi}$ is a Killing vector field for g_G.

Conversely, it is obvious that a Killing vector field for g_G is a conformal vector field of (M, c). This concludes the proof of Theorem 1.1.

Acknowledgments This work was partly supported by the PNRR-III-C9-2023-I8 grant CF 149/31.07.2023 *Conformal Aspects of Geometry and Dynamics* and by the Procope Project No. 57650868 (Germany) / 48959TL (France).

References

1. A. Andrada, V. del Barco, A. Moroianu, Locally conformally product structures on solvmanifolds. Ann. Mat. Pura Appl. **203**, 2425–2456 (2024)
2. F. Belgun, B. Flamencourt, A. Moroianu, Weyl structures with special holonomy on compact conformal manifolds. arXiv:2305.06637 (2023)
3. F. Belgun, A. Moroianu, On the irreducibility of locally metric connections. J. Reine Angew. Math. **714**, 123–150 (2016)
4. V. del Barco, A. Moroianu, The structure of locally conformally product Lie algebras. arxiv:2404.17956 (2024)
5. B. Flamencourt, Locally conformally product structures. Int. J. Math. **35**(5), 2450013 (2024)
6. B. Flamencourt, The characteristic group of conformally product structures. Ann. Mat. **204**, 189–211 (2025)
7. P. Gauduchon, La 1-forme de torsion d'une variété hermitienne compacte. Math. Ann. **267**, 495–518 (1984)
8. M. Kourganoff, Similarity structures and de Rham decomposition. Math. Ann. **373**, 1075–1101 (2019)
9. V. Matveev, Y. Nikolayevsky, A counterexample to Belgun–Moroianu conjecture. C. R. Math. Acad. Sci. Paris **353**, 455–457 (2015)
10. V. Matveev, Y. Nikolayevsky, Locally conformally Berwald manifolds and compact quotients of reducible manifolds by homotheties. Ann. Inst. Fourier (Grenoble) **67**(2), 843–862 (2017)
11. A. Moroianu, M. Pilca, Conformal vector fields on lcK manifolds. Math. Res. Lett. **30**(5), 1591–1608 (2023)
12. A. Moroianu, M. Pilca, Adapted metrics on locally conformally product manifolds. Proc. Am. Math. Soc. **152**, 2221–2228 (2024)
13. Y. Tashiro, K. Miyashita, Conformal transformations in complete product Riemannian manifolds. J. Math. Soc. Jpn. **19**, 328–346 (1967)

On Some Properties of Hopf Manifolds

Nicolina Istrati and Alexandra Otiman

Dedicated to Professor Paul Gauduchon for his 80th birthday.

Abstract We review old and new properties of Hopf manifolds from the point of view of their analytic and metric structure.

Keywords Contractions · Locally conformally Kähler metric · Algebraic dimension · Deformations · Automorphism group

1 Introduction

Hopf manifolds constitute the main examples of compact complex manifolds not admitting any compatible Kähler metric. These manifolds date back to Hopf [14], who first studied complex manifolds having the diffeomorphism type of $\mathbb{S}^1 \times \mathbb{S}^{2n-1}$. Later on, motivated by the classification of compact complex surfaces, Kodaira [20] reviewed his definition and proposed a more general one, which is the one we adopt

N.I. was partially supported by the PNRR-III-C9-2023-I8 grant CF 149/31.07.2023 "Conformal Aspects of Geometry and Dynamics" and by a CNRS PEPS JCJC project. A.O. is supported by a DFF Sapere Aude grant "Conformal geometry: metrics and cohomology" and by a grant of the Ministry of Research and Innovation, CNCS-UEFISCDI, project no. P1-1.1-PD-2021-0474

N. Istrati (✉)
Département de mathématiques, Faculté des Sciences, Université d'Angers, Angers, France
e-mail: nicolina.istrati@univ-angers.fr

A. Otiman
Institut for Matematik and, Aarhus Institute of Advanced Studies, Aarhus University, Aarhus C, Denmark

Institute of Mathematics "Simion Stoilow" of the Romanian Academy, Bucharest, Romania
e-mail: alexandra.otiman@imar.ro; aiotiman@aias.au.dk

L. Ornea (ed.), *Real and Complex Geometry*,
https://doi.org/10.1007/978-3-031-92297-8_9

here, namely Hopf manifolds are compact complex manifolds of dimension $n \geq 2$ whose universal cover is $\mathbb{C}^n \setminus \{0\}$. Equivalently, they are finite quotients of $W_\gamma := \mathbb{C}^n \setminus \{0\}/\langle\gamma\rangle$ for $\gamma \in \mathrm{Aut}(\mathbb{C}^n, 0)$ a contraction.

The geometry of Hopf manifolds is very rich and they have been intensely studied, particularly as model spaces bearing locally conformally Kähler metrics. The aim of this short survey is to discuss some of the more fundamental properties of Hopf manifolds, concerning both the analytic invariants and the metric structure.

The paper is organized as follows. In Sect. 2, we review the general theory of Hopf manifolds, their classification according to the Poincaré-Dulac normal form of the contracting germ γ and several properties concerning their analytic deformations. This part mostly follows [12]. In Sect. 3, we discuss in more detail the structure of the automorphism group of a Hopf manifold W_γ. In particular, this part allows us to determine the Lee group of a Vaisman metric when γ is diagonal, and to show that when γ is not diagonal, no such Lee group, and therefore no Vaisman metric can exist (Proposition 3.2). This result is possibly new, but generalizes Belgun's proof [3] in dimension 2.

In Sect. 4 we revisit analytical and cohomological properties, such as the Dolbeault and the Bott-Chern cohomology and the Kodaira dimension of Hopf manifolds of the form W_γ. Moreover, for the case when γ is diagonal, we determine the algebraic dimension of W_γ in terms of the relations satisfied by the eigenvalues of $d_0\gamma$ (Proposition 4.4). This generalizes [20, Thm. 31], which determines the algebraic dimension of all Hopf surfaces.

Finally, Sect. 5 is concerned with the metric properties of Hopf manifolds, with a focus on locally conformally Kähler (lcK) metrics. Gauduchon and Ornea [10] gave the first proof that Hopf surfaces admit lcK metrics, and in the diagonal case, even Vaisman ones. Since then, multiple generalizations have been given, see for instance [3, 5, 18, 23]. In the last section, we briefly recall the history of the problem and choose to present Brunella's proof [5] adapted to the Hopf setting, which is the most general one in some sense, as it does not distinguish among the special froms of the corresponding contraction. We end with some non-existence results concerning other distinguished Hermitian metrics, following [7, 25].

2 Normal Forms of Contractions and the Classification of Hopf Manifolds

Definition 2.1 A *Hopf manifold* of dimension $n \geq 2$ is a compact complex manifold with universal cover biholomorphic to $W := \mathbb{C}^n \setminus \{0\}$. It is called *primary* if it has infinite cyclic fundamental group, and *secondary* otherwise.

Hopf's initial definition [14], which is also sometimes the one used in the literature, is the one corresponding to primary Hopf manifolds. In fact, a general Hopf manifold is very close to a primary one, by the following result:

Theorem 2.2 *(see [2] in dimension 2, [13, Thm 2.1, Cor 2.2] in general dimension) Let X be a Hopf manifold. Then $\pi_1(X)$ is a finite extension of an infinite cyclic subgroup generated by a contraction. In particular, X admits a primary Hopf manifold as a finite unramified normal cover.*

In [19], Kato moreover classifies the posibilities for the finite groups giving rise to secondary Hopf surfaces, as well as the diffeomorphism types of these. Hasegawa [13] partially generalises Kato's results to any dimension.

In any case, an equivalent definition of Hopf manifolds is:

Definition 2.3 A primary Hopf manifold of dimension $n \geq 2$ is a complex manifold W_γ obtained as the quotient of $W := \mathbb{C}^n \setminus \{0\}$ by the $\mathbb{Z}$-action generated by a holomorphic contraction $\gamma \in \mathrm{Aut}(\mathbb{C}^n \setminus \{0\})$. A secondary Hopf manifold is a finite quotient of a primary Hopf manifold.

Recall that a *contraction* γ of W is a biholomorphism of W such that γ^m converges uniformly to 0 on compact neighbourhoods of 0 when m goes to infinity. Note that by Hartogs' lemma, any $\gamma \in \mathrm{Aut}(\mathbb{C}^n \setminus \{0\})$ extends naturally to an automorphism of $\mathbb{C}^n$ fixing 0, and via Schwarz lemma it can be shown that γ is a contraction precisely when the eigenvalues of $d_0\gamma$ are all in the punctured unit disk $\mathbb{D}^*$ of $\mathbb{C}$.

As explained in [12, Appendix] following [2], any contraction can be put into normal form after a holomorphic change of coordinates of W. In order to explain this precisely, let us first introduce some terminology.

Definition 2.4 Let $\lambda = (\lambda_1, \ldots, \lambda_n) \in (\mathbb{C}^*)^n$. **A resonance relation** for λ is a relation of the form

$$\lambda_s = \lambda^m := \lambda_1^{m_1} \cdots \lambda_n^{m_n}, \quad s \in \{1, \ldots, n\}, m = (m_1, \ldots, m_n) \in \mathbb{N}^n \setminus \{e_s\}.$$

We will also denote this by $(s; m)$. Here $e_1, \ldots, e_n$ denotes the standard basis of $\mathbb{C}^n$. If λ has no resonance relation, we say that λ is **non-resonant**.

Remark 2.5 Note that if $\lambda \in (\mathbb{D}^*)^n$, then there is only a finite number of resonance relations for λ. Indeed, without loss of generality we can assume that $|\lambda_1| \leq |\lambda_2| \leq \ldots \leq |\lambda_n| < 1$ and then note that for each $s \in \{1, \ldots n\}$, for a relation $\lambda_s = \lambda^m$ with $m \in \mathbb{N}^n$, the length $|m| := \sum_{k=1}^n m_k \in \mathbb{N}$ is bounded:

$$\frac{\ln |\lambda_s|}{\ln |\lambda_1|} \leq |m| \leq \frac{\ln |\lambda_s|}{\ln |\lambda_n|}.$$

Definition 2.6 Let $\lambda \in (\mathbb{C}^*)^n$. An analytic map $\gamma = (\gamma_1, \ldots, \gamma_n)\colon \mathbb{C}^n \to \mathbb{C}^n$ is called λ**-resonant** if each component γ_i expresses as

$$\gamma_i(z) = \alpha_i z_i + \sum_m a_{m,i} z^m \tag{2.1}$$

with $\alpha_i, a_{m,i} \in \mathbb{C}$ and the sum is taken over each $m \in \mathbb{N}^n$ corresponding to a resonance relation $(i; m)$ for λ. We denote by R_λ the subalgebra of λ-resonant analytic maps, and by $G_\lambda \subset R_\lambda$ the group of invertible elements.

Remark 2.7 According to Remark 2.5, if $\lambda \in (\mathbb{D}^*)^n$ then there are only finitely many resonance relations, therefore R_λ consists only in polynomial maps and G_λ, being open in R_λ, is a finite dimensional connected complex Lie group. Its Lie algebra is given by

$$\mathfrak{g}_\lambda = \langle z_1 \frac{\partial}{\partial z_1}, \dots z_n \frac{\partial}{\partial z_n} \rangle \oplus \langle z^m \frac{\partial}{\partial z_s} \mid (s; m) \rangle$$

where the second family is taken over all resonance relations $(s; m)$ for λ, and the sum is a vector space sum.

Remark 2.8 If we denote by

$$d_\lambda \colon \mathbb{C}^n \to \mathbb{C}^n, \quad z \mapsto (\lambda_1 z_1, \dots, \lambda_n z_n)$$

then it is easy to check that

$$R_\lambda = \{\gamma \colon \mathbb{C}^n \to \mathbb{C}^n \mid \gamma \circ d_\lambda = d_\lambda \circ \gamma\}.$$

Thus the above remark implies that $\mathfrak{g}_\lambda = \mathfrak{aut}(W_{d_\lambda})$ and $G_\lambda / \langle d_\lambda \rangle = \mathrm{Aut}_0(W_{d_\lambda})$.

We have the following classification result:

Theorem 2.9 (Poincaré-Dulac Normal Form, cf. [2, p. 187] and [12, Appendix]) *Let $\gamma \in \mathrm{Aut}(\mathbb{C}^n, 0)$ be a contraction and let $\lambda = (\lambda_1, \dots, \lambda_n) \in (\mathbb{D}^*)^n$ be the eigenvalues of $d_0\gamma$. Then there exists a biholomorphism $h \in \mathrm{Aut}(\mathbb{C}^n, 0)$ and a contraction $\tilde{\gamma} \in G_\lambda$ so that $\tilde{\gamma} = h\gamma h^{-1}$. Moreover, one can pick $\tilde{\gamma}$ in upper triangular form, i.e. so that each component of $\tilde{\gamma}$ is of the form*

$$\tilde{\gamma}_j(z) = \lambda_j z_j + g_j(z_{j+1}, \dots, z_n), \quad \alpha_j \in \mathbb{C}^*, j \in \{1, \dots, n\}.$$

Such a $\tilde{\gamma}$ is called a normal form for γ.

Definition 2.10 A Hopf manifold of the form W_{d_λ} for some $\lambda \in (\mathbb{D}^*)^n$ is called *a diagonal Hopf manifold*. A primary Hopf manifold W_γ for which γ does not admit a diagonal normal form is also called a *a non-diagonal Hopf manifold.*

2.1 *Moduli Spaces and Deformations*

Note at this point that while $\lambda \in (\mathbb{D}^*)^n$ is an invariant of the Hopf manifold W_γ, the normal form $\tilde{\gamma}$ of γ is not unique. Indeed, G_λ acts on itself by conjugation and every element in the orbit $G_\lambda.\tilde{\gamma}$ gives a different representative of a normal form for the same Hopf manifold. In any case, as a consequence we obtain that G_λ parametrizes all small deformations of a Hopf manifold W_γ, more specifically:

Theorem 2.11 (see [12] for the full statement, but also [29], [4] for particular cases) *Let W_γ be a Hopf manifold with $\gamma \in G_\lambda$ in normal form and let $S \subset G_\lambda$ be a complex analytic slice passing through γ and complementary to the orbit $G_\lambda.\gamma$. Then the family $(W_s)_{s\in S}$ forms a versal deformation of W_γ. Moreover we have:*

$$H^0(W_\gamma, TW_\gamma) \cong H^1(W_\gamma, TW_\gamma) \cong \ker(id - \gamma_* \in \mathrm{End}(\mathfrak{g}_\lambda)),$$
$$H^k(W_\gamma, TW_\gamma) = 0, \quad k > 1.$$

For primary Hopf manifolds we have the following result, typically very useful in order to extend properties of diagonal Hopf manifolds to non-diagonal ones via upper-semicontinuity.

Theorem 2.12 (See for Instance [12, p. 242]) *Let W_γ be a Hopf manifold with $\gamma \in G_\lambda$ in normal form.There exits a holomorphic family of Hopf manifolds $p\colon \mathcal{X} \to \mathbb{C}$ such that $p^{-1}(0) = W_{d_\lambda}$ and for any $t \in \mathbb{C}^*$ we have $p^{-1}(t) \cong W_\gamma$.*

Proof The proof of this result comes down to noticing that when the components of γ are in the form (2.1), by conjugating with diagonal automorphisms, which live in G_λ, one can simultaneously make all the coefficients $a_{m,i}$ as small as one wishes. More specifically, let

$$\gamma = d_\lambda + (g_1, \ldots, g_n), \quad g_i(z) = \sum_m a_{m,i} z^m,$$

be in upper triangular normal form, so that for each coefficient $a_{m,i}$ that does not vanish we have $m_j = 0$ as soon as $1 \leq j \leq i$, and in particular $n_{i,m} := \sum_{j=1}^n j m_j - i > 0$.

Define the family $\mathcal{X} := W \times \mathbb{C}/\mathbb{Z}$ where $1 \in \mathbb{Z}$ acts on $(z, t) \in W \times \mathbb{C}$ by

$$1.(z, t) = (d_\lambda(z) + g_t(z), t)$$

with the i-th component of g_t being

$$\sum_m t^{n_{i,m}} a_{m,i} z^m.$$

The map p is the naturally induced projection, so that $p^{-1}(0) = W_{d_\lambda}$.

For any $\mu = (\mu_1, \dots, \mu_n) \in (\mathbb{C}^*)^n$ we have $d_\mu \in G_\lambda$ and the i-th component of $d_\mu \gamma d_\mu^{-1}$ is given by

$$\lambda_i z_i + \sum_m a_{m,i} \frac{\mu_i}{\mu^m} z^m.$$

In particular, for $t \in \mathbb{C}^*$ and $\mu := (t^{-1}, t^{-2}, \dots, t^{-n})$ we obtain that $d_\mu \gamma d_\mu^{-1} = d_\lambda + g_t$ and so $W_\gamma \cong p^{-1}(t)$, which concludes the proof. ■

3 Automorphisms

Given $\lambda \in (\mathbb{D}^*)^n$ and $\gamma \in G_\lambda$ with eigenvalues $\lambda = (\lambda_1, \dots, \lambda_n)$, let us denote by

$$\mathfrak{g}_\gamma := \mathfrak{g}_\lambda^{\langle \gamma_* \rangle} = H^0(W_\gamma, TW_\gamma)$$

where the last equality holds by Theorem 2.11. In particular, $\mathrm{Aut}_0(W_\gamma) = G_\gamma / \langle \gamma \rangle$, where G_γ is the exponential of $\mathfrak{g}_\gamma$ in the group G_λ.

Let us now make more explicit the Lie group structure of G_λ. Recall that by Remarks 2.7 and 2.8, the group G_λ has Lie algebra

$$\mathfrak{g}_\lambda = \langle z_1 \frac{\partial}{\partial z_1}, \dots z_n \frac{\partial}{\partial z_n} \rangle \oplus \langle z^m \frac{\partial}{\partial z_s} \mid (s; m) \rangle = \mathfrak{a} \oplus \mathfrak{r}.$$

Note that the first summand $\mathfrak{a}$ corresponds to the Lie algebra of the standard torus $(\mathbb{C}^*)^n \subset \mathrm{Aut}(W)$. A vector field of the form $\xi_{s,m} = z^m \frac{\partial}{\partial z_s} \in \mathfrak{r}$ generates the effective action of $\mathbb{C} \subset \mathrm{Aut}(W)$ given by

$$\Phi^t_{\xi_{s,m}}(z) = z + t z^m e_s, \quad t \in \mathbb{C}, z \in W.$$

Moreover, $\mathfrak{r}$ has a natural finite $\mathbb{N}$-grading given by

$$\mathfrak{r}_l := \langle \xi_{s,m} | (s; m), l = |m| \rangle \quad l \in \mathbb{N}, l \geq 1.$$

Denoting by $Z_j = z_j \frac{\partial}{\partial z_j} = \xi_{j,e_j}$, $j \in \{1, \dots, n\}$, we have the following relations:

$$[Z_j, \xi_{s,m}] = (m_j - \delta_{j,s}) \xi_{s,m}; \quad [\xi_{s,m}, \xi_{t,n}] = n_s \xi_{t,m+n-e_s} - m_t \xi_{s,m+n-e_t}. \qquad (3.1)$$

In particular, $[\mathfrak{r}_a, \mathfrak{r}_b] \subset \mathfrak{r}_{a+b-1}$ every time $(a, b) \neq (1, 1)$, so that $\mathfrak{n}_\lambda := \oplus_{l \geq 2} \mathfrak{r}_l$ is a nilpotent ideal of $\mathfrak{g}_\lambda$ and

$$\mathfrak{g}_\lambda^{red} := \mathfrak{a} \oplus \mathfrak{r}_1 \subset \mathfrak{gl}(n, \mathbb{C})$$

is a sub-algebra of $\mathfrak{g}_\lambda$. Moreover, the equivalence relation $\sim$ on $\{1, \dots, n\}$ defined by $i \sim j \Leftrightarrow \lambda_i = \lambda_j$ gives us a partition $\{1, \dots, n\} = \bigsqcup_{i=1}^k I_i$ so that for each orbit I_i, denoting by l_i its cardinal, we readily find, via (3.1), isomorphisms of Lie algebras

$$\mathfrak{g}_i := \langle \xi_{j,e_k} | j, k \in I_i \rangle \cong \mathfrak{gl}(l_i, \mathbb{C}), \quad \mathfrak{g}_\lambda^{red} \cong \bigoplus_{i=1}^k \mathfrak{g}_i.$$

In particular, $\mathfrak{g}_\lambda^{red} = \mathfrak{g}_\lambda / \mathfrak{n}_\lambda$ is reductive, implying that $\mathfrak{n}_\lambda$ is the nilradical of $\mathfrak{g}_\lambda$ and we have a semi-direct decomposition

$$\mathfrak{g}_\lambda = \mathfrak{n}_\lambda \rtimes \mathfrak{g}_\lambda^{red}. \tag{3.2}$$

Denoting by $N := \exp(\mathfrak{n}_\lambda) \subset \mathrm{Aut}(W)$ and by $G_\lambda^{red} := \exp(\mathfrak{g}_\lambda^{red}) \subset \mathrm{GL}(n, \mathbb{C}) \subset \mathrm{Aut}(W)$, this lifts to

$$G_\lambda = N \rtimes G_\lambda^{red}, \quad G_\lambda^{red} = \prod_{i=1}^k \mathrm{GL}(l_i, \mathbb{C}) \tag{3.3}$$

where the product group is identified with a subgroup of $\mathrm{GL}(n, \mathbb{C})$ via the standard block-diagonal embedding.

Note that the group N has no compact subgroup, since any vector field of $\mathfrak{n}_\lambda$ generates an effective $\mathbb{C}$-action. In particular, the exponential map $\mathfrak{n}_\lambda \to N$ is a biholomorphism and N is simply connected.

Remark 3.1 If λ has only resonance relations $(s; m)$ of degree $|m| \geq 2$, i.e. if $\mathfrak{n}_\lambda = \mathfrak{r}$, then $\mathfrak{n}_\lambda = [\mathfrak{g}_\lambda, \mathfrak{g}_\lambda]$ and $\mathfrak{g}_\lambda$ has the structure of a complex *splittable* solvable Lie algebra, i.e. $\mathfrak{g}_\lambda = \mathfrak{a} \ltimes_\phi \mathfrak{n}$ for $\phi : \mathfrak{a} \to \mathfrak{aut}(\mathfrak{n}_\lambda)$ given by the adjoint representation. In particular, we have an exact sequence of complex Lie groups:

$$0 \longrightarrow N \longrightarrow G_\lambda \longrightarrow (\mathbb{C}^*)^n \longrightarrow 0.$$

Proposition 3.2 *Let W_γ be a primary Hopf manifold. If W_γ is non-diagonal, there exists no compact torus $H \subset \mathrm{Aut}_0(W_\gamma)$ so that $J\mathfrak{h} \cap \mathfrak{h} \neq \{0\}$. If $\gamma = d_\lambda$, then the only complex one-dimensional Lie group G sitting in a compact torus $H \subset \mathrm{Aut}_0(W_\gamma)$ is generated by*

$$Z := \sum_{j=1}^n \ln |\lambda_j| z_j \frac{\partial}{\partial z_j}.$$

Its action on W_γ is induced by the action of $\mathbb{C}^*$ via

$$\mu.[z_1, \dots, z_n] = [\mu^{\ln |\lambda_1|} z_1, \dots, \mu^{\ln |\lambda_n|} z_n], \quad \mu \in \mathbb{C}^*, [z] \in W_\gamma.$$

Proof Let $\gamma \in G_\lambda$ be in normal form, let us denote by $G := G_\gamma/\langle\gamma\rangle \subset G_\lambda/\langle\gamma\rangle$, let

$$q\colon G_\lambda \to G_\lambda^{red}, \quad q'\colon G_\lambda/\langle\gamma\rangle \to G_\lambda^{red}/\langle q(\gamma)\rangle$$

be the natural projections and we denote also by $G' := G_\lambda^{red}/\langle q(\gamma)\rangle$. Note that since $N \cap \langle\gamma\rangle = \{\mathrm{id}\}$, we have $N = \ker q'$.

Let $H \subset G$ be a torus as in the statement. By eventually replacing it with a smaller torus, we can assume that $H_c := \exp(J\mathfrak{h} \cap \mathfrak{h}) \subset H$ is dense in H. Since the group N has no compact subgroup, $H \cap N$ is trivial and therefore H projects isomorphically via q' to a torus H' in G' whose Lie algebra $\mathfrak{h}'$ also satisfies $J\mathfrak{h}' \cap \mathfrak{h}' \neq \{0\}$.

By Eq. (3.3), the exponential map of G_λ^{red} is surjective, therefore we can write $\gamma' := q(\gamma) = \exp(\xi'_\gamma)$ for some vector field $\xi'_\gamma \in \mathfrak{g}_\lambda^{red}$. Let us denote by $\mathfrak{t} \subset \mathfrak{a}$ the Lie algebra of the real compact subtorus of $(\mathbb{C}^*)^n \subset G_\lambda$. Since any maximal torus in G_λ^{red} is conjugated to $\exp(\mathfrak{t})$, we find that any maximal torus in G' up to conjugation has Lie algebra either $\mathfrak{t}$ or $\mathfrak{t} \oplus \mathbb{R}\xi'_\gamma$. Since such a torus must moreover contain the non-purely real Lie group H', we must be in the second case, so that $\mathfrak{h}' \subset \mathfrak{t} \oplus \mathbb{R}\xi'_\gamma$.

Note however that if $\xi'_\gamma = \xi_a + \xi_\mathfrak{r}$ with $\xi_\mathfrak{a} \in \mathfrak{a}, \xi_\mathfrak{r} \in \mathfrak{r}$, then the condition $J\mathfrak{h}' \cap \mathfrak{h}' \neq \{0\}$ implies that $\xi_\mathfrak{r} = 0$, since else we would have that $J\xi_r$ and ξ_r span the same real line. This implies thus that $\gamma' \in (\mathbb{C}^*)^n$, and so $\gamma' = d_\lambda$ and $\gamma = \gamma_N \circ d_\lambda$ with $\gamma_N \in N$. Note that since γ is central in G_γ, γ_N is central in $N_\gamma := N \cap G_\gamma$.

Consider now the holomorphic map $\phi\colon G_\gamma \to N_\gamma$, $g \mapsto q(g)^{-1}g$ and the induced holomorphic map $\phi'\colon G \to N_\gamma/\langle\gamma_N\rangle$. Then $\phi'|_{H_c}\colon H_c \to N_\gamma/\langle\gamma_N\rangle$ is holomorphic and bounded, since $\phi'(H_c) \subset \phi'(H)$. On the other hand, as $\gamma_N \in \mathcal{Z}(N_\gamma)$, we have that if $\gamma_N \neq \mathrm{id}$ then $N_\gamma/\langle\gamma_N\rangle$ is biholomorphic to $\mathbb{C}^* \times \mathbb{C}^{\dim N_\gamma - 1}$ via the exponential map, therefore we find by Liouville's theorem that $\phi'|_{H_c}$ is constant, and hence also $\phi'|_H$ is constant by density. In particular, as $\hat{H} := \exp\mathfrak{h} \subset G_\gamma$ is connected, we also have that $\phi|_{\hat{H}}$ is constant and so $\hat{H} \subset G_\lambda^{red}$. At the same time, the argument from before shows that $\hat{H}$ intersects $\langle\gamma\rangle$ non trivially, therefore $\gamma = d_\lambda \in G_\lambda^{red}$ is indeed diagonal.

When $\gamma = d_\lambda$ is diagonal, writing $\lambda_j = r_j e^{i\theta_j}$ for $j \in \{1, \ldots, n\}$, we find that $\mathfrak{h} \cap J\mathfrak{h}$ is generated over $\mathbb{C}$ by

$$Z := \sum_{j=1}^{n} \ln r_j z_j \frac{\partial}{\partial z_j}.$$

This concludes the proof. ∎

4 Cohomology and Analytic Invariants

We present in this section cohomological properties of primary Hopf manifolds. Since these have the tolological type of $\mathbb{S}^1 \times \mathbb{S}^{2n-1}$, de Rham cohomology is readily computable. On the other hand, the complex cohomologies, such as Dolbeault and Bott-Chern, are more intricate and can be detected via the analytic family constructed by Theorem 2.12.

For the next result we follow the ideas of proof of Mall [21] for which we provide a few more details. Note that different proofs are possible, by using for instance the existence of Vaisman metrics on diagonal Hopf manifolds.

Theorem 4.1 ([21, Thm. 3]) *The Hodge numbers of an n-dimensional primary Hopf manifold W_γ are given by*

$$h_{\overline{\partial}}^{p,q} = \begin{cases} 1, & \text{if } (p,q) \in \{(0,0),(0,1),(n-1,n),(n,n)\} \\ 0, & \text{otherwise.} \end{cases}$$

In particular, Hopf manifolds do not satisfy the $\partial\overline{\partial}$-lemma, however their Frölicher spectral sequence degenerates at first page.

Proof We compute first the Hodge numbers of a diagonal Hopf manifold $X = W_{d_\lambda}$, where $\lambda \in (\mathbb{D}^*)^n$. We will take a spectral sequence approach, since this will simplify some computations from [21]. By [11, Subsection 5.2], the Dolbeault cohomology of X can be computed as $\mathbb{Z}$-equivariant cohomology of W. More precisely, for each fixed $k \in \mathbb{N}$, one can consider a spectral sequence $E_r^{\bullet,\bullet}$ whose second page is given by

$$E_2^{p,q} = H^p(\mathbb{Z}, H^q(W, \Omega_W^k)) \Rightarrow H^{p+q}(X, \Omega_X^k),$$

where $\mathbb{Z} = \langle d_\lambda \rangle$ acts on a class $[\omega] \in H^q(W, \Omega_W^k)$ by $d_\lambda \cdot [\omega] = [d_\lambda^* \omega]$.

Since $\mathbb{Z}$ is of rank 1, $E_2^{p,\bullet} = 0$ for $p \geq 2$, therefore $E_\infty^{\bullet,\bullet} = E_2^{\bullet,\bullet}$ and we get

$$\begin{aligned} h_{\overline{\partial}}^{k,l} := \dim_{\mathbb{C}} H^l(X, \Omega_X^k) &= \dim_{\mathbb{C}} H^0(\mathbb{Z}, H^l(W, \Omega_W^k)) \\ &\quad + \dim_{\mathbb{C}} H^1(\mathbb{Z}, H^{l-1}(W, \Omega_W^k)). \end{aligned}$$

Furthermore, since Ω_W^k is free over $\mathcal{O}_W$ and since by [12, Lemma 3.2] we have that $H^p(W, \mathcal{O}_W) = 0$ for $p \neq 0, n-1$, we readily infer that $h_{\overline{\partial}}^{k,l} = 0$ for $1 < l < n-1$ and we are left with computing

$$h_{\overline{\partial}}^{k,0} = \dim_{\mathbb{C}} H^0(\mathbb{Z}, H^0(W, \Omega_W^k)) = \dim_{\mathbb{C}} \operatorname{Ker}(\mathrm{Id} - d_\lambda^*)_{|H^0(W, \Omega_W^k)}$$

$$h_{\overline{\partial}}^{k,1} = \dim_{\mathbb{C}} H^1(\mathbb{Z}, H^0(W, \Omega_W^k)) = \dim_{\mathbb{C}} \operatorname{CoKer}(\mathrm{Id} - d_\lambda^*)_{|H^0(W, \Omega_W^k)}$$

$$h_{\bar{\partial}}^{k,n-1} = \dim_{\mathbb{C}} H^0(\mathbb{Z}, H^{n-1}(W, \Omega_W^k)) = \dim_{\mathbb{C}} \mathrm{Ker}(\mathrm{Id} - d_\lambda^*)_{|H^{n-1}(W,\Omega_W^k)}$$

$$h_{\bar{\partial}}^{k,n} = \dim_{\mathbb{C}} H^1(\mathbb{Z}, H^{n-1}(W, \Omega_W^k)) = \dim_{\mathbb{C}} \mathrm{CoKer}(\mathrm{Id} - d_\lambda^*)_{|H^{n-1}(W,\Omega_W^k)}.$$

Clearly $h_{\bar{\partial}}^{0,0} = 1$. We start by showing that $h_{\bar{\partial}}^{k,0} = 0$ for all $k \neq 0$. Indeed, take a d_λ-invariant holomorphic k-form on W

$$\omega = \sum_I \sum_{\alpha \in \mathbb{N}^n} a_\alpha^I z^\alpha dz_I, \tag{4.1}$$

where $I = \{(i_1, \ldots, i_k) | 1 \leq i_1 < \ldots < i_k \leq n\}$, $dz_I = dz_{i_1} \wedge \ldots \wedge dz_{i_k}$, $z^\alpha = z_1^{\alpha_1} \ldots z_n^{\alpha_n}$ and $a_\alpha \in \mathbb{C}$. The condition $d_\lambda^* \omega = \omega$ reads as

$$a_\alpha^I \lambda^\alpha \cdot \lambda_{i_1} \ldots \lambda_{i_k} = a_\alpha^I \tag{4.2}$$

for any I, where $\lambda^\alpha = \lambda_1^{\alpha_1} \ldots \lambda_n^{\alpha_n}$. If $k \neq 0$, since $|\lambda_{i_1}| \leq \ldots \leq |\lambda_{i_p}| < 1$, (4.2) can take place if and only if $a_\alpha^I = 0$, implying $\omega = 0$.

In order to show $h_{\bar{\partial}}^{0,1} = 1$, we notice that

$$\mathrm{CoKer}(\mathrm{Id} - d_\lambda^*)_{|H^0(W,\mathcal{O}_W)} = \frac{H^0(W, \mathcal{O}_W)}{\mathrm{Im}(\mathrm{Id} - d_\lambda^*)} \simeq \mathbb{C},$$

since any holomorphic function f on W of the form

$$f = \sum_{\alpha \in \mathbb{N}^n \setminus \{0\}} f_\alpha z^\alpha = g - d_\lambda^* g \text{ for } g := \sum_{\alpha \in \mathbb{N}^n \setminus \{0\}} \frac{f_\alpha}{1 - \lambda^\alpha} z^\alpha$$

and the series defining g converges, since $(1/(1 - \lambda^\alpha))_{\alpha \in \mathbb{N}^n \setminus \{0\}}$ is bounded.

For $k \neq 1$, we have $\frac{H^0(W, \Omega_W^k)}{\mathrm{Im}(\mathrm{Id} - d_\lambda^*)} = 0$ since for any ω presented as in (4.1), we have

$$\omega = (\mathrm{Id} - d_\lambda^*)(h), \quad h := \sum_I \sum_{\alpha \in \mathbb{N}^n} \frac{a_\alpha^I}{1 - \lambda^\alpha \cdot \lambda_{i_1} \ldots \lambda_{i_p}} z^\alpha dz_I \in H^0(W, \Omega_W^k).$$

This implies that $h_{\bar{\partial}}^{k,1} = 0$ for $k \neq 1$.

By Serre duality, the only Dolbeault numbers left to compute are $h_{\bar{\partial}}^{0,n-1}$, $h_{\bar{\partial}}^{1,n-1}$, $h_{\bar{\partial}}^{0,n}$ and $h_{\bar{\partial}}^{1,n}$. By [12, Lemma 3.2], we have

$$H^{n-1}(W, \mathcal{O}_W) \cong \{ \sum_{\alpha \in (\mathbb{Z}_{<0})^n} a_\alpha z^\alpha \text{ convergent on } (\mathbb{C}^*)^n \}$$

and the induced $\mathbb{Z}$-action on the right hand-side is generated by the natural action of d_λ on holomorphic functions on $(\mathbb{C}^*)^n$. Then the same reasoning as above gives us the remaining Dolbeault numbers.

For the general case, we use that any primary Hopf manifold W_γ with $\gamma \in G_\lambda$ is arbitrarily close to the diagonal one W_{d_λ} via Theorem 2.12. Then, by the upper semi-continuity of the Hodge numbers (see [6, Ch. 3, Thm. 4.12]), we get $h^{p,q}_{\overline{\partial}}(W_\gamma) \leq h^{p,q}_{\overline{\partial}}(W_{d_\lambda})$ for any degree (p, q). In particular, the numbers $h^{p,q}_{\overline{\partial}}(W_\gamma)$ are lower than 1 for $(p, q) \in \{(0, 0), (0, 1), (n-1, n), (n, n)\}$ and vanish in any other degree. The final result is obtained by applying the Frölicher inequality $h^{0,1}_{\overline{\partial}} \geq b_1(W_\gamma) = 1$. ■

For the computation of the Bott-Chern numbers, we will in fact use the existence of Vaisman metrics, cf. Sect. 5.

Theorem 4.2 *The Bott-Chern numbers of a primary Hopf manifold are given by*

$$h^{p,q}_{BC} = \begin{cases} 1, & \text{if } (p,q) \in \{(0,0), (1,1), (n-1,n), (n,n-1), (n,n)\} \\ 0, & \text{otherwise.} \end{cases}$$

Proof A diagonal Hopf manifold admits a Vaisman metric, following the discussion of Sect. 5. For such manifolds, the Bott-Chern numbers can be recovered from the Dolbeault cohomology, as shown in [17, Thm 3.1, Cor. 5.2], retrieving the numbers in the statement. As a general primary Hopf manifold is arbitrarily close to a diagonal one, and since [17, Cor. 5.4] states that the Bott-Chern numbers remain constant for a small deformation of a compact Vaisman manifold, we infer the statement for any primary Hopf manifold. ■

Proposition 4.3 *The Kodaira dimension of an arbitrary Hopf manifold is* $-\infty$.

Proof Since any Hopf manifold is finitely covered by a primary one, it is enough to show the result for any Hopf manifold of the form W_γ, with $\gamma \in G_\lambda$ in normal form. We first show that for the diagonal Hopf W_{d_λ}, all plurigenera vanish. Indeed, the meromorphic form on W

$$\frac{dz_1 \wedge \ldots \wedge dz_n}{z_1 z_2 \ldots z_n}$$

is invariant under d_λ, hence descends to a form ω on W_{d_λ}. The canonical divisor $K_{W_{d_\lambda}}$ is then given by $div(\omega) = -W_1 - \ldots - W_n$, where W_i is the complex hypersurface defined by $\{z_i = 0\}$. Consequently, the canonical bundle $K_{W_{d_\lambda}}$ is anti-effective and so the plurigenera vanish for W_{d_λ}.

Consider the family of deformation $p : \mathcal{X} \to \mathbb{C}$ in Theorem 2.12 such that $p^{-1}(0) \simeq W_{d_\lambda}$ and $p^{-1}(t) \simeq W_\gamma$, for any $t \neq 0$. By [6, Ch. 3, Thm. 4.12], for any $l > 0$, the plurigenera $P_t := \dim_{\mathbb{C}} H^0(p^{-1}(t), K^l_{p^{-1}(t)})$ give upper semi-continuous functions $t \mapsto P_t$, therefore the plurigenera of W_γ vanish as well and $\mathrm{Kod}(W_\gamma) = -\infty$. ■

The following result computes the algebraic dimension of a diagonal Hopf manifold in terms of the relations between the $\lambda_j's$. We note in particular that all possible values strictly lower than the complex dimension can be attained as algebraic dimensions. Of course there is no Moishezon Hopf manifold, as Hopf manifolds do not satisfy the $\partial\bar{\partial}$-lemma. We also note that since the algebraic dimension is not upper-semicontinuous in general [8], we cannot directly infer anything about the algebraic dimension of a non-diagonal Hopf manifold.

For any $\lambda \in (\mathbb{C}^*)^n$, let us introduce the group homomorphism

$$\psi_\lambda \colon \mathbb{Z}^n \to \mathbb{C}^*, \quad m \mapsto \lambda^m.$$

Proposition 4.4 *Let W_{d_λ} be an n-dimensional diagonal Hopf manifold. Then* $a(W_{d_\lambda}) = \mathrm{rank}(\ker \psi_\lambda)$.

Proof Let us first assume that $\ker \psi_\lambda = \{0\}$ and show that W_{d_λ} has no meromorphic function, whence vanishing algebraic dimension. Assume, by contradiction, that f is the pullback on W of a non-constant meromorphic function on W_{d_λ}. Since $n \geq 2$, f can be extended as a meromorphic function on $\mathbb{C}^n$ and be written as $\frac{g}{h}$, where g and h are holomorphic functions on $\mathbb{C}^n$ and $d_\lambda^* f = f$. Both g and h can be expressed as power series

$$g(z) = \sum_{m\in\mathbb{N}^n} a_m z^m, \quad h(z) = \sum_{l\in\mathbb{N}^n} b_l z^l$$

satisfying $g(\lambda_1 z_1, \ldots, \lambda_n z_n)h(z) = h(\lambda_1 z_1, \ldots, \lambda_n z_n)g(z)$. This implies

$$\sum_{m\in\mathbb{N}^n} a_m \lambda^m z^m \sum_{l\in\mathbb{N}^n} b_l z^l = \sum_{m\in\mathbb{N}^n} a_m z^m \sum_{l\in\mathbb{N}^n} b_l \lambda^l z^l.$$

Take now the lowest degrees k_1 and k_2 in the expansions of g and h, respectively and take $m_0 = (m_{01}, \ldots, m_{0n})$ and $l_0 = (l_{01}, \ldots, l_{0n})$ such that m_{01} and l_{01} are the smallest among the multi-indices m and l of degree k_1 and k_2, respectively. Then

$$a_{m_0} b_{l_0} \lambda^{m_0} z^{m_0+l_0} = a_{m_0} b_{l_0} \lambda^{l_0} z^{m_0+l_0},$$

which gives $\lambda^{m_0} = \lambda^{l_0}$, contradicting the assumption. Thus indeed $a(W_{d_\lambda}) = 0$.

In the general case, assuming that $k := \mathrm{rank}(\ker \psi_\lambda)$, any $m \in \ker \psi_\lambda$ gives rise to the d_λ-invariant meromorphic function z^m on W, and so to a meromorphic function f_m on W_{d_λ}. In particular, we have $\mathbb{C}(f_m | m \in \ker \psi_\lambda) \subset \mathbb{C}(W_{d_\lambda})$. Now note that $\ker \psi_\lambda$ is a subgroup of $\mathbb{Z}^n$, therefore torsion free and so there exist $m_1, \ldots, m_k \in \ker \psi_\lambda$ so that $\ker \psi_\lambda = \mathbb{Z}\langle m_1, \ldots, m_k\rangle \cong \mathbb{Z}^k$. We claim that the associated k meromorphic functions $f_i = f_{m_i}$, $i \in \{1, \ldots, k\}$, are algebraically independent. Indeed, if there is some polynomial $0 \neq P = \sum a_l X^l \in \mathbb{C}[X_1, \ldots, X_k]$ with $P(f_1, \ldots, f_k) = 0$ then, for each $l \in \mathbb{N}^k$ with $a_l \neq 0$ we must have $\sum_{i=1}^k l_i m_i = 0$, which then implies that $l = 0$ which is a contradiction.

Therefore, we have $k \leq a(W_{d_\lambda})$. If $k = n - 1$, this automatically implies that $a(W_{d_\lambda}) = n - 1$ since $a(W_{d_\lambda}) < n$.

For the case $k < n - 1$, we still need to show the converse inequality. In order to do so, note that there exists a subset $J \subset \{1, \ldots, n\}$ of cardinal $n - k$ so that $\ker \psi_\lambda \cap \langle e_j | j \in J\rangle = \{0\}$. After eventually reordering coordinates, we can assume that $J = \{k + 1, \ldots, n\}$. By defining $\mu := (\lambda_{k+1}, \ldots, \lambda_n) \in (\mathbb{D}^*)^{n-k}$, it follows that $\ker \psi_\mu = \{0\}$. At the same time, $Y := \{z_1 = \ldots = z_k = 0\} \subset W_{d_\lambda}$ is a complex submanifold of W_{d_λ} of codimension k which is biholomorphic to W_{d_μ}. Since $\dim_{\mathbb{C}} Y > 1$, the above discussion implies that Y has algebraic dimension 0. Applying [27, Thm 3.8 (ii)], we get $a(W_{d_\lambda}) \leq a(Y) + \mathrm{codim}(Y) = k$. We conclude therefore that $a(W_{d_\lambda}) = k$. ■

Remark 4.5 Note that for a generic $\lambda \in (\mathbb{D}^*)^n$, ψ_λ is injective and therefore $a(W_{d_\lambda}) = 0$.

Remark 4.6 The simplest Hopf manifolds of positive algebraic dimension k correspond to $\lambda \in (\mathbb{D}^*)^n$ with $\lambda_1 = \ldots = \lambda_{k+1}$ and so that ψ_μ is injective, where $\mu = (\lambda_{k+1}, \ldots, \lambda_n) \in (\mathbb{D}^*)^{n-k}$. In this case, the algebraic reduction is given by $\mathbb{CP}^k$. Indeed, consider the meromorphic map induced by the linear projection to the first $k + 1$-coordinates:

$$r \colon W_{d_\lambda} \dashrightarrow \mathbb{P}^k, \quad [z_1, \ldots, z_n] \mapsto (z_1 : \cdots : z_{k+1}),$$

where $[z]$ denotes the image of $z \in W$ in W_{d_λ}. Then $Z := \{z_1 = \ldots = z_{k+1} = 0\} \subset W_{d_\lambda}$ is the indeterminacy locus of r and r lifts to a holomorphic map r^* on the blow-up of W_{d_λ} along Z

$$\mathrm{Bl}_Z W_{d_\lambda} = \{([z], (u_1 : \cdots u_{k+1})) \mid z_i u_j = z_j u_i, \forall i, j \in \{1, \ldots, k+1\}\} \subset W_{d_\lambda} \times \mathbb{P}^k$$

which coincides with the projection to the second factor. In particular, the fibers of r^* are connected, all isomorphic to the $n - k$-dimensional diagonal Hopf manifold W_{d_μ}. Since $a(W_{d_\lambda}) = k$ by the above, it follows that $(r^*, \mathbb{P}^k)$ is the algebraic reduction of W_{d_λ}.

5 Hermitian Metrics

Due to their topological restrictions, Hopf manifolds cannot support a Kähler metric. Therefore, a relevant question for Hermitian non-Kähler geometry is what kind of special metrics Hopf manifolds possess. We recall briefly some of the definitions of these special geometries and we shall focus more on the one which has been greatly linked to their study, which is *locally conformally Kähler (lcK) geometry*.

5.1 lcK and Vaisman Metrics

A Hermitian metric ω on a complex manifold X is called *locally conformally Kähler* (lcK) if $d\omega = \theta \wedge \omega$ for a closed one-form θ, called the Lee form of the metric. If moreover θ is parallel with respect to the Levi-Civita connection of the metric, ω is called *Vaisman*.

The existence of lcK metrics on Hopf manifolds has a long history, going back to Vaisman [28], who proved than on a diagonal Hopf manifold W_{d_λ} with $|\lambda_1| = \ldots = |\lambda_n|$, the Hermitian metric induced by

$$\omega = \frac{i\partial\overline{\partial}||z||^2}{||z||^2}$$

is Vaisman. In fact, this is the only case where an explicit lcK metric can be written down on a Hopf manifold. Next, Gauduchon and Ornea [10] proved that any Hopf surface admits lcK metrics. On the general diagonal Hopf surface, they construct an lcK metric of the form $\omega = f^{-1}i\partial\overline{\partial}f$, where the potential f is given as the implicit unique solution of some functional equation, and then by direct computation they infer that this metric is Vaisman. On non-diagonal Hopf surfaces, the existence of the metric is inferred via a deformation argument. A different deformation argument for constructing lcK metrics on Hopf surfaces was used by Belgun [3], who also showed that non-diagonal Hopf surfaces are lcK, but never Vaisman.

Since then, many different arguments for the existence of lcK metrics on primary Hopf manifolds of any dimension have been given. For instance the direct generalisation of [10] to any diagonal Hopf manifolds appears in [18] (see also [15, Sect. 2.6.1]), where the authors also show that the constructed metric is Vaisman. One can then use a deformation argument of [22] and the isotrivial family of Theorem 2.12 in order to deduce the existence of lcK metrics on any non-diagonal Hopf manifold. Other works of Ornea-Verbitsky also construct such metrics using more involved arguments (see [23] for linear Hopf manifolds and [24] for non-linear Hopf manifolds).

A unifying point of view for the existence of lcK metrics on any primary Hopf manifold, diagonal or not, appears in an argument of Brunella [5] formulated for Kato surfaces. This construction uses the fact that a conformal class of an lcK metric on X is equivalent to a conformal class of a Kähler metric on the universal cover $\tilde{X}$ on which the deck group acts by homotheties. Without going into the definition of a Kato manifold, let us make explicit the argument from [5] for Hopf manifolds:

Theorem 5.1 (Brunella) *Every primary Hopf manifold carries lcK metrics.*

Proof Let W_γ be the n-dimensional Hopf manifold in question, and for any $r > 0$, let us denote by $\mathbb{B}_r = \{z \in \mathbb{C}^n, ||z||^2 < r\}$. Since γ is a contraction, we can pick $0 < s < 1$ so that $\gamma(\overline{\mathbb{B}}_1) \subset \mathbb{B}_s$.

Let $f_0(z) := ||z||^2$, $f(z) := f_0 \circ \gamma^{-1}$. Note that as f is strictly plurisubharmonic, via the maximum principle we have that $m := \min_{\partial\mathbb{B}_s} f < M := \max_{\partial\mathbb{B}_1} f$, thus for any $A > \frac{M-m}{1-s^2} > 0$ there exists $c \in (M-A, m-As^2)$ and we have $A \cdot f_0 + c > f$ on $\partial\mathbb{B}_1$ and $A \cdot f_0 + c < f$ on $\partial\mathbb{B}_s$. Thus the function ψ defined as the regularised maximum between $A \cdot f_0 + c$ and f on $\mathbb{B}_1 - \mathbb{B}_s$ and equal to f on $\mathbb{B}_s$ is smooth and defines a Kähler metric $\omega := i\partial\bar{\partial}\psi$. Moreover, since $\gamma^*\psi|_{\gamma(\partial\mathbb{B}_1)} = f_0|_{\partial\mathbb{B}_1}$, we have that

$$\gamma^*\omega|_{\gamma(\partial\mathbb{B}_1)} = \frac{1}{A}\omega|_{\partial\mathbb{B}_1}.$$

Since $\mathbb{B}_1 - \gamma(\mathbb{B}_1) \subset W$ is a fundamental domain for the action of $\Gamma := \langle\gamma\rangle$ on W, it follows that the metric ω corresponds to the conformal class of an lcK metric on W_γ. ■

In order to detect whether a given lcK manifold can be Vaisman or not, a useful tool is the so called Lee group. For a Vaisman metric (ω, θ) on X, the smooth vector field which is metric dual to θ, $B := \theta^{\sharp_g} \in \mathcal{C}^\infty(TX)$, together with JB, generate a complex one-dimensional Lie group acting by biholomorphic isometries

$$G = \exp\langle B, JB\rangle \subset \mathrm{Aut}_0(X, \omega)$$

called *the Lee group* of X. If X is compact, then $\mathrm{Aut}_0(X, \omega)$ is a compact real Lie group, so that we can also define the compact Lie group $H := \overline{G} \subset \mathrm{Aut}_0(X, \Omega)$ as the closure of G. As G is abelian, H is a real compact torus of dimension at least 2. Let us denote by $\mathfrak{h}$ the Lie algebra of H. Note that $\mathfrak{g} = \langle B, JB\rangle \subset \mathfrak{h} \cap J\mathfrak{h} \neq \{0\}$ and in fact we have equality here, by [16, Prop. 3]. More specifically, a compact lcK manifold bears a Vaisman metric if and only if it bears a complex Lie group $G \subset \mathrm{Aut}_0(X)$ which is contained in a real compact Lie subgroup $H \subset \mathrm{Aut}_0(X)$. Moreover, if it exists, such a group is unique.

This, together with Proposition 3.2, implies:

Proposition 5.2 *A primary Hopf manifold W_γ bears a Vaisman metric if and only if γ is equivalent to a diagonal automorphism d_λ. Moreover, in this case, the Lee vector field is a positive multiple of*

$$B = \mathrm{Re}\left(\sum_{j=1}^{n} \ln|\lambda_j| z_j \frac{\partial}{\partial z_j}\right).$$

Proof The first part is contained in Proposition 3.2. In particular, in the diagonal case it was shown that $\mathfrak{g} = \mathfrak{h} \cap J\mathfrak{h} = \langle Z\rangle_{\mathbb{C}}$, where

$$Z = \sum_{j=1}^{n} \ln|\lambda_j| z_j \frac{\partial}{\partial z_j}.$$

Since a Lee vector field B must satisfy $\theta(B) > 0$ and $\theta(JB) = 0$ for a one-form θ cohomologuous to $-d\ln|z|^2$, the only possibility for B is the one given. ■

Remark 5.3 The above results only discuss the existence of lcK metrics on primary Hopf manifolds. However, the deformation result of [22] together with Theorem 2.12 ensure the existence of *lcK metrics with potential* on these manifolds, i.e. metrics of the form $\omega = f^{-1}i\partial\overline{\partial}f$ on $\tilde{X} = W$ so that the deck group acts by homotheties on f. Since for a general Hopf manifold X one has, by [13, Thm. 2.3], that $\pi_1(X) \cong \Gamma \ltimes F$, with $\Gamma < \mathrm{Aut}(W)$ infinite cyclic and $F \triangleleft \pi_1(X)$ finite, one can first construct an lcK metric with potential on W/Γ and then average the smooth, positive, strictly plurisubharmonic function f to an F-invariant one of the same kind h on W. This then allows one to descend the lcK metric $h^{-1}i\partial\overline{\partial}h$ from W directly to X.

5.2 Non-existence of Other Distinguished Metrics

If a Hermitian metric ω satisfies $\partial\overline{\partial}\omega = 0$, then ω is called *pluriclosed*. This condition coincides with the notion of *Gauduchon* metric in complex dimension 2, therefore such metrics always exist on surfaces by [9]. A Hermitian metric ω on an n-dimensional complex manifold is called *balanced* if $d\omega^{n-1} = 0$. More generally, ω is called *strongly Gauduchon* if $\partial\omega^{n-1}$ is $\overline{\partial}$-exact.

The problem of existence of several special metrics can be solved easily with a negative answer if the Hopf manifold is Vaisman, as shown in [1]. Even though most of the arguments cannot be transfered in general from the Vaisman case to any Hopf manifold, there are still neat results about Hopf manifolds in general. The following two results rule out the existence of pluriclosed and strongly Gauduchon metrics.

Theorem 5.4 *Hopf manifolds of complex dimension larger than* 2 *do not admit pluriclosed metrics.*

Proof The proof rests purely on cohomological properties. Moreover, it is clearly enough to establish the result on primary Hopf manifolds. We reproduce briefly the argument in [7, Thm. 5.16]. By Theorem 4.1, any primary Hopf manifold X satisfies $H^{2,1}_{\overline{\partial}}(X) = H^{3,0}_{\overline{\partial}}(X) = 0$. If a pluriclosed metric ω existed, then $\partial\omega = \overline{\partial}\sigma$ for some $\sigma \in \Omega^{2,0}(X)$ with $\partial\sigma \in H^{3,0}_{\overline{\partial}}(X) = 0$. This amounts to $\Omega := \omega - \sigma - \overline{\sigma}$ being a closed 2-form. Moreover, as $\Omega(V, JV) > 0$ for any vector field V, Ω is non-degenerate, hence a symplectic form. However this cannot exist on X, as $H^2(X) = 0$. ■

Theorem 5.5 *Hopf manifolds do not admit strongly Gauduchon metrics. In particular, they cannot be balanced.*

Proof Again it is enough to show this for primary Hopf manifolds. The argument is given in [25, Section 2], by using the characterization of existence of a strongly Gauduchon metric in terms of currents. As proven in [26, Prop. 3.3], a complex manifold carries a strongly Gauduchon metric if and only if there exists no non-zero current T of type $(1, 1)$ such that $T \geq 0$ and T is d-exact. However, such a current exists on a primary Hopf manifold $X = W_\gamma$. Indeed, if γ is in normal form and $n = \dim_{\mathbb{C}} X$, then $\{z_n = 0\}$ defines a hypersurface in W_γ, therefore its current of integration gives precisely a current obstructing the existence of strongly Gauduchon metrics, as $H^2(X) = 0$.

References

1. D. Angella, A. Otiman, *A Note on Compatibility of Special Hermitian Structures*. arXiv:2306.02981
2. V.I. Arnol'd, *Geometrical Methods in the Theory of Ordinary Differential Equations*. Grundlehren der mathematischen Wissenschaften, vol. 250 (Springer, Berlin, 1996)
3. F. Belgun, On the metric structure of the non-Kähler complex surfaces. Math. Ann. **317**, 1–40 (2000)
4. C. Borcea, Some remarks on deformations of Hopf manifolds. Rev. Roum. Math. pures et appl. **2**(6), 1287–1294 (1981)
5. M. Brunella, Locally conformally Kähler metrics on Kato surfaces. Nagoya Math. J **202**, 77–81 (2011)
6. C. Bănică, O. Stănăşilă, *Méthodes algébrique dans la théorie globale des espaces complexes* (Gauthier-Villars, Paris, 1977)
7. G. Cavalcanti, *Hodge Theory and Deformations of SKT Manifolds*. arXiv:1203.0493
8. A. Fujiki, M. Pontecorvo, Non-upper-semicontinuity of algebraic dimension for families of compact complex manifolds. Math. Ann. **348**, 593–599 (2010)
9. P. Gauduchon, Le théorème de l'excentricité nulle. C.R. Acad. Sci. Paris **285**, 387–390 (1977)
10. P. Gauduchon, L. Ornea, Locally conformally Kähler metrics on Hopf surfaces. Ann. Inst. Fourier **48**, 1107–1127 (1998)
11. A. Grothendieck, Sur quelques points d'algèbre homologique. Tôhoku Math. J. **9**(2), 119–221 (1957)
12. A. Haefliger, Deformations of transversely holomorphic flows on spheres and deformations of Hopf manifolds. Compos. Math. **55**, 241–251 (1985)
13. K. Hasegawa, Deformations and diffeomorphism types of generalized Hopf manifolds. Ill. J. Math. **37**(7), 643–651 (1993)
14. H. Hopf, Zur Topologie der komplexen Mannigfaltigkeiten, in *Studies and Essays presented to R. Courant* (Interscience, New York, 1948), pp. 167–185
15. N. Istrati, *Conformal Structures on Compact Complex Manifolds*. Ph.D. Thesis, IMJ-PRG, 2018
16. N. Istrati, Existence criteria for special locally conformally Kähler metrics. Ann. Mat. Pura Appl. **198**(2), 335–353 (2019)
17. N. Istrati, A. Otiman, Bott-Chern cohomology of compact Vaisman manifolds. Trans. Am. Math. Soc. **376**, 3919–3936 (2023)
18. Y. Kamishima, L. Ornea, Geometric flow on compact locally conformally Kähler manifolds. Tohoku Math. J. **57**, 201–221 (2005)
19. M. Kato, Topology of Hopf surfaces. J. Math. Soc. **27**, 222–238 (1975)
20. K. Kodaira, On the structure of compact complex surfaces, II. Am. J. Math. **88**, 682–721 (1966)
21. D. Mall, The cohomology of line bundles on Hopf manifolds. Osaka J. Math. **28**, 999–1015 (1991)
22. L. Ornea, M. Verbitsky, Locally conformally Kähler manifolds with potential. Math. Ann. **248**(1), 25–33 (2010)

23. L. Ornea, M. Verbitsky, Locally conformally Kähler metrics obtained from pseudoconvex shells. Proc. Am. Math. Soc. **144**, 325–335 (2016)
24. L. Ornea, M. Verbitsky, Non-linear Hopf manifolds are locally conformally Kähler. J. Geom. Anal. **33**, Article number 201 (2023)
25. D. Popovici, Deformation openness and closedness of various classes of compact complex manifolds; examples. Ann. Sc. Norm. Super. Pisa Cl. Sci. (5) **13**, 255–305 (2014)
26. D. Popovici, *Limits of Projective Manifolds under Holomorphic Deformations*. arXiv:0910.2032
27. K. Ueno, *Classification Theory of Algebraic Varieties and Compact Complex Spaces*. Lecture Notes in Mathematics, vol. 439 (Springer, Berlin/Heidelberg/New York, 1975)
28. I. Vaisman, On locally conformal almost Kähler manifolds. Isr. J. Math. **24**, 338–351 (1976)
29. J. Wehler, *Versal deformation of Hopf surfaces*. J. Reine. Angew. Math. **328**, 22–32 (1981)

Einstein Constants and Smooth Topology

Claude LeBrun

To my dear friend and esteemed collaborator Paul Gauduchon, in celebration of his eightieth birthday.

Abstract It was first shown in Catanese and LeBrun (Math. Res. Lett. **4**, 843–854, 1997) that certain high-dimensional smooth closed manifolds admit pairs of Einstein metrics with Ricci curvatures of opposite sign. After reviewing subsequent progress that has been made on this topic, we then prove various related results, with the ultimate goal of stimulating further research on associated questions.

Keywords Einstein metric · Einstein constant · h-Cobordism · Kähler-Einstein · Ricci-flat · Sasaki-Einstein · G_2 manifold · Fano manifold · Calabi-Yau

1 Introduction

A Riemannian metric g is said to be *Einstein* if its Ricci curvature, considered as a function on the unit tangent bundle, is constant. This is of course equivalent to requiring that the Ricci tensor r of g satisfy

$$r = \lambda g \tag{1}$$

for some real number λ. In the original context of general relativity, where the metric is taken to be Lorentzian rather than Riemannian, Eq. (1) is called the Einstein vacuum equation with cosmological constant. In fact, Einstein did briefly consider

Supported in part by NSF grant DMS-2203572.

C. LeBrun (✉)
Department of Mathematics, Stony Brook University, Stony Brook, NY, USA
e-mail: claude@math.stonybrook.edu

L. Ornea (ed.), *Real and Complex Geometry*,
https://doi.org/10.1007/978-3-031-92297-8_10

the possibility that the constant λ in (1) might be non-zero, but he eventually [22, p. 44] came to consider this idea to be "the biggest blunder of his life." It must therefore surely be considered to be amusingly ironic that we mathematicians have acquired the habit of calling λ the *Einstein constant* of a Riemannian metric g satisfying (1).

In this article, the term *Einstein manifold* will mean a compact connected manifold (always without boundary) that is equipped with a Riemannian metric satisfying (1). Of course, the modern prominence of the study of Einstein manifolds is partly attributable to Besse's celebrated book on the subject [7], and many questions raised there by the pseudonymous A.L. Besse were therefore quickly elevated to the status of folk-conjectures. One of these questions [7, p. 19] asked whether the sign of the Einstein constant is somehow determined by the diffeomorphism-type of an Einstein manifold; that is, if a smooth closed manifold M admits two Einstein metrics g_1 and g_2, do the Einstein constants λ_1 and λ_2 of these two metrics always have the same sign? In general, the answer to this question [10] is actually a resounding, "No!" However, if we refine the question by also specifying the dimension n of the manifold M, the answer certainly depends on n, since, for example, Einstein manifolds of dimension $n = 2$ or 3 automatically have constant sectional curvature, and their Einstein constants must therefore have a specific, fixed sign that is predetermined by the fundamental group. This article will emphasize the degree to which we still do not know whether or not this phenomenon really occurs in most dimensions n, while also highlighting some interesting related questions that surely merit further investigation.

Even though there is no reason to believe that this phenomenon actually occurs in dimension $n = 4$, all of the currently available examples curiously hinge on a key 4-dimensional differential-topological fact. Suppose that X and Y are two smooth closed oriented simply-connected non-spin 4-manifolds that have the same Euler characteristic and signature. In light of Donaldson's thesis [17] and the classification of symmetric bilinear forms over the integers [28], a classic result of Terry Wall [55] implies that X and Y are then in fact *h-cobordant*, meaning that there is a smooth compact oriented 5-manifold-with-boundary W with $\partial W = \overline{X} \sqcup Y$ such that the inclusions $X, Y \hookrightarrow W$ are homotopy equivalences. But this implies that $X \times X$ and $Y \times Y$ are also h-cobordant, via the 9-manifold-with-boundary $W' = (W \times X) \cup_{Y \times X} (Y \times W)$. Since the simply connected-manifold $X \times X$ has dimension $8 > 4$, Smale's h-cobordism theorem [41, 51] therefore implies that $X \times X$ and $Y \times Y$ are diffeomorphic. Similarly, the h-cobordism theorem also implies that the k-fold Cartesian products $X^{\times k} = X \times \cdots \times X$ and $Y^{\times k} = Y \times \cdots \times Y$ are diffeomorphic for every $k \geq 2$. In the examples that will interest us, however, X and Y will actually [21, 35] have different Seiberg-Witten invariants, implying that they are non-diffeomorphic, and thereby clearly illustrating the failure of the h-cobordism theorem in dimension 4. Nevertheless, the h-cobordism between X and Y does play an important role in 4-dimensional topology, because it provides the starting point for Freedman's proof [20] that X and Y are nonetheless *homeomorphic*.

Applying this observation to Besse's Einstein-constant question was first made possible by the discovery of a remarkable complex surface of general type by

Rebecca Barlow [5]. Her non-singular compact complex surface B is simply connected, with $c_1^2 = 1$ and $h^{2,0} = 0$, and it then follows that its underlying 4-manifold is non-spin, with Euler characteristic 11 and signature -7. Barlow's surface is minimal, meaning that it does not contain any rational curves of self-intersection -1, but it nonetheless contains four rational curves of self-intersection -2, and for this reason it does not have $c_1 < 0$. However, Fabrizio Catanese and I showed in [10] that generic small deformations of the complex structure of B do in fact have $c_1 < 0$, and the Aubin/Yau theorem [4, 58] therefore implies that these generic deformations all carry compatible Kähler-Einstein metrics with $\lambda < 0$. In particular, the underlying smooth 4-manifold of the Barlow surface B carries some Einstein metric g^- with $\lambda < 0$, and all of the Cartesian products $B^{\times k}$ therefore admit an Einstein product metric $g^- \oplus \cdots \oplus g^-$ with the same negative Einstein constant. But the punchline is that there is also a family of *del Pezzo* surfaces S that have $c_1 > 0$, are simply-connected, and also have $c_1^2 = 1$ and $h^{2,0} = 0$. These specific del Pezzo surfaces are diffeomorphic to the connected sum $\mathbb{CP}_2\#8\overline{\mathbb{CP}}_2$, and some of them had been shown by Tian and Yau [54] to admit compatible $\lambda > 0$ Kähler-Einstein metrics. In particular, the smooth compact 4-manifold S admits an Einstein metric g^+ with $\lambda > 0$, and the Cartesian products $S^{\times k}$ therefore admit product Einstein metrics $g^+ \oplus \cdots \oplus g^+$ with the same Einstein constant $\lambda > 0$. But since the above h-cobordism argument also shows that $B^{\times k}$ and $S^{\times k}$ are actually diffeomorphic, the following result [10] therefore becomes an immediate consequence:

Theorem 1.1 (Catanese-LeBrun) *For each $k \geq 2$, there is a closed simply-connected $4k$-manifold M that admits a pair of Einstein metrics with Einstein constants of opposite signs.*

A series of breakthroughs in complex surface theory eventually allowed Răsdeaconu and Şuvaina [49] to prove a beautiful improvement of this result. First, Lee and Park [38] had discovered that Park's surprising rational-blow-down construction [46] of an exotic symplectic 4-manifold homeomorphic to $\mathbb{CP}_2\#7\overline{\mathbb{CP}}_2$ could be refined to produce a family of simply-connected minimal complex surface of general type with $c_1^2 = 2$ and $h^{2,0} = 0$. Building on these ideas, Park et al. [47] then produced a family of simply-connected complex surface of general type with $c_1^2 = 3$ and $h^{2,0} = 0$ by an analogous construction. In both cases, one first produces an orbifold complex surface by collapsing disjoint chains of rational curves on a rational complex surface, and then shows that the problem of smoothing all these singularities is unobstructed. What Răsdeaconu and Şuvaina then showed was that the generic surface in each smoothing family has $c_1 < 0$. Park et al. [48] then went on to produce a similar family of simply-connected minimal complex surface of general type with $c_1^2 = 4$ and $h^{2,0} = 0$, and then checked that the Răsdeaconu-Şuvaina argument also shows that the generic surface in their family has $c_1 < 0$. By once again invoking the Aubin-Yau theorem, and remembering our previous discussion of the Barlow case, we thus obtain four smooth simply-connected closed 4-manifolds X_ℓ, $\ell = 1, \ldots, 4$, each of which carries an Einstein metric g_ℓ^- with

$\lambda = -1$ that is Kähler with respect to some complex structure with $c_1^2 = \ell$ and $h^{2,0} = 0$. On the other hand, Wall's theorem tells us that each X_ℓ is h-cobordant to the smooth oriented manifold $Y_\ell := \mathbb{CP}_2\#(9-\ell)\overline{\mathbb{CP}}_2$ via some compact oriented 5-manifold-wth-boundary W_ℓ with $\partial W_\ell = \overline{X}_\ell \sqcup Y_\ell$. It then follows that the products $X_{\ell_1} \times \cdots \times X_{\ell_k}$ and $Y_{\ell_1} \times \cdots \times Y_{\ell_k}$ are h-cobordant via the manifold-with-boundary obtained by gluing the k different h-cobordisms

$$X_{\ell_1} \times \cdots \times X_{\ell_{j-1}} \times W_{\ell_j} \times Y_{\ell_{j+1}} \times \cdots \times Y_{\ell_k}$$

end-to-end, and the h-cobordism theorem therefore guarantees that these products are actually diffeomorphic, for any choice of $\ell_1, \ldots, \ell_k$, where $k \geq 2$. However, Tian and Yau [54] had also shown that each such Y_ℓ is diffeomorphic to a del Pezzo surface that admits a compatible Kähler-Einstein metric g_ℓ^+ with $\lambda = +1$. (Indeed, we now know [45] that all del Pezzo surfaces of these diffeotypes actually admit Kähler-Einstein metrics.) It thus follows that any product $X_{\ell_1} \times \cdots \times X_{\ell_k}$ admits a $\lambda = -1$ Einstein metric $g_{\ell_1}^- \oplus \cdots \oplus g_{\ell_k}^-$, while any product $Y_{\ell_1} \times \cdots \times Y_{\ell_k}$ admits a $\lambda = +1$ Einstein metric $g_{\ell_1}^+ \oplus \cdots \oplus g_{\ell_k}^+$. But since these products have already been observed to be diffeomorphic, we therefore obtain [49] the following result:

Theorem 1.2 (Răsdeaconu-Şuvaina) *For every $k \geq 2$, there are at least $\binom{k+3}{3}$ distinct smooth closed simply-connected $4k$-manifolds that admit both $\lambda > 0$ and $\lambda < 0$ Einstein metrics.*

Proof If we choose any k smooth 4-manifolds $Y_{\ell_j} = \mathbb{CP}_2\#(9-\ell_j)\overline{\mathbb{CP}}_2$, where the integers ℓ_j satisfy $1 \leq \ell_j \leq 4$ for every $j = 1, \ldots, k$, we have just seen that their Cartesian product is a smooth compact simply-connected $4k$-manifold that admits both $\lambda > 0$ and $\lambda < 0$ Einstein metrics. But the chosen integers ℓ_j are also determined, up to permutations, by the Betti numbers of the product manifold. Indeed, the Poincaré polynomial $\sum b_i t^i$ is multiplicative under Cartesian products, while the Poincaré polynomial $1+(10-\ell_j)t^2+t^4$ of each $Y_{\ell_j} = \mathbb{CP}_2\#(9-\ell_j)\overline{\mathbb{CP}}_2$ is irreducible in $\mathbb{Z}[t]$. The fact that $\mathbb{Z}[t]$ is a unique factorization domain thus implies that the ℓ_j are determined, up to permutations, by the Betti numbers of the Cartesian product. □

These examples clearly show that $\lambda > 0$ and $\lambda < 0$ Einstein metrics can definitely coexist on various smooth compact manifolds. But the list of manifolds where this phenomenon has actually been seen to occur remains surprisingly limited, and consists of spaces that share many rare and unusual features. We can therefore probably expect that many of the patterns seen in these known examples will be nothing but artifacts of the method of construction. For example, all of these known examples live in dimensions $\equiv 0 \bmod 4$. Do analogous things also happen in odd dimensions? Or in dimensions $\equiv 2 \bmod 4$? The known constructions also never involve any Einstein metrics with $\lambda = 0$. What can actually be proved about the coexistence of Ricci-flat metrics with Einstein metrics for which $\lambda \neq 0$? The bulk of this article will actually concern the narrower problem of finding closed manifolds that admit both Ricci-flat metrics and $\lambda > 0$ Einstein metrics, with a particular emphasis on the situation in dimensions 6 or 7. One motivation for this focus is that there are many simply-connected Ricci-flat manifolds of special holonomy in these dimensions, and yet every simply-connected closed manifolds of dimension 6 or 7 also admits [24] metrics of positive scalar curvature. This narrower problem turns out to be surprisingly delicate, however, and many of the results we will prove will be negative in nature. Nonetheless, it is hoped that describing some of these results here will have a salutary effect, both by encouraging other researchers to join the search, and by charting various reefs and shoals that future explorers might do well to avoid.

2 G_2 Manifolds and Sasaki-Einstein Metrics

One plausible-sounding strategy for finding odd-dimensional avatars of the phenomenon under discussion might be to look for a simply-connected compact 7-manifold that admitted both a Sasaki-Einstein metric g_1 and a metric g_2 of holonomy G_2. Indeed, if this were possible, g_2 would then be Ricci-flat, while g_1 would be an Einstein metric with positive Einstein constant. Since established methods are available [9, 15, 31] for manufacturing vast numbers of compact 7-dimensional Einstein manifolds of both these types, one might therefore be tempted to begin a search for diffeomorphic pairs of Einstein 7-manifolds of these different types. However, any such search would be doomed to fail. Indeed, in this section, we will prove the following result:

Theorem 2.1 *No smooth compact 7-manifold can admit both a Sasaki-Einstein metric g_1 and a metric g_2 with holonomy contained in G_2.*

Our proof of the incompatibility of these two types of Einstein metric is based on the following topological fact:

Proposition 2.2 *If M is a smooth compact 7-manifold that carries a Sasaki-Einstein metric g_1, then $p_1(M) \in H^4(M, \mathbb{Z})$ is a torsion class.*

Before we prove this in full generality, let us first see why it holds for the prototypical examples that were originally discovered [32, Theorem 5] by Shoshichi Kobayashi. These are exactly the Sasaki-Einstein manifolds (M, g_1) which are *regular* [9], in the sense that the Reeb vector field ξ is periodic, with all orbits of the same minimal period. This periodicity assumption forces M to be a principal circle bundle over a compact 6-manifold

$$\begin{array}{ccc} S^1 & \to & M^7 \\ & & \downarrow^{\varpi} \\ & & X^6 \end{array}$$

and the quotient space $X = M/S^1$ then inherits a uniquely-determined metric h that makes $\varpi : M \to X$ into a Riemannian submersion. With standard normalization conventions, the Sasaki-Einstein metric g_1 then has Einstein constant 6, while the Riemannian-submersion metric h induced on X by via ϖ is Kähler-Einstein, with Einstein constant 8.

By passing to a finite cover if necessary, we may now arrange for M to be simply connected. The circle bundle $M \to X$ can then be identified with the unit-circle bundle of $K^{-1/m}$, where $K = \Lambda^{3,0}_X$ is the canonical line-bundle for an h-compatible complex structure J on X, and where the Fano index $m \in \{1, 2, 3, 4\}$ of (X, J) is by definition the largest integer that divides $c_1(X) \in H^2(X, \mathbb{Z})$. With standard normalization conventions, each fiber of ϖ is then a closed geodesic in (M, g_1) of total length $m\pi/2$.

We next compare the cohomologies of M and X by invoking the Gysin exact sequence [42, §6]

$$\cdots \to H^k(M) \to H^{k-1}(X) \xrightarrow{\mathsf{L}} H^{k+1}(X) \xrightarrow{\varpi^*} H^{k+1}(M) \to \cdots$$

where L is given by cup product with the Euler class $c_1(X)/m$ of the oriented circle bundle $M \to X$. Notice, however, that $c_1(X)/m$ is also the Kähler class of $4h/m\pi$. In deRham cohomology, this means that L actually becomes the *Lefschetz operator* of a Kähler metric on the compact complex 3-fold (X, J). The Hard Lefschetz theorem [23, §0.7] therefore guarantees that

$$\mathsf{L} : H^2(X, \mathbb{R}) \to H^4(X, \mathbb{R})$$

is a vector-space isomorphism. The real-coefficient version

$$\cdots \to H^2(X, \mathbb{R}) \xrightarrow{\mathsf{L}} H^4(X, \mathbb{R}) \xrightarrow{\varpi^*} H^4(M, \mathbb{R}) \to \cdots$$

of the Gysin sequence consequently guarantees that the pull-back map

$$\varpi^* : H^4(X, \mathbb{R}) \to H^4(M, \mathbb{R})$$

must vanish identically.

In particular, the pull-back $\varpi^* p_1(X) \in H^4(M, \mathbb{Z})$ of the first Pontrjagin class of X must be a torsion class, since its image $\varpi^* p_1^{\mathbb{R}}(X) \in H^4(M, \mathbb{R})$ in deRham cohomology necessarily vanishes. But since $\ker \varpi_* \subset TM$ is just the trivial rank-1 bundle spanned by the Reeb vector field, $TM \cong \mathbb{R} \oplus \varpi^* TX$, and it therefore follows that

$$p_1(M) = p_1(TM) = p_1(\mathbb{R} \oplus \varpi^* TX) = p_1(\varpi^* TX) = \varpi^* p_1(TX) = \varpi^* p_1(X)$$

is therefore a torsion class, as claimed. ◇

To prove the general case of Proposition 2.2, we will now replace the cohomology of X with the basic cohomology of the Reeb foliation of M.

Proof of Proposition 2.2 Recall that a Sasaki-Einstein manifold (M, g_1) is an Einstein manifold that is equipped [8, §2.1] with a unit-length Killing field ξ such that (M, g_1) has sectional curvature $K(P) = 1$ for every 2-plane $P \subset TM$ with $\xi \in P$, and where the simple 2-form representing any such oriented 2-plane P is moreover an eigenvector of the curvature operator. The 1-form $\eta = g_1(\xi, \cdot)$ is then a contact form, and ξ is then the Reeb vector field of the scaled contact manifold (M, η). Setting $\omega = d\eta/2$, and letting $\Pi : TM \to \ker \eta$ denote projection modulo the span of ξ, we then have

$$g_1 = \eta \otimes \eta + \omega(\cdot, (J \circ \Pi) \cdot)$$

for a unique CR structure $J \in \mathrm{End}(\ker \eta)$, $J^2 = -I$, that is moreover invariant under the flow of ξ. Letting $\mathfrak{F}$ denote the foliation of M by the flow-lines of ξ, the transverse metric h locally induced on the leaf space $M/\mathfrak{F}$ is then Kähler-Einstein, with complex structure induced by J. We will henceforth assume that M has real dimension 7, which then implies that g_1 has Einstein constant 6, and that h has Einstein constant 8.

Relative to the Reeb foliation $\mathfrak{F}$, a differential p-form $\alpha \in \mathcal{E}^p(M)$ is said to be *basic* if $\xi \lrcorner \alpha = 0$ and $\xi \lrcorner d\alpha = 0$. The basic forms $\mathcal{E}^*_B(M, \mathfrak{F})$ then constitute a sub-complex

$$\cdots \xrightarrow{d} \mathcal{E}^{p-1}_B(M, \mathfrak{F}) \xrightarrow{d} \mathcal{E}^p_B(M, \mathfrak{F}) \xrightarrow{d} \mathcal{E}^{p+1}_B(M, \mathfrak{F}) \xrightarrow{d} \cdots$$

of the deRham complex $\mathcal{E}^*(M)$, and the basic cohomology of $(M, \mathfrak{F})$ can therefore be defined to be

$$H^p_B(M, \mathfrak{F}) := \frac{\ker\left[d : \mathcal{E}^p_B(M, \mathfrak{F}) \to \mathcal{E}^{p+1}_B(M, \mathfrak{F})\right]}{\mathrm{image}\left[d : \mathcal{E}^{p-1}_B(M, \mathfrak{F}) \to \mathcal{E}^p_B(M, \mathfrak{F})\right]}.$$

If a basic p-form α is closed, we will use $[\![\alpha]\!]$ to denote its equivalence class in $H^p_B(M, \mathfrak{F})$, while instead using $[\alpha] \in H^p_{dR}(M)$ to denote its (larger) equivalence class in ordinary deRham cohomology.

Because M is compact and $\mathfrak{F}$ is a Riemannian foliation, the basic cohomology $H^*_B(M, \mathfrak{F})$ is [19] automatically finite-dimensional. Moreover, because the complex dimension of $M/\mathfrak{F}$ is 3, and the transverse metric h is Kähler, with the basic form ω as its Kähler form, El Kacimi-Alaoui's Hard Lefschetz Theorem [18, §3.4.7(i)] guarantees that the linear map

$$\mathsf{L} : H^2_B(M, \mathfrak{F}) \to H^4_B(M, \mathfrak{F})$$
$$[\![\alpha]\!] \mapsto [\![\omega \wedge \alpha]\!]$$

is an isomorphism.

On the other hand, notice that the first Pontrjagin class of $TM/T\mathfrak{F}$ is represented in deRham cohomology by the closed basic 4-form

$$\wp_1(TM/T\mathfrak{F}) = -\frac{1}{8\pi^2}\operatorname{Tr}(\mathscr{R}\wedge\mathscr{R})$$

where $\mathscr{R}$ denotes the Riemann curvature tensor of the transverse metric h. We can thus define a basic version of the first Pontrjagin class $TM/T\mathfrak{F}$ by

$$p^{\mathbb{R}}_{1,B}(TM/T\mathfrak{F}) := [\![-\frac{1}{8\pi^2}\operatorname{Tr}(\mathscr{R}\wedge\mathscr{R})]\!] \in H^4_B(M,\mathfrak{F}).$$

By the above version of Hard Lefschetz, we must therefore have

$$p^{\mathbb{R}}_{1,B}(TM/T\mathfrak{F}) = [\![\omega\wedge\alpha]\!]$$

for some closed basic 2-form α. It therefore follows that

$$-\frac{1}{8\pi^2}\operatorname{Tr}(\mathscr{R}\wedge\mathscr{R}) = \omega\wedge\alpha + d\beta$$

for some basic 3-form β on M. But since we can identify $TM/T\mathfrak{F}$ with $\ker\eta$, the usual first Pontrjagin class of $\ker\eta$ in deRham cohomology must be

$$p^{\mathbb{R}}_1(\ker\eta) = p^{\mathbb{R}}_1(TM/T\mathfrak{F}) = [-\frac{1}{8\pi^2}\operatorname{Tr}(\mathscr{R}\wedge\mathscr{R})] = [\omega\wedge\alpha] = [d(\tfrac{1}{2}\eta\wedge\alpha)] = 0,$$

and it therefore follows that

$$p^{\mathbb{R}}_1(M) = p^{\mathbb{R}}_1(TM) = p^{\mathbb{R}}_1(\mathbb{R}\oplus\ker\eta) = p^{\mathbb{R}}_1(\ker\eta) = 0 \in H^4_{dR}(M) = H^4(M,\mathbb{R}).$$

Hence $p_1(M)\in H^4(M,\mathbb{Z})$ must be a torsion class, as claimed. □

With this result in hand, it now becomes straightforward to prove the main result of this section:

Proof of Theorem 2.1 If the closed 7-manifold M admits a Sasaki-Einstein metric g_1, the positivity of the Ricci curvature of g_1 forces $\pi_1(M)$ to be finite by Myers' Theorem. On the other hand, Proposition 2.2 tells us that $p_1(M)\in H^4(M,\mathbb{Z})$ must be a torsion class.

But now suppose that M also admits a metric g_2 with holonomy contained in G_2. Any g_2-compatible choice of torsion-free G_2-structure then determines a calibrating 3-form φ that is parallel with respect to g_2. In particular, this φ is closed, and so endows M with a preferred deRham class $[\varphi] \in H^3_{dR}(M)$. However, a remarkable formula of Joyce [31, Proposition 10.2.7] then tells us that

$$\langle p_1(M) \cup [\varphi], [M] \rangle = -\frac{1}{8\pi^2} \int_M |\mathcal{R}|^2 d\mu_{g_2}, \tag{2}$$

where $\mathcal{R}$ denotes the Riemann curvature tensor of g_2. But since $p_1(M)$ has been shown to be a torsion class in $H^4(M, \mathbb{Z})$, both sides of (2) must therefore vanish. Hence g_2 must be flat, and the universal cover of (M, g_2) must be $\mathbb{R}^7$. Since M is compact, this implies that $\pi_1(M)$ must be infinite. But we have already shown that $\pi_1(M)$ is finite, so this is a contradiction! Thus, the purported coexistence of the metrics g_1 and g_2 is actually impossible. □

While this does not by any means prove that a G_2 manifold (M^7, g_2) can never carry an Einstein metric g_1 with $\lambda > 0$, it does at least rule out one way that g_1 could be related to metrics of special holonomy. We now close this section by eliminating various other simple ways in which such a metric g_1 might arise from a special-holonomy construction.

Proposition 2.3 *If a smooth compact connected 7-manifold M admits a $\lambda > 0$ Einstein metric g_1, and also admits a metric g_2 with holonomy in G_2, then the holonomy group of (M, g_1) must be exactly $SO(7)$.*

Proof By applying Myers' theorem to the universal cover $(\widetilde{M}, \tilde{g}_1)$ of (M, g_1), we immediately see that $\widetilde{M}$ is also compact. It will therefore suffice to prove the claim when M is simply connected, since the holonomy group $\mathrm{Hol}(\widetilde{M}, \tilde{g}_1) = \mathrm{Hol}^0(M, g_1)$ would in any case be a subgroup of $\mathrm{Hol}(M, g_1) \subset O(7)$, while the existence of g_2 would guarantee that M is orientable, thereby implying that $\mathrm{Hol}(M, g_1) \subset SO(7)$.

With this simplification, Berger's classification theorem [6, 7, 50] implies that our simply-connected, positive-λ Einstein 7-manifold (M, g_1) must have holonomy G_2 or $SO(7)$ unless it is either a non-trivial Riemannian product or an irreducible symmetric space of compact type. But, aside from the 5-dimensional special-Lagrangian Grassmannian $SU(3)/SO(3)$, any simply-connected, compact-type irreducible symmetric space of prime dimension p is a round p-sphere, with holonomy $SO(p)$; cf. [7, §10.K, Tables 2 & 4] or [26]. Likewise, we can immediately rule out G_2 as a candidate for $\mathrm{Hol}(M, g_1)$, since g_1 is assumed to have non-zero Ricci curvature. To prove the claim, it will therefore suffice to prove that (M, g_1) cannot be a non-trivial Riemannian product.

However, a non-trivial Riemannian-product decomposition of the simply-connected positive-λ Einstein 7-manifold (M, g_1) would express the space as a product of two positive-λ Einstein manifolds, one of which would necessarily have dimension < 4. But the only simply-connected positive-λ Einstein manifolds of dimension < 4 are the round spheres of dimension 2 and 3. Thus, such a decomposition would necessarily take the form

$$(M^7, g_1) = \begin{cases} (S^2, g_S) \times (X^5, g_X), & \text{or} \\ (S^3, g_S) \times (Y^4, g_Y), \end{cases}$$

where g_S is some constant multiple of the unit-sphere metric, and where g_X or g_Y would be a positive-λ Einstein metric. On the other hand, we have assumed that M also admits a metric g_2 with holonomy in G_2, so X or Y would necessarily be spin. Moreover, since M is assumed to be simply connected, g_2 cannot be flat, and (2) therefore implies that M must admit a cohomology class $[\varphi] \in H^3_{dR}(M)$ such that $[\varphi] \cup p_1(M) \neq 0$. We will now use these observations to obtain a contradiction in either case.

First, if $M = S^2 \times X^5$, the fact that TS^2 is stably trivial immediately implies that $p_1(M) = \varpi^* p_1(X)$, where $\varpi : M \to X$ is the second-factor projection. But since the simply-connected 5-manifold X certainly satisfies $H^1_{dR}(X) = 0$, the Künneth formula $H^*_{dR}(M) = H^*_{dR}(S^2) \otimes H^*_{dR}(X)$ tells us that $\varpi^* : H^3_{dR}(X) \to H^3_{dR}(M)$ is an isomorphism. Consequently, every $[\varphi] \in H^3_{dR}(M)$ can be expressed as $[\varphi] = \varpi^*[\phi]$ for some $[\phi] \in H^3_{dR}(X)$. It then follows that $p_1(M) \cup [\varphi] = \varpi^*(p_1(X) \cup [\phi]) = \varpi^* 0 = 0$. Since this holds for any $[\varphi] \in H^3_{dR}(M)$, no such product $S^2 \times X$ could possibly admit a metric g_2 with holonomy in G_2.

On the other hand, if $M = S^3 \times Y^4$, recall that the existence of a metric g_2 with holonomy in G_2 forces M, and hence Y, to be spin. But since TS^3 is trivial, we also have $\varpi^* p_1(Y) = p_1(M) \neq 0$, where $\varpi : M \to Y$ denotes the second-factor projection. Hence $p_1(Y) \neq 0$, and it therefore follows that the closed spin 4-manifold Y has $\widehat{A}(Y) = -\frac{1}{24}\langle p_1(Y), [Y]\rangle \neq 0$. But since $\widehat{A}(Y)$ is actually the index of the chiral Dirac operator $\not{D} : \Gamma(\mathbb{S}_+) \to \Gamma(\mathbb{S}_-)$ on Y, the Lichnerowicz Weitzenböck formula then guarantees that the spin 4-manifold Y cannot admit a metric g_Y of positive scalar curvature [27, 34, 39]. This contradiction now shows that the positive-λ Einstein metric g_1 must in fact be holonomy irreducible, and so must actually have holonomy $SO(7)$. □

3 Fano Manifolds and Calabi-Yau 3-Folds

In dimension 7, we have just seen that Ricci-flat metrics of special holonomy cannot coexist with $\lambda > 0$ Einstein metrics of Sasakian type. But where else, besides dimension 7, might we look for a smooth compact manifold M^n that admitted both a Ricci-flat metric g_0 and a positive-λ Einstein metric g_1? If, for ease of construction, we also required the Ricci-flat metric g_0 to have special holonomy and to be holonomy irreducible, then any candidate for M would consequently have dimension $n \equiv 6 \bmod 8$. Indeed, after arranging for M to be simply connected by, if necessary, passing to a finite cover, the Ricci-flat metric g_0 would have holonomy

$SU(m)$ (with $n = 2m$), $Sp(k)$ (with $n = 4k$), or $Spin(7)$ (with $n = 8$). Because this reduces the structure-group of M to a simply-connected Lie group, our manifold would therefore be spin, and, since $\pi_1(M) = 0$, its spin structure would moreover be unique. Because $n = \dim M$ is even, we therefore obtain a chiral Dirac operator

$$\not{D} : \Gamma(\mathbb{S}_+) \to \Gamma(\mathbb{S}_-)$$

associated with each metric on the manifold. However, because a section of $\mathbb{S}_+ \oplus \mathbb{S}_-$ on a compact, even-dimensional, Ricci-flat spin manifold (M, g_0) belongs to $\ker \not{D} \oplus \ker \not{D}^*$ if and only if it is parallel [27, 39], the generalized index of $\not{D}$ can therefore be read off by counting the trivial factors of the spin representation of the holonomy group $\mathrm{Hol}(M, g_0)$. When $n \equiv 0 \bmod 4$, the classical index $\widehat{A}(M) = \mathrm{Ind}(\not{D}) = \dim \ker(\not{D}) - \dim \ker(\not{D}^*)$ is thus

$$\widehat{A}(M) = \begin{cases} 2, & \text{if } \mathrm{Hol}(g_0) = SU(2m) \text{ and } n = 2m, \\ k+1, & \text{if } \mathrm{Hol}(g_0) = Sp(k) \text{ and } n = 4k, \\ 1, & \text{if } \mathrm{Hol}(g_0) = Spin(7) \text{ and } n = 8. \end{cases}$$

On the other hand, for $n = 8\ell + 2$, our Ricci-flat metric g_0 would have holonomy $SU(4\ell+1)$, and the same reasoning then yields $\dim \ker \not{D} = 1$; and since Hitchin's α-invariant is exactly the mod-2 reduction of $\dim \ker \not{D}$ when $n \equiv 2 \bmod 8$, we similarly conclude that $\alpha(M) \neq 0$ in this case, too. In each of these cases, it therefore follows [27] that the manifold cannot admit metrics of positive scalar curvature, which in particular precludes the existence of a $\lambda > 0$ Einstein metric. This leaves us with only the case where the Ricci-flat metric has holonomy $SU(4\ell + 3)$, and lives on a manifold of real dimension $8\ell + 6$. On the other hand, this remaining case certainly has a much more encouraging feel to it, because the cobordism-based results of Gromov-Lawson [24] and Stolz [53] show that any simply-connected manifold of dimension 6 mod 8 does actually admit Riemannian metrics of positive scalar curvature.

While these arguments provide only very rough guidance that is far from watertight, they at least suggest that we might do well to begin our search in real dimension 6. The specific problem we will therefore focus on in this section is that of finding a smooth closed 6-manifold M that admits both a Ricci-flat Kähler metric g_0 and a positive-λ Kähler-Einstein metric g_1, where it is to be understood from the outset that the complex structures associated with these two Kähler structures would be entirely unrelated. To get a handle on the problem, we will also invoke a beautiful result of Terry Wall [56, Theorem 5] that reveals the non-existence of exotic differential structures in dimension 6 by determining the diffeomorphism-type of a large class of 6-manifolds in terms of a small number of classical invariants.

Now recall that a compact complex manifold (X, J_X) is said to be *Fano* if $c_1(X, J_X) > 0$. By Yau's proof [59] of the Calabi conjecture, this is equivalent to saying that (X, J_X) admits a compatible Kähler metric of positive Ricci curvature. Since the Kodaira vanishing theorem implies that the Todd genus $\mathrm{Td}(X) =$

$\chi(X, \mathcal{O})$ of any Fano manifold must equal 1, and since the Todd genus is multiplicative under finite covers, it follows [32] that any Fano manifold is necessarily simply connected. On the other hand, while any complex manifold that admits a compatible $\lambda > 0$ Kähler-Einstein metric is necessarily Fano, the converse is certainly false. Instead, a landmark result of Chen et al. [11–14] says that a Fano manifold admits a compatible ($\lambda > 0$) Kähler-Einstein metric if and only if it satisfies the subtle algebro-geometric condition of *K-polystability*. Fortunately, though, Fano manifolds of complex dimension 3 have been completely classified [29, 30, 43, 44]. This has become a springboard for systematic studies of their K-stability [1–3, 16, 40, 52], and these studies will provide us with all the information about existence that will be required for our purposes.

Of course, we will only be interested here in those smooth 6-manifolds underlying Fano 3-folds that might also simultaneously support a Calabi-Yau metric g_0 that is compatible with some other complex structure J_0. Since we've just seen that the existence of a positive-λ Kähler-Einstein metric g_1 forces M to be simply connected, the flat canonical line-bundle of (M, g_0, J_0) must therefore be trivial, and the mod-2 reduction $w_2(M) \in H^2(M, \mathbb{Z}_2)$ of $c_1(M, J_0) = 0$ consequently vanishes. Thus M must necessarily be spin. If J_1 denotes a complex structure compatible with g_1, the mod-2 reduction of $c_1(M, J_1)$ therefore also vanishes, and the exact sequence

$$\cdots \to H^2(M, \mathbb{Z}) \stackrel{2\cdot}{\to} H^2(M, \mathbb{Z}) \to H^2(M, \mathbb{Z}_2) \to \cdots$$

then tells us that $2|c_1(M, J_1)$. Thus, the *Fano index* of (M, J_1) must be even, where the Fano index of a Fano manifold is by definition the largest integer that divides the first Chern class. On the other hand, a general result of Kobayashi-Ochiai [33] implies that the Fano index of a Fano 3-fold (M, J_1) is necessarily ≤ 4, with equality iff $(M, J_1) = \mathbb{CP}_3$. However, since (M, J_0) has already been observed to have trivial canonical line bundle, we then have $h^{3,0}(M, J_0) = 1$, and hence $b_3(M) \geq 2$ by the Hodge decomposition. Thus M certainly cannot be diffeomorphic to $\mathbb{CP}_3$, and the Fano index of any Fano candidate (M, J_1) is therefore exactly 2.

But these same ideas also rule out several other possibilities. Indeed, by the classification results of [43, 44], or by a more general result of Wiśniewski [57], the only index-2 Fano 3-folds with $b_2 \geq 2$ are $\mathbb{P}(T^*\mathbb{CP}_1)$, the one-point blow-up of $\mathbb{CP}_3$, and $\mathbb{CP}_1 \times \mathbb{CP}_1 \times \mathbb{CP}_1$; these are often identified by the standard catalog numbers 2–32, 2–35, and 3–27, respectively, where the digit before the dash records the second Betti number b_2. Since these are all $\mathbb{CP}_1$-bundles over $\mathbb{CP}_2$ or $\mathbb{CP}_1 \times \mathbb{CP}_1$, the Serre spectral sequence shows that they all satisfy $b_3 = 0$, and this of course rules out the existence of Calabi-Yau metrics g_0 on all of their underlying 6-manifolds.

In addition, there is one last deformation-type of degree-2 Fano 3-folds that is also ruled out by this same argument. This candidate, which bears the conventional catalog number 1–15, is the transverse intersection of a linear $\mathbb{CP}_6 \subset \mathbb{CP}_9$ with the image of the Plücker embedding $\mathrm{Gr}(2, 5) \hookrightarrow \mathbb{P}(\Lambda^2\mathbb{C}^5) = \mathbb{CP}_9$. Since $\mathrm{Gr}(2, 5) = U(5)/[U(2) \times U(3)]$ has $H_2 = \pi_2 = \mathbb{Z}$ by the homotopy exact sequence of a fibration, the Lefschetz hyperplane section theorem shows that this Fano 3-fold must indeed have $b_2 = 1$, as is implicitly claimed by the first digit of its catalog number. However, a Chern-class computation also reveals that this Fano manifold has Euler characteristic $c_3 = 4$, and we therefore deduce that it also has $b_3 = 0$. Thus, the existence of a Calabi-Yau metric on its underlying 6-manifold is automatically excluded.

Thus, we have failed to exclude exactly four deformation-types of Fano 3-folds, namely those with Fano index 2, $b_2 = 1$, and $b_3 \neq 0$. Moreover, every Fano 3-fold in each of these four families is actually K-stable by the results of [1, 3, 16, 40, 52], and therefore admits a Kähler-Einstein metric; see [2] for an excellent overview, and further references. The following Proposition articulates this conclusion more explicitly, and in further detail:

Proposition 3.1 *If (M, J_1, g_1) is a $\lambda > 0$ Kähler-Einstein manifold of complex dimension 3 whose underlying 6-manifold M also admits a Calabi-Yau metric g_0 (compatible with some other complex structure J_0), then the Fano 3-fold (M, J_1) belongs to one of the four following families:*

ID#	c_1^3	$h^{1,2}$	$h^{1,1}$	Index	Description
1–11	8	21	1	2	Sextic hypersurface in $\mathbb{CP}(1, 1, 1, 2, 3)$
1–12	16	10	1	2	Double cover of $\mathbb{CP}_3$ branched over a smooth quartic
1–13	24	5	1	2	Cubic hypersurface in $\mathbb{CP}_4$
1–14	32	2	1	2	Complete intersection of two quadrics in $\mathbb{CP}_5$

Moreover, every Fano manifold in any of these families is K-stable, and therefore admits a compatible $\lambda > 0$ Kähler-Einstein metric g_1.

Inspection now also reveals the following useful topological fact:

Corollary 3.1 *Every 6-manifold M in Proposition 3.1 satisfies*

$$\mathfrak{T}_2 = \mathfrak{T}^3 = \mathfrak{T}_3 = \mathfrak{T}^4 = 0,$$

where $\mathfrak{T}_j$ denotes the torsion subgroup of $H_j(M, \mathbb{Z})$, and where $\mathfrak{T}^j$ denotes the torsion subgroup of $H^j(M, \mathbb{Z})$.

Proof Applications of the Lefschetz hyperplane-section theorem imply that $H_2(M, \mathbb{Z}) = \mathbb{Z}$ in each case. On the other hand, $\mathfrak{T}^3 \cong \mathfrak{T}_2$ by the universal coefficients theorem, while $\mathfrak{T}^3 \cong \mathfrak{T}_3$ and $\mathfrak{T}^4 \cong \mathfrak{T}_2$ by Poincaré duality. □

Wall's classification theorem [56, Theorem 5] thus allows one to easily determine whether any other given smooth compact 6-manifold is diffeomorphic to one of these Fano manifolds. This result specifically concerns closed, spin, simply-connected 6-manifolds M with $\mathfrak{T}_2 = 0$. The remaining invariants of such a space are then

- the even integer $b_3(M) = \dim H^3(M, \mathbb{R})$;
- the symmetric trilinear form

$$H^2(M, \mathbb{Z}) \times H^2(M, \mathbb{Z}) \times H^2(M, \mathbb{Z}) \to \mathbb{Z}$$

  defined by $(\mathsf{a}, \mathsf{b}, \mathsf{c}) \mapsto \langle \mathsf{a} \cup \mathsf{b} \cup \mathsf{c}, [M] \rangle$; and
- the first Pontrjagin class $p_1(M)$, as conveniently encoded by the linear map $H^2(M, \mathbb{Z}) \to \mathbb{Z}$ defined by $\mathsf{a} \mapsto p_1 \cdot \mathsf{a} := \langle p_1(M) \cup \mathsf{a}, [M] \rangle$.

Wall's theorem then says that two closed simply-connected spin 6-manifolds with torsion-free homology are diffeomorphic iff they have the same such invariants relative to some isomorphism of their second cohomology groups.

For our purposes, it now suffices to focus on the special class of simply-connected 6-manifolds with $H_2(M, \mathbb{Z}) \cong \mathbb{Z}$ and non-trivial trilinear form. Given an orientation of the manifold, there is then a preferred generator $\mathsf{H} \in H^2(M, \mathbb{Z})$ such that $\mathsf{H}^3 > 0$. The invariants in Wall's theorem are then encoded by the three integers b_3, H^3, and $p_1 \cdot \mathsf{H}$. For example, one has $\mathsf{H} = c_1/2$, and hence $\mathsf{H}^3 = c_1^3/8$, for each of the index-2 Fano manifolds of Proposition 3.1. At the same time, since any Fano 3-fold has Todd genus $1 = \chi(M, \mathcal{O}) = c_1 c_2/24$, the first Pontrjagin class $p_1 = c_1^2 - 2c_2$ must therefore satisfy

$$p_1 \cdot \mathsf{H} = \frac{c_1 \cdot p_1}{2} = \frac{c_1^3}{2} - c_1 c_2 = 4\mathsf{H}^3 - 24.$$

Comparing invariants with putative Calabi-Yau manifolds, we thus obtain:

Proposition 3.3 *A simply-connected Calabi-Yau 3-fold $(M, J, [\omega])$ is diffeomorphic to one of the Fano 3-folds of Proposition 3.1 if and only if it fulfills the list of properties on the corresponding row of the following table,*

Partner ID	H^3	$c_2 \cdot \mathsf{H}$	$h^{1,1}$	$h^{1,2}$	$\mathfrak{T}_2$
I-11	1	10	1	20	0
I-12	2	8	1	9	0
I-13	3	6	1	4	0
I-14	4	4	1	1	0

where $\mathsf{H} \in H^2(M, \mathbb{Z})$ is indivisible over $\mathbb{Z}$, and is a positive multiple of the Kähler class $[\omega] \in H^2(M, \mathbb{R})$.

Proof For any such Calabi-Yau, one has $p_1 = -2c_2$, and $b_3 = 2h^{1,2} + 2$. □

There does not seem to be any obvious reason why these Calabi-Yaus could not exist. However, while there are constructions [25, 36, 37] of many Calabi-Yau 3-folds with $h^{1,1} = 1$, the ones sought by Proposition 3.3 cannot, for example, arise as complete intersections or branched covers. Indeed, Hirzebruch-Riemann-Roch would predict that $h^0(M, \mathcal{O}(\mathsf{H})) = \chi(M, \mathcal{O}(\mathsf{H})) = (2\mathsf{H}^3 + c_2 \cdot \mathsf{H})/12 = 1$, so the linear system $|\mathsf{H}|$ would therefore consist of a single divisor. Entirely new ideas would therefore seem to be required to prove or disprove the existence of the Calabi-Yau 3-folds in question.

Acknowledgments The author warmly thanks Johannes Nordström for some illuminating conversations that helped put him on the road to the results of Sect. 2. He would also like to thank Charles Boyer, Rareş Răsdeaconu, and Ioana Şuvaina for a number of stimulating discussions of relevant topics.

References

1. H. Abban, Z. Zhuang, K-stability of Fano varieties via admissible flags, Forum Math. Pi **10**, pp. Paper No. e15, 43 (2022)
2. C. Araujo, A.-M. Castravet, I. Cheltsov, K. Fujita, A.-S. Kaloghiros, J. Martinez-Garcia, C. Shramov, H. Süß, N. Viswanathan, *The Calabi Problem for Fano Threefolds*, London Mathematical Society Lecture Note Series, vol. 485 (Cambridge University Press, Cambridge, 2023)
3. C. Arezzo, A. Ghigi, G.P. Pirola, Symmetries, quotients and Kähler-Einstein metrics. J. Reine Angew. Math. **591**, 177–200 (2006)
4. T. Aubin, Équations du type Monge-Ampère sur les variétés kähleriennes compactes. C. R. Acad. Sci. Paris Sér. A-B **283**, pp. Aiii, A119–A121 (1976)
5. R. Barlow, A simply connected surface of general type with $p_g = 0$. Inv. Math. **79**, 293–301 (1985)
6. M. Berger, Sur les groupes d'holonomie homogène des variétés à connexion affine et des variétés riemanniennes. Bull. Soc. Math. France **83**, 279–330 (1955)
7. A.L. Besse, *Einstein Manifolds*, Ergebnisse der Mathematik und ihrer Grenzgebiete (3), vol. 10 (Springer, Berlin, 1987)
8. C. Boyer, K. Galicki, 3-Sasakian manifolds, in *Surveys in Differential Geometry: Essays on Einstein Manifolds*, vol. 6 of Surv. Differ. Geom. (Int. Press, Boston, MA) (1999), pp. 123–184
9. C.P. Boyer, K. Galicki, *Sasakian Geometry*, Oxford Mathematical Monographs (Oxford University Press, Oxford, 2008)
10. F. Catanese, C. LeBrun, On the scalar curvature of Einstein manifolds. Math. Res. Lett. **4**, 843–854 (1997)
11. X. Chen, S. Donaldson, S. Sun, Kähler-Einstein metrics and stability. Int. Math. Res. Not. IMRN **2014**, 2119–2125 (2014)
12. X. Chen, S. Donaldson, S. Sun, Kähler-Einstein metrics on Fano manifolds. I: Approximation of metrics with cone singularities. J. Amer. Math. Soc. **28**, 183–197 (2015)
13. X. Chen, S. Donaldson, S. Sun, Kähler-Einstein metrics on Fano manifolds. II: Limits with cone angle less than 2π. J. Amer. Math. Soc. **28**, 199–234 (2015)
14. X. Chen, S. Donaldson, S. Sun, Kähler-Einstein metrics on Fano manifolds. III: Limits as cone angle approaches 2π and completion of the main proof. J. Amer. Math. Soc. **28**, 235–278 (2015)
15. A. Corti, M. Haskins, J. Nordström, T. Pacini, G_2-manifolds and associative submanifolds via semi-Fano 3-folds. Duke Math. J. **164**, 1971–2092 (2015)
16. R. Dervan, On K-stability of finite covers. Bull. Lond. Math. Soc. **48**, 717–728 (2016)

17. S.K. Donaldson, An application of gauge theory to four-dimensional topology. J. Differential Geom. **18**, 279–315 (1983)
18. A. El Kacimi-Alaoui, Opérateurs transversalement elliptiques sur un feuilletage riemannien et applications. Compos. Math. **73**, 57–106 (1990)
19. A. El Kacimi-Alaoui, V. Sergiescu, G. Hector, La cohomologie basique d'un feuilletage riemannien est de dimension finie. Math. Z. **188**, 593–599 (1985)
20. M. Freedman, On the topology of 4-manifolds. J. Differential Geom. **17**, 357–454 (1982)
21. R. Friedman, J. Morgan, Algebraic surfaces and Seiberg-Witten invariants. J. Alg. Geom. **6**, 445–479 (1997)
22. G. Gamow, *My World Line; An Informal Autobiography* (Viking Press, New York, 1970)
23. P. Griffiths, J. Harris, *Principles of Algebraic Geometry* (Wiley-Interscience, New York, 1978)
24. M. Gromov, H.B. Lawson, The classification of simply connected manifolds of positive scalar curvature. Ann. Math. **111**, 423–434 (1980)
25. M. Gross, Primitive Calabi-Yau threefolds. J. Differential Geom. **45**, 288–318 (1997)
26. S. Helgason, *Differential Geometry, Lie Groups, and Symmetric Spaces*, vol. 80 of Pure and Applied Mathematics (Academic Press [Harcourt Brace Jovanovich Publishers], New York, 1978)
27. N.J. Hitchin, Harmonic spinors. Adv. Math. **14**, 1–55 (1974)
28. D. Husemoller, J. Milnor, *Symmetric Quadratic Forms* (Springer, 1973)
29. V.A. Iskovskih, Fano threefolds. I. Izv. Akad. Nauk SSSR Ser. Mat. **41**, 516–562, 717 (1977)
30. V.A. Iskovskih, Fano threefolds. II. Izv. Akad. Nauk SSSR Ser. Mat. **42**, 506–549 (1978)
31. D.D. Joyce, *Compact Manifolds with Special Holonomy*, Oxford Mathematical Monographs (Oxford University Press, Oxford, 2000)
32. S. Kobayashi, Topology of positively pinched Kaehler manifolds. Tohoku Math. J. (2) **15**, 121–139 (1963)
33. S. Kobayashi, T. Ochiai, Characterizations of complex projective spaces and hyperquadrics. J. Math. Kyoto Univ. **13**, 31–47 (1973)
34. H.B. Lawson, M. Michelsohn, *Spin Geometry* (Princeton University Press, 1989)
35. C. LeBrun, On the scalar curvature of complex surfaces. Geom. Funct. Anal. **5**, 619–628 (1995)
36. N.-H. Lee, Calabi-Yau coverings over some singular varieties and new Calabi-Yau 3-folds with Picard number one. Manuscripta Math. **125**, 531–547 (2008)
37. N.-H. Lee, Calabi-Yau double coverings of Fano-Enriques threefolds. Proc. Edinb. Math. Soc. (2) **62**, 107–114 (2019)
38. Y. Lee, J. Park, A simply connected surface of general type with $p_g = 0$ and $K^2 = 2$. Invent. Math. **170**, 483–505 (2007)
39. A. Lichnerowicz, Spineurs harmoniques. C.R. Acad. Sci. Paris **257**, 7–9 (1963)
40. Y. Liu, C. Xu, K-stability of cubic threefolds. Duke Math. J. **168**, 2029–2073 (2019)
41. J. Milnor, *Lectures on the h-Cobordism Theorem* (Princeton University Press, 1965)
42. J.W. Milnor, J.D. Stasheff, *Characteristic Classes*, Annals of Mathematics Studies, No. 76 (Princeton University Press, Princeton, 1974)
43. S. Mori, S. Mukai, Classification of Fano 3-folds with $B_2 \geq 2$. Manuscripta Math. **36**, 147–162 (1981/1982)
44. S. Mori, S. Mukai, Erratum: "Classification of Fano 3-folds with $B_2 \geq 2$" [Manuscripta Math. **36** (1981/82), no. 2, 147–162; MR0641971 (83f:14032)]. Manuscripta Math. **110**, 407 (2003)
45. Y. Odaka, C. Spotti, S. Sun, Compact moduli spaces of del Pezzo surfaces and Kähler-Einstein metrics. J. Differential Geom. **102**, 127–172 (2016)
46. J. Park, Simply connected symplectic 4-manifolds with $b_2^+ = 1$ and $c_1^2 = 2$. Invent. Math. **159**, 657–667 (2005)
47. H. Park, J. Park, D. Shin, A simply connected surface of general type with $p_g = 0$ and $K^2 = 3$. Geom. Topol. **13**, 743–767 (2009)
48. H. Park, J. Park, D. Shin, A simply connected surface of general type with $p_g = 0$ and $K^2 = 4$. Geom. Topol. **13**, 1483–1494 (2009)

49. R. Răsdeaconu, I. Şuvaina, Smooth structures and Einstein metrics on $\mathbb{CP}^2\#5,\ 6,\ 7\overline{\mathbb{CP}^2}$. Math. Proc. Camb. Philos. Soc. **147**, 409–417 (2009)
50. J. Simons, On the transitivity of holonomy systems. Ann. Math. (2) **76**, 213–234 (1962)
51. S. Smale, On the structure of manifolds. Ann. Math. **87**, 387–399 (1964)
52. C. Spotti, S. Sun, Explicit Gromov-Hausdorff compactifications of moduli spaces of Kähler-Einstein Fano manifolds. Pure Appl. Math. Q. **13**, 477–515 (2017)
53. S. Stolz, Simply connected manifolds of positive scalar curvature. Ann. Math. (2) **136**, 511–540 (1992)
54. G. Tian, S.-T. Yau, Kähler-Einstein metrics on complex surfaces with $C_1 > 0$. Comm. Math. Phys. **112**, 175–203 (1987)
55. C.T.C. Wall, On simply connected 4-manifolds. J. Lond. Math. Soc. **39**, 141–149 (1964)
56. C.T.C. Wall, Classification problems in differential topology. V. On certain 6-manifolds. Invent. Math. **1**, 355–374 (1966); corrigendum, ibid, **2**, 306 (1966)
57. J.A. Wiśniewski, On Fano manifolds of large index. Manuscripta Math. **70**, 145–152 (1991)
58. S.T. Yau, Calabi's conjecture and some new results in algebraic geometry. Proc. Nat. Acad. Sci. U. S. A. **74**, 1798–1799 (1977)
59. S.T. Yau, On the Ricci curvature of a compact Kähler manifold and the complex Monge-Ampère equation. I. Comm. Pure Appl. Math. **31**, 339–411 (1978)

The Lee–Gauduchon Cone on Complex Manifolds

Liviu Ornea and Misha Verbitsky

To our friend and teacher, Paul Gauduchon, at his 80th birthday

Abstract Let M be a compact complex n-manifold. A Gauduchon metric is a Hermitian metric whose fundamental 2-form ω satisfies the equation $dd^c(\omega^{n-1}) = 0$. Paul Gauduchon has proven that any Hermitian metric is conformally equivalent to a Gauduchon metric, which is unique (up to a constant multiplier) in its conformal class. Then $d^c(\omega^{n-1})$ is a closed $(2n-1)$-form; the set of cohomology classes of all such forms, called the Lee-Gauduchon cone, is a convex cone, superficially similar to the Kähler cone. We prove that the Lee-Gauduchon cone is a bimeromorphic invariant, and compute it for several classes of non-Kähler manifolds.

Keywords Positive current · Pseudo-effective cone · Bott–Chern cohomology · Aeppli cohomology

2010 Mathematics Subject Classification 53C55, 32H04

Liviu Ornea is partially supported by the PNRR-III-C9-2023-I8 grant CF 149/31.07.2023 Conformal Aspects of Geometry and Dynamics.

Misha Verbitsky is partially supported by FAPERJ SEI-260003/000410/2023 and CNPq - Process 310952/2021-2.

L. Ornea (✉)
Faculty of Mathematics and Informatics, University of Bucharest, Bucharest, Romania

Institute of Mathematics "Simion Stoilow" of the Romanian Academy, Bucharest, Romania
e-mail: lornea@fmi.unibuc.ro

M. Verbitsky
Instituto Nacional de Matemática Pura e Aplicada (IMPA), Rio de Janeiro, RJ, Brazil
e-mail: verbit@impa.br

L. Ornea (ed.), *Real and Complex Geometry*,
https://doi.org/10.1007/978-3-031-92297-8_11

1 Introduction

This paper was inspired by our work [24] on the following conjecture.

Conjecture 1.1 A compact locally conformally Kähler manifold never admits a balanced Hermitian metric.

Both of these notions are defined properly in Sect. 5.

It turns out that this question is essentially a question about the shape of the Gauduchon cone, the cone which plays in non-Kähler complex geometry the same role as the Kähler cone plays in the conventional (Kähler) complex geometry.

A Hermitian metric ω on a complex n-manifold is called **Gauduchon** if $dd^c\omega^{n-1} = 0$ (see Sect. 2.4 for a proper introduction to the geometry of Gauduchon forms). The **Gauduchon form** ω^{n-1} can be used to reconstruct ω in an unambiguous way. The set of all Gauduchon forms is clearly a convex cone in the infinite-dimensional space of all dd^c-closed $(n-1, n-1)$-forms.

To make sense of the cohomological properties of the Gauduchon forms, one uses the Aeppli cohomology, defined as $\frac{\ker dd^c}{\operatorname{im} d + \operatorname{im} d^c}$ (Sect. 2.2). The Gauduchon metrics define a convex cone, called **the Gauduchon cone**, in the Aeppli cohomology group $H^{n-1,n-1}_{AE}(M)$.

For a general complex manifold, the Aeppli cohomology is hard to compute; this group is not a topological invariant, and not constant in holomorphic families of complex manifolds. In this paper we introduce a more accessible counterpart of the Gauduchon cone, called the Lee-Gauduchon cone, which lives in the de Rham cohomology group $H^{2n-1}(M, \mathbb{R})$. We introduce this object and give its proper definition in Sect. 3.1. From the definition of a Gauduchon form, it is clear that the form $d^c(\omega^{n-1})$ is closed. The Lee-Gauduchon cone is the set of all cohomology classes of this form.

In this paper we show that the Lee-Gauduchon cone is a bimeromorphic invariant (Theorem 4.5) and compute it for several important classes of complex manifolds. We also characterize the Lee-Gauduchon cone in terms of positive currents (Theorem 3.11).

2 Bott–Chern and Aeppli Cohomologies

We present briefly the two main cohomology theories which are most useful on non-Kähler manifolds. We refer to [4] for more details and references.

2.1 Bott–Chern Cohomology

Definition 2.1 Let M be a complex manifold, and $H^{p,q}_{BC}(M)$ the space of closed (p,q)-forms modulo $dd^c(\Lambda^{p-1,q-1}M)$. Then $H^{p,q}_{BC}(M)$ is called **the Bott–Chern cohomology** of M.

Remark 2.2 A (p,q)-form η is closed if and only if $\partial\eta = \overline{\partial}\eta = 0$. Since $2\sqrt{-1}\,\partial\overline{\partial} = dd^c$, an equivalent definition of the Bott–Chern cohomology is $H^*_{BC}(M) := \frac{\ker\partial\cap\ker\overline{\partial}}{\operatorname{im}\partial\overline{\partial}}$.

Remark 2.3 While there exist no morphisms between the de Rham and the Dolbeault cohomology, there are natural and functorial maps from the Bott–Chern cohomology to both the Dolbeault cohomology $H^*(\Lambda^{*,*}M, \overline{\partial})$ and the de Rham cohomology. However, there exists no multiplicative structure on the Bott–Chern cohomology.

Denote by $H^{1,0}_d(M)$ the space of closed holomorphic forms on M. Since any exact holomorphic 1-form on a compact complex manifold vanishes, $H^{1,0}_d(M)$ can be considered as a subspace in de Rham (or Dolbeault) cohomology. Clearly, $H^{1,0}_d(M)\oplus\overline{H^{1,0}_d(M)}$ is the space of all forms $\alpha\in\Lambda^1(M,\mathbb{C})$ such that $d\alpha = d^c\alpha = 0$; this space is the complexification of $\ker d\oplus\ker d^c$.

Proposition 2.4 *The sequence*

$$0 \longrightarrow \operatorname{Re}(H^{1,0}_d(M)) \longrightarrow H^1(M,\mathbb{R}) \xrightarrow{d^c} H^{1,1}_{BC}(M,\mathbb{R}) \tag{2.1}$$

is exact.

Proof This sequence is clearly exact in the first term: an exact holomorphic form is a differential of a global holomorphic function on M, and all such functions are constant. To prove that it is exact in the second term, let x be a closed 1-form, and $[x]\in H^1(M)$ its cohomology class. The cohomology class of d^cx vanishes in $H^{1,1}_{BC}(M,\mathbb{R})$ if and only if $d^cx = dd^cf$, for some function $f\in C^\infty M$. However, $d^cx = dd^cf$ means that $x+df$ is d-closed and d^c-closed, hence $[x]$ belongs to the image of $H^{1,0}_d(M)\oplus\overline{H^{1,0}_d(M)}$. ∎

2.2 Aeppli Cohomology

Definition 2.5 Let M be a complex manifold, and $H^{p,q}_{AE}(M)$ the space of dd^c-closed (p,q)-forms modulo $\partial(\Lambda^{p-1,q}M)+\overline{\partial}(\Lambda^{p,q-1}M)$. Then $H^{p,q}_{AE}(M)$ is called **the Aeppli cohomology** of M.

Remark 2.6 The spaces $H^{p,p}_{AE}(M)$ and $H^{p,p}_{BC}(M)$ are preserved by the complex conjugation. We denote the subspaces fixed by the complex conjugation by $H^{p,p}_{AE}(M,\mathbb{R}) \subset H^{p,p}_{AE}(M)$ and $H^{p,p}_{BC}(M,\mathbb{R}) \subset H^{p,p}_{BC}(M)$.

Theorem 2.7 (A. Aeppli, [2]) *Let M be a compact complex n-manifold. Then the Aeppli cohomology is finite-dimensional. Moreover, the natural pairing*

$$H^{p,q}_{BC}(M) \times H^{n-p,n-q}_{AE}(M) \longrightarrow H^{2n}(M) = \mathbb{C},$$

$$(x, y) \mapsto \int_M x \wedge y,$$

is non-degenerate and identifies $H^{p,q}_{BC}(M)$ with the dual $(H^{n-p,n-q}_{AE}(M))^*$.

2.3 *The Lee-Gauduchon Space $W \subset H^{2n-1}(M)$*

We define a subspace in the de Rham cohomology which will play a central role in the sequel. The motivation for its name will be apparent after we define the Lee-Gauduchon cone.

Definition 2.8 Let $W \subset H^{2n-1}(M,\mathbb{R})$ be the space of all cohomology classes α such that $\int_M \alpha \wedge \rho = 0$ for all closed holomorphic forms $\rho \in \Lambda^{1,0}(M,\mathbb{R})$. We call W **the Lee-Gauduchon space**.

Remark 2.9 Since $\int_M d^c u \wedge \rho = \int_M u \wedge d^c \rho = 0$, the image of the natural map $d^c : H^{n-1,n-1}_{AE}(M) \longrightarrow H^{2n-1}(M)$ belongs to W.

Proposition 2.10 *Let M be a compact complex n-manifold. Then*

$$W = d^c(H^{n-1,n-1}_{AE}(M,\mathbb{R})).$$

Proof Consider the exact sequence (2.1) constructed above. Dualizing it, we obtain

$$H^{n-1,n-1}_{AE}(M,\mathbb{R}) \xrightarrow{d^c} H^{2n-1}(M,\mathbb{R}) \longrightarrow H^{2n-1}(M,\mathbb{R})/W \longrightarrow 0,$$

because W is the annihilator of $H^{1,0}_d \oplus \overline{H^{1,0}_d(M)} \subset H^1(M,\mathbb{R})$. ∎

2.4 *The Gauduchon Cone of a Complex Manifold*

Definition 2.11 A Hermitian metric on a complex n-manifold is called a **Gauduchon metric** if its fundamental form satisfies $dd^c(\omega^{n-1}) = 0$. In this case, ω^{n-1} is called **a Gauduchon form**.

Remark 2.12 Recall that on a compact complex manifold, a Gauduchon metric exists in any conformal class of Hermitian metrics, and it is unique up to constant multiplier, [10].

Definition 2.13 **The Gauduchon cone** of a compact complex n-manifold is the set of all classes $\omega^{n-1} \in H_{AE}^{n-1,n-1}(M, \mathbb{R})$ of all Gauduchon forms.

Remark 2.14 Since any strictly positive $(n-1, n-1)$-form is the $(n-1)$-th power of a Hermitian form [19], the Gauduchon cone is open and convex in $H_{AE}^{n-1,n-1}(M, \mathbb{R})$.

2.5 *Positive Currents*

In this subsection we express the Gauduchon cone in terms of currents. We recall the definition of positive currents and refer the reader to [7] for an introduction on currents on complex manifolds.

Definition 2.15 A $(1, 1)$-current α is called **positive** if $\int_M \alpha \wedge \tau \geqslant 0$ for any positive $(n-1, n-1)$-form τ with compact support.

Remark 2.16 The cone of positive $(1, 1)$-currents is generated by $-\sqrt{-1}\,\alpha u \wedge \overline{u}$, where α is a positive generalized function (that is, a measure), and u a (1,0)-form.

Definition 2.17 The **pseudo-effective cone** $P \subset H_{BC}^{1,1}(M)$ is the set of Bott–Chern classes of all positive, closed (1,1)-currents.

The next result gives a characterization of the Gauduchon cone in terms of currents.

Theorem 2.18 *On a compact complex manifold, the Gauduchon cone is dual to the pseudo-effective cone.*

Proof *[18, Lemme 3.3]; See also [27]* We include the proof of Theorem 2.18 in Sect. 3.2. ∎

3 Lee Forms and Lee Classes of a Gauduchon Form

3.1 *The Lee-Gauduchon Cone*

The following is but one of the equivalent definitions of the Lee form of a Hermitian metric. For another approach, see [10].

Definition 3.1 **The Lee form** of a Hermitian metric ω is $\frac{1}{n-1} * (d^c \omega^{n-1})$, where $*$ denotes the Hodge star operator.

Remark 3.2 Clearly, the Lee form is d^*-closed if and only if ω is Gauduchon.

Definition 3.3 Let ω^{n-1} be a Gauduchon form. The corresponding **Lee-Gauduchon form** is $d^c(\omega^{n-1})$. This form is clearly closed; its cohomology class $[d^c(\omega^{n-1})] \in H^{2n-1}(M, \mathbb{R})$ is called **the Lee-Gauduchon class**.

Remark 3.4 As in Sect. 2.3, let $W \subset H^{2n-1}(M, \mathbb{R})$ be the space of all cohomology classes α such that $\int_M \alpha \wedge \rho = 0$ for all closed holomorphic forms $\rho \in \Lambda^{1,0}(M)$. Since $\int_M d^c(\omega^{n-1}) \wedge \rho = \int_M \omega^{n-1} \wedge d^c\rho = 0$, all Lee-Gauduchon classes belong to W, called **the Lee-Gauduchon space** (Sect. 2.3).

Definition 3.5 **The Lee-Gauduchon cone** $LG(M) \subset W$ is the set of all Lee-Gauduchon classes of Gauduchon forms.

Claim 3.6 The Lee-Gauduchon cone is a convex, open cone in the Lee-Gauduchon space W.

Proof The set of Lee-Gauduchon forms is open, because it is the image of the Gauduchon cone, which is open, and the projection from dd^c-closed forms to W is surjective by Proposition 2.10. ■

Remark 3.7 The Lee-Gauduchon space is trivial on all compact complex manifolds for which the dd^c-lemma holds, for example, on all projective, Moishezon, Kähler or Fujiki class C manifolds.

3.2 *Exact Pseudoeffective Bott–Chern Classes*

In this subsection we express the Lee-Gauduchon cone in terms of exact pseudoeffective Bott–Chern classes.

We use the Hahn–Banach theorem (on the model of [12, 28]) in the following formulation:

Theorem 3.8 (Hahn–Banach) *Let V_1 be a locally convex topological vector space, $V \subset V_1$ a closed subspace, and $A \subset V_1$ an open, convex subset, not intersecting A. Then there exists a continuous linear functional $\xi \in V_1^*$ vanishing on V and positive on A.*

We start by proving Lamari's theorem (Theorem 2.18).

Definition 3.9 Let $\alpha \in \Lambda^{p,q}(M)$ be a dd^c-closed form. We say that α is **Aeppli exact** if it is the (p, q)-part of an exact form.

Theorem 3.10 *On a compact complex manifold, the Gauduchon cone is dual to the pseudo-effective cone.*

Proof Let A be the set of all strictly positive $(n-1, n-1)$-forms, $u \in H_{AE}^{n-1,n-1}(M, \mathbb{R})$, and $V = u + V'$, where V' is the space of Aeppli exact dd^c-closed $(n-1, n-1)$-forms. Then $V \cap A = \emptyset$ if and only if there exists a functional on $\Lambda^{n-1,n-1}(M)$ (that is, a (1,1)-current) ξ such that $\langle \xi, A\rangle > 0$ and $\langle \xi, V\rangle = 0$. Since V is an affine space and V' its linearization, $\langle \xi, V\rangle = 0$ implies $\langle \xi, V'\rangle = 0$.

The condition $\langle \xi, A\rangle > 0$ means precisely that ξ is a non-zero positive current. The condition $\langle \xi, V'\rangle = 0$ is equivalent to $\int \xi \wedge dw = 0$ for all $(2n-3)$-forms w, because V' is the space of $(n-1, n-1)$-parts of exact forms, and ξ is a (1,1)-current. However, $\int \xi \wedge dw = 0$ for all w is equivalent to ξ being closed. Then, a class $u \in H_{AE}^{n-1,n-1}(M, \mathbb{R})$ belongs to the Gauduchon cone if and only if $\int_M \xi \wedge u > 0$ for all positive, closed (1,1)-currents ξ. ■

We may now prove:

Theorem 3.11 *Let M be a compact complex manifold, and $\mathcal{C} \subset H^1(M)$ the set of classes $\rho \in H^1(M, \mathbb{R})$ such that $d^c(\rho) \in H_{BC}^{1,1}(M)$ is pseudo-effective. Then the Lee-Gauduchon cone $LG(M) \subset W$ is the dual cone to $\mathcal{C}$. In other words, $\alpha \in LG(M)$ if and only if $\int_M \alpha \wedge \rho > 0$ for any closed 1-current ρ such that $d^c\rho$ is positive.*

Proof **Step 1:** If $\alpha \in LG(M)$, then $\alpha = d^c\omega^{n-1}$ which gives $\int_M \alpha \wedge \rho = \int_M \omega^{n-1} \wedge d^c\rho$. For any non-zero positive current, the integral $\int_M \omega^{n-1} \wedge d^c\rho$ (known as "the mass" of the current) is positive. This means that $LG(M)$ is evaluated positively on all elements of $\mathcal{C}$. It remains to prove the converse inclusion, that is, to show that for all $\alpha \in W\backslash LG(M)$, there exists a closed 1-current ξ such that $d^c\xi$ is positive, but $\langle \alpha, \xi\rangle = 0$.

Step 2: Fix $u \in W$ and apply the Hahn–Banach theorem to the closed affine space

$$V = u + d^c(\text{Aeppli exact } (n-1, n-1)\text{–forms}) + \text{exact } 1-\text{forms}$$

and the open cone $d^c(\text{strictly positive } (n-1, n-1) - \text{forms})$. By Hahn–Banach, these spaces do not intersect if there exists a 1-current ξ such that:

$$\begin{aligned} \langle -\xi, u + d^c(\text{Aeppli exact } (n-1, n-1) - \text{forms})\rangle &= 0, \text{ and} \\ \langle -\xi, d^c(\text{strictly positive } (n-1, n-1) - \text{forms})\rangle &> 0, \text{ and} \\ \langle \xi, \text{exact } 1 - \text{forms}\rangle &= 0. \end{aligned} \tag{3.1}$$

Step 3: The condition $\langle \xi, \text{exact } 1 - \text{forms}\rangle = 0$ means that ξ is d-closed; indeed, $\langle \xi, d\zeta\rangle = \pm\langle d\xi, \zeta\rangle$. However, $d^c(\text{closed } 1 - \text{forms}) \subset \Lambda^{1,1}(M)$, because locally a closed 1-form is exact, and $d^c df \in \Lambda^{1,1}(M)$ for any function f.

Step 4: Using integration by parts, the second condition of (3.1) translates to

$$\langle d^c\xi, \text{strictly positive } (n-1, n-1) - \text{forms}\rangle > 0,$$

which is equivalent to $d^c\xi$ being a positive current.

Step 5: The first condition of (3.1) implies that

$$\langle -\xi, d^c(\text{Aeppli exact } (n-1, n-1) - \text{forms})\rangle = 0$$

that is, $\langle d^c\xi, \text{Aeppli exact } (n-1, n-1)\text{–forms}\rangle = 0$. This is equivalent to $d^c\xi$ being a closed (1,1)-form; it follows from the same argument that proves Lamari's theorem (Theorem 3.10). ∎

4 The Lee-Gauduchon Cone Is Bimeromorphically Invariant

Definition 4.1 Let X, Y be complex manifolds, and $Z \subset X \times Y$ a closed subvariety such that the projections of Z to X and Y are proper and generically bijective. Then Z is called **a bimeromorphic map,** and X and Y are called **bimeromorphic.**

The structure of bimeromorphisms is given in the following fundamental result:

Theorem 4.2 (Weak Factorization Theorem, [1]) *Any bimeromorphism can be decomposed into a composition of blow-ups and blow-downs with smooth centers (in arbitrary order).*

Remark 4.3 This immediately implies that bimeromorphic manifolds have the same fundamental group. Also, the spaces of global holomorphic forms on bimeromorphic manifolds are naturally isomorphic.

This implies the following

Corollary 4.4 *Let M_1, M_2 be compact complex n-manifolds which are bimeromorphic. Then $W(M_1) = W(M_2)$, where $W \subset H^{2n-1}(M)$ is the Lee-Gauduchon space (Sect. 2.3).*

Theorem 4.5 *Let M_1, M_2 be compact complex n-manifolds which are bimeromorphic. Then $LG(M_1) = LG(M_2)$.*

Proof **Step 1:** Let $[\theta_1], [\theta_2]$ classes in $H^1(M_i)$ which are identified by the natural isomorphism $H^1(M_1) = H^1(M_2)$. We are going to prove that $[\theta_1] \in \mathcal{C}(M_1) \Leftrightarrow [\theta_2] \in \mathcal{C}(M_2)$, where $\mathcal{C}(M_i)$ is the dual cone to the LG-cone in M_i (Theorem 3.11). By definition, $\mathcal{C}(M_i)$ is the cone of classes $[\theta_i] \in H^1(M, \mathbb{R})$ such that the Bott-Chern class of $d^c([\theta_i])$ is pseudoeffective. Denote by θ_i the closed currents representing $[\theta_i]$ such that the currents $d^c\theta_i$ are positive.

Step 2: The pushforward of a positive current is always positive; the pullback of a current is, in general, not defined. This makes it difficult to identify the pseudoeffective cones of bimeromorphic manifolds. However, the 1-currents θ_i are exact on the universal cover $\tilde{M}_i$: $\tilde{\theta}_i = df_i$, where f_i are generalized functions on $\tilde{M}_i$. If $d^c\theta_i$ is positive, f_i is plurisubharmonic; however, a plurisubharmonic function is always L^1_{loc}-integrable, and can be extended over a closed analytic subset, [7, Theorem 5.4, p. 45]. Therefore, the pullback *and* the pushforward of a plurisubharmonic function are plurisubharmonic, which implies that f_1 can be lifted to the graph of the bimeromorphic correspondence and pushed forward to a plurisubharmonic function on $\tilde{M}_2$. ■

5 The Lee–Gauduchon Cone: Examples

We end this note with results on the Lee–Gauduchon cone on special classes of compact complex manifolds: Oeljeklaus-Toma, strongly Gauduchon, balanced and LCK.

5.1 Strongly Gauduchon and Balanced Manifolds

Definition 5.1 ([26]) A Gauduchon form ω^{n-1} is called **strongly Gauduchon** if $\partial(\omega^{n-1})$ is $\overline{\partial}$-exact.

Remark 5.2 This is equivalent to ω^{n-1} being the $(n-1, n-1)$-part of a closed form. Indeed, $\partial(\omega^{n-1}) = \overline{\partial}(x^{n,n-2})$ implies that $d(x^{n,n-2}) - d\omega^{n-1} + d(\overline{x^{n,n-2}}) = 0$.

Remark 5.3 The strongly Gauduchon property is implied by $d\omega^{n-1} = 0$; such Hermitian form are known as **balanced**, [19]. The class of balanced manifolds is bimeromorphically invariant, [3].

Proposition 5.4 *Let ω be a strongly Gauduchon form. Then its Lee-Gauduchon class vanishes.*

Proof Let $x^{n,n-2} + \omega^{n-1} + \overline{x^{n,n-2}}$ be a closed form with ω^{n-1} a Gauduchon form. Then $I(x^{n,n-2} + \omega^{n-1} + \overline{x^{n,n-2}}) = -x^{n,n-2} + \omega^{n-1} - \overline{x^{n,n-2}}$, hence $d^c\omega^{n-1} = -d(\omega^{n-1})$. Therefore, the Lee-Gauduchon class of any strongly Gauduchon metric vanishes. ■

Remark 5.5 If $0 \in LG(M)$, then $LG(M) = W$. Indeed, this cone is open, convex and $\mathbb{R}^{>0}$-invariant. Therefore, $LG(M) = W$ for the class of strongly Gauduchon manifolds, which includes balanced manifolds, which includes all twistor spaces (of ASD manifolds, [13, 19], of conformally flat Riemannian manifolds of even dimension, [11], of quaternion-Kähler manifolds, [25], of hyperkähler manifolds, [16], of hypercomplex manifolds, [29]).

Remark 5.6 The condition "$d\omega^{n-1}$ is d^c-exact" is a weaker form of the strongly Gauduchon condition. A manifold is strongly Gauduchon if and only if all exact positive (1,1)-currents vanish [26], and $d\omega^{n-1}$ is d^c-exact if and only if all positive (1,1)-currents with Bott–Chern cohomology classes in $d^c(H^1(M))$ vanish (Theorem 3.11).

Example 5.7 Let M be a Calabi-Eckmann manifold [5], a complex manifold diffeomorphic to $S^3 \times S^3$ and fibered over $\mathbb{C}P^1 \times \mathbb{C}P^1$ with fiber an elliptic curve. The Bott-Chern group $H^{1,1}_{BC}(M)$ is non-zero; indeed, let $\omega \in \Lambda^{1,1}(M)$ be the pullback of the Kähler form on $\mathbb{C}P^1 \times \mathbb{C}P^1$. The Bott-Chern class $[\omega] \in H^{1,1}_{BC}(M)$ of this form is pseudoeffective, and it is exact, because $H^2(M) = 0$. Moreover, $[\omega] \neq 0$, because a plurisubharmonic function on a compact manifold is constant. Therefore, M is not strongly Gauduchon; however, the Lee-Gauduchon cone of M is trivial, because $H^5(M, \mathbb{R}) = 0$.

Remark 5.8 We obtained that for a Calabi-Eckmann manifold, the cone of positive exact (1,1)-currents and the cone of positive (1,1)-currents with Bott–Chern cohomology classes in $d^c(H^1(M))$ are distinct.

5.2 LCK Manifolds

Definition 5.9 A Hermitian manifold of complex dimension $> 1 (M, I, g, \omega)$ is called **locally conformally Kähler** (LCK) if there exists a closed 1-form θ such that $d\omega = \theta \wedge \omega$. The 1-form θ is called the **Lee form**.

Remark 5.10 We recall that all known compact LCK manifolds belong to one of three classes: blow-ups of LCK with potential, blow-ups of Oeljeklaus–Toma and Kato manifolds. We refer to [22] for LCK geometry.

Recently, we proved the following result:

Theorem 5.11 ([24]) *Let (M, ω, θ) be a compact LCK manifold in any of these classes. Then $d^c\theta$ is exact pseudoeffective.*

Corollary 5.12 *For a compact LCK manifold in any of the above classes, $LG(M) \not\supseteq 0$ and M does not admit a strongly Gauduchon metric.*

This leads to the following conjecture which was the starting point for writing this note:

Conjecture 5.13 The two-form $d^c\theta$ is pseudoeffective on all compact LCK manifolds.

The Lee-Gauduchon cone for these three classes of LCK manifolds can be computed explicitly.

The following proposition was proven in [23].

Proposition 5.14 *Let (M, ω, θ) be an LCK manifold with potential [22]. Then $H^1(M, \mathbb{C}) = H^{1,0}_d(M) \oplus \overline{H^{1,0}_d(M)} \oplus \langle\theta\rangle$.*

Proof [23, Theorem 6.1]. ■

From this theorem it is clear that $\dim W = 1$, and since $d^c\theta$ is pseudo-effective on any LCK manifold with potential, its Lee-Gauduchon cone is a half-line $\{x \in W \mid \langle x, \theta\rangle > 0\}$.

The same is true for any Kato manifold M [14, 15]. Indeed, $b_1(M) = 1$, because M is a deformation of a blown-up Hopf manifold, and $H^{1,0}_d(M) = 0$ because $H^{1,0}_d(M) \oplus \overline{H^{1,0}_d(M)}$ contribute an even number to $b_1(M)$. Therefore, $\dim W = 1$. Since $d^c\theta$ is pseudo-effective on a Kato manifold [24], its Lee-Gauduchon cone is a half-line $\{x \in W \mid \langle x, \theta\rangle > 0\}$.

We deal with all OT-manifolds, locally conformally Kähler or not, in the next subsection.

5.3 *Oeljeklaus-Toma Manifolds*

Oeljeklaus-Toma manifolds were introduced in [21], and were much studied since then. These are manifolds associated with number fields. Their geometry and topology ultimately depends on two invariants associated with a number field: its number of real embeddings, denoted s, and the number of complex embeddings, denoted $2t$ (complex embeddings go in pairs, which are complex conjugate). When $t = 1$, an OT manifold is LCK [21], and otherwise it is not [6]. A dimension 2 OT manifold is called **Inoue surface of type** S^0. OT manifolds are complex solvmanifolds [17], that is, they can be obtained as a left quotient of a solvable Lie group G with a left-invariant complex structure by a cocompact lattice Γ.

One of the main tools which are applied to OT-manifolds is the averaging procedure, originally defined by F. Belgun for Inoue surfaces, and then extended to nilmanifolds in [8] (see also [17], [9]). Averaging is defined on any quotient manifold of form $X := G/\Gamma$, where G is a Lie group and $\Gamma \subset G$ a discrete subgroup of finite covolume. The averaging is a map $\mathrm{Av} : \Lambda^k(X) \longrightarrow \Lambda^k(\mathfrak{g}^*)$, where $\Lambda^k(\mathfrak{g}^*)$ denotes the space of antisymmetric k-forms on its Lie algebra. It takes a k-form α to a functional $T \mapsto \int_{G/\Gamma} \langle T, \alpha\rangle \,\mathrm{Vol}$, where Vol is a left-invariant Haar volume, and $T \in \Lambda^k(\mathfrak{g})$ an antisymmetric k-vector. Since $\Lambda^k(\mathfrak{g})^* = \Lambda^k(\mathfrak{g}^*)$, the averaging takes values in $\Lambda^k(\mathfrak{g}^*)$.

The integral $\int_{G/\Gamma} T \wedge \alpha$ is well defined for any k-current and any $\alpha \in \Lambda^{n-k}(\mathfrak{g}^*)$, hence the averaging is naturally extended to currents [9].

We will identify $\Lambda^k(\mathfrak{g}^*)$ and the space of left-invariant forms on G. The de Rham differential on $\Lambda^k(\mathfrak{g}^*)$ is known as the Chevalley-Eilenberg differential. It is not hard to see that the averaging commutes with the de Rham differential; when G is equipped with a left-invariant complex structure, averaging preserves the Hodge decomposition. By Nomizu theorem [20], for nilmanifolds, the averaging map induces an isomorphism in cohomology. This theorem is false for solvmanifolds; however, H. Kasuya [17, Lemma 2.1] has shown that for OT-manifolds the averaging induces isomorphism on H^1.

The averaging map is very convenient to deal with the positive currents: it takes a positive, closed current to a non-zero, positive, closed invariant form.

The following claim trivially follows from these observations.

Claim 5.15 Let M be an OT-manifold, and $[\alpha] \in H^1(M, \mathbb{R})$ a cohomology class; using [17], we represent $[\alpha]$ by a closed, invariant 1-form $\alpha \in \Lambda^1(\mathfrak{g}^*)$. Then $[\alpha]$ can be represented by a current α_1 such that $d^c\alpha_1$ is positive if and only if $d^c\alpha$ is positive.

Proof The form $\operatorname{Av}(\alpha_1)$ is equal to α, because these forms are cohomologous, and $d : \Lambda^0(\mathfrak{g}^*) \longrightarrow \Lambda^1(\mathfrak{g}^*)$ is zero. Since the averaging preserves positivity, the (1,1)-form $d^c\alpha = \operatorname{Av}(d^c(\alpha_1))$ is also positive. ■

To proceed, we need to describe the complex structure on the Lie algebra of the solvable group G associated with the OT-manifold explicitly.

An OT-manifold is constructed from a number field K which has $2t$ complex embeddings and s real ones (in other words, $K \otimes_{\mathbb{Q}} \mathbb{R} = \mathbb{C}^{2t} \oplus \mathbb{R}^s$). For such a number field, the group of units $\mathcal{O}_K^*$ has rank $t + s - 1$ by Dirichlet unit theorem. Oeljeklaus and Toma construct a torsion-free subgroup $U \subset \mathcal{O}_K^*$ of rank s, such that the quotient described below is compact.

Let $\tilde{M}_0 = \mathbb{C}^t \times \mathbb{R}^s$; we identify $\tilde{M}_0$ with a subspace of $K \otimes_{\mathbb{Q}} \mathbb{R} = \mathbb{C}^{2t} \oplus \mathbb{R}^s$ obtained by taking only one $\mathbb{C}$ in each pair of complex conjugate components. The additive group $\mathcal{O}_K = \mathbb{Z}^{2s+t}$ acts on $\tilde{M}_0$ by translations; this action is cocompact, totally discontinuous and the quotient $\frac{\tilde{M}_0}{\mathcal{O}_K}$ is a torus.

Now, let $\mathbb{H}$ denote the upper half-plane. Consider the manifold $\tilde{M} := \mathbb{C}^t \times \mathbb{H}^s$, identified with $\tilde{M}_0 \times \mathbb{R}^s_+$. We equip the quotient manifold $\frac{\tilde{M}}{\mathcal{O}_K}$ with the action of U as follows. The action of U on $\tilde{M}_0 = K \otimes_{\mathbb{Q}} \mathbb{R}$ is the standard multiplicative action; it clearly commutes with the $\mathcal{O}_K$-action on the same manifold. To describe the action of U on each $\mathbb{H}$-component, we index these components by the set $\{\sigma_1, ...\sigma_s\}$ of real embeddings of K. Then $u \in U$ acts on the $\mathbb{H}$-component number k as a multiplication by the real number $\sigma_k(u)$.

It is not hard to see that if we choose generators u_i of U in such a way that $\sigma_k(u_i) > 1$ for each k, i, then the quotient $M = \frac{\tilde{M}}{\mathcal{O}_K \rtimes U}$ is compact; the action of $\mathcal{O}_K \rtimes U$ on $\mathbb{C}^t \times \mathbb{H}^s \subset \mathbb{C}^{s+t}$ is by construction complex affine. Clearly, M is a cocompact quotient of the Lie group $G := (\mathcal{O}_K \otimes_{\mathbb{Z}} \mathbb{R}) \rtimes (U \otimes_{\mathbb{Z}} \mathbb{R})$ obtained as a semidirect product of the abelian Lie groups identified with the vector spaces $\mathcal{O}_K \otimes_{\mathbb{Z}} \mathbb{R}$ and $U \otimes_{\mathbb{Z}} \mathbb{R}$.

Following [17, Section 6], we fix the basis of the dual space to the Lie algebra $\mathfrak{g}$ of G. Let γ_i be generators associated with the $\mathbb{C}$-factors, with the complex structure I taking γ_{2i-1} to γ_{2i}, and α_i, β_i be the generators associated with the $\mathbb{H}$-factors, with $I(\alpha_i) = \beta_i$. The standard coordinate system on $\mathbb{H}$ is $x = \operatorname{Re}(z)$, $y = \log(\operatorname{Im}(z))$; then β_i correspond to the real components, $\beta_i = \frac{dx}{y}$ and α_i correspond to the imaginary component, $\alpha_i = \frac{dy}{y}$.

In this basis, the Chevalley-Eilenberg differential (the operator dual to the Lie bracket) is written as in [17, Section 6]: $d\alpha_i = 0, d\beta_i = -\alpha_i \wedge \beta_i$, and

$$d\gamma_{2i-1} = \psi_i \wedge \gamma_{2i-1} + \varphi_i \wedge \gamma_{2i},\quad d\gamma_{2i} = -\varphi_i \wedge \gamma_{2i-1} + \psi_i \wedge \gamma_{2i},$$

where ψ_i, φ_i are linear combinations of α_i.

Since I exchanges α_i with β_i, the intersection $d\mathfrak{g}^* \cap I(d\mathfrak{g}^*)$ is generated by $d\beta_i$, with $d\gamma_i$ contributing nothing. However, $-d\beta_i$ are positive for all i. This brings the following proposition.

Proposition 5.16 *Let $M = \frac{\mathbb{C}^t \times \mathbb{H}^s}{\mathcal{O}_K \rtimes U}$ be an OT-manifold, z_i the complex coordinates on the $\mathbb{H}$-components, and $d^c d(\log(\operatorname{Im} z_i))$ the Kähler form associated with the Poincaré metric on this $\mathbb{H}$-component. Then the cone A of pseudo-effective Bott–Chern cohomology classes which belong to $d^c(H^1(M, \mathbb{R}))$ is generated by positive linear combinations of $d^c d(\log(\operatorname{Im} z_i))$.*

Proof By Claim 5.15, the cone A is generated by positive forms $d^c\rho$, where $\rho \in \Lambda^1(\mathfrak{g}^*)$ is closed. However, the intersection $d\Lambda^1(\mathfrak{g}^*) \cap I(d\Lambda^1(\mathfrak{g}^*))$ is generated by $\alpha_i \wedge \beta_i$, as indicated above, hence the only positive (1,1)-forms in $d\Lambda^1(\mathfrak{g}^*)$ are linear combinations of $-d\beta_i$.

By definition, β_i is the G-invariant form on the i-th $\mathbb{H}$-component of $\mathbb{C}^t \times \mathbb{H}^s$, corresponding to $d^c \log \operatorname{Im}(z_i)$, hence $d^c d(\log(\operatorname{Im} z_i)) = -d\beta_i$. ■

Using Proposition 5.16 in conjunction to Theorem 3.11, we can compute the Lee-Gauduchon cone of an OT-manifold M explicitly. Recall that $H^1(M)$ is generated by $d\log \operatorname{Im} z_i$, where z_i are coordinates on the $\mathbb{H}$-components of $\mathbb{C}^t \times \mathbb{H}^s$ [17, 21]. However, the forms $d^c\alpha_i$ are positive. By Theorem 3.11, the Lee-Gauduchon cone of an OT-manifold is dual to the cone of cohomology classes α in $H^1(M)$ such that $d^c\alpha$ is positive. This brings the following result.

Theorem 5.17 *Let M be an OT-manifold, and $[\alpha_1]$, ..., $[\alpha_s]$ generators of the first cohomology constructed above, $\alpha_i = d \log \operatorname{Im} z_i$. Then the LG-cone of M is dual to the convex cone generated by α_i in $H^1(M, \mathbb{R})$.*

Acknowledgments We are grateful to Dan Popovici for an important email conversation and consultations about the strong Gauduchon metrics.

References

1. D. Abramovich, K. Karu, K. Matsuki, J. Włodarczyk, Torification and factorization of birational maps. J. Amer. Math. Soc. **15**, 531–572 (2002)
2. A. Aeppli, Some exact sequences in cohomology theory for Kähler manifolds. Pacific J. Math. **12** (1962)
3. L. Alessandrini, G. Bassanelli, Metric properties of manifolds bimeromorphic to compact Kähler spaces. J. Differential Geometry **37**, 95–121 (1993)

4. D. Angella, *Cohomological Aspects in Complex non-Kähler Geometry* (Springer, 2014)
5. E. Calabi, B. Eckmann, A class of compact complex manifolds which are not algebraic. Ann. Math. **58**, 494–500 (1953)
6. S. Deaconu, V. Vuletescu, On Locally Conformally Kähler metrics on Oeljeklaus-Toma Manifolds. Manuscr. Math. **171**, 643–647 (2023)
7. J.-P. Demailly, Complex analytic and differential geometry, https://www-fourier.ujf-grenoble.fr/~demailly/manuscripts/agbook.pdf
8. A. Fino, G. Grantcharov, On some properties of the manifolds with skew-symmetric torsion and holonomy $SU(n)$ and $Sp(n)$. Adv. Math. **189**(2), 439–450 (2004)
9. A. Fino, L. Grantcharov, M. Verbitsky, Algebraic dimension of complex nilmanifolds. J. Math. Pures Appl. (9) **118**, 204–218 (2018)
10. P. Gauduchon, Le théorème de l'excentricité nulle. C. R. Acad. Sci. Paris Sér. A-B **285**(5), A387–A390 (1977)
11. P. Gauduchon, Structures de Weyl et théorèmes d'annulation sur une variété conforme autoduale. Ann. Sc. Norm. Sup. Pisa XVII **I**, 563–629 (1991)
12. R. Harvey, H.B. Lawson Jr., An intrinsic characterization of Kähler manifolds. Invent. Math. **74**(2), 169–198 (1983)
13. N. Hitchin, Kählerian twistor spaces. Proc. Lond. Math. Soc. **43**, 133–150 (1981)
14. N. Istrati, A. Otiman, M. Pontecorvo, On a class of Kato manifolds. Int. Math. Res. Not. **7**, 5366–5412 (2021)
15. N. Istrati, A. Otiman, M. Pontecorvo, M. Ruggiero Toric Kato manifolds. J. Éc. Polytech. Math. **9**, 1347–1395 (2022)
16. D. Kaledin, M. Verbitsky, Non-Hermitian Yang-Mills connections. Sel. Math. New Series **4**, 279–320 (1998)
17. H. Kasuya, Vaisman metrics on solvmanifolds and Oeljeklaus-Toma manifolds. Bull. Lond. Math. Soc **45**(1), 15–26 (2013)
18. A. Lamari, Le cône Kählérien d'une surface. J. Math. Pures Appl. IX. Sér. **78**(3), 249–263 (1999)
19. M.L. Michelsohn, On the existence of special metrics in complex geometry. Acta Math. **149**, 261–295 (1982)
20. K. Nomizu, On the cohomology of compact homogeneous spaces of nilpotent Lie groups. Ann. Math. **2**(59), 531–538 (1954)
21. K. Oeljeklaus, M. Toma, Non-Kähler compact complex manifolds associated to number fields. Ann. Inst. Fourier **55**(1), 1291–1300 (2005)
22. L. Ornea, M. Verbitsky, *Principles of Locally Conformally Kähler Geometry*, Progress in Math., vol. 354 (Birkhäuser, 2024)
23. L. Ornea, M. Verbitsky, Lee classes on LCK manifolds with potential. Tohoku. Math. J. **76**, 105–125 (2024)
24. L. Ornea, M. Verbitsky, Balanced metrics and Gauduchon cone of locally conformally Kähler manifolds, Int. Math. Res. Not. **2025**(3) (2025). https://doi.org/10.1093/imrn/rnaf014
25. M. Pontecorvo, Complex structures on quaternionic manifolds. Diff. Geom Appl. **4**, 163–177 (1994)
26. D. Popovici, Deformation limits of projective manifolds: hodge numbers and strongly gauduchon metrics. Invent. Math. **194**, 515–534 (2013)
27. D. Popovici, L. Ugarte, Compact complex manifolds with small Gauduchon cone. Proc. Lond. Math. Soc. **116**, 1161–1186 (2018)
28. D. Sullivan, Cycles for the dynamical study of foliated manifolds and complex manifolds. Invent. Math. **36**, 225–255 (1976)
29. A. Tomberg, Twistor spaces of hypercomplex manifolds are balanced. Adv. Math. **280**, 282–300 (2015)

Bi-Hermitian and Locally Conformally Kähler Surfaces

Massimiliano Pontecorvo

To Paul Gauduchon, master of Geometry

Abstract We present work by various authors on a classification of four-dimensional Riemannian manifolds (M^4, g) admitting two complex structures which are g-orthogonal, inducing the same orientation.

Keywords Compact complex surfaces · Bi-Hermitian

1 Introduction

We intend to report on some work which has been going on for about 20 years on bi-Hermitian surfaces, by which we mean Riemannian four–manifolds (M^4, g) equipped with two *integrable* complex structures J_+ and J_- which are both g-orthogonal inducing the same orientation.

Definition 1.1 An (almost) complex structure $J \in End(TM)$ with $J^2 = -id.$ on an even-dimensional Riemannian manifold (M, g) is said to be orthogonal if $g(JX, JY) = g(X, Y)$ for all tangent vectors $X, Y \in TM$.

It is important to stress here that in the framework of bi-Hermitian metrics each $J_\pm$ is assumed to be integrable, in the usual sense that induces local holomorphic coordinates. This condition imposes constrains on the curvature of the Riemannian

The author is a member of GNSAGA—INdAM and of the research project PRIN: "Real and Complex Manifolds: Geometry and Holomorphic Dynamics".

M. Pontecorvo (✉)
Dipartimento di Matematica e Fisica, Roma Tre University, Roma, Italy
e-mail: massimiliano.pontecorvo@uniroma3.it

L. Ornea (ed.), *Real and Complex Geometry*,
https://doi.org/10.1007/978-3-031-92297-8_12

metric g; in fact, by conformal invariance, it imposes conditions on the Weil tensor W of the metric.

The following *local* result of Salamon provided the initial motivation for studying bi-Hermitian metrics in four-dimension:

Theorem 1.2 ([40]) *Each point $p \in (M^4, g, or.)$ of an oriented Riemannian four-manifold belongs to a neighborhood in which there are either zero, one, two or infinitely many complex structures (up to sign) which are $g-$orthogonal, inducing the given orientation.*

Notice that the last alternative occurs if and only if g is anti-self-dual. Soon after the above local result the following question appeared to be quite natural:

Question 1.3 (Salamon 90's) What about Riemannian four-manifolds admitting exactly two (distinct, up to sign) orthogonal complex structures inducing the same orientation? In what follows, such a structure $(g, J_{\pm})$ will be called a bi-Hermitian metric.

The first local results are due to Piotr Kobak [30] who explicitly constructed bi-Hermitian metrics on compact tori. Then came the remarkable work of Apostolov-Gauduchon-Grantcharov [6] where the local theory was developed along with global constructions coming from deformations of hyperkähler metrics.

The present paper is devoted to discuss the following **global version** of Salamon question:

Question 1.4

(a) Which compact four-manifolds support bi-Hermitian metrics ?
(b) More precisely, which compact complex surfaces (S, J_+) admit a Riemannian metric g and a different complex structure J_- such that $J_{\pm}$ are g-orthogonal and induce the same orientation ?

Remark 1.5 The precise meaning of *different* in this question is that there is $p \in S$ where, as endomorphisms of the tangent bundle we have, $J_+(p) \neq \pm J_-(p)$. It will then turn up that the two complex structures may or may not be biholomorphic, in any case they will be in the same class of Enriques-Kodaira classification. This is the framework we intend to work in, because it soon appeared that in order to answer question (b) is enough to assume (S, J_+) to be a *minimal* complex surface [20] [13]. This means, by Castelnuovo criterium, free from smooth rational curves of self-intersection -1. △

We end up this introduction with a quick look at dimensions greater than four.

High Dimensions We mention here briefly about two possible higher dimensional generalizations of this setting in the Riemannian and in the quaternionic case. The Weil tensor W of a *Riemann* metric becomes irreducible as soon as the real dimension $n \geq 5$ and certainly the condition $W = 0$ implies, for n is even, the existence of an abundance of Orthogonal Complex Structures, simply because the twistor space is a complex manifold. Conversely, Salamon showed [41] that each orthogonal complex structure imposes linear conditions on the conformal curvature W, which however has a big number of components, difficult to control.

A different behavior was observed in the higher dimensional *quaternionic* setting where the natural analogue of an oriented conformal structure $(M^4, [g], or.)$ is a $GL(n, \mathbb{H})Sp(1)$-structure on the tangent bundle of M^{4n}. It was shown in [1] that for every quaternionic dimension $4k \geq 8$ the local existence of two compatible complex structures forces the almost quaternionic structure to be a quaternionic structure. By the integrability of the corresponding twistor space, this is equivalent to say that in this situation we always have infinitely many complex structures compatible with the quaternionic structure, locally. We conclude that there is no analogue notion of bi-Hermitian metric in this setting.

2 Fundamental Divisor and Flat Bundles

2.1 The Fundamental Line Bundle of a Bi-Hermitian Metric

In what follows we will denote by S the compact complex surface $S = (M, J_+)$ where M is a four-manifold with bi-Hermitian structure given by two complex structures $J_\pm$ which are orthogonal to the same Riemannian metric and induce the same orientation.

Our goal is to give informations about the complex geometry of S. To this end we start to introduce here what will be called the *fundamental divisor* of S. It is the effective divisor T geometrically described below as the intersection of smooth complex surfaces in the almost complex, real 6-dimensional twistor space Z of $(M, [g], or.)$. The space Z only depends on the conformal class of the metric which we denote by $[g]$.

The standard notation $[D]$ for the holomorphic line bundle of the divisor D will be used here together with additive notation for divisors and multiplicative for line bundle so that $[D + D'] = [D] \otimes [D']$. We start by saying that the *support* $\mathcal{T}$ of the fundamental divisor T is easily seen, by the description below, to be the set of points where the two complex structures agree, up to sign:

$$\mathcal{T} = \{p \in M \mid J_+(p) = \pm J_-(p)\}.$$

As a matter of fact $\mathcal{T}$ turns out to be the closure of a union of smooth 2-dimensional real submanifolds [21]. However, being a non-necessarily transverse intersection, the divisor T will typically be singular and highly non-reduced, see [22] for a computation of all its multiplicities.

Setting $\mathbf{t} := b_0(\mathcal{T})$ it follows from surface classification that the number of connected components can only be

$$\mathbf{t} = 0, 1, 2 \tag{1}$$

in other words T can have at most two connected components and may well be empty.

We end up this subsection with some considerations involving the twistor space.

The bi-Hermitian structure $([g], J_{\pm})$ defines two complex surfaces in the twistor space $Z = \{J \in \text{End}(TM) \mid J^2 = -id.\}$ of all (locally defined) almost complex structures compatible with the metric and orientation of M, for which we will use the following notation $S_{\pm} := J_{\pm}(M) \subset Z$. Similarly we set $X_{\pm} := S_{\pm} \amalg \sigma(S_{\pm})$ where $\sigma : Z \to Z$ is the involution $J \mapsto -J$. Notice that there is a tautological biholomorphism $(M, J_{\pm}) \cong S_{\pm}$.

In this notation $T \subset S = (M, J_+)$ can be identified with either of the following two subsets in the twistor space: $X_+ \cap S_-$ or $S_+ \cap X_- \subset Z$; this exhibits T as an almost complex subvariety of either S_+ or S_-; T is therefore a complex curve in each of the two smooth surfaces, in particular closed in the analytic Zariski topology and nowhere dense. A first Chern class computation in Z then yields the following:

Proposition 2.1 *Let $t : Z \to M$ denote the twistor projection, the image $t(T) = t(X_+ \cap S_-) = t(S_+ \cap X_+) \subset M$ has a natural structure of divisor T in both surfaces $S_{\pm}$ with the property*

$$c_1(T) = c_1(S_{\pm})$$

For each compact complex surface $S_{\pm}$ there is a holomorphic line bundle $F_{\pm}$ such that

1. $c_1(F_{\pm}) = 0$
2. $[T] = K_{\pm}^{-1} \otimes F_{\pm}$ *where $K_{\pm}$ is the canonical line bundle of the surface $S_{\pm}$.*

Definition 2.2 We will call T the *fundamental* divisor of the bi-Hermitian surface and $F = [T] \otimes K$ the *fundamental flat line bundle.*

We summarize this discussion with the following statement which gives an important obstruction to the existence of bi-Hermitian surfaces

Corollary 2.3 *For every bi-Hermitian surface $S = (M, J_+)$ the fundamental divisor T is what is called a numerically anti-canonical (NAC) divisor: simply because its Chern class is opposite to that of the canonical bundle: $[T] = F \otimes K_S^{-1}$ with $c_1(F) = 0$.*

2.2 Lee Bundle of a LCK Metric

For dimensional reasons, the fundamental $(1, 1)$-form ω of an Hermitian surface $S = (M^4, g, J)$ always satisfies the equation

$$d\omega = \omega \wedge \theta$$

for a unique 1-form θ called the *Lee form* of the Hermitian metric such that:

1. The metric is Kähler if and only if $\theta = 0$
2. The Lee form $\theta = df$ is exact when the metric $e^{-f} g$ is Kähler—i.e. (g, J) is a globally conformally Kähler metric.
3. The Lee form is closed—i.e. locally exact—precisely when the Hermitian metric is conformal to Kähler, locally. We abbreviate this condition, which was introduced by Vaisman in the 70's and is still a topic of active research in every complex dimension [36], by LCK metric.

Definition 2.4 We say that a compact complex Hermitian surface is LCK if it admits a LCK metric which is not globally conformally Kähler.

In the above situation, the closed Lee form θ defines a flat line bundle $L_\theta \in \check{H}^1(M, \mathbb{R}^{>0})$ which we call here the *Lee bundle* —or also called *weight bundle* by Gauduchon et al.—with local trivializations $e^{-(f_i)}$ and transition functions given by $\{g_{ij} = \frac{e^{-(f_i)}}{e^{-(f_j)}}\}$ which are *positive constants* because the locally exact Lee form satisfies $df_i = \theta = df_j$ on double intersections while the cocycle condition on triple intersections holds automatically.

Remark 2.5 These same considerations show that the correspondence between closed Lee forms and Lee bundles extends to a linear map in cohomology

$$\exp : H^1_{dR}(S) \to \check{H}^1(S, \mathbb{R}^{>0})$$

from the first deRham cohomology group to the multiplicative group of line bundles with positive constant transition functions (and therefore flat) on M. △

For the purpose of the present work, the most important instance of this correspondence is given by the following result of Apostolov-Gauduchon-Grantcharov:

Proposition 2.6 ([6][Lemma 1, Prop. 2]) *For any bi-Hermitian metric $(M^4, g, J_\pm)$ consider the respective two Lee forms $\theta_\pm$ of $(g, J_\pm)$ which of course are real and need not be closed individually. However, the sum $\theta_+ + \theta_-$ is always closed and corresponds to the fundamental flat bundle F of the bi-Hermitian structure: $F := L_{\frac{1}{2}(\theta_+ + \theta_-)}$.*

2.3 Gauduchon Degree

Remark 2.7 *By a strong result of Gauduchon [24]*, in every conformal class c of Hermitian metrics on a compact complex surface S there is a unique representative $g \in c$ whose Lee form θ is coclosed or equivalently whose fundamental $(1, 1)$-form ω is $\partial\bar{\partial}$-closed. Furthermore the Lee form is coexact—$\theta = \delta G$—if and only if the second Betti number is even [24].

Such a Riemannian metric g will be called the *Gauduchon metric* of the conformal class and this result has many useful applications among which we mention the notion of *degree* of a holomorphic line bundle on the Hermitian surface $S = (M, g, J)$.

Because we will only be interested in *the special case of flat line bundles* $L_a \in H^1(M, \mathbb{R}^{>0})$ associated to closed 1-form $a \in H^1_{dR}(M)$ the formula in the definition simplifies to [4]

$$2\pi \deg(\mathcal{L}_a) = -\langle a, \theta\rangle = -\langle a_h, \theta_h\rangle \tag{2}$$

where θ is the Lee form of the Gauduchon metric g with g-harmonic part denoted by θ_h while a_h denotes the g-harmonic part of the 1-form a and $\langle , \rangle$ is the global inner product with respect to g; the last equality holds because a is closed and θ is coclosed.

We will make the following use of degree: let $\mathcal{L}$ denote the holomorphic line bundle which is the complexification of L, then deg measures the g-volume with sign of a virtual meromorphic section of $\mathcal{L}$—even if $\mathcal{L}$ might not admit non-trivial meromorphic sections. In particular $\deg(\mathcal{L}) > 0$ whenever $H^0(S, \mathcal{L}) \neq 0$ and $\mathcal{L} \neq \mathcal{O}$. Finally, because the space of Gauduchon metrics is connected and $\mathbb{R} \setminus 0$ is not, the sign of $\deg(L)$ is independent of the particular choice of Gauduchon metric on (S, J). It may, however, depend on J [22]. △

The above mentioned result of Apostolov-Grantcharov-Gauduchon allows to compute the degree of the fundamental flat line bundle with respect to the J_+–Gauduchon metric, see [2]

$$2\pi \deg_{J_+}(F) = -\langle \theta_+, \frac{1}{2}(\theta_+ + \theta_-)\rangle = -\frac{1}{4}||\theta_+ + \theta_-||^2$$

because the equality $||\theta_+||^2 = ||\theta_-||^2$ holds for any metric in the conformal class [38]. From this we get the result that the fundamental flat bundle F of a bi-Hermitian surface has

$$\deg(F) \leq 0 \tag{3}$$

with equality if and only if F is trivial

We end this section by observing that the Gauduchon degree of a Lee bundle $\mathcal{L}_\theta$—i.e. the holomorphic flat line bundle of a LCK metric—is always strictly negative, because from Eq. (2) it follows that $\deg(\mathcal{L}_\theta) = -||\theta_h||^2 < 0$ otherwise g would be conformally Kähler and $b_1(S)$ even.

3 Bi-Hermitian Surfaces

Getting back to the main question: which compact complex surfaces (S, J) admit a bi-Hermitian structure $(g, J_\pm)$ with $J = J_+$? The aim of this section is to present all surfaces which are known to admit such metrics. It is convenient to split our presentation according to the parity of the first Betti number which, in complex dimension 2, is equivalent to the existence of a Kähler metric.

3.1 Käherian Case

It turns out to be convenient to start with the case $b_1(S)$ even, because of the following

Proposition 3.1 *[6, Lemma 4] When $(M^4, g, J_\pm)$ is compact bi-Hermitian with* even *first Betti number the two Lee forms always satisfy the condition that*

$$\textit{the sum} \quad \theta_+ + \theta_- \quad \textit{is exact.} \tag{4}$$

Proof Let g be the J_+-Gauduchon metric, as $b_1(M)$ is even, we can assume $\theta_+ = \delta\alpha$. On the other hand $d(\theta_+ + \theta_-) = 0$ so that the orthogonal decomposition of θ_- into harmonic, coexact and coclosed part is of the form $\theta_- = \theta_-^h - \delta\alpha + df$. Because the two Lee forms have the same global L^2-norm we conclude that $\theta_-^h = 0 = df$ so that $\theta_+ + \theta_- = 0$ and g is J_--Gauduchon too. □

The fundamental equation of Proposition 2.1 (2.) reduces to

$$[T] = K^{-1} \tag{5}$$

and the necessary condition to admit bi-Hermitian metrics is then $H^0(S, K^{-1}) \neq 0$.

As a consequence, the canonical bundle of S is trivial or else S must be a ruled surface. In the first case, explicit examples where constructed by Kobak [30] and Apostolov-Grantcharov-Gauduchon [6] on tori and K3 surfaces, while Enriques or hyperelliptic surfaces cannot be bi-Hermitian because the flat bundle F is not real.

The remaining case is also of very high interest because the above proposition says that bi-Hermitian surfaces are indeed *generalized Kähler* for b_1 even. Important contributions by Hitchin [28], Gualtieri [27], Goto [25] et al. proved the converse statement that indeed any ruled surface with anti-canonical divisor is bi-Hermitian, when b_1 is even.

3.2 Non-Kählerian Surfaces

This case is where we gave a contribution and where some questions are still open. When $b_1(S)$ is odd, the flat line bundle F may not be trivial, equivalently the closed 1-form $\theta_1 + \theta$ may or may not be exact. In any case S cannot be a surface of Kodaira class VI:

Proposition 3.2 *[2] The canonical bundle of a compact bi-Hermitian surface S of odd first Betti number has negative degree, as a consequences all plurigenera must vanish and* $Kod(S) = -\infty$. *In other words the surface S belongs to class VII of Kodaira classification.*

Proof By the fundamental equation $T = K^{-1} \otimes F$ and recall that T is effective or zero while F has negative degree or is trivial Proposition 3.2. It follows that deg $K =$ deg F- vol $T < 0$ unelse F and T are both trivial. However, since $b_1(S)$ is odd, a result of Apostolov-Gauduchon-Grantcharov [6, Prop. 4] says that when F is trivial the fundamental divisor T must have two non-empty connected components, therefore cannot be trivial. □

From now on the bi-Hermitian surface S belongs to class VII_0 which exactly means: $\mathrm{Kod}(S) = -\infty$, S is minimal with odd first Betti number. Recall that indeed all such surfaces satisfy $b_1(S) = 1$ and have negative definite intersection form so that the topological signature $\sigma(S) = -b_2(S)$ agrees with the Chern number $c_1^2(S) = 2\chi(S) + 3\sigma(S) = -b_2(S)$.

Although there is no complete classification of surfaces in this class, we now present a few results which are particularly useful for discussing bi-Hermitian surfaces. We start from the case of vanishing first Chern class.

3.2.1 Surfaces in class VII_0 and $b_2(S) = 0$

A Theorem of Bogomolov [9], later clarified by Teleman [43] and Yau et al. [33] asserts that there exist only two different types of surfaces $S \in \mathrm{VII}_0$ with $b_2(S) = 0$.

The first class is given by Hopf surfaces which are defined to be compact quotients of $\mathbb{C}^2 \setminus \{0\}$; they have fundamental group isomorphic to $\mathbb{Z} \oplus (\mathbb{Z}/\mathbb{Z}_k)$ for $k \geq 1$ and are diffeomorphic to $S^1 \times S^3$, up to finite coverings.. The easiest example is when the action is generated by the contraction $(z, w) \mapsto \frac{1}{2}(z, w)$. The complete list of contractions was described by Kodaira and shows that anti-canonical divisors are always effective, consisting of two elliptic curves in the diagonal case and of one curve with multiplicity in the other case.

It follows that every Hopf surface satisfies the necessary condition (2.1) for existence of bi-Hermitian metrics.

Indeed, it has been shown in [38], [3] and [42], that Hopf surfaces admit an abundance of bi-Hermitian structures, with all possible values of $\mathbf{t} = b_0(T) = 0, 1, 2$.

By Bogomolov theorem the only other case satisfying $S \in$VII and $b_2(S) = 0$ is that S is a so called Inoue-Bombieri surface [31] [10]. Such surfaces are always discrete quotients of $\mathcal{H} \times \mathbb{C}$, the product of the upper-half plane with a complex line; they have no curves at all. Inoue-Bombieri surfaces however cannot admit bi-Hermitian metrics by Proposition 3.2 because the degree of their canonical bundle is always positive by a result of Teleman [44].

3.2.2 Surfaces in Class VII_0^+

In order to investigate bi-Hermitian surfaces we are now left with the non-Kählerian case of negative Kodaira dimension and positive second Betti number: $\mathrm{Kod}(S) = -\infty$ and $b_2(S) > 0$. Recall furthermore that we can assume that S is *minimal;* the standard notation for this class of surfaces is VII_0^+. Althought the Kodaira classification of these surfaces is still an open problem, thanks to the work of Dloussky all the bi-Hermitian surfaces in this class belong to a well known class of surfaces which will be presented in the last section.

Proposition 3.3 ([15]) *Every bi-Hermitian surface $S \in VII_0^+$ is a Kato surface.*

Proof The fundamental divisor T of any bi-Hermitian surface is either effective or zero with the same Chern class of S: $c_1(T) = c_1(K_S^{-1}) = c_1(S)$ so that $c_1^2(S) = b_2(S) > 0$ implies that T is a non-trivial NAC (numerical (pluri-) anti-canonical) divisor on $S \in \mathrm{VII}_0^+$ which must therefore be a Kato surface by a strong result of Dloussky [15]. □

4 Bi-Hermitian and LCK Metrics, an Interplay

The first examples of bi-Hermitian Kato surfaces, are due to Lebrun [32] on so called Parabolic Kato surfaces. Their construction is a byproduct of the LeBrun ansatz for self-dual metrics on the connected sum of arbitrary many copies of $\mathbb{CP}_2$. As such these bi-Hermitian metrics are also LCK, by a result of [11].

This circle of ideas motivated the construction of Fujiki-Pontecorvo which was aiming at producing LCK metrics on Kato surfaces. Although the relevant log-deformation theory of singular twistor spaces turned out to be obstructed in the general case, it was proved in [20] that the construction produces bi-Hermitian surfaces on all Kato surfaces admitting a disconnected anti-canonical divisor, of real type. This result completed the classification of bi-Hermitian Kato surfaces under the hypothesis that the flat line bundle F is actually trivial.

Theorem 4.1 *[20] Every Hyperbolic Inoue surface and every Parabolic Inoue surface of real type admits bi-Hermitan metrics and these are the only possibilities in class VII_0^+, when $\mathbf{t} = 2$.*

Indeed, coming from twistor theory, these metrics are automatically anti-self-dual and therefore LCK. Indeed a strong motivation for this work was to give new inputs to the following old interesting

Question 4.2 (Vaisman 1970's) Which compact complex surfaces admit LCK metrics?

We refer to [39] for an overview of results and just mention here that a consequence of [20] was the first proof that all Kato surfaces with Dloussky number equal to $2b$ or $3b$ are LCK; see the next section for the definition of Dloussky number.

Soon after, this result was greatly generalized by the remarkable work of Marco Brunella which was able to show that *all* Kato surfaces are LCK [12].

So that, after Brunella's work, the state of the art about Vaisman question is: all *known* compact complex surfaces are LCK, except for the Inoue-Bombieri surfaces of Belgun [8].

A different, powerful interplay between LCK and bi-Hermitian metrics is given by the following result of Apostolov-Bailey-Dloussky which, in the same way as LCK geometry can be thought of as a twisted version of Kähler geometry, appears as a twisted version of a Hamiltonian construction for generalized Kähler metrics of Gualtieri [26].

Theorem 4.3 *[7] Let $S = (M^4, J)$ be a compact complex surface in class VII with a locally conformally Kähler metric and denote by F its Lee bundle. Then, M also admits a bi-Hermitian metric $(g, J_{\pm})$ whose fundamental divisor is T and $J = J_+$ if and only if there is a* connected *NAC divisor T of index 1—i.e. $[T] = K^{-1} \otimes F$— with the same real flat bundle F of the LCK metric*

This theorem was used by Apostolov-Bailey-Dloussky to produce the first bi-Hermitian metrics on Hopf surfaces with $\mathbf{t} = 1$ [7]; and then by Fujiki-Pontecorvo to prove the same result for parabolic Inoue surfaces [23]. In fact, surprisingly enough, the Lee bundle of the LCK bi-Hermitian metric turns out to be exactly what is needed for the Apostolov-Bailey-Dloussky theorem to work. The conclusion is that Hopf surfaces admit bi-Hermitian metrics with all possible values of $\mathbf{t} = 0, 1, 2$ while Parabolic Inoue are bi-Hermitian with $\mathbf{t} = 1, 2$; it is not hard to see that these are the only surfaces which can possibly be bi-Hermitian for different values of $\mathbf{t}$.

5 Kato Surfaces

The classification of compact complex surfaces in class VII_0^+ has received much attention in the last 40 years from the Japanese school of Nakamura, Inoue, Enoki as well from the French-based school of Dloussky, Oeljeklaus, Toma and Teleman; we present some of their results in this section. For the moment however, the only known surfaces in this class can be described by the following geometric construction of Kato [29], who also showed that they can be deformed into blown-up Hopf surfaces and are therefore always diffeomorphic to $(S^1 \times S^3)\#b\overline{\mathbb{CP}}_2$.

A *Kato surface* S is geometrically obtained by successive blow ups of the unit ball in $\mathbb{C}^2$ starting from the origin and continuing to blow up a point on the last constructed exceptional curve until, in order to get a *minimal* compact complex surface, one takes the quotient by a biholomorphism σ which identifies a small ball centered at a point p in the last exceptional curve with a small ball around the origin $0 \in \mathbb{C}^2$ in such a way that $\sigma(p) = 0$, see the picture on [16, p.5]. This process constructs $b := b_2(S)$ rational curves $D_1, \ldots, D_b$ of self-intersection ≤ -2 which are the only rational curves in S.

In other words $S = S_{\pi,\sigma}$ is obtained by identifying, via $\sigma \circ \pi$, the two boundary components of $\hat{B} \setminus \sigma(B)$—denoted by $\partial \hat{B} \amalg \sigma(\partial B)$—each of which is a 3-sphere. This shows that S has a global spherical shell, briefly called GSS, and $\pi_1(S) = \mathbb{Z}$.

Conjecture 5.1 The **GSS conjecture** of Nakamura [34] asserts that every surface in class VII_0^+ is a Kato surface.

As a matter of fact it is not even known if every surface in class VII_+^0 must have curves and, since the algebraic dimension vanishes, these surfaces can have at most finitely many curves. However, by a very relevant result of Dloussky-Oeljekluas-Toma [18] a surface $S \in \mathrm{VII}_0^+$ with $b_2(S)$ rational curves must be a Kato surface and the configuration of the rational curves is encoded by the Dloussky sequence $\mathrm{Dl}(S)$ which we now present.

The blow-up construction of S produces exactly $b_2(S) =: b$ rational curves most of which will have self-intersection number -2 because a curve C with self-intersection number $C^2 = -(k+2) \leq -3$ can only be obtained by repeatedly blow-up at a fixed node of previously created exceptional curves and will therefore come along with a chain of -2-curves of length $(k-1)$. All other curves in S are either obtained in this way or else by blowing up a general point of the previously created exceptional divisor, in particular their self-intersection number is -2. Therefore, if $(a_1, \ldots, a_b)$ denotes the string of *opposite* self-intersection numbers of all rational curves in S we can separate it into *singular* sequences $s_k := ((k+2)2\ldots 2)$ of length k and *regular* sequences $r_m := (2\ldots 2)$ of length m.

This notation was introduced by Dloussky in [14] and the *Dloussky sequence* of a Kato surface S is a string of $b = b_2(S)$ positive integers describing the opposite self-intersection number of the rational curves, following their *order of creation,* and grouped into singular sequences s_k and regular sequences of maximal length r_m.

We refer to [22] for a picture of the dual graph of curves in a Kato surface S with *simple* Dloussky sequence

$$\mathrm{DlS} = [s_{k_1} s_{k_2} \ldots s_{k_p} r_m] \tag{6}$$

A weaker biholomorphic invariant of Kato surfaces is the Dloussky number $\sigma(S)$ which is defined to be the sum of opposite self-intersection numbers of the rational curves $D_1, \ldots, D_b$ in S so that $\sigma(S) := -\sum_{i=1}^{n} D_i^2$. It follows from the previous discussion that

$$2b \leq \sigma(S) \leq 3b \tag{7}$$

We now give a quick presentation of these surfaces using the terminology of the Japanese school. Kato surfaces with $2b = \sigma(S)$—or equivalently $\mathrm{Dl}(S) = [r_b]$—are called Enoki surfaces and are geometrically characterized by the property of having a cycle C, of b rational curves and self-intersection $C^2 = 0$, which therefore represents a flat line bundle. The versal family of Enoki surfaces is b-dimensional and has a 1-dimensional subfamily of surfaces containing an elliptic curve. These special Enoki surfaces are called Parabolic Inoue. By the previously mentioned results, every Parabolic Inoue surface with line bundle $[C]$ of real type has bi-Hermitian metrics with $\mathbf{t} = 2$ as well as different bi-Hermitian metrics with $\mathbf{t} = 1$. To the contrary, a general Enoki surface cannot be bi-Hermitian simply because there are no anti-canonical divisors.

The other extreme case $\sigma(S) = 3b$ is when all blow-ups occur at intersection points so that $\mathrm{Dl}(S) = [s_{k_1} s_{k_2} \ldots s_{k_p}]$ presents no regular sequences. As a consequence, every rational curve belongs to a cycle and there are at most two cycles in S. The corresponding surface is then called Hyperbolic Inoue when there are two cycles (p is even). Every such surface is bi-Hermitian by [20], while in the other case p =odd there is no bi-Hermitian metric because the only NAC divisor is not of real type.

All other Kato surfaces satisfy the strict inequalities $2b < \sigma(S) < 3b$ and are therefore called *intermediate*. In this case the union of all rational curves is connected and contains a unique cycle with branches departing from it. The number of branches equals the number of regular sequences in $\mathrm{Dl}(S)$.

There are no examples of bi-Hermitian intermediate Kato surfaces but in fact there are some serious known obstructions which we now would like to present.

The first obstruction is the index. This is defined to be the least positive integer for which $H^0(S, K^{-m} \otimes F) \neq 0$, for some flat line bundle $F \in \mathrm{Pic}_0(S)$.

Therefore, if an intermediate Kato surface S is bi-Hermitian it needs to have index 1.

This (index=1-)condition only depends on $\mathrm{Dl}(S)$ and as such poses severe restrictions on the configuration of curves, like for example that the number of irreducible components in the cycle must be bigger than that of the branches, see [22].

More obstructions come from the complex structure of intermediate Kato surfaces. Before getting into the relevant complex deformations theory, we need to mention the following convenient way of Dloussky-Oeljeklaus to parametrize flat line bundles.

Following [17][§1.2], for any Kato surface S the group $\mathrm{Pic}_0(S)$ of flat holomorphic line bundles is non-compact and isomorphic to $\mathbb{C}^*$. Furthermore, being a surface with a global spherical shell, S has a Leray cover by two open sets whose intersection has two connected components (like in the circle) so that every $F \in \mathrm{Pic}_0(S)$ is determined by a pair of transition functions $(1, e^f)$ with f a complex constant. The notation used to indicate this correspondence is as follows

$$\begin{array}{ccc} \mathrm{Pic}_0(S) & \longleftrightarrow & \mathbb{C}^* \\ F & \longmapsto & L^{e^f} \end{array}$$

and notice that *real* flat line bundles have $f \in \mathbb{R}$ while $f = 0$ if and only if F is trivial.

Now we come to *logarithmic deformations.* More precisely, we consider the versal family $\mathcal{S}$ of deformations of the pair (S, D) where S is intermediate Kato surface and D is the union of all its rational curves. D is always a connected and reduced normal crossing divisor with b irreducible components. The complex deformation family of (S, D) is unobstructed of complex dimension $h^1(S, \Theta(\log(-D))) = m$, where m is the total length of the regular sequences in $\mathrm{Dl}(S)$. Notice that the deformation preserves the divisor D and even more so $\mathrm{Dl}(S)$. In particular, there exists an integer $k \geq 2$ which can be defined in various ways [16] [22] and explicitly computed as a polynomial of degree p in the variables $(k_1, \ldots, k_p)$. So that k remains fixed in any logarithmic deformation.

The following result says that every intermediate Kato surface S has a unique numerically anti-canonical divisor. Furthermore, S is deformation equivalent to a surface S_u with a holomorphic vector field and $\mathrm{Dl}(S_u)$ $=\mathrm{Dl}(S)$, if and only if the index is 1. Finally, recall that the sign of degree is independent of the Gauduchon metric.

Theorem 5.2 *[17][4.5 and 5.2] Let $\mathcal{S} \to U$ be a versal family of deformations of the pair (S, D) with (fixed) integer $k \geq 2$, as above. Then, there is a surjective holomorphic map $\mu : U \to \mathbb{C}^*$ such that $\mu(u)$ is the unique complex number for which $H^0(S_u, K^{-1} \otimes L^{\mu(u)}) \neq 0$. Furthermore, μ can be chosen to satisfy the following properties:*

1. *The flat bundle $L^{\mu(u)}$ is of real type if and only if $\mu(u) \in \mathbb{R}^{>0}$; the sign of its degree is positive for $\mu(u) > 1$ and therefore negative when $0 < \mu(u) < 1$.*
2. *S_u has a holomorphic anti-canonical section if and only if index$(S) = 1 = \mu(u)$.*
3. *S_u has a holomorphic vector field if and only if index$(S) = 1$ and $\mu(u) = \frac{1}{k}$.*

The above versal family can be described using contracting germ of the biholomorphism σ in the GSS construction of the Kato surface S. This germ is a polynomial map $\phi : \mathbb{C}^2 \to \mathbb{C}^2$ which has been put in the following normal form by Favre [19].

$$\phi(z, \xi) = (\lambda \xi^s z + Q(\xi), \xi^k) \tag{8}$$

The degree of the second component is exactly the integer $k \geq 2$ of the Theorem and can be shown to regulate existence of vector fields. The function μ depends on the first coefficient $\lambda \in \mathbb{C}^*$ of the germ.

The relevance of the germ technique and the role of the integer k shows up strongly in the following recent result of Apostolov-Dloussky who used Favre germ (8) to construct a negative plurisubharmonic function on the universal cover of any intermediate Kato surface (of arbitrary index) with automorphic factor exactly equal to k. In particular, by their general result [5][3.7], $L^{\mu(u)}$ cannot be the Lee bundle of a LCK metric on S_u, unless $\mu(u) > k$.

This is an obstruction for the Lee bundle of a LCK metric on S which induces an obstruction for bi-Hermitian metrics via the Apostolov-Bailey-Dloussky Theorem [7].

We close the paper with some simple concluding remarks. Every compact bi-Hermitian four-manifolds with $b_1(M)$ is always diffeomorphic to a blown up Hopf surface $(S^1 \times S^3)\#b\overline{\mathbb{CP}}_2$. However the finer question of determining the complex structure remains open in the intermediate case, only:

Question 5.3 Can intermediate Kato surfaces support bi-Hermitian metrics ?

As we have seen, this question is very much related to the study of possible Lee classes of LCK metrics on intermediate Kato surfaces. This is in fact a very interesting question, and an active field of research for LCK manifolds of any dimension [35]. In complex dimension 2 it has been extended by Apostolov-Dloussky to the study of locally conformally symplectic structures, see [4], [5].

We conclude with a final remark about Vaisman question: first of all, the answer is not always positive, as pointed out by Belgun [8]. Indeed, in his thesis at Ècole Polytechnique he showed that some of the Inoue-Bombieri surfaces are not LCK, proving at the same time that the LCK property is not stable under deformations of the complex structure.

A simple remark is then that a solution of the GSS conjecture would also imply that Belgun examples are the only possible non-LCK surfaces and this would provide a complete answer to Vaisman question. This has been observed for example by Ornea-Verbitsky-Vuletescu in [37].

Acknowledgments Most of our results have been obtained in collaboration with Prof. Akira Fujiki which we hearthfully thank for the longstanding collaboration and friendship. It is a great pleasure to say that I have been learning a lot about the subject from Simon Salamon, Piotr Kobak, Gueo Grantcharov, Paul Gauduchon, Vestislav Apostolov, Georges Dloussky, Ryushi Goto and Liviu Ornea.

References

1. D. Alekseevsky, S. Marchiafava, M. Pontecorvo, Compatible complex structures on almost quaternionic manifolds. Trans. A.M.S. **351**, 997–1014 (1999)
2. V. Apostolov, Bihermitian surfaces with odd first Betti number. Math. Z. **238**, 555–568 (2001)
3. V. Apostolov, G. Dloussky, Bihermitian metrics on Hopf surfaces. Math. Res. Lett. **15**(5), 827–839 (2008)

4. V. Apostolov, G. Dloussky, Locally conformally symplectic structures on compact non-Kähler complex surfaces. Int. Math. Res. Not. IMRN **9**, 2717–2747 (2016)
5. V. Apostolov, G. Dloussky, Twisted differentials and Lee classes of locally conformally sympletic complex surfaces. Math. Z. **303**, 76 (2023)
6. V. Apostolov, P. Gauduchon, G. Grantcharov, Bi-Hermitian structures on complex surfaces. Proc. Lond. Math. Soc. (3) **79**(2), 414–428 (1999); Proc. Lond. Math. Soc. (3) **92**(1), 200–202 (2006)
7. V. Apostolov, M. Bailey, G. Dloussky, From locally conformally Kähler to bi-Hermitian structures on non-Kähler complex surfaces. Math. Res. Lett. **22**, 317–336 (2015)
8. F.A. Belgun, On the metric structure of non-Kähler complex surfaces. Math. Ann. **317**, 1–40 (2000)
9. F.A. Bogomolov, Classification of surfaces of class VII_0 and affine geometry. Math.USSr-Izv **10**, 255–269 (1976)
10. E. Bombieri, *Letter to Kodaira* (1973)
11. C.P. Boyer, Conformal duality and compact complex surfaces. Math. Ann. **274**, 517–526 (1986)
12. M. Brunella, Locally conformally Kähler metrics on Kato surfaces. Nagoya Math. J. **202**, 77–81 (2011)
13. G. Cavalcanti, M. Gualtieri, Blowing up generalized Kähler 4-manifolds. Bull. Braz. Math. Soc. (N.S.) **42**(4), 537–557 (2011)
14. G. Dloussky, Structure des Surfaces de Kato Memoire Soc. Mat. Fr. **14**(14), 1–122 (1984)
15. G. Dloussky, On surfaces of class VII_0 with numerically anticanonical divisor. Am. J. Math. **128**(3), 639–670 (2006)
16. G. Dloussky, From non-Kählerian surfaces to Cremona group of $\mathbb{CP}_2$. Complex Manifolds **1** 1–33 (2014)
17. G. Dloussky, K. Oeljeklaus, Vector fields and foliations an compact surfaces of class VII_0. Ann. Inst. Fourier **49**, 1503–1545 (1999)
18. G. Dloussky, K. Oeljeklaus, M. Toma, Class VII_0 surfaces with b_2 curves. Tohoku Math. J. (2) **55**, 283–309 (2003)
19. C. Favre, Classification of 2-dimensional contracting rigid germs. J. Math. Pures Appl. **79**, 475–514 (2000)
20. A. Fujiki, M. Pontecorvo, Anti-self-dual bihermitian structures on Inoue surfaces. J. Differ. Geom. **85**(1), 15–71 (2010)
21. A. Fujiki, M. Pontecorvo, Twistors and bi-Hermitian surfaces of non-Kähler type. Symmetry Integrability Geometry Methods Appl. **10**, Paper 042, 13 pp. (2014)
22. A. Fujiki, M. Pontecorvo, Numerically anti-canonical divisors on Kato surfaces. J. Geom. Phys. **91**, 117–130 (2015)
23. A. Fujiki, M. Pontecorvo, Bi-Hermitian metrics on Kato surfaces. J. Geom. Phys. **138**, 33–43 (2019)
24. P. Gauduchon, La 1-forme de torsion d'une variété hermitienne compacte. Math. Ann **267**, 495–518 (1984)
25. R. Goto, Unobstructed K-deformations of generalized complex structures and bi-Hermitian structures. Adv. Math. **231**, 1041–1067 (2012)
26. M. Gualtieri, Branes on Poisson varieties. *The Many Facets of Geometry* (Oxford University Press, Oxford, 2010), pp. 368–394
27. M. Gualtieri, Generalized Kähler geometry. Commun. Math. Phys. **331**, 297–331 (2014). arXiv:1007.3485
28. N.J. Hitchin, Instantons, Poisson structures and generalized Kähler geometry. Commun. Math. Phys. **265**, 131–164 (2006)
29. M. Kato, Compact complex manifolds containing "global" spherical shells. *I. Proceedings of the International Symposium on Algebraic Geometry (Kyoto University, Kyoto, 1977)* (Kinokuniya Book Store, Tokyo, 1978), pp. 45–84
30. P. Kobak, Explicit doubly-Hermitian metrics. Differ. Geom. Appl. **10**, 179-185 (1999)
31. M. Inoue, On Surfaces of Class VII_0. Inv. Math. **24**, 269–310 (1974)

32. C.R. Lebrun, Anti-self-dual Hermitian metrics on blown-up Hopf surfaces. Math. Ann. **289**, 383–392 (1991)
33. J. Li, Yau, S.-T., F. Zheng, On projectively flat Hermitian manifolds. Commun. Anal. Geom. **2**, 103–109 (1994)
34. I. Nakamura, Classification of non-Kähler complex surfaces. Sugaku Expositions **2**, 209–229 (1989). Translated in Japanese Sugaku **36**(2), 110–124 (1984)
35. Ornea L., M. Verbitsky, Lee classes on LCK manifolds with potential. Tohoku Math. J.**76**(2), 105–125 (2024)
36. Ornea L., M. Verbitsky, *Principles of Locally Conformally Käheler Geometry*. Progress in Mathematics (Birkhäuser, Cham, 2024)
37. L. Ornea, M. Verbitsky, V. Veletescu, Classification of non-Kähler surfaces and locally conformally Kähler geometry Russ. Math. Surveys. **76**, 261–289 (2021)
38. M. Pontecorvo, Complex structures on Riemannian four-manifolds. Math. Ann. **309**(1), 159–177 (1997)
39. M. Pontecorvo, On a question of Vaisman concerning complex surfaces. Ann. Mat. Pura Appl. **193**, 1283–1293 (2014)
40. S. Salamon, Special structures on four-manifolds. Riv. Mat. Univ. Parma (4) **17**, 109–123 (1994)
41. S. Salamon, Orthogonal complex structures. *Differential Geometry and Applications (Brno, 1995)* (Masaryk University, Brno, 1996), pp. 103–117
42. J. Streets, Y. Ustinov, Classification of generalized Kähler-Ricci solitons on complex surfaces. Commun Pure Appl. Math. **74**, 1896–1914 (2021)
43. A. Teleman, Projectively flat surfaces and Bogomolov's theorem on class VII_0 surfaces. Int. J. Math. **5**, 253–264 (1994)
44. A. Teleman, The pseudo-effective cone of a non-Kählerian surface and applications. Math. Ann. **335**, 965–989 (2006)

Revisiting 3-Sasakian and G_2-structures

Simon Salamon and Ragini Singhal

On the occasion of Paul Gauduchon's 80th birthday

Abstract The algebra of exterior differential forms on a regular 3-Sasakian 7-manifold is investigated, with special reference to nearly-parallel G_2 3-forms. This is applied to the study of 3-forms invariant under cohomogeneity-one actions by $SO(4)$ on the 7-sphere and on Berger's space $SO(5)/SO(3)$.

Keywords Self-dual Einstein · Cohomogeneity one · Nearly parallel · Berger space · Nearly half-flat

1 Introduction

Let M be an oriented self-dual Einstein 4-manifold M with positive scalar curvature. The bundle of 2-forms

$$\Lambda^2 T^*M = \Lambda^+ \oplus \Lambda^- \tag{1.1}$$

splits into self-dual (SD) and anti-self-dual (ASD) subbundles, in line with the double covering

$$SO(4) \to SO(3)^+ \times SO(3)^-$$

S. Salamon (✉)
King's College London, Strand, London, UK
e-mail: simon.salamon@kcl.ac.uk

R. Singhal
University of Münster, Münster, Germany
e-mail: rsinghal@uni-muenster.de

L. Ornea (ed.), *Real and Complex Geometry*,
https://doi.org/10.1007/978-3-031-92297-8_13

of Lie groups. Our hypothesis implies that Λ^- has constant curvature in the sense of (2.4) below. The reason for focussing on Λ^- is that (with its standard orientation and Fubini-Study metric) the projective plane $\mathbb{CP}^2$ satisfies the hypothesis, as do the metrics featuring in [4]. Much of the theory extends to orbifolds, such as quaternion-Kähler quotients [15, 21], and the $SO(3)$-invariant orbifolds discovered by Hitchin [26]. The latter are parametrized by an integer $k \geqslant 3$ and the first two examples are S^4 and $\mathbb{CP}^2/\mathbb{Z}_2$.

The (total space of the) $SO(3)^-$ principal bundle P^- associated to Λ^- is 3-Sasakian [30], and therefore admits an Einstein metric and a family of nearly-parallel G_2 (for short, NP_2) structures. An equivalent way of expressing the 3-Sasakian condition is to say that the Riemannian cone $P^- \times \mathbb{R}^+$ is hyperkähler, meaning that it has a metric with holonomy $Sp(2)$ [8]. However, it is well known that a squashing of the fibres of P^- produces a second Einstein metric and canonical NP_2 structure whose cone has holonomy *equal* to *Spin*(7).

Now suppose (restricting to an open set if necessary) that M is spin, so that its structure group lifts to $Spin(4) \cong SU(2)^2$. The negative spin bundle Σ^- can be regarded as a quaternionic line bundle over M, and admits a family of metrics with *Spin*(7) holonomy, all asymptotic to the cone over (a 2 : 1 cover of) the squashed NP_2 manifold [13]. An initial purpose of this article is to relate well-known constructions of the relevant differential forms on these spaces, one via a *Spin*(7) 4-form on a vector bundle, the other via 3-Sasakian 1-forms on a principal bundle. This is motivated by the situation in which M itself admits an action by $SO(3)$ with 3-dimensional orbits. In this case, P^- or its double cover $\tilde{P}^-$, will admit an action by $SO(4)$ or $SU(2)^2$ of cohomogeneity one, preserving the G_2 structures. This happens when M is S^4 or $\mathbb{CP}^2$, and more generally for the Hitchin orbifolds.

We shall make this explicit for the 7-sphere $S^7 \cong \tilde{P}^- S^4$, in order to express its squashed NP_2 structure in canonical form. The result is consistent with a known expression for the round metric on S^7 subject to $SO(4)$ and $\mathbb{Z}_3$ symmetry [38]. The same technique is applied to the isotropy irreducible space $B^7 = SO(5)/SO(3)$, which Berger showed has positive sectional curvature [10]. This space also has a homogeneous NP_2 structure [35], a cohomogeneity-one action by $SO(4)$ [34], and is diffeomorphic to $\tilde{P}^+$ for a non-trivial Hitchin orbifold [25]. Although not our main concern, exhaustive classifications of associative submanifolds in B^7 and in squashed 3-Sasakian manifolds are undertaken in [6, 7] using twistor correspondences, pursuing work in [27, 32] for S^7.

Paying debts to this previous mountain of work, our survey highlights the distinction between the homogeneous NP_2 structures on S^7 and B^7, when expressed in terms of the local differential geometry on an interval times $SO(4)$. The choice of geodesics orthogonal to the orbits allows one to construct bases (X_i) and (Y_i) of invariant 1-forms that enable the respective 3-forms to be built up using upper triangular 2×2 matrices; see Corollaries 4.4 and 5.4. These bases ultimately arise from different conjugacy classes of 3-dimensional subgroups of $SO(4)$, and different group diagrams for the cohomogeneity-one actions.

Just as a hypersurface of a G_2 holonomy manifold acquires a so-called half-flat $SU(3)$ structure, so a hypersurface of a NP_2 manifold has an analogous nearly-half-flat (NHF) property. There is a sense in which a nearly-Kähler 6-manifold has both properties. We conclude with a brief investigation of induced structures on hypersurfaces of S^7 and B^7, and we show that constants defining the homogeneous NP_2 structures are to some extent determined by the NHF condition. The results of this section form part of a broader project to describe exceptional structures in dimensions 6 and 7.

2 Notation

In the first part of this paper, we shall adopt the technique of [13], which relies on quaternion-valued differential forms on the total spaces of Λ^- and Σ^-, though only the latter will concern us.

The Riemannian metric of M^4 will be denoted by g_4 and its positive scalar curvature by 12κ, where κ is another constant. The sphere $S^n(r)$ of radius r has scalar curvature $n(n-1)/r^2$, so $\kappa = 1$ for $S^4(1)$. **From now on, we shall scale g_4 so that $\kappa = 1$.**

The induced connection on Σ^- is defined by a vector-valued 1-form

$$\phi = i\phi_1 + j\phi_2 + k\phi_3$$

on the total space of Σ^-. We shall refer to a differential form with values in $\operatorname{Im}\mathbb{H}$ as 'vector valued', as opposed to 'quaternion-valued' (for $\mathbb{H}$). Thus, $\bar{\phi} = -\phi$, where a bar denotes quaternionic conjugate. The curvature of Σ^- is

$$\begin{aligned}\Phi &= d\phi + \phi \wedge \phi \\ &= i(d\phi_1 + 2\phi_{23}) + j(d\phi_2 + 2\phi_{31}) + k(d\phi_3 + 2\phi_{12})\end{aligned} \tag{2.1}$$

where $\phi_{23} = \phi_2 \wedge \phi_3$. See [13, p. 841].

The choice of a unit section of Σ^- determines a local orthonormal basis of ASD 2-forms via the isomorphism $S^2\Sigma^- \cong (\Lambda^2_-)_c$ of complex vector spaces or bundles. These 2-forms are the components of the tautological vector-valued 2-form

$$\tfrac{1}{2}(\overline{\omega} \wedge \omega) = i\Omega_1 + j\Omega_2 + k\Omega_3 = \Omega.$$

We use a Greek letter without an index to stand for a quantity with values in $\mathbb{H}$ that encodes all the individual components. Scalings are fixed by the existence pointwise of an orthonormal basis $\{\omega_0, \omega_1, \omega_2, \omega_3\}$ of 1-forms for g_4 such that

$$\Omega_k = \omega_{0k} - \omega_{ij} \tag{2.2}$$

for $(ijk) = (123)$.

It follows from (2.1) that

$$d\Phi = -2\phi \wedge \Phi \tag{2.3}$$

where $\wedge$ combines multiplication of imaginary quaternions (cross product) with wedging. With the notation above, the Einstein ASD condition becomes

$$\Phi = \tfrac{1}{2}\kappa\Omega \tag{2.4}$$

as stated in [13, p. 842], although we are assuming throughout that $\kappa = 1$.

Now let

$$a = a_0 + a_1 i + a_2 j + a_3 k \tag{2.5}$$

be a quaternionic coordinate on the fibre of Σ^-. Note that $\overline{a}a = a\overline{a} = R$ is the radius squared (denoted r in [13, p. 846]). Set

$$\alpha = da - a\phi = \alpha_0 + i\alpha_1 + j\alpha_2 + k\alpha_3. \tag{2.6}$$

At each point, the annihilator of α is the horizontal subspace determined by the connection. We can then determine the effect of the action $a \rightsquigarrow aq$ on differential forms. It yields the transformations $\phi \rightsquigarrow \overline{q}\phi q$ and $\Omega \rightsquigarrow \overline{q}\Omega q$.

Next, consider the quaternion-valued 1-form

$$\beta \;=\; \overline{a}\alpha \;=\; \beta_0 + i\beta_1 + j\beta_2 + k\beta_3 \tag{2.7}$$

as defined in [13, p. 846]. In particular,

$$\beta_1 \;=\; a_0\alpha_1 - a_1\alpha_0 - a_2\alpha_3 + a_3\alpha_2$$

is the result of applying a fibrewise ASD almost complex structure to $\beta_0 = \tfrac{1}{2}dR$. However, from now on, we work on $\tilde{P}^-$, so that $R = 1$, and β becomes a vector-valued 1-form. In this setting,

Lemma 2.1 $d\beta + \beta \wedge \beta = -2\,\mathrm{Im}\,(\phi \wedge \beta) - \Phi.$

Proof Using (2.1), (2.4), and (2.6),

$$d\alpha \;=\; -(\alpha + a\phi) \wedge \phi - a d\phi \;=\; -\alpha \wedge \phi - a\Phi. \tag{2.8}$$

Therefore

$$\begin{aligned} d\beta \;=\; d(\overline{a}\alpha) &= d\overline{a} \wedge \alpha + \overline{a}\, d\alpha \\ &= d\overline{a} \wedge \alpha + \overline{a}(-\alpha \wedge \phi - a\Phi) \\ &= \overline{\alpha} \wedge \alpha - (\phi \wedge \beta + \beta \wedge \phi) - \Phi. \end{aligned}$$

Now restrict to $R = 1$, so that β (and, as always, ϕ) have values in $\mathrm{Im}\,\mathbb{H}$. Because of the wedging, the real parts of $\phi \wedge \beta$ and $\beta \wedge \phi$ cancel out. Moreover,

$$\overline{\alpha} \wedge \alpha = \overline{\alpha}a \wedge \overline{a}\alpha = \overline{\beta} \wedge \beta = -\beta \wedge \beta \tag{2.9}$$

and the result follows. □

The first component of the equation in Lemma 2.1 is

$$d\beta_1 + 2\beta_{23} \;=\; -2(\phi_2 \wedge \beta_3 - \phi_3 \wedge \beta_2) - \Phi_1. \tag{2.10}$$

Note that β_{23} is totally vertical whereas $\Phi_1 = \frac{1}{2}\kappa\Omega_1$ 'lives' on M, so their exterior derivatives cannot be proportional, which forces the appearance of the mixed terms on the right-hand side. The fibrewise ASD property of $d\beta$ is indicated by the presence of $\overline{\alpha} \wedge \alpha$ rather than the self-dual $\alpha \wedge \overline{\alpha}$.

3 Spin(7) and G_2-Structures

One of the main results in [13, §4], by now well known, is that the total space of Σ^- admits a family of metrics with holonomy equal to $Spin(7)$ and asymptotically conical. The induced Einstein metric h_7 on the link of this conical limit therefore underlies a so-called nearly-parallel G_2 structure (our abbreviation is NP_2). The latter comprises a 3-form φ with stabilizer G_2 for which

$$d\varphi = \mu * \varphi \tag{3.1}$$

where $21\mu^2/8$ is known to equal the scalar curvature of h_7 [12]. The metric h_7 is also said to have weak holonomy G_2 [24]. We proceed to write down φ explicitly.

In order to extract useful real-valued differential forms from those defined in Sect. 1, we need to exploit the obvious invariants for the action of $SO(3)$ on $\mathrm{Im}\,\mathbb{H} \cong \mathbb{R}^3$, namely inner product and determinant. In this way, we can define the real 3-forms

$$\begin{aligned}
\Upsilon &= \beta_{123} &&= -\tfrac{1}{6}\beta \wedge \beta \wedge \beta\\
\Theta &= \Phi_1 \wedge \beta_1 + \Phi_2 \wedge \beta_2 + \Phi_3 \wedge \beta_3 &&= -\mathrm{Re}\,[\Phi \wedge \beta]
\end{aligned}$$

and the real 4-forms

$$\begin{aligned}
-d\Upsilon &= \Phi_1 \wedge \beta_{23} + \Phi_2 \wedge \beta_{31} + \Phi_3 \wedge \beta_{12} &&= -\tfrac{1}{2}\mathrm{Re}\,[\Phi \wedge \beta \wedge \beta]\\
X &= \Phi_1 \wedge \Phi_1 + \Phi_2 \wedge \Phi_2 + \Phi_3 \wedge \Phi_3 &&= -\mathrm{Re}\,[\Phi \wedge \Phi].
\end{aligned}$$

Using these, we can build G_2 structures. In particular:

Proposition 3.1 *Let $\lambda = 2/\sqrt{5}$. The 3-form*

$$\varphi = 2\lambda\,\Theta + \lambda^3\Upsilon \tag{3.2}$$

defines an NP$_2$ structure on $\tilde{P}^-$ with $\mu = 6/\sqrt{5}$. The associated Einstein metric is

$$h_7 = \tfrac{4}{5}\sum_{l=1}^{3}\beta_i\otimes\beta_i + \pi^* g_4. \tag{3.3}$$

Proof The calculation of $d\Theta$ is a little delicate:

$$\begin{aligned} d\Theta &= -\mathrm{Re}\,[d\Phi\wedge\beta + \Phi\wedge d\beta] \\ &= -\mathrm{Re}\,[-2\phi\wedge\Phi\wedge\beta] - \mathrm{Re}\,[\Phi\wedge(-\beta\wedge\beta - 2\phi\wedge\beta - \Phi)] \\ &= 2\,d\Upsilon - X. \end{aligned}$$

Terms in $\phi\wedge\Phi$ and $\Phi\wedge\phi$ cancel out because wedging is commutative but imaginary quaternion multiplication is not.

Recall from (2.2) that $\Phi_i = \frac{1}{2}(\omega_{0i} - \omega_{jk})$ is a horizontal ASD 2-form, so $X = -\frac{3}{2}\omega_{0123}$, the volume form of (M, g_4) being ω_{0123}. It follows that (3.2) defines a G_2 structure whose associated metric has an orthonormal basis

$$\omega_0,\ \omega_1,\ \omega_2,\ \omega_3,\ \lambda\beta_1,\ \lambda\beta_2,\ \lambda\beta_3 \tag{3.4}$$

so

$$*\varphi = 2\lambda^2 d\Upsilon - \tfrac{2}{3}X.$$

On the other hand,

$$d\varphi = \lambda(4+\lambda^2)d\Upsilon - 2\lambda X$$

so the NP$_2$ condition is satisfied when

$$\mu = \tfrac{1}{2}\lambda^{-1}(4+\lambda^2) = 3\lambda$$

and $\lambda^2 = 4/5$. The sign of λ and μ is merely a matter of orientation, so we choose them to be positive for this proposition. The form of h_7 follows from (3.4). □

The complete metric g_Φ with holonomy *Spin*(7) appears on [13, p. 848]. If we take the asymptotic limit and set $r = 1$, we obtain $\frac{1}{5}h_7$. It is well known that (3.3) is the so-called squashed Einstein metric that exists on the S^3 (or $\mathbb{RP}^3$) bundle over M^4. The more standard Einstein metric (the round metric in the case of $\pi : S^7 \to S^4$) has the same form, but without the factor $1/5$.

The proof above is modelled on [22, Prop. 2.5] from the theory of 3-Sasakian 7-manifolds. See also [2, 28]. Indeed, we shall explain next that the 3-form (3.2) coincides with one that can be defined using globally-defined 1-forms η_1, η_2, η_3 dual to the Lie algebra of $SU(2)$ acting on the fibre of a regular 3-Sasakian 7-manifold. This is accomplished by means of a gauge transformation that converts the geometry of a vector bundle to that of a principal bundle.

Modify (2.7) by conjugation and (partly for convention) a sign change:

$$\eta \;=\; -a\beta\overline{a} \;=\; -\alpha\overline{a} = a\overline{\alpha} \tag{3.5}$$

where a is a unit quaternion. Retaining the condition $R = 1$, we can write

$$\eta \;=\; i\eta_1 + j\eta_2 + k\eta_3.$$

In particular,

$$\eta_1 \;=\; -a_0\alpha_1 + a_1\alpha_0 - a_2\alpha_3 + a_3\alpha_2.$$

Observe that, as a direct consequence of the conjugation, η is *invariant* under the action $a \rightsquigarrow aq$ of $SU(2)^-$. Moreover, η, like β, annihilates the horizontal distribution on $\tilde{P}^-$. It follows that each real component η_i is globally defined on $\tilde{P}^-$. Moreover,

$$\alpha \wedge \overline{\alpha} = \alpha\overline{a} \wedge a\overline{\alpha} = -\eta \wedge \eta$$

in analogy to (2.9), but with a fibrewise self-dual property. Converting from β to η furnishes a more succinct version of Lemma 2.1 on $\tilde{P}^-$, namely

Lemma 3.2 $d\eta + \eta \wedge \eta = a\Phi\overline{a}$.

Proof Applying (2.6) and (2.8),

$$\begin{aligned} d\eta \;=\; -d(\alpha\overline{a}) &= -d\alpha\,\overline{a} + \alpha \wedge d\overline{a} \\ &= (\alpha \wedge \phi + a\Phi)\overline{a} + \alpha \wedge (\overline{\alpha} - \phi\overline{a}) \\ &= \alpha \wedge \overline{\alpha} + a\Phi\overline{a}. \end{aligned}$$

As has to be, the individual terms in the last line are unchanged when a, α, Φ are replaced by $aq, \alpha q, \overline{q}\Phi q$, respectively. □

It is instructive to compare Lemmas 2.1 and 3.2. In analogy to (2.1), set

$$H \;=\; a\Phi\overline{a} \;=\; i(d\eta_1 + 2\eta_{23}) + j(d\eta_2 + 2\eta_{31}) + k(d\eta_3 + 2\eta_{12}). \tag{3.6}$$

Then H is '3-horizontal', i.e. its individual real components are pointwise the pullbacks of forms on the 4-manifold M. However, they are not invariant by

$SU(2)^-$, only their quaternionic combination enjoys this property, the price to pay for avoiding analogues of the terms $\phi \wedge \beta$ in Lemma 2.1. If we replace the symbol H_i by $\overline{\phi}^i$ (the overline is just notation), this coincides with [22, p. 50].

Remark 3.3 Conjugation of an imaginary quaternion by the unit quaternion (2.5) describes the double covering $SU(2)^- \longrightarrow SO(3)^-$ by mapping (2.5) to the orthogonal matrix

$$(P_{i,j}) = \begin{pmatrix} a_0^2 + a_1^2 - a_2^2 - a_3^2 & 2(-a_0a_3 + a_1a_2) & 2(a_0a_2 + a_3a_1) \\ 2(a_0a_3 + a_1a_2) & a_0^2 - a_1^2 + a_2^2 - a_3^2 & 2(-a_0a_1 + a_2a_3) \\ 2(-a_0a_2 + a_3a_1) & 2(a_0a_1 + a_2a_3) & a_0^2 - a_1^2 - a_2^2 + a_3^2 \end{pmatrix} \tag{3.7}$$

that features in many articles on the quaternions, such as [1, Prop. 23.12]. The nine entries of (3.7) are tracefree quadratics, elements of $S_0^2(\mathbb{R}^4) \cong \Lambda_+^2 \otimes \Lambda_-^2$. Indeed, $P_{i,j}$ equals minus the contraction of standard basis elements of Λ^+ and Λ^-. The triple $(P_{1,i})$ is the hyperkähler moment map $\mathbb{R}^4 \to \mathfrak{so}(3)$ for the action of $U(1)$ on $\mathbb{R}^4$ tangent to I, and is familiar from the Gibbons-Hawking construction. In components,

$$d\eta_1 + 2\eta_{23} = 2\sum_p P_{1,p}\Omega_p. \tag{3.8}$$

It is easy to compute dH using (3.6) twice. In particular,

$$dH_1 \;=\; 2(d\eta_2 \wedge \eta_3 - \eta_2 \wedge d\eta_3) \;=\; -2(\eta_2 \wedge H_3 - \eta_3 \wedge H_2)$$

which (as it has to) matches (2.3):

$$d\Phi_1 \;=\; -2(\phi_2 \wedge \Phi_3 - \phi_3 \wedge \Phi_2)$$

All the factors of the right-hand side of the last line are 3-horizontal.

Proposition 3.4 *Let $\lambda = 2/\sqrt{5}$. Define*

$$\varphi' \;=\; 2\lambda \sum_{i=1}^{3} H_i \wedge \eta_i + \lambda^3 \eta_{123}.$$

Then φ' coincides with minus the 3-form (3.2)*, and therefore gives an equivalent description of the squashed NP_2 structure on $\tilde{P}^-$.*

Proof The definitions of η and H, and some telescoping, imply that

$$\varphi' = -a\varphi\bar{a} = -\varphi$$

since both φ and φ' are real differential forms. □

The definition of φ' in Proposition 3.4 is modelled on that of φ, but is perhaps more familiar. However, there is a significant difference between Lemmas 2.1 and 3.2, either of which can be used to compute $d\varphi' = -d\varphi$. The necessity to change the sign of λ arises from a minus sign that we inserted in the definition (3.5) of η to preserve the combination $d\eta + \eta^2$ to match $d\beta + \beta^2$.

As explained in [22, p. 51], in the 3-Sasakian context, one can define a hyperkähler triple of exact 2-forms

$$\varpi_i = d(2R\eta_i) = 4\eta_{0i} - 4R\eta_{jk} + 2RH_i \tag{3.9}$$

on the cone over $\tilde{P}^-$. Pointwise, each $2H_i = \omega_{0i} - \omega_{jk}$ is a horizontal ASD 2-form that matches the other terms, see (3.4) with $\lambda = 2$. The existence of such a hyperkähler structure is well known in the setting of twistor space and quaternionic geometry [37], though we are here more interested in induced G_2 structures on the 3-sphere bundle $\tilde{P}^-$.

The metric associated to (3.9) is the cone over $\tilde{P}^-$, and this is one way of defining a 3-Sasakian structure [8, 11]. Since the holonomy of the cone lies in $Sp(2)$, it also lies in *Spin*(7) and $\tilde{P}^-$ must have an NP_2 structure. However, such a subgroup *Spin*(7) is conjugate to the stabilizer of a 4-form $-\varpi_1^2 + \varpi_2^2 + \varpi_3^2$, so one needs to break the $SO(3)$ symmetry of the hyperkähler triple, whereupon there emerges a family of NP_2 structures.

We can replace each β_i by η_i in (3.3) without affecting the sum. More general families of metrics of the form

$$dt^2 + \sum_{k=1}^{3} A_i(t)^2 \eta_i \otimes \eta_i + B(t)^2 g_4,$$

with holonomy *Spin*(7) were constructed by Bazaikin by deforming 3-Sasakian structures [9]. This has guided our own presentation of (or rather, hope to generalize) the homogeneous NP_2 structures, in which trigonometric functions of t appear on subspaces $\langle \beta_k, \omega_k \rangle$ for $k = 1, 2, 3$, see Corollaries 4.4 and 5.4.

4 The Cohomogeneity-One Action of SO(4) on S^7

The group $Sp(n)$ acts transitively on the sphere S^{4n-1} in quaternionic space $\mathbb{H}^n$ and, in particular,

$$S^7 \cong \frac{Sp(2)}{Sp(1)} \cong \frac{SO(5)}{SU(2)}. \tag{4.1}$$

This is the homogeneous model of 3-Sasakian geometry in dimension 7, with its triple of invariant 1-forms. From the point of view of Sect. 2, it is more appropriate to start from the description

$$S^7 \cong \frac{Sp(2)\times Sp(1)}{Sp(1)\times Sp(1)} = \frac{Sp(2)Sp(1)}{Sp(1)Sp(1)}, \tag{4.2}$$

which features in the classification [20] of homogeneous NP_2 spaces. The second factor of the isotropy group $Sp(1)Sp(1)\cong SO(4)$ acts diagonally, and juxtaposition indicates a quotient by ± 1 across both factors. In quaternion-Kähler jargon, S^7 is here regarded as the unit sphere in the $Sp(2)Sp(1)$ module $\mathfrak{m}$, with complexification $\mathfrak{m}_c = E\otimes H$, where $Sp(2)$ acts on $E\cong\mathbb{C}^4$ and $Sp(1)$ acts on $H\cong\mathbb{C}^2$.

Let $\Sigma^+\oplus\Sigma^-$ denote the spin representation of $Spin(4)\cong Sp(1)\times Sp(1)$, so that $Sp(1)^+$ acts on $\Sigma^+\cong\mathbb{C}^2$ and (as before) $Sp(1)^-$ acts on $\Sigma^-\cong H$. The isotopy subgroup in (4.2) then decomposes $E=\Sigma^+\oplus\Sigma^-$, so that

$$\mathfrak{m}_c \cong (\Sigma^+\oplus\Sigma^-)\otimes\Sigma^-.$$

Since $S^2\Sigma^-$ is the complexification of the space Λ^- of ASD 2-forms on $\mathbb{R}^4$, the isotropy subgroup $SO(4)$ acts on the real tangent space to S^7 as $\mathbb{R}^4\oplus\Lambda^-$. It is well known that this action factors through G_2. Moreover there are independent $SO(4)$ invariant 3-forms φ_1,φ_2 on S^7, a linear combination φ of which must satisfy (3.1) with $\mu = 6/\sqrt{5}$, in accordance with Proposition 3.1.

To obtain an action by $SO(4)\cong Sp(1)^+Sp(1)^-$ of cohomogeneity one on S^7, we move from isotropy subgroups to isometry groups and set $E=S^3\Sigma^+$, so that

$$\mathfrak{m}_c \cong S^3\Sigma^+\otimes\Sigma^-$$

is the isotropy representation of the Wolf space $G_2/SO(4)$, as in [38]. Then

$$\begin{aligned} S^2\mathfrak{m}_c &\cong (S^2E\otimes S^2H)\oplus(\Lambda^2E\otimes\Lambda^2H)\\ &\cong (S^6\Sigma^+\otimes S^2\Sigma^-)\oplus(S^2\Sigma^+\otimes S^2\Sigma^-)\oplus S^4\Sigma^+\oplus\mathbb{C}. \end{aligned} \tag{4.3}$$

The penultimate summand is the complexification of the space $S^2_0(\mathbb{R}^3)\cong\mathbb{R}^5$ of 3×3 tracefree symmetric matrices. Projection to this module of $v\otimes v$ (where v is a non-zero vector in $\mathfrak{m}=\mathbb{R}^8$) defines the $Sp(2)Sp(1)$-equivariant fibration $S^7\to S^4$.

Remark 4.1 The 9-dimensional summand of (4.3) is isomorphic to $S^2_0(\mathbb{R}^4)$ (cf. Remark 3.3), and is the model for a geometrical structures on 9-manifolds investigated in [18]. In the present context, it gives rise to an embedding of $\mathbb{RP}^7$ as a hypersurface of S^8.

Let $A\in S^4\subset S^2_0(\mathbb{R}^3)$ be a 3×3 matrix with distinct eigenvalues. Then the orbit of $SO(3)=SU(2)^+/\mathbb{Z}_2$ containing A has dimension 3 and principal isotropy group $\mathbb{Z}_2^2$. There are two singular orbits, each copies of $\mathbb{RP}^2$, corresponding to repeated eigenvalues, which can be negative or positive. To study the round metric g_4 on S^4 relative to the action by $SO(3)$ of cohomogeneity one, set $\vartheta_k = 2\pi(k-1)/3$, and

$$\begin{aligned} c_k &= c_k(t) = \cos(t+\vartheta_k) \\ s_k &= s_k(t) = \sin(t+\vartheta_k) \end{aligned} \tag{4.4}$$

for $k = 1, 2, 3$. These trigonometric quantities satisfy

$$c_1 + c_2 + c_3 \;=\; 0 \;=\; s_1 + s_2 + s_3 \tag{4.5}$$

as well as the identity

$$c_j c_k + s_j s_k = -\tfrac{1}{2}, \tag{4.6}$$

for $(ijk) = (123)$.

Now choose left-invariant 1-forms on $SO(3)$ satisfying

$$de_1 = -2e_{23}, \quad de_2 = -2e_{31}, \quad de_3 = -2e_{12}. \tag{4.7}$$

This convention is chosen to match expressions such as (2.10). Combined with the constraint on κ, it leads us to adopt an orthonormal basis of 1-forms on S^4 formed from dt and $4s_k e^k$ $(k = 1, 2, 3)$, so that

$$g_4 = dt^2 + 16\sum_{k=1}^{3} s_k(t)^2 e_k \otimes e_k \tag{4.8}$$

away from the singular orbits. The associated basis of ASD 2-forms consists of

$$\begin{aligned} \omega_1 &= 4s_1 dt \wedge e_1 - 16 s_2 s_3 e_{23} \\ \omega_2 &= 4s_2 dt \wedge e_2 - 16 s_3 s_1 e_{31} \\ \omega_3 &= 4s_3 dt \wedge e_3 - 16 s_1 s_2 e_{12}. \end{aligned}$$

Lemma 4.2 *With this choice, the connection 1-form ϕ for $\Lambda^2_- T^*S^4$ is given by*

$$-\phi = i(2c_2+1)e_1 + j(2c_2+1)e_2 + k(2c_3+1)e_3, \tag{4.9}$$

and $\kappa = 1$.

Proof It suffices to verify (2.3) and (2.4), since these equations have a unique solution by the theories of Cartan or Levi-Civita.

First observe that

$$d\omega_1 \;=\; 8(s_1 - 2(c_2 s_3 + s_2 c_3))dt \wedge e_{23},$$

whereas

$$\begin{aligned}-2(\phi_2 \wedge \omega_3 - \phi_3 \wedge \omega_2) &= 8(ac_2 - 1)s_3 dt \wedge e_{23} + 8(ac_3 - 1)s_2 dt \wedge e_{23} \\ &= 8(s_1 - 2(c_2 s_3 + c_3 s_2))dt \wedge e_{23}.\end{aligned}$$

This verifies (2.3) after cyclic permutations.

As regards (2.4), we have

$$\begin{aligned}d\phi_1 + 2\phi_{23} &= 2s_1 dt \wedge e_1 + 2(2c_1 + 1)e_{23} + 2(2c_2 + 1)(2c_3 + 1)e_{23} \\ &= 2s_1 dt \wedge e_1 + (4 + 8c_2 c_3)e_{23} \\ &= 2s_1 dt \wedge e_1 - 8s_2 s_3 e_{23} \\ &= \tfrac{1}{2}\omega_1,\end{aligned}$$

which is consistent with (2.4) and $\kappa = 1$. □

Next, consider the locally-defined 1-form

$$f = \overline{a}\, da = \overline{a}(\alpha + a\phi) = \beta + \phi \tag{4.10}$$

on Σ^- over a self-dual Einstein 4-manifold. It is vector valued on the 3-sphere bundle, and transforms as $f \rightsquigarrow \overline{q} f q$. It arises from the Maurer-Cartan form of the fibre, and

$$df = d(\overline{a}\, da) = d\overline{a} \wedge da = (d\overline{a}\, a) \wedge (\overline{a}\, da) = -f \wedge f$$

so in components,

$$df_1 = -2f_{23}, \quad df_2 = -2f_{31}, \quad f_3 = -2f_{12}. \tag{4.11}$$

Regard $S^7 = \tilde{P}^- S^4$ as the S^3 bundle over S^4. Consider the 1-forms (e_k) and (f_k) satisfying (4.11) and (4.7). Lemma 4.2 yields

Theorem 4.3 *For $k = 1, 2, 3$, we have $\beta_k = (2c_k + 1)e_k + f_k$.*

These equations relate the 1-forms from Lemma 2.1 to the action of $SO(4)$, which is a significant step in understanding the cohomogeneity-one setup. The trigonometric terms arise from the action of the Weyl group in the cohomogeneity-one theory [25], and is suggestive of a discrete Fourier transform. One cannot usefully substitute (β_k) by the 3-Sasakian 1-forms (η_k) in the theorem, since this would yield $a\phi\overline{a}$ in (4.10).

An orthonormal basis of 1-forms for the round metric g_7 on S^7 is given by

$$dt, \; 2\beta_1, \; 2\beta_2, \; 2\beta_3, \; 4s_1 e_1, \; 4s_2 e_2, \; 4s_3 e_3.$$

Set $\tau = t/2$. Then

$$\begin{aligned} g_7 &= dt^2 + 4\sum_{k=1}^{3}(\beta_k)^2 + \pi^* g_4 \\ &= dt^2 + 4\sum_{k=1}^{3}(\beta_k)^2 + 16\sum_{k=1}^{3}(s_k e_k)^2 \\ &= dt^2 + 4\sum_{k=1}^{3}\left[f_k^2 + \gamma_k(e_k f_k + f_k e_k) + \delta_k e_k^2\right], \end{aligned}$$

where

$$\gamma_k = 2c_k + 1 = -1 + 4\cos^2\tau, \quad \delta_k = (2c_k+1)^2 + (2s_k)^2 = 1 + 8\cos^2\tau.$$

It follows that the round metric can be expressed as dt^2 plus a 6×6 matrix with 2×2 symmetric blocks, of which the first is

$$4\begin{pmatrix} 1 & -1+4\cos^2\tau \\ -1+4\cos^2\tau & 1+8\cos^2\tau \end{pmatrix}. \tag{4.12}$$

This can be reconciled with [38, p. 7] up to scale, in which the period of t in (4.4) has been halved to $\pi/6$. A change of off-diagonal sign results from our ASD convention.

We now turn to the squashed metric (3.3) on S^7 and its NP_2 structure. Consider the orthonormal basis 1-forms (X_i) defined by $X_{2k-1} = \lambda\beta_k$, $X_{2k} = 4s_k e_k$, and $X_7 = -dt$. Theorem 4.3 implies

Corollary 4.4 *Let $\kappa = 1$ and $\lambda = 2/\sqrt{5}$. Then*

$$\begin{pmatrix} X_{2k-1} \\ X_{2k} \end{pmatrix} = \begin{pmatrix} \lambda & \lambda(2c_k+1) \\ 0 & 4s_k \end{pmatrix}\begin{pmatrix} f_k \\ e_k \end{pmatrix}$$

for $k = 1, 2, 3$.

The behaviour of both g_7 and h_7 at the two singular orbits of $SO(4)$ can be investigated by setting $t = 0$ and $t = \pi/3$, so that $X_2 = 0$ and $X_4 = 0$, respectively.

Referring to (3.2), we now see that $\lambda^3\Upsilon = X_{135}$ and

$$2\lambda\Theta = (X_{72} - X_{46}) \wedge X_1 + (X_{74} - X_{62}) \wedge X_3 + (X_{76} - X_{24}) \wedge X_5.$$

Proposition 3.1 therefore yields

Theorem 4.5 *The natural NP_2 structure on S^7 has*

$$\varphi = dt \wedge (X_{12} + X_{34} + X_{56}) + X_{135} - X_{146} - X_{236} - X_{245}.$$

The theorem asserts that in the stated basis, φ assumes a standard form

$$\begin{aligned}\varphi &= dt \wedge \xi + \mathrm{Re}\, \Xi \\ *\varphi &= -dt \wedge \mathrm{Im}\, \Xi + \tfrac{1}{2}\xi^2\end{aligned}$$

relative to the $SO(4)$ orbits, where

$$\begin{aligned}\xi &= X_{12} + X_{34} + X_{56} \\ \Xi &= (X_1 + iX_2) \wedge (X_3 + iX_4) \wedge (X_5 + iX_6).\end{aligned} \tag{4.13}$$

To sum up: we have described the 'squashed' 3-form φ from Propositions 3.1 and 3.4 in terms of $SO(4)$ invariant 1-forms, via Corollary 4.4. This provides a model for the construction of NP_2 structures in the cohomogeneity-one situation of an interval times (a finite quotient of) $SO(4)$. In the next section, we shall discover a strikingly similar description on the principal part of the isotropy irreducible space $SO(5)/SO(3)$.

5 The Cohomogeneity-One Action of SO(4) on B^7

Berger discovered two homogeneous spaces of positive sectional curvature, of respective dimensions 7 and 13, namely

$$B^7 = \frac{SO(5)}{SO(3)}, \qquad B^{13} = \frac{SU(5)}{Sp(2)U(1)},$$

which he studied in [10]. The former is the subject of this section; it is the coset space that parametrizes principal subalgebras $\mathfrak{so}(3)$ of $\mathfrak{so}(5)$, equivalently Veronese surfaces in S^4 [7].

The subgroup $SO(3)$ of $SO(5)$ acts irreducibly on $\mathbb{R}^5$, which can be identified with the space $S^2_0(\mathbb{R}^3)$, equivalently the space of traceless symmetric 3×3 matrices. The isotropy subgroup of B^7 is a principal 3-dimensional subgroup that acts irreducibly on $\mathbb{R}^7$. Indeed,

$$\mathfrak{so}(5) = \mathfrak{so}(3) \oplus \mathfrak{m}, \tag{5.1}$$

where $\mathfrak{m}$ can be identified with the space $S^3_0(\mathbb{R}^3)$ of primitive cubic forms on $\mathbb{R}^3$. After complexification,

$$\mathfrak{so}(3)_c \cong S^2(\mathbb{C}^2), \quad (\mathbb{R}^5)_c \cong S^4(\mathbb{C}^2), \quad \mathfrak{m}_c \cong S^6(\mathbb{C}^2) \tag{5.2}$$

where $\mathbb{C}^2$ denotes the faithful representation of the double-cover $SU(2)$ of $SO(3)$.

It is known that the action of $SO(3)$ on $\mathfrak{m}$ factors through G_2. Indeed, there is an $SU(2)$ equivariant isomorphism

$$\Lambda^3(S^6(\mathbb{C}^2)) \cong S^{12} \oplus S^8 \oplus S^6 \oplus S^4 \oplus S^0 \tag{5.3}$$

(with obvious abbreviation), and the trivial 1-dimensional submodule $S^0 \cong \mathbb{C}$ is spanned by a G_2-admissible 3-form φ. The latter also arises from the projection to $\mathfrak{m}$ of the Lie bracket $[\mathfrak{m}, \mathfrak{m}]$ within (5.1). The fact that $[\mathfrak{m}, \mathfrak{m}]$ does not lie in $\mathfrak{so}(3)$ implies that B^7 is not a symmetric space, and its holonomy group equals $SO(7)$.

Now, φ determines an $SO(5)$-invariant 3-form on B^7 (which we denote with the same symbol), and an invariant 4-form again unique up to scaling (which we can denote $*\varphi$). Since there is no non-zero invariant 2-form, $d\varphi$ must be a nonzero multiple of $*\varphi$, which is the NP$_2$ condition (3.1). Hence, the well-known

Proposition 5.1 *B^7 has an $SO(5)$-invariant NP$_2$ structure.*

It follows that the Riemannian cone $B^7 \times \mathbb{R}^+$ has holonomy *Spin*(7). It is not known whether this singularity can be resolved in the manner of [9], but it motivates the study of metrics with holonomy *Spin*(7) admitting a codimension-two action by $SO(4)$.

Remark 5.2 The space B^{13} is not isotropy irreducible. There is a fibration $B^{13} \to \mathbb{CP}^4$, with fibre $SU(4)/Sp(2) \cong S^5$. Its complexified isotropy representation is $(V \oplus \overline{V}) \oplus \Lambda^2_0 E$, where $V = E \otimes L \cong \mathbb{C}^4$ is a standard representation of $Sp(2) \times U(1)$. The space B^{13} has a 3-dimensional space of invariant 4-forms, which are relevant to the study of 'strongly positive curvature' [3].

The 3-form φ on B^7 was computed explicitly in [7, 14, 23], but in different ways. The calculations below are equivalent, but our approach emphasizes the underlying 3-fold symmetry.

Denote by S_{ij} the symmetric 3×3 matrix with (i, j) and (j, i) entry equal to 1 and 0 elsewhere. We then adopt the following basis of traceless symmetric matrices:

$$\begin{gathered} \overline{S}_1 = \tfrac{1}{\sqrt{6}}\mathrm{diag}(1, 1, -2), \qquad \overline{S}_2 = \tfrac{1}{\sqrt{2}}\mathrm{diag}(1, -1, 0), \\ \overline{S}_3 = S_{12}, \quad \overline{S}_4 = S_{13}, \quad \overline{S}_5 = S_{23}. \end{gathered} \tag{5.4}$$

We represent elements of the Lie algebra (5.2) by skew-symmetric 5×5 matrices. Let E_{ij} denote the matrix with 1 at (i, j), -1 at (j, i), and 0 elsewhere. For the subalgebra $\mathfrak{so}(3)$, we may choose a basis $\{\varepsilon_1, \varepsilon_2, \varepsilon_3\}$, where

$$\begin{aligned} \varepsilon_1 &= \tfrac{1}{s}(2E_{23} + E_{45}) \\ \varepsilon_2 &= \tfrac{1}{s}(E_{24} + E_{35} + \sqrt{3}E_{14}) \\ \varepsilon_3 &= \tfrac{1}{s}(-E_{25} + E_{34} + \sqrt{3}E_{15}). \end{aligned}$$

One can check that $\mathrm{tr}(\varepsilon_i\varepsilon_j) = -2\delta_{ij}$, and that $[\varepsilon_i\varepsilon_j] = -\lambda\varepsilon_k$, where $(ijk) = (123)$.

Now set

$$R(t) = \begin{pmatrix} \cos t & \sin t & 0\,0\,0 \\ -\sin t & \cos t & 0\,0\,0 \\ 0 & 0 & 0\,1\,0 \\ 0 & 0 & 0\,0\,1 \\ 0 & 0 & 1\,0\,0 \end{pmatrix}.$$

We have chosen the lower block to be a 3×3 permutation matrix (rather than the identity) because $\rho = R(\frac{2\pi}{3})$ has special significance; it is derived from the effect on (5.4) of permuting the basis of $\mathbb{R}^3$. It therefore lies in the intersection of the principal $SO(3)$ with the subgroup $SO(2) \times SO(3)$ of $SO(5)$. Conjugation by this element of order 3 cycles the ε_i, so that $\rho^{-1}\varepsilon_1\rho = \varepsilon_2$, etc.

Let $a = \cos(2\pi/3)$ and $b = \sin(2\pi/3)$. A basis for $\mathfrak{m}$ is given by $\{\gamma_1, \ldots, \gamma_7\}$ where

$$\begin{aligned} \gamma_1 &= \tfrac{1}{s}(E_{23} - 2E_{45}) \\ \gamma_2 &= -E_{13} \\ \gamma_3 &= \tfrac{1}{s}(bE_{15} + aE_{25} - 2E_{34}) \\ \gamma_4 &= -aE_{15} + bE_{25} \\ \gamma_5 &= \tfrac{1}{s}(-bE_{14} + aE_{24} + 2E_{35}) \\ \gamma_6 &= -aE_{14} - bE_{24} \\ \gamma_7 &= E_{12}. \end{aligned} \tag{5.5}$$

One can check that $\mathrm{tr}(\gamma_i\gamma_j) = -2\delta_{ij}$, and that $\mathrm{tr}(\gamma_i\varepsilon_j) = 0$ for all i, j. Observe that ρ cycles the pairs $(\gamma_{2k-1}, \gamma_{2k})$ for $k = 1, 2, 3$, in the sense that $\rho^{-1}(\gamma_1, \gamma_2)\rho = (\gamma_3, \gamma_4)$ etc. In fact, it was only necessary to choose the orthogonal basis γ_1, γ_2 orthogonal to $\mathfrak{so}(3)$; the rest were generated by ρ. On the other hand, γ_7 commutes with ρ.

Having made these choices, the $SO(5)$-invariant 3-form φ must be proportional to

$$\frac{1}{10} \sum_{i<j<k} \mathrm{tr}([\gamma_i, \gamma_j]\gamma_k)\, \gamma_i^* \wedge \gamma_j^* \wedge \gamma_k^*,$$

where (γ_i^*) is the dual basis of invariant 1-forms. With the given normalization, we discover

Proposition 5.3 $\varphi = \gamma_{127}^* + \gamma_{135}^* - \gamma_{146}^* - \gamma_{236}^* - \gamma_{245}^* + \gamma_{347}^* + \gamma_{567}^*.$

This is in canonical form for a G_2 structure. Since B^7 has a unique $SO(5)$-invariant Riemannian metric g_B (up to an overall constant), it is a corollary of the proposition that (γ_i^*) is an orthonormal basis for g_B.

There is a cohomogeneity-one action of $SO(4)$ on B^7 [34]. The principal orbits are $SO(4)/\mathbb{Z}_2^2$, and the singular orbits are both isomorphic to $SO(4)/O(2)$.

To analyse this, we start with bases (e_k), (f_k), whose exterior derivatives are normalized as in (4.7), (4.11). We then define positive and negative combinations $p_k = \frac{1}{2}(f_k + e_k)$ and $n_k = \frac{1}{2}(-f_k + e_k)$, so that

$$\begin{pmatrix} p_k \\ n_k \end{pmatrix} = L \begin{pmatrix} f_k \\ e_k \end{pmatrix}, \qquad L = \frac{1}{2}\begin{pmatrix} 1 & 1 \\ -1 & 1 \end{pmatrix} \tag{5.6}$$

for $k = 1, 2, 3$. This switch of bases will play a significant role in simplifying subsequent expressions, as we are about to see. A coefficient of $1/2$ is chosen in the light of the differential relations (5.10) below.

The Maurer-Cartan form of the subgroup $SO(4)$ in $SO(5)$ can be expressed as

$$A = 2\begin{pmatrix} 0 & 0 & 0 & 0 & 0 \\ 0 & 0 & n_3 & n_2 & n_1 \\ 0 & -n_3 & 0 & -p_1 & p_2 \\ 0 & -n_2 & p_1 & 0 & -p_3 \\ 0 & -n_1 & -p_2 & p_3 & 0, \end{pmatrix} \tag{5.7}$$

so that e_1 corresponds to the combination $-E_{25} - E_{34}$, and f_1 to $E_{25} - E_{34}$, etc. These choices match the conventions (4.7),(4.11) and (the subsequent) (5.10).

The pullback of A on $[0, \frac{\pi}{3}] \times SO(4)$ is given by

$$R(t)^{-1}AR(t) + \gamma_7^* \wedge dt = \begin{pmatrix} 0 & dt & -sn_1 & -sn_3 & -sn_2 \\ -dt & 0 & cn_1 & cn_3 & cn_2 \\ sn_1 & -cn_1 & 0 & -p_2 & p_3 \\ sn_2 & -cn_3 & p_2 & 0 & -p_1 \\ sn_3 & -cn_2 & -p_3 & p_1 & 0 \end{pmatrix}$$

where $c = \cos t$ and $s = \sin t$ as temporary abbreviations.

Let Y_i denote the pullback of $2\gamma_i^*$. These 1-forms satisfy

$$\begin{aligned} Y_1 &= \lambda(cn_1 + 2p_1) \\ Y_2 &= 2sn_1 \\ Y_3 &= \lambda((ca - sb)n_2 + 2p_2) \\ Y_4 &= 2(sa + cb)n_2 \\ Y_5 &= \lambda((ca + sb)n_3 + 2p_3) \\ Y_6 &= 2(sa - cb)n_3 \\ Y_7 &= 2dt \end{aligned}$$

where $\lambda = 2/\sqrt{5}$ and (as before) $a = \cos(2\pi/3)$ and $b = \sin(2\pi/3)$.

These equations manifest the $\mathbb{Z}_3$ symmetry arising from the action of the Weyl group for the action of $SO(4)$ [25], and can be restated as

Corollary 5.4 *Let* $\kappa = 1$ *and* $\lambda = 2/\sqrt{5}$. *Then*

$$\begin{pmatrix} Y_{2k-1} \\ Y_{2k} \end{pmatrix} = \begin{pmatrix} 2\lambda & \lambda c_k \\ 0 & 2s_k \end{pmatrix} \begin{pmatrix} p_k \\ n_k \end{pmatrix} \tag{5.8}$$

for $k = 1, 2, 3$.

The 3-form from Proposition 5.3 defining the NP_2 structure on B^7 can be written

$$\varphi = \xi \wedge Y_7 + \mathrm{Re}\,\Xi,$$

where Ξ is defined by (4.13) with Y_i in place of X_i. If we denote the 2×2 matrix in Corollary 5.4 by $M_k = M_k(t)$, the analogue of (4.12) for the term of g_B involving the plane $\langle f_1, e_1 \rangle$ equals

$$L^\top M_1^\top M_1 L = \frac{1}{5}\begin{pmatrix} 5+4\sin^2 t+4\cos t & 1-4\cos^2 t \\ 1-4\cos^2 t & 5+4\sin^2 t-4\cos t \end{pmatrix}. \tag{5.9}$$

The change of basis (5.6) reveals a striking similarity between Corollaries 4.4 and 5.4, which we explore next.

The 1-forms in (5.6) satisfy

$$\begin{cases} dp_k = -2(p_{ij} + n_{ij}) \\ dn_k = -2(p_i n_j - p_j n_i) \end{cases} \tag{5.10}$$

for $(ijk) = (123)$. If ∇ is a connection on $T^*SO(4)$ that satisfies

$$\nabla n_k = p_i \otimes n_j - p_j \otimes n_i$$

then we may regard the p_i as connection 1-forms on $\mathcal{D}$, whose curvature 2-forms are

$$dp_k + 2p_{ij} = -2n_{ij}.$$

The second equation of (5.10) implies that $\mathcal{D} = \langle Y_2, Y_4, Y_6 \rangle$ is a differential ideal in $SO(4)$. Regarding $\mathcal{D}$ as a subspace of $\mathfrak{so}(4)^*$, its annihilator is the diagonal subalgebra

$$\mathfrak{so}(3)_d \subset \mathfrak{so}(3)_+ + \mathfrak{so}(3)_- = \mathfrak{so}(4)$$

represented by the bottom 3×3 block in (5.7). This subalgebra corresponds to the choice of isometry $\sigma: \langle e_1, e_2, e_3\rangle \to \langle f_1, f_2, f_3\rangle$ defined by our indexing, or (up to sign) the vector in $\mathbb{R}^4$ that is fixed by our choice of geodesic.

Inside each generic $SO(4)$ orbit, the integral submanifolds of $\mathcal{D}$ consist of $\mathbb{Z}_2^2$ quotients of conjugacy classes of a diagonal $SO(3)_d$, or cosets parametrized by $SO(4)/SO(3)_d \cong S^3$. Moreover,

$$\operatorname{Im}\Xi = -Y_{246} + Y_{235} + Y_{145} + Y_{136}$$

vanishes on these submanifolds, and they are associative (see (4.13)). They play a role analogous to that of the orbits of $SU(2)^-$ acting on S^7, insofar as the homogeneous NP_2 metrics are concerned. However, the foliation on B^7 degenerates over the singular $SO(4)$ orbits. The complete theory of associative submanifolds of B^7 is very rich because of the many different orbit types of a principal $SO(3)$ subgroup of G_2 acting on the space $G_2/SO(4)$ of associative subspaces of $\mathbb{R}^7$ [7].

Remark 5.5 In the wider homogeneous context, conjugacy classes of 3-dimensional subalgebras of $\mathfrak{so}(5)$ (namely $\mathfrak{su}(2)^-$, $\mathfrak{so}(3)_d$ and principal $\mathfrak{so}(3)$) correspond to critical submanifolds (namely S^4, $\mathbb{G}r_3(\mathbb{R}^5)$ and B^7) for the natural Morse-Bott functional on the 21-dimensional Grassmannian $\mathbb{G}r_3(\mathfrak{so}(5))$. The product $B^7 \times \mathbb{R}^5$ admits an incomplete quaternion-Kähler metric by the theory in [29, 31], and it is tempting to tie this in with the correspondences of [7] relating the critical submanifolds.

The behaviour of the 1-forms Y_i at $t = 0$ and $\pi/3$ gives some information about the action of $SO(4)$ on the two special orbits. For example, Y_2 vanishes at $t = 0$, and at our distinguished point of the corresponding orbit $B_0 \cong SO(4)/O(2)$ (which we can take to be the identity coset), the connected component $SO(2)$ of the isotropy rotates $\langle E_4, E_5\rangle$ by θ and $\langle E_2, E_3\rangle$ by 2θ. If we orient $\mathbb{R}^4$ by the basis $\langle E_4, E_5, E_2, E_3\rangle$, the induced action on (1.1) fixes $E_{45} \pm E_{23}$, but rotates $\langle E_{42} + E_{35}, E_{43} + E_{52}\rangle$ by 3θ, and $\langle E_{42} - E_{35}, E_{43} - E_{52}\rangle$ by θ. It follows that B_0 is covered by a circle bundle in $\mathcal{O}(3, 1) \to \mathbb{CP}^1 \times \mathbb{CP}^1$.

An element of $SO(2)$ that lies in the subgroup $SU(2)^+$ of $SO(4)$ must act trivially on $T = \langle E_{42} - E_{35}, E_{43} - E_{52}\rangle$, and (given the weight 1 above) is necessarily the identity. Therefore $SU(2)^+$ acts freely on B_0, and we can identify the (co)tangent space of $B_0/SU(2)^+ \cong \mathbb{RP}^2$ with T. By contrast, $SU(2)^-$ has stabilizer $\mathbb{Z}_3$ on B_0, as does $SU(2)^+$ at $t = \pi/3$. It follows that $B^7/SU(2)^+$ is smooth over $[0, \pi/3)$, but not at $\pi/3$. The commutator $SU(2)^-$ acts on the orbifold $B^7/SU(2)^+$ as $SO(4)/SU(2)^+ \cong SO(3)$ with generic stabilizer $V = \mathbb{Z}_2^2$. This observation leads to the $SO(4)$-equivariant diffeomorphism $B^7 \cong \tilde{P}^+\mathcal{O}_5$ established by Grove et al. [25], where $\mathcal{O}_k$ is Hitchin's orbifold with a $\mathbb{Z}_{k-2}$ singularity transverse to $\mathbb{RP}^2$ (we use $k \geqslant 3$ as in [26]). We outline this setup next because of its implications for NP_2 structures with an $SO(4)$ symmetry, working backwards from $\mathcal{O}_k$. Recall our convention that a self-dual conformal structure has vanishing Weyl component W_-, cf. [4].

The irreducible action of $SO(3) = G$ on the round 4-sphere S^4 has generic orbit G/V, and singular orbits $\mathcal{L} \cong G/L$ and $\mathcal{R} \cong G/R$ where L and R are copies of $O(2)$ in different blocks of $SO(3)$. Topologically, the orbifold O_k is constructed by gluing disk bundles over $\mathcal{L}$ and $\mathcal{R}$, each a copy of $\mathbb{RP}^2$. If we choose $\mathcal{R}$ to be the singular locus, then

$$O_k \approx (G \times_L D^2) \bigcup_{G/V} (G \times_{R/\ell} (D^2/\ell),$$

where $\ell = k-2$, and this space is still homeomorphic to S^4. The subgroups L, R act on the tangent and normal spaces T, $T^\perp$ as $(e^{i\theta}, e^{2i\theta})$ at $x \in \mathcal{L}$ and as $(e^{\ell i\theta}, e^{-2i\theta})$ at $y \in \mathcal{R}$; factors of 2 are needed to match the generic stabilizer V. The actions on $\Lambda^+(T \oplus T^\perp)$ are therefore $e^{3i\theta}$ at x and $e^{(\ell-2)i\theta}$ at y. The inclusion

$$U(1) \subset R \subset SO(3) \times SU(2)^+ SU(2)^- \tag{5.11}$$

is given by $(e^{\ell i\theta}, e^{(\ell-2)i\theta}, e^{(-\ell-2)i\theta})$ at y, and almost coprime weights imply that the $SO(3)^+$ principal bundle P^+O_k is smooth. The same is true of P^-O_k. The SD/ASD distinction can be verified by considering $O_4 = \mathbb{CP}^2/\mathbb{Z}_2$ for which $\Lambda^+ = \langle\omega\rangle \oplus [\![\Lambda^{2,0}]\!]$; the singular locus $\mathcal{R}$ of is totally real, reflecting the weight $\ell - 2 = 0$. When $\ell = 3$, we can obtain the group diagram for $\tilde{P}^+O_5$ by ignoring the third factor of (5.11) and letting $SO(3)$ play the role of $SU(2)^-$ acting on B^7. This gives weights $(1, 3)$ over $\mathcal{L}$ and $(3, 1)$ over $\mathcal{R}$ that coincide with those of $SO(4)$ on B^7, whence the equivalence $\tilde{P}^+O_5 \cong B^7$.

Remark 5.6 Although B^7 cannot be the total space of a smooth principal S^3 bundle, it is known to be diffeomorphic to a smooth S^3 bundle over S^4 with Euler class ± 10 and (in particular) $p_1 = \pm 16$ [23].

The orbifolds O_k arise from those on the moduli spaces of monopoles of charge 2 and mass 1 on hyperbolic space $\mathcal{H}^3(-16/(k-2)^2)$ for $k \geqslant 3$. Their metrics can (in theory, all) be constructed from solutions of Painlevé VI, and approach the hyperkähler metric on the Atiyah-Hitchin space M_0^2 [5, 26] as $k \to \infty$. Explicit algebraic expressions and fascinating graphs for $\ell = 3, 4, 6$ can be found in [38]. The Konishi bundle P^-O_k inherits an $SO(4)$-invariant 3-Sasakian metric. Therefore both P^-O_5 and P^+O_5 admit (albeit, very different) NP_2 structures, and it is natural to ask whether the same could be true for $k > 5$. There is an analogous question for positive sectional curvature, since modification of the 3-Sasakian metric on P^-O_5 has this property for $k = 5$ [16, 25].

6 Hypersurface Calculations

In the setup of (4.13), we resort to the description of an NP_2 manifold by means of the induced $SU(3)$ structures on the hypersurfaces for which $t \in (0, \pi/3)$ is constant, so the generic $SO(4)$ orbits. The condition (3.1) is equivalent to the next two equations:

$$d\xi - \tfrac{\partial}{\partial t}\mathrm{Re}\,\Xi = \mu\,\mathrm{Im}\,\Xi \tag{6.1}$$

$$d(\mathrm{Re}\,\Xi) = \mu\,\tfrac{1}{2}\xi^2. \tag{6.2}$$

The second characterizes the so-called nearly-half-flat (NHF) condition introduced by [17], and studied in the present context by the second author [36].

Recall the NP_2 structure on S^7 determined by the basis (X_i) and the 2×2 matrices that feature in Corollary 4.4. Let us just consider 6-dimensional orbits of $SO(4)$ with slightly more general $SU(3)$ structures. The next result shows that it is to some extent possible to reverse engineer the constants occurring in Corollary 4.4 by merely applying the NHF condition for all values of t:

Proposition 6.1 *Define ξ and Ξ by* (4.13) *with*

$$\begin{pmatrix} X_{2k-1} \\ X_{2k} \end{pmatrix} = \begin{pmatrix} \lambda & \lambda(ac_k + b) \\ 0 & 4s_k \end{pmatrix} \begin{pmatrix} f_k \\ e_k \end{pmatrix},$$

where λ, a, b are (unconstrained) constants. Then (6.2) *holds for all t iff*

(i) $(a, b) = (0, -1)$ *or* $(0, 0)$, *in which case* $\mu = -2/\lambda$, *or*
(ii) $(a, b) = (\pm 2, 1)$, *in which case* $\mu = -\frac{1}{2}(\lambda^2 + 4)/\lambda$.

In this formulation, the coefficient '4' is retained as a normalization.

Proof Equations (4.13) yield $\xi = -4\lambda(s_1e_1f_1 + s_2e_2f_2 + s_3e_3f_3)$ so

$$\tfrac{1}{2}\xi^2 = -16\lambda^2(s_2s_3e_{23}f_{23} + s_3s_1e_{31}f_{31} + s_1s_2e_{12}f_{12}).$$

Moreover

$$\begin{aligned} d\,\Xi = (&-2\lambda f_{23} - 2\lambda(ac_1{+}b)e_{23} - 8is_1e_{23}) \wedge \\ &(\lambda f_2 + \lambda(ac_2{+}b)e_2 + 4is_2e_2) \wedge (\lambda f_3 + \lambda(ac_3{+}b)e_3 + 4is_3e_3) + \cdots \end{aligned}$$

Therefore the $e_{23}f_{23}$ component, call it Γ_1, of $d(\mathrm{Re}\,\Xi)$ equals

$$\begin{aligned} &-2\lambda[\lambda^2(ac_2{+}b)(ac_3{+}b) - 16s_2s_3] - 2\lambda^3(ac_1{+}b) \\ &= -2\lambda^3[a^2c_2c_3 - abc_1 + b^2] + 32\lambda s_2s_3 - 2\lambda^3(ac_1 + b) \\ &= -2\lambda^3(b + b^2) - 2\lambda^3a^2c_2c_3 + 32\lambda s_2s_3 + 2\lambda^3(ab - a)c_1 \\ &= -2\lambda^3(b + b^2) - 16\lambda - 2\lambda(16 + a^2\lambda^2)c_2c_3 + 2\lambda^3(ab - a)c_1, \end{aligned}$$

the last line by (4.6). This must equal μ times $-16\lambda^2 s_2 s_3 = 8\lambda^2(1+2c_2c_3)$ for all t. To eliminate the coefficient of c_1 above, we need $a = 0$ or $b = 1$.

If $a = 0$, we obtain the system

$$\begin{cases} -32\lambda = 16\lambda^2\mu \\ -2\lambda^3 b(1+b) - 16\lambda = 8\lambda^2\mu, \end{cases}$$

which implies that $\lambda\mu = -2$ and $b = 0$ or -1. If $b = 1$, we obtain

$$-4\lambda^3 - 2\lambda(8 + a^2\lambda^2)c_2c_3 - 8\lambda \;=\; \mu(8a^2\lambda^2 + 16c_2c_3),$$

which provides the system

$$\begin{cases} \lambda^2 + 2\mu\lambda + 4 = 0 \\ a^2\lambda^2 + 2\mu\lambda + 4 = 0, \end{cases}$$

and $a = \pm 2$. The unique quadratic then gives the constraint between λ and μ. □

As a corollary, any positive definite linear combination of g_7 and h_7 on S^7 is associated to an NHF structure. Note however that $\lambda = 2/\sqrt{5}$ implies that $\mu = -6/\sqrt{5}$, as in Proposition 3.1 modulo orientation.

The analysis of (6.1) requires more extensive calculation. Turning to B^7, we have established that the solution in Corollary 5.4 is essentially unique within a 3-parameter family. This is part (i) of

Theorem 6.2 *Define ξ and Ξ by* (4.13) *with Y_i in place of X_i, where now*

$$\begin{pmatrix} Y_{2k-1} \\ Y_{2k} \end{pmatrix} = \begin{pmatrix} 2\lambda & 2\lambda a c_k \\ b & 2s_k \end{pmatrix} \begin{pmatrix} p_k \\ n_k \end{pmatrix},$$

and λ, a, b are (unconstrained) constants. Then

(i) (6.1) *and* (6.2) *both hold iff* $(\lambda, a, b) = (\pm\frac{2}{\sqrt{5}}, \frac{1}{2}, 0)$;

(ii) (6.2) *holds if either* $\lambda^2(4 - a^2) = 3$ *and* $b = 0$, *or* $(\lambda, a, b) = (\frac{1}{2}, 2, \pm\sqrt{3})$.

In all cases, μ is determined by λ.

We omit the proof of (i), since analysis of (6.1) is somewhat fruitless. The proof of (ii) is analogous to that of Proposition 6.1, but more involved since it is based upon (5.10). Both parts were checked by hand and using *Maple*, but in different orders.

The significance of the condition $b = 0$ is that it guarantees smooth extension to the singular orbits. This justifies our presentation of the S^7 and B^7 solutions using upper triangular 2×2 matrices, and contrasting geometry at the singular orbits. The special NHF solutions in this section will be analysed elsewhere.

There is an obvious generalization of the hypotheses of Theorem 6.2 that will also capture the NP_2 solution on the principal part $SO(4) \times (0, \pi/3)$ of B^7. Namely one can saturate the first row of the 2×2 matrix by linear combinations 1 and c_k, and its second row by terms proportional to s_k. For that matter, one could insert less trivial Fourier series in each slot. The prospect of adapting the methods to [19] to find new compact NP_2 manifolds of cohomogeneity one appears remote. Nonetheless, the results of this section motivate a computational search for other NP_2 solutions of the above type. Another approach arises from the study of NP_2 structures admitting a cohomogeneity-one action by $Sp(1)^3$ [33].

Acknowledgments We thank the referee for careful reading of the first version. Both authors were supported by the Simons Foundation (#488635 Simon Salamon), and the second author was also funded by the DFG under Germany's Excellence Strategy EXC 2044–390685587, Mathematics Münster: Dynamics–Geometry–Structure. Some of the results of this paper were presented at the conference Special Holonomy and Geometric Structures on Complex Manifolds at IMPA in March 2024.

References

1. E. Abbena, S. Salamon, A. Gray, *Modern Differential Geometry of Curves and Surfaces with Mathematica*, 3rd edn. (Chapman & Hall/CRC, 2006)
2. I. Agricola, T. Friedrich, 3-Sasakian manifolds in dimension seven, their spinors and G_2-structures. J. Geom. Phys. **60**, 326–332 (2010)
3. I. Agricola, G. Dileo, L. Stecker, Curvature properties of 3-(α, δ)-Sasaki manifolds. Ann. Mat. Pura Appl. **202**, 2007–2033 (2023)
4. V. Apostolov, P. Gauduchon, Selfdual Einstein hermitian four-manifolds. Ann. Scuola Norm. Sup. Pisa **1**, 203–243 (2002)
5. M.F. Atiyah, N.J. Hitchin, *The Geometry and Dynamics of Magnetic Monopoles*, Porter Lectures (Princteon University Press, 1988)
6. G. Ball, J. Madnick, Associative submanifolds of squashed 3-Sasakian manifolds. arXiv:2208.10622
7. G. Ball, J. Madnick, Associative Submanifolds of the Berger Space. Commun. Anal. Geom. **31**, 1125–1175 (2023)
8. C. Bär, Real killing spinors and holonomy. Commun. Math. Phys. **154**, 509–521 (1993)
9. Ya.V. Bazaikin, Noncompact Riemannian spaces with the holonomy group *Spin*(7) and 3-Sasakian manifolds. Proc. Steklov Inst. Math. **263**, 2–12 (2008)
10. M. Berger, Les variétés riemanniennes homogènes normales simplement connexes à courbure strictement positive. Ann. Scuola Norm. Sup. Pisa **15**, 179–246 (1961)
11. C. Boyer, K. Galicki, *Sasakian Geometry*, Oxford Mathematical Monographs (Oxford University Press, 2008)
12. R.L. Bryant, Some remarks on G_2 structures, in *Proc. Gokova Geometry-Topology Conference 2005*, ed. by S. Akbulut, T. Onder, R.J. Stern (International Press, 2006), pp. 75–109
13. R.L. Bryant, S.M. Salamon, On the construction of complete metrics with exceptional holonomy. Duke Math. J. **58**, 829–850 (1989)
14. I. Chrysikos, Killing and twistor spinors with torsion. Ann. Glob. Anal. Geom. **49**, 105–141 (2016)
15. O. Dearricott, Positively curved self-dual Einstein metrics on weighted projective planes. Ann. Glob. Anal. Geom. **27**, 79–86 (2005)
16. O. Dearricott, A 7-manifold with positive curvature. Duke Math. J. **158**, 307–346 (2011)

17. M. Fernández, S. Ivanov, V. Muñoz, L. Ugarte, Nearly hypo structures and compact nearly Kähler 6-manifolds with conical singularities. J. Lond. Math. Soc. **78**, 580–604 (2008)
18. A. Fino, P. Nurowski, Analog of selfduality in dimension nine. J. Reine Angew. Math. **699**, 67–110 (2015)
19. L. Foscolo, M. Haskins, New G_2 holonomy cones and exotic nearly Kähler structures on S^6 and $S^3 \times S^3$. Ann. Math. **185**, 59–130 (2017)
20. Th. Friedrich, I. Kath, A. Moroianu, U. Semmelmann, On nearly parallel G_2-structures. J. Geom. Phys. **23**, 259–286 (1997)
21. K. Galicki, H.B. Lawson Jr., Quaternionic reduction and quaternionic orbifolds. Math. Ann. **282**, 1–22 (1988)
22. K. Galicki, S. Salamon, Betti numbers of 3-Sasakian manifolds. Geom. Dedicata **63**, 45–68 (1996)
23. S. Goette, N. Kitchloo, K. Shankar, Diffeomorphism type of the Berger space $SO(5)/SO(3)$. Amer. J. Math. **126**, 395–416 (2004)
24. A. Gray, The structure of nearly Kähler manifolds. Math. Ann. **223**, 233–248 (1976)
25. K. Grove, B. Wilking, W. Ziller, Positively curved cohomogeneity one manifolds and 3-Sasakian geometry. J. Differ. Geom. **78**, 33–111 (2008)
26. N.J. Hitchin, A new family of Einstein metrics, in *Manifolds and Geometry (Pisa, 1993). Sympos. Math.*, vol. XXXVI (1996), pp. 190–222
27. K. Kawai, Some associative submanifolds of the squashed 7-sphere. Q. J. Math. **66**, 861–893 (2015)
28. A. Kennon, J. Lotay, Geometric flows of G_2-structures on 3-Sasakian 7-manifolds. J. Geom. Phys. **187**, 104793 (2023)
29. P.Z. Kobak, A. Swann, Classical nilpotent orbits as hyperkähler quotients. Int. J. Math. **7**, 193–210 (1996)
30. M. Konishi, On manifolds with Sasakian 3-structure over quaternion Kaehler manifolds. Kodai Math. Sem. Rep. **26**, 194–200 (1975)
31. P.B. Kronheimer, Instantons and the geometry of the nilpotent variety. J. Differ. Geom. **32**, 473–490 (1990)
32. J.D. Lotay, Associative submanifolds of the 7-sphere. Proc. Lond. Math. Soc. **105**, 1183–1214 (2012)
33. F. Podestà, Nearly parallel G_2-structures with large symmetry group. Canad. J. Math. **73**, 339–359 (2021)
34. F. Podestà, L. Verdiani, Positively curved 7-dimensional manifolds. Q. J. Math. **50**, 497–504 (1999)
35. S. Salamon, *Riemannian Geometry and Holonomy Groups* (Longman Scientific and Technical, Harlow, Essex, 1989)
36. R. Singhal, Nearly half-flat $SU(3)$-structures on $S^3 \times S^3$. Differ. Geom. Appl. **97**, 102187 (2024)
37. A. Swann, HyperKähler and quaternionic Kähler geometry. Math. Ann. **289**, 421–450 (1991)
38. W. Ziller, On the geometry of cohomogeneity one manifolds with positive curvature, in *Riemannian Topology and Geometric Structures on Manifolds*, Progress Math., vol. 271 (2009), pp. 233–262

Kodaira Dimension of SU(m)-structures

Lorenzo Sillari and Adriano Tomassini

Dedicated to Paul Gauduchon, in occasion of his 80th birthday.

Abstract We study the Kodaira dimension of almost complex manifolds admitting an SU(m)-structure. We introduce the notion of almost complex structure of splitting type and of associated SU(m)-structure. When the latter is pseudoholomorphic, we provide two constructions that allow to obtain non-invariant almost complex structures with Kodaira dimension 0, resp. with Kodaira dimension $-\infty$. Our results apply, in particular, to complex structures of splitting type and to several almost complex manifolds already well-studied in the literature.

Keywords Almost complex structure · Canonical bundle · Kodaira dimension · Pseudoholomorphic section · Solvmanifold · SU(m)-structure

2020 Mathematics Subject Classification Primary: 32Q60, 53C10; Secondary: 32L05, 53C15

The authors are partially supported by GNSAGA of INdAM. The second author is partially supported by the Project PRIN 2022 "Real and Complex Manifolds: Geometry and Holomorphic Dynamics" (code 2022AP8HZ9)

L. Sillari · A. Tomassini (✉)
Dipartimento di Scienze Matematiche, Fisiche e Informatiche, Unità di Matematica e Informatica, Università degli Studi di Parma, Parma, Italy
e-mail: lorenzo.sillari@unipr.it; adriano.tomassini@unipr.it

L. Ornea (ed.), *Real and Complex Geometry*,
https://doi.org/10.1007/978-3-031-92297-8_14

1 Introduction

The Kodaira dimension is a classical complex invariant that measures the growth rate of the space of sections of the pluricanonical bundle of a complex manifold. It becomes especially useful in low dimension, where it plays a significant role in Enriques–Kodaira's classification of complex surfaces.

Recently, Chen and Zhang gave a definition of *Kodaira dimension of almost complex manifolds* that, in the integrable case, reduces to the usual one, see [9] and [10]. They study the basic properties of the Kodaira dimension (in particular on 4-manifolds) and of the spaces of pseudoholomorphic sections of *bundle almost complex structures* [12]. Furthermore, they provide several explicit examples and computations and they ask for other explicit computations of Kodaira dimension, possibly on manifolds with a certain structure (e.g., fibrations). So far, the available explicit computations are:

- on the 4-torus and on 4-manifolds that admit a pseudoholomorphic elliptic fibration over Riemann surfaces of genus $g \geq 2$ [9];
- on the Kodaira–Thurston 4-manifold, see [5] and [9], or on its higher-dimensional generalization [17];
- on 4-dimensional solvmanifolds without complex structures [7];
- on the completely solvable Nakamura manifold [6];
- on the Lie groups $\mathrm{SU}(2) \times \mathrm{SU}(2)$, $\mathrm{SO}(4)$, $\mathrm{Sp}(2)$, $\mathrm{SU}(3)$, and on Lie groups of the form $G \times G$, where G is a connected Lie group [8];
- on the 6-sphere with the standard almost complex structure [9].

All of the mentioned computations, with the exception of the 4-torus, are performed for left-invariant structures on Lie groups or for invariant structures on compact quotients of Lie groups. Even though determining the Kodaira dimension of an invariant structure is already a hard task, finding that of a non-invariant one usually requires solving a difficult PDE whose coefficients are smooth functions (or showing that such a PDE has no solution).

Since the almost complex Kodaira dimension can be computed in terms of sections of the canonical bundle, it is natural to relate it to $\mathrm{SU}(m)$-structures, see, e.g., [4] or [22]. An almost complex $2m$-manifold admits an associated $\mathrm{SU}(m)$-structure when its canonical bundle is trivial as a complex (not necessarily pseudoholomorphic) line bundle. Lie groups and their compact quotients by lattices are parallelizable, hence they always admit an $\mathrm{SU}(m)$-structure, and it becomes natural to exploit its existence to help in the computations of the Kodaira dimension.

In this paper, we focus on the Kodaira dimension of $\mathrm{SU}(m)$-structures. We have two main goals. The first one, is to study general properties of the Kodaira dimension of such structures. We are able to relate the vanishing of the Kodaira dimension to the triviality of tensor powers of the canonical bundle, see Sect. 2. Our results apply to parallelizable manifolds and to compact quotients of Lie groups, and allow to recover results from [1] and [8] as special cases.

The second goal is to partially answer the problem posed by Chen and Zhang by providing two constructions that allow to build *non-invariant* almost complex structures for which we can compute the Kodaira dimension. To do so, in Sect. 2 we introduce *almost complex structures of splitting type* on Lie groups and their compact quotients, see Definition 2.5, a natural non-integrable broad generalization of complex structures of splitting type on solvmanifolds, see [2] and [19]. Compact quotients of these Lie groups have the structure of a pseudoholomorphic fibration over a complex torus, and they inherit a standard SU(m)-structure compatible with the fibration. Section 3 is where we prove our main results: when the standard SU(m)-structure has pseudoholomorphically trivial canonical bundle, we use the fibration to build many non-invariant almost complex structures with Kodaira dimension 0 or $-\infty$. More precisely, we have the following.

Theorem 3.1 *Let $G = \mathbb{C}^k \ltimes_\eta H$ be a $2m$-dimensional Lie group endowed with an almost complex structure of splitting type whose standard* SU(m)*-structure is pseudoholomorphic, and let $M = \Gamma\backslash G$ be its compact quotient. Then there exists a family of non-invariant almost complex structures J_F on M with $\kappa_{J_F} = 0$, parametrized by $F \in \mathcal{F}^{m-k}$.*

Theorem 3.2 *Let $G = \mathbb{C}^k \ltimes_\eta H$ be a $2m$-dimensional Lie group endowed with an almost complex structure of splitting type whose standard* SU(m)*-structure is pseudoholomorphic, and let $M = \Gamma\backslash G$ be its compact quotient. Then there exists a family of non-invariant almost complex structures J_F on M with $\kappa_{J_F} = -\infty$, parametrized by $F \in \mathcal{G} \times C^\infty(M)^{k-1}$.*

In the statement of the theorems, we denoted by Γ a lattice in G and by $\mathcal{F}$ and $\mathcal{G}$ suitable subspaces of functions in $C^\infty(M)$. We are also excluding the trivial case $k = m$, that is discussed in Example 4.1. The assumptions of the theorems hold, in particular, for *complex* structures of splitting type, see Remark 2.6.

We point out that all the examples that we mentioned, on which the Kodaira dimension has been computed (except for the 6-sphere), can be seen as Lie groups endowed with an almost complex structure of splitting type or as their compact quotients. The same is true for other relevant manifolds present in the literature, whose Kodaira dimension has not been computed before. In Sect. 4, we show some application of our results to those examples. In particular, our construction can be performed on $2m$-dimensional tori, on the Iwasawa manifold, on the holomorphically parallelizable Nakamura manifold, on the 6-dimensional solvmanifold without invariant complex structures built by Fernández, de León and Saralegui [15], and on the 8-dimensional solvmanifold built by Sferruzza and the second author [23].

2 Preliminaries

Notation Given a basis of complex forms ϕ^j, $j = 1, \dots, m$, we abbreviate $\phi^j \wedge \phi^k$ to ϕ^{jk} and $\bar{\phi}^j$ to $\phi^{\bar{j}}$.

In this section, we recall some preliminaries on almost complex structures and their Kodaira dimension, on SU(m)-structures, and on the canonical bundle of compact quotients of Lie groups.

2.1 Almost Complex Kodaira Dimension

Let (M, J) be a compact almost complex $2m$-manifold. Denote by $\Lambda^{p,q}$ the bundle of complex (p, q)-forms and by $A^{p,q}$ the space of its smooth sections, that is, the space of complex (p, q)-forms.

The *canonical bundle* of (M, J) is the bundle $K_J := \Lambda^{m,0}$. A (p, q)-form $\alpha \in A^{p,q}$ is called *pseudoholomorphic* if $\bar{\partial}\alpha = 0$. In particular, a pseudoholomorphic section of the canonical bundle is a $\bar{\partial}$-closed complex $(m, 0)$-form Ω. If J is integrable, pseudoholomorphic forms are actually holomorphic forms. Note that, in the complex case, the term *canonical bundle* usually refers to the bundle of *holomorphic* $(m, 0)$*-forms*, see [18]. Even though in practice we will often consider pseudoholomorphic $(m, 0)$-forms, we do not require $\bar{\partial}$-closedness in the definition of K_J.

Denote by P_l the space of pseudoholomorphic sections of $K_J^{\otimes l}$, for $l \geq 1$. The space P_1 is just the space of pseudoholomorphic $(m, 0)$-forms. Chen and Zhang recently introduced the notion of *Kodaira dimension of an almost complex manifold*, see [9] and [10]. The almost complex Kodaira dimension is the number

$$\kappa_J := \begin{cases} -\infty & \text{if } P_l = \{0\} \text{ for all } l \geq 1, \\ \limsup\limits_{l \to \infty} \dfrac{\log \dim_{\mathbb{C}} P_l}{\log l} & \text{otherwise,} \end{cases}$$

and it generalizes the usual notion of Kodaira dimension of a complex manifold.

Remark 2.1 In addition to the definition given above, Chen and Zhang gave another, more algebraic, definition of Kodaira dimension of almost complex manifolds. In the complex case, the two definitions are equivalent to the usual definition of Kodaira dimension. In the non-integrable case, it is not known if the two definitions are equivalent. Due to the analytical approach of this paper, we deal with the definition given in terms of sections of tensor powers of the canonical bundle that we recalled above.

2.2 SU(m)-structures

Let M be a compact $2m$-manifold. We say that M admits an SU(m)-structure if the structure group of its tangent bundle can be reduced to SU(m). The choice of a reduction corresponds to the choice of an SU(m)-structure, which determines an almost complex structure J whose canonical bundle K_J is trivial, and the choice of a trivialization Ω of K_J, that is, a never-vanishing $(m, 0)$-form. This also provides a preferred volume form on M, given by $\Omega \wedge \bar{\Omega}$. If, in addition, we can take Ω to be pseudoholomorphic, then K_J is pseudoholomorphically trivial. By Proposition 3.6 in [8], if the canonical bundle is pseudoholomorphically trivial, then the Kodaira dimension of J vanishes. We show that this is also the case when a tensor power of the canonical bundle is pseudoholomorphically trivial.

Lemma 2.2 *Let (M, J) be a compact almost complex manifold admitting an* SU(m)*-structure. If the canonical bundle of J or one of its tensor powers is pseudoholomorphically trivial, then $\kappa_J = 0$.*

Proof Let Ω be a (not necessarily pseudoholomorphic) trivialization of K_J given by an SU(m)-structure, and let γ be the $(0, 1)$-form such that $\bar{\partial}\Omega = \gamma \wedge \Omega$. The trivialization Ω is pseudoholomorphic if and only if γ vanishes. Our claim for the case $\gamma = 0$ is already proved in Proposition 3.6 in [8] for the case of parallelizable manifolds. The proof for SU(m)-manifolds is very similar, therefore we omit it, and we can suppose $\gamma \neq 0$. A section $f\Omega^{\otimes l}$ of $K_J^{\otimes l}$ is pseudoholomorphic if and only if f is a solution of the equation

$$D_l(f) := \bar{\partial} f + l f \gamma = 0,$$

cf. [8] or [9]. By assumption, there exists a minimum integer l_0 such that $K_J^{\otimes l_0}$ is pseudoholomorphically trivial. We show that

$$\dim_{\mathbb{C}} P_l = \begin{cases} 1, & \text{if } l \text{ is a multiple of } l_0, \\ 0, & \text{otherwise}, \end{cases}$$

which readily implies that $\kappa_J = 0$.

Let $F\Omega^{\otimes l_0}$ be a trivialization for $K_J^{\otimes l_0}$, where F is a never-vanishing smooth function. Then F solves the equation $D_{l_0}(F) = 0$. Let $g\Omega^{\otimes l_0}$ be another pseudoholomorphic section of $K_J^{\otimes l_0}$, where g is not necessarily never-vanishing. Then

$$0 = \bar{\partial}(g\Omega^{\otimes l_0}) = \bar{\partial}(\frac{g}{F} F\Omega^{\otimes l_0}) = \bar{\partial}(\frac{g}{F}) \wedge (F\Omega^{\otimes l_0}).$$

Since $F\Omega^{\otimes l_0}$ is never-vanishing, this implies that $\bar{\partial}(g/F) = 0$ and, since every $\bar{\partial}$-closed function on a compact almost complex manifold must be constant, see e.g., Corollary 4.9 in [11], that g/F is constant. Hence, we have that $\dim_{\mathbb{C}} P_{l_0} = 1$.

Similarly, if $l = pl_0$, then we have that $D_l(F^p) = 0$, that $F^p\Omega^{\otimes l}$ is never-vanishing and pseudoholomorphic, and that every other pseudoholomorphic section of $K_J^{\otimes l}$ is a constant multiple of $F^p\Omega^{\otimes l}$, which gives $\dim_{\mathbb{C}} P_l = 1$.

Now suppose, by contradiction, that for $p < l_0$ there exists a function g such that $g\Omega^{\otimes p}$ is pseudoholomorphic. Then $g^{l_0}\Omega^{\otimes pl_0}$ and $F^p\Omega^{\otimes pl_0}$ are both pseudoholomorphic sections of $K_J^{\otimes pl_0}$. By the argument of the first part of the proof, we have that g^{l_0} is a constant multiple of F^p, which is never-vanishing. In particular, g^{l_0} is never-vanishing, and so is g. This contradicts the minimality assumption on l_0 and proves that $K_J^{\otimes p}$ has no pseudoholomorphic sections for $p < l_0$, giving $\dim_{\mathbb{C}} P_p = 0$.

To conclude the proof, let p be any integer that is not a multiple of l_0, and let q be the integer such that $ql_0 < p < (q+1)l_0$. Suppose that $g\Omega^{\otimes p}$ is a pseudoholomorphic section of $K_J^{\otimes p}$. Then, we have that

$$
\begin{aligned}
0 &= \bar{\partial}(g\Omega^{\otimes p}) = \bar{\partial}(\frac{g}{F^q}\Omega^{\otimes(p-ql_0)} \otimes F^q\Omega^{\otimes ql_0}) = \bar{\partial}(\frac{g}{F^q}\Omega^{\otimes(p-ql_0)}) \otimes (F^q\Omega^{\otimes ql_0}) \\
&= (\bar{\partial}(\frac{g}{F^q}) + (p-ql_0)\frac{g}{F^q}\gamma) \wedge (F^q\Omega^{\otimes p}),
\end{aligned}
$$

which is equivalent to $D_{p-ql_0}(g/F^q) = 0$. Since $p - ql_0 < l_0$, there are no non-zero solutions to such equation, and we get $\dim_{\mathbb{C}} P_p = 0$. □

2.3 *Compact Quotients of Lie Groups with Trivial Canonical Bundle*

Let G be a connected, simply connected Lie group admitting a co-compact lattice Γ, and denote by $M := \Gamma\backslash G$ their compact quotient. If G is nilpotent, resp. solvable, we say that M is a *nilmanifold*, resp. *solvmanifold*. If G is completely solvable, we say that M is a solvmanifold *of completely solvable type*.

Since G admits a lattice, it must be unimodular. This implies that the quotient M is parallelizable. In particular, it admits invariant almost complex structures and invariant $SU(m)$-structures induced by structures on $\mathfrak{g}$, the Lie algebra of G. In addition, it will also admit genuine almost complex or $SU(m)$-structures that are not invariant.

Let J be an invariant almost complex structure on M, and let ϕ^j, $j = 1, \ldots, m$, be a basis of invariant $(1,0)$-forms for J. Then, the invariant $(m,0)$-form $\phi^{1\ldots m}$ induces a trivialization of the canonical bundle K_J, hence a canonical invariant $SU(m)$-structure on M. By choosing a different basis of $(1,0)$-forms, the trivialization changes by a constant.

As a direct application of Lemma 2.2, we have the following.

Proposition 2.3 *Let $M = \Gamma\backslash G$ be a compact quotient of a Lie group and let J be an invariant almost complex structure on M. If the canonical bundle of J or one of its tensor powers is pseudoholomorphically trivial, then $\kappa_J = 0$.*

We recall that if (M, J) is a nilmanifold endowed with an invariant *complex* structure, then its canonical bundle is always holomorphically trivial [3], so that $\kappa_J = 0$. Moreover, the trivialization is necessarily provided by an invariant form. In the solvable case, there are examples of invariant *complex* structures whose canonical bundle is holomorphically trivial, but the trivialization cannot be invariant, see [1] and [24]. We point out that this difference amounts to understanding if a given invariant complex structure admits or not a compatible invariant SU(m)-structure with holomorphically trivial canonical bundle.

Finally, we give the almost complex version of Corollary 4.2 in [1].

Lemma 2.4 *Let $M = \Gamma\backslash G$ be a compact quotient of a Lie group and let J be an invariant complex structure on M. Then pseudoholomorphic trivializations of tensor powers of its canonical bundle are either all invariant or all non-invariant (if they exist).*

Proof Let Ω be a (not necessarily pseudoholomorphic) trivialization of K_J. Suppose that the bundle $K_J^{\otimes l}$ admits a never-vanishing pseudoholomorphic section $F\Omega^{\otimes l}$, $F \in C^\infty(M)$, and let $\hat{F}\Omega^{\otimes l}$, $\hat{F} \in C^\infty(M)$, be another pseudoholomorphic section of $K_J^{\otimes l}$. Since $F\Omega^{\otimes l}$ trivializes $K_J^{\otimes l}$, there exists a function g such that $\hat{F}\Omega^{\otimes l} = gF\Omega^{\otimes l}$. Then, we have that

$$0 = \bar{\partial}(\hat{F}\Omega^{\otimes l}) = \bar{\partial}(gF\Omega^{\otimes l}) = F\bar{\partial}g \wedge \Omega^{\otimes l} + g\bar{\partial}(F\Omega^{\otimes l}) = F\bar{\partial}g \wedge \Omega^{\otimes l},$$

where in the last equality we used that $F\Omega^{\otimes l}$ is pseudoholomorphic. Since F is never-vanishing and $\Omega^{\otimes l}$ induces an isomorphism between $A^{0,1}$ and $A^{0,1} \wedge K_J^{\otimes l}$, we conclude that $\bar{\partial}g = 0$ and, by compactness of M, that g is constant. In particular, every pseudoholomorphic section of $K_J^{\otimes l}$ is a constant multiple of $F\Omega^{\otimes l}$. The trivialization $F\Omega^{\otimes l}$ is invariant if and only if F is constant, if and only if g is constant. Similarly, F is non-constant if and only if g is. In this case every trivialization is non-invariant. □

2.4 Almost Complex and SU(m)-structures of Splitting Type

We introduce a special class of Lie groups that split as the semi-direct product of $\mathbb{C}^k$ and a Lie group endowed with a left-invariant almost complex structure, with the action of $\mathbb{C}^k$ compatible with the almost complex structure.

Let G be a connected, simply connected, $2m$-dimensional Lie group.

Definition 2.5 Let G be a Lie group that can be written as a semi-direct product $G = \mathbb{C}^k \ltimes_\eta H$, $k \geq 1$, such that:

(i) H is a connected, simply connected, $2(m-k)$-dimensional Lie group endowed with a left-invariant almost complex structure J_H;
(ii) G admits a co-compact lattice Γ that splits as $\Gamma = \Gamma_{\mathbb{C}^k} \ltimes_\eta \Gamma_H$ and such that, for all $z \in \Gamma^k_{\mathbb{C}}$, we have $\eta(z) \in \mathrm{Aut}(\Gamma_H)$.

Then G admits a standard left-invariant almost complex structure built as follows. Let dz^j, $j = 1, \ldots, k$, be the standard basis of $(1,0)$-forms for $\mathbb{C}^k$, and let ψ^j, $j = k+1, \ldots, m$, be a basis of left-invariant $(1,0)$-forms on H for the almost complex structure J_H. Let $g = (z, h) \in G$ and $(Z, W) \in \mathrm{Lie}(G)$. For $j = k+1, \ldots, m$, we consider the left-invariant 1-forms ϕ^j on G, defined by

$$(\phi^j(Z, W))(z, h) := (\psi^j(d\eta_z^{-1}(W)))(h).$$

For brevity, we write $\phi^j = \eta^{-1}\psi^j$. The *almost complex structure of splitting type* induced on G by J_H is the one defined by the co-frame of $(1,0)$-forms

$$\{dz^1, \ldots, dz^k, \phi^{k+1}, \ldots, \phi^m\}. \tag{2.1}$$

Note that the construction depends only on the choice of co-frame of $(1,0)$-forms for the almost complex structure J_H. In this case, we say that G is an *almost complex Lie group of splitting type*.

Every almost complex structure of splitting type naturally defines an $\mathrm{SU}(m)$-structure given by the left-invariant $(m,0)$-form

$$\Omega := dz^{1\ldots k} \wedge \phi^{k+1\ldots m},$$

that we call the *standard left-invariant* $\mathrm{SU}(m)$*-structure* associated to J. In general, the form Ω will not be pseudoholomorphic since

$$\bar\partial\Omega = (-1)^k dz^{1\ldots k} \wedge \bar\partial(\phi^{k+1\ldots m}).$$

If $\bar\partial\Omega = 0$, we say that the standard left-invariant $\mathrm{SU}(m)$-structure on G is *pseudoholomorphic*. Note that pseudoholomorphic $\mathrm{SU}(m)$-structures have already been considered, under different names, by de Bartolomeis and the second author in [14] and [13].

Remark 2.6 The reader should compare our definition of almost complex structure of splitting type with the one of complex structure of splitting type given in [19], see also [2]. In addition to the assumptions given in Definition 2.5, when dealing with complex structure of splitting type one assumes that H is nilpotent, that J_H is integrable, and that the action of η on the Lie algebra of H is semi-simple. This guarantees that the resulting structure on G is integrable and that G is solvable. The original definition also requires some special cohomological properties on H. We drop the assumption since in this paper we are not concerned with the cohomology of almost complex manifolds.

Lemma 2.7 *Let $G = \mathbb{C}^k \ltimes_\eta H$ be an almost complex Lie group endowed with a **complex** structure of splitting type. Then the standard left-invariant* SU(m)-*structure on G is holomorphic.*

Proof The basis of $(1, 0)$-forms for J_H defines a left-invariant $(m - k, 0)$-form

$$\hat{\Omega} := \psi^{k+1...m}$$

on H. By [3], if H is nilpotent and J_H is integrable, then $\bar{\partial}\hat{\Omega} = 0$. In particular, since the action of η is holomorphic with respect to J_H and it can be diagonalized over $\mathbb{C}$ by semi-simplicity, we have that

$$\bar{\partial}(\eta^{-1}\hat{\Omega}) = 0,$$

proving the lemma. □

Almost complex structures of splitting type on G and the associated left-invariant SU(m)-structures naturally descend to invariant structures on the compact quotient $M := \Gamma\backslash G$. By assumption, the lattice Γ splits as $\Gamma = \Gamma_{\mathbb{C}^k} \ltimes_\eta \Gamma_H$. Hence, there is a well-defined projection

$$\pi : M \longrightarrow T^{2k} := \Gamma_{\mathbb{C}^k}\backslash\mathbb{C}^k, \tag{2.2}$$

that gives to M the structure of a fiber bundle over T^{2k}, with fiber diffeomorphic to $\Sigma := \Gamma_H\backslash H$. By the expression of (2.1), the fibration is pseudoholomorphic with respect to the almost complex structure of splitting type on M and the complex structure on T^{2k}.

Example 2.8 Tori, as well as the direct products by $\mathbb{C}^k$ by almost complex Lie groups endowed with a lattice, trivially admit almost complex structures of splitting type. The Iwasawa manifold and the Nakamura manifold, which are two widely studied solvmanifolds endowed with complex and almost complex structures, both have many almost complex structures of splitting type, and provide non-trivial examples. We refer to Sect. 4 for more non-trivial examples.

Finally, due to the importance it plays in Theorem 3.1, we denote by $\mathcal{F}$ the space of smooth functions on M that are constant along the fibers of its fibration structure. By pushing down along the projection, such functions correspond to functions on the torus T^{2k}. In particular, if we denote by ξ_j, $j = 1, \ldots, m$, a basis of $(1, 0)$-vector fields dual to (2.1), then we have that

$$\mathcal{F} = C^\infty(M) \cap \ker \xi_{k+1} \cap \ldots \cap \ker \xi_m.$$

The inclusion of $\mathcal{F}$ into the kernel of each operator ξ_j, $j = k+1, \dots, m$, is clear since functions in $\mathcal{F}$ are constant on the fibers. Denote by ∂_Σ the operator on the fibers of π such that

$$\partial f = \sum_{j=1}^{k} \partial_{z_j}(f) dz^j + \partial_\Sigma f,$$

and let $f \in C^\infty(M) \cap \ker \xi_{k+1} \cap \dots \cap \ker \xi_m$. This implies that $\partial_\Sigma f = 0$. The operator $\bar\partial_\Sigma \partial_\Sigma$ is elliptic on Σ, which is compact. Hence, by the maximum principle, the function f is constant on each fiber, proving the opposite inclusion.

3 Kodaira Dimension of Almost Complex Structures of Splitting Type

Let G be an almost complex Lie group of splitting type and let J_0 be an almost complex structure of splitting type on it. If the associated $\mathrm{SU}(m)$-structure is pseudoholomorphic, then K_{J_0} admits an invariant pseudoholomorphic trivialization that descends to the quotient $M = \Gamma \backslash G$. In particular, the structure J_0 induces on M an invariant structure J that has Kodaira dimension 0.

In the next theorem, we describe a construction that allows to produce non-invariant structures with Kodaira dimension 0.

Theorem 3.1 *Let $G = \mathbb{C}^k \ltimes_\eta H$ be a $2m$-dimensional Lie group endowed with an almost complex structure of splitting type whose standard $\mathrm{SU}(m)$-structure is pseudoholomorphic, and let $M = \Gamma \backslash G$ be its compact quotient. Then there exists a family of non-invariant almost complex structures J_F on M with $\kappa_{J_F} = 0$, parametrized by $F \in \mathcal{F}^{m-k}$.*

Proof We first prove the theorem for the case $k = 1$. Let ϕ^j, $j = 1, \dots, m$, be a basis of invariant $(1,0)$-forms for the standard almost complex structure J_0 on G, with differentials

$$d\phi^j = \sum_{p<q} (C^j_{pq} \phi^{pq} + C^j_{\bar p \bar q} \phi^{\bar p \bar q}) + \sum_{p,q} C^j_{p\bar q} \phi^{p\bar q}, \quad j = 1, \dots, m, \tag{3.1}$$

with C^j_{pq}, $C^j_{\bar p \bar q}$ and $C^j_{p\bar q} \in \mathbb{C}$. Since G is of splitting type, we can suppose $d\phi^1 = 0$. Since the invariant $\mathrm{SU}(m)$-structure associated to J_0 is pseudoholomorphic, we also have that

$$0 = \bar\partial(\phi^{1\dots m}) = \sum_j (-1)^{j+1} \bar\partial \phi^j \wedge \phi^{1\dots\hat{j}\dots m} = -(\sum_{j,q} C^j_{j\bar q} \phi^{\bar q}) \wedge \phi^{1\dots m}.$$

In particular, this gives the relations

$$\sum_j c^j_{j\bar{q}} = 0, \quad \text{for } q = 1, \ldots, m. \tag{3.2}$$

Let $f_j \in \mathcal{F}$, $j = 2, \ldots, m$, be non-constant functions, and consider the non-invariant almost complex structure J_F on M defined by the $(1,0)$-forms

$$\omega^1 := \phi^1 \quad \text{and} \quad \omega^j := \phi^j + f_j \phi^{\bar{1}} \quad \text{for } j = 2, \ldots, m,$$

with $F = (f_2, \ldots, f_m) \in \mathcal{F}^{m-1}$. The inverse of the transformation is

$$\phi^1 = \omega^1, \quad \phi^j = \omega^j - f_j \omega^{\bar{1}} \quad \text{for } j = 2, \ldots, m.$$

Moreover, if we denote by ξ_j, $j = 1, \ldots, m$, the $(1,0)$-vector fields dual to the ϕ^j, then the $(1,0)$-vector fields dual to the ω^j are

$$\psi_1 = \xi_1 - \sum_{j \geq 2} \bar{f}_j \xi_{\bar{j}} \quad \text{and} \quad \psi_j = \xi_j, \quad \text{for } j = 2, \ldots, m.$$

To prove the theorem, we show that, for our choice of the functions f_j, we have that

$$\bar{\partial}_F(\omega^{1\ldots m}) = \sum_j (-1)^{j+1}\, \bar{\partial}_F \omega^j \wedge \omega^{1\ldots\hat{j}\ldots m} = 0,$$

where $\bar{\partial}_F$ denotes the operator $\bar{\partial}$ associated to the structure defined by the ω^j. To measure the difference of the action of $\bar{\partial}_F$ and $\bar{\partial}$, it is useful to define

$$\bar{\partial}_0 \phi^j := \sum_{p,q} c^j_{p\bar{q}} \omega^{p\bar{q}},$$

obtained by formally replacing ϕ^j by ω^j in the expression of $\bar{\partial}\phi^j$ given in (3.1). In particular, it follows from (3.2) that

$$\sum_j (-1)^{j+1}\, \bar{\partial}_0 \phi^j \wedge \omega^{1\ldots\hat{j}\ldots m} = 0. \tag{3.3}$$

Then, we have that

$$\bar{\partial}_F \omega^1 = 0 \quad \text{and} \quad \bar{\partial}_F \omega^j = \bar{\partial}_0 \phi^j + \epsilon_j + \partial_F f_j \wedge \omega^{\bar{1}}, \quad \text{for } j = 2, \ldots, m, \tag{3.4}$$

where ϵ_j measures the difference between $\bar{\partial}_F$ and $\bar{\partial}_0$. So we have that

$$\begin{aligned}\bar{\partial}_F(\omega^{1\dots m}) &= \sum_{j\geq 2}(-1)^{j+1}\,\bar{\partial}_F\omega^j \wedge \omega^{1\dots\hat{j}\dots m}\\ &= \sum_{j\geq 2}(-1)^{j+1}\,(\bar{\partial}_0\phi^j + \epsilon_j + \partial_F f_j \wedge \omega^{\bar{1}}) \wedge \omega^{1\dots\hat{j}\dots m}\\ &= \sum_{j\geq 2}(-1)^{j+1}\,\epsilon_j \wedge \omega^{1\dots\hat{j}\dots m},\end{aligned}$$

where in the second equality we used (3.4), while in the last equality we used (3.3) and the fact that $f_j \in \mathcal{F}$ gives

$$\partial_F f_j \wedge \omega^{1\dots\hat{j}\dots m} = \psi_1(f_j)\,\omega^1 \wedge \omega^{1\dots\hat{j}\dots m} = 0. \tag{3.5}$$

To obtain the explicit expression of the ϵ_j, we write (3.1) in terms of the ω^j and take the (1, 1)-bidegree part. First, we have that

$$\sum_{p<q} C^j_{pq}\,\phi^{pq} = \sum_{q>1} C^j_{1q}\,\omega^1 \wedge (\omega^q - f_q\omega^{\bar{1}}) + \sum_{1<p<q} C^j_{pq}(\omega^p - f_p\omega^{\bar{1}}) \wedge (\omega^q - f_q\omega^{\bar{1}}).$$

Taking bidegree (1, 1), the remaining terms are

$$\sum_{q>1} -f_q\,C^j_{1q}\,\omega^{1\bar{1}} + \sum_{1<p<q} C^j_{pq}\,\omega^{\bar{1}} \wedge (f_q\omega^p - f_p\omega^q). \tag{3.6}$$

Similarly, when developing $\sum_{p<q} C^j_{\bar{p}\bar{q}}\,\phi^{\bar{p}\bar{q}}$, we get the extra terms

$$\sum_{q>1} \bar{f}_q\,C^j_{1\bar{q}}\,\omega^{1\bar{1}} + \sum_{1<p<q} C^j_{\bar{p}\bar{q}}\,\omega^1 \wedge (\bar{f}_q\omega^{\bar{p}} - \bar{f}_p\omega^{\bar{q}}). \tag{3.7}$$

Finally, we have that

$$\begin{aligned}\sum_{p,q} C^j_{p\bar{q}}\,\phi^{p\bar{q}} &= \sum_q (C^j_{1\bar{q}}\,\omega^1 \wedge (\omega^{\bar{q}} - \bar{f}_q\omega^1) + C^j_{p\bar{1}}(\omega^p - f_p\omega^{\bar{1}}) \wedge \omega^{\bar{1}})\\ &\quad + \sum_{p,q\neq 1} C^j_{p\bar{q}}(\omega^p - f_p\omega^{\bar{1}}) \wedge (\omega^{\bar{q}} - \bar{f}_q\omega^1).\end{aligned}$$

Taking bidegree (1, 1) and using (3.6) and (3.7), we have that

$$\begin{aligned}\bar{\partial}_F\phi^j &= \sum_{p,q} C^j_{p\bar{q}}\,\omega^{p\bar{q}} + \sum_{q>1}(\bar{f}_q\,C^j_{1\bar{q}} - f_q\,C^j_{\bar{1}q})\omega^{1\bar{1}} \\ &\quad + \sum_{1<p<q} C^j_{pq}\,\omega^{\bar{1}}\wedge(f_q\omega^p - f_p\omega^q) + \sum_{1<p<q} C^j_{\bar{p}\bar{q}}\,\omega^1\wedge(\bar{f}_q\omega^{\bar{p}} - \bar{f}_p\omega^{\bar{q}}).\end{aligned}$$

By definition, the first summand is $\bar{\partial}_0\phi^j$, while the second and fourth summand are irrelevant when wedging with $\omega^{1\ldots\hat{j}\ldots m}$, since they all contain the term ω^1. This leaves us with

$$\begin{aligned}\bar{\partial}_F(\omega^{1\ldots m}) &= \sum_{j\geq 2}(-1)^{j+1}\,\epsilon_j\wedge\omega^{1\ldots\hat{j}\ldots m} \\ &= \sum_{j\geq 2}(-1)^{j+1}\sum_{1<p<q} C^j_{pq}\,\omega^{\bar{1}}\wedge(f_q\omega^p - f_p\omega^q)\wedge\omega^{1\ldots\hat{j}\ldots m} \\ &= \sum_{j\geq 2}\Big(\sum_{q>j} C^j_{jq}f_q - \sum_{1<p<q} C^j_{pj}f_p\Big)\omega^{\bar{1}1\ldots m}.\end{aligned}$$

By rearranging the indices in a suitable way, we can write

$$\sum_{j\geq 2}\Big(\sum_{q>j} C^j_{jq}f_q - \sum_{1<q<j} C^j_{qj}f_q\Big) = \sum_{j\geq 2} A_j f_j,$$

where $A_j\in\mathbb{C}$ has the expression

$$A_j = \sum_{2\leq q<j} C^q_{qj} - \sum_{j<q\leq m} C^q_{jq}, \quad j = 2,\ldots,m.$$

To conclude the proof, we show that $A_j = 0$ for all j. Using the duality pairing between forms and vector fields on $\mathfrak{g}$, and the fact that the differential of invariant forms is, up to sign, the dual of the Lie bracket, we have that

$$\begin{aligned}A_j &= \sum_{2\leq q<j} d\phi^q(\xi_q,\xi_j) - \sum_{j<q\leq m} d\phi^q(\xi_j,\xi_q) \\ &= \sum_{q=1}^m d\phi^q(\xi_q,\xi_j) = -\sum_{q=1}^m \phi^q([\xi_q,\xi_j]) = \mathrm{tr}\big(\mathrm{ad}_{\phi^j}\big),\end{aligned}$$

where in the second equality we used that $[\xi_j,\xi_j] = 0$ and that $[\xi_1,\xi_q] = 0$ for all $q = 1,\ldots,m$. Since G admits a lattice, it is unimodular and the trace of the adjoint representation vanishes. By equation (3.2), we also have that the trace of $J\circ\mathrm{ad}$ vanishes, giving $A_j = 0$ for all $j = 1,\ldots,m$.

The proof for the case $k > 1$ follows similarly, choosing $f_j = 0$ for $j = 1, \dots, k$. The only difference is in (3.5), that becomes

$$\partial_F f_j \wedge \omega^{1\dots\hat{j}\dots m} = \sum_p \psi_p(f_j)\,\omega^p \wedge \omega^{1\dots\hat{j}\dots m} = (-1)^j \psi_j(f_j)\omega^{1\dots m}.$$

The function $\psi_j(f_j)$ vanishes for every j since we are taking $f_j = 0$ when $j \leq k$, and $f_j \in \mathcal{F}$ when $j \geq k+1$, proving the theorem for $k > 1$. □

In the second result, we describe a construction that allows to produce non-invariant structures with Kodaira dimension $-\infty$. We suppose that $k \neq m$. The case $k = m$, corresponding to almost complex tori, is discussed in Example 4.1.

Theorem 3.2 *Let $G = \mathbb{C}^k \ltimes_\eta H$ be a $2m$-dimensional Lie group endowed with an almost complex structure of splitting type whose standard* SU(m)*-structure is pseudoholomorphic, and let $M = \Gamma\backslash G$ be its compact quotient. Then there exists a family of non-invariant almost complex structures J_F on M with $\kappa_{J_F} = -\infty$, parametrized by $F \in \mathcal{G} \times C^\infty(M)^{k-1}$.*

Proof Let ϕ^j, $j = 1, \dots, m$, be a basis of invariant $(1,0)$-forms for an almost complex structure of splitting type J_0 on G, with differentials given as (3.1). Since J_0 is of splitting type, we can suppose $d\phi^j = 0$ for $j = 1, \dots, k$. Since the invariant SU(m)-structure associated to J_0 is pseudoholomorphic, we also have that (3.2) holds.

Let $f_j \in C^\infty(M)$, $j = 1, \dots, k$, be non-constant functions, and consider the non-invariant almost complex structure J_F on M defined by the $(1,0)$-forms

$$\omega^j := \phi^j, \quad \text{for } j = 1, \dots, m-1, \quad \text{and} \quad \omega^m := \phi^m + \sum_{j=1}^k f_j \phi^{\bar{j}},$$

with $F = (f_1, \dots, f_k) \in C^\infty(M)^k$. The inverse of the transformation is

$$\phi^j = \omega^j, \quad \text{for } j = 1, \dots, m-1, \quad \text{and} \quad \phi^m = \omega^m - \sum_{j=1}^k f_j \omega^{\bar{j}}.$$

Moreover, if we denote by ξ_j, $j = 1, \dots, m$, the vector fields dual to the ϕ^j, then the vector fields dual to the ω^j are

$$\psi_j = \xi_j - \bar{f}_j \xi_{\bar{m}}, \quad \text{for } j = 1, \dots, m-1, \quad \text{and} \quad \psi_m = \xi_m.$$

To prove the theorem, we start by showing that, for a suitable choice of the functions f_j, we have that the equation

$$\bar{\partial}_F(g\,\omega^{1\dots m}) = 0,$$

has no solution $g \in C^\infty(M)$, where $\bar\partial_F$ denotes the operator $\bar\partial$ associated to the structure defined by the ω^j. With the same notation of the proof of Theorem 3.1, we have that

$$\bar\partial_F \omega^j = \bar\partial_0 \phi^j + \epsilon_j, \quad \text{for } j = 1, \ldots, m-1, \quad \text{and}$$

$$\partial_F \omega^m = \bar\partial_0 \phi^m + \epsilon_m + \sum_{j=1}^{k} \partial_F f_j \wedge \omega^{\bar j} + \sum_{j=1}^{k} f_j \partial_F \omega^{\bar j}.$$

Since $d\phi^j = d\omega^j = 0$ for $j = 1, \ldots, k$, the last summand in the differential $\partial_F \omega^m$ is zero, so that

$$\begin{aligned}
\bar\partial_F(\omega^{1\ldots m}) &= \sum_{j=1}^{m} (-1)^{j+1}\, \bar\partial_F \omega^j \wedge \omega^{1\ldots\hat{j}\ldots m} \\
&= \sum_{j=1}^{m} (-1)^{j+1} \,(\bar\partial_0 \phi^j + \epsilon_j) \wedge \omega^{1\ldots\hat{j}\ldots m} + (-1)^{m+1} \sum_{j=1}^{k} \partial_F f_j \wedge \omega^{\bar j 1\ldots m-1} \\
&= \sum_{j=1}^{m} (-1)^{j+1}\, \epsilon_j \wedge \omega^{1\ldots\hat{j}\ldots m} - \sum_{j=1}^{k} \psi_m(f_j) \omega^{\bar j} \wedge \omega^{1\ldots m}
\end{aligned}$$

where in the third equality we used (3.3). To obtain the explicit expression of the ϵ_j, we proceed as in the proof of Theorem 3.1. The contribution of the terms $C^j_{pq}\, \phi^{pq}$ to the differential $\bar\partial_F \omega^j$ is

$$-\sum_{p<m} \sum_{q=1}^{k} C^j_{pm} f_q\, \omega^{p\bar q}.$$

Similarly, the terms $C^j_{\bar p \bar q}\, \phi^{\bar p \bar q}$ contribute as

$$\sum_{p<m} \sum_{q=1}^{k} C^j_{\bar p \bar m} \bar f_q\, \omega^{q\bar p}.$$

Finally, the last contribution is that of $C^j_{m\bar m}\, \phi^{m\bar m}$, that gives

$$-\sum_{p,q=1}^{k} C^j_{m\bar m} f_q \bar f_p \omega^{p\bar q}$$

With a slight abuse of notation, and to write the sums in a more compact way, we set $f_j = 0$ for $j \geq k+1$. This leaves us with

$$\begin{aligned}\bar{\partial}_F(\omega^{1\dots m}) &= \sum_{j=1}^{m}(-1)^{j+1}\Big(\sum_{p,q=1}^{m-1}(-C^j_{pm}f_q + C^j_{\bar{q}\bar{m}}\bar{f}_p - C^j_{m\bar{m}}f_q\bar{f}_p)\omega^{p\bar{q}}\Big)\wedge\omega^{1\dots\hat{j}\dots m} \\ &\quad - \sum_{j=1}^{k}\psi_m(f_j)\omega^{\bar{j}}\wedge\omega^{1\dots m} \\ &= -\sum_{j=1}^{m}\Big(\sum_{q=1}^{m-1}(-C^j_{jm}f_q + C^j_{\bar{q}\bar{m}}\bar{f}_j - C^j_{m\bar{m}}f_q\bar{f}_j)\omega^{\bar{q}}\Big)\wedge\omega^{1\dots m} \\ &\quad - \sum_{j=1}^{k}\psi_m(f_j)\omega^{\bar{j}}\wedge\omega^{1\dots m} \\ &= \sum_{q=1}^{m} H_q\,\omega^{\bar{q}}\wedge\omega^{1\dots m}\end{aligned}$$

where in the second equality we used that the only summands that matter are those for which $p = j$, and where the $H_q \in C^\infty(M)$ are given by

$$H_q = \sum_{j=1}^{m}(C^j_{jm}f_q - C^j_{\bar{q}m}\bar{f}_j + C^j_{m\bar{m}}f_q\bar{f}_j) - \psi_m(f_q) \quad \text{for } q = 1,\dots,m-1,$$

and $H_m = -\psi_m(f_m)$. The first summand vanishes since $\sum_{j=1}^{m} C^j_{jm}$ is the trace of the adjoint of ϕ^m, see the proof of Theorem 3.1. The second and third summands vanish since either $j \leq k$, so that $C^j_{\bar{q}m} = 0$, or $j \geq k+1$, so that $f_j = 0$. Recalling that we set $f_j = 0$ for $j \geq k+1$, we conclude that

$$H_q = -\psi_m(f_q), \quad \text{for } q = 1,\dots,k \quad \text{and} \quad H_q = 0, \quad \text{for } q = k+1,\dots,m.$$

The equation $\bar{\partial}_F(g\omega^{1\dots m})$ has a solution $g \in C^\infty(M)$ if and only if g solves the system

$$\begin{cases}\psi_q(g) = g\psi_m(f_q), & \text{for } q \leq k,\\ \psi_q(g) = 0, & \text{for } q \geq k+1.\end{cases} \tag{3.8}$$

To show that it has no solution, observe that, by the last equation, we have that $\psi_m(g) = 0$, that implies $\psi_q(g) = \xi_q(g)$ for $q = 1, \dots, m$. Consider the fibration $M \to T^{2k}$ defined by the splitting structure on G. Then the operator

$$\Delta_{k+1,\dots,m} := \sum_{j=k+1}^{m} \xi_j \xi_{\bar{j}}$$

is elliptic on the fibers and any solution g of (3.8) lies in its kernel. By the maximum principle, the function g must be constant on the fibers, and it descends to a function on the base T^{2k}. By the equations of (3.8) for $q \leq k$, if we choose the f_q in a way that $f_q \in \mathcal{G} := (\mathbb{C}^\infty(M) \setminus \ker \xi_m)$ for at least one index $q \leq k$, then the system has no solution g. This implies that the canonical bundle has no pseudoholomorphic section.

To show that $\kappa_{J_F} = -\infty$, consider the equation $\bar{\partial}_F(g(\omega^{1\dots m})^{\otimes l}) = 0$, for $l \geq 2$. Then g is a solution of the equation if and only if it solves the system

$$\begin{cases} \psi_q(g) = l\, g \psi_m(f_q), & q \leq k, \\ \psi_q(g) = 0, & q \geq k+1. \end{cases}$$

With the same reasoning as above, we conclude that there are no solutions and that the Kodaira dimension of J_F is $-\infty$. □

Corollary 3.3 *Every almost complex Lie group of splitting type whose standard* SU(m)*-structure is pseudoholomorphic admits non-invariant almost complex structures of Kodaira dimension* 0 *and of Kodaira dimension* $-\infty$.

Remark 3.4 By Remark 2.6 and Lemma 2.7, the claims of Theorems 3.1 and 3.2, and of Corollary 3.3 hold for complex structures of splitting type. They also hold for compact quotients of complex parallelizable Lie groups [25], for the 6-dimensional solvmanifolds with holomorphically trivial canonical bundle classified by Fino et al. [16], and for the compact quotients of the 6-dimensional unimodular Lie groups with holomorphically trivial canonical bundle described by Otal and Ugarte [21].

4 Examples and Applications

In this section we present several examples of Lie groups endowed with almost complex structures of splitting type and of their compact quotients. All the examples come from manifolds that, more or less extensively, have already been considered in the literature, showing the abundance of this kind of structure.

Example 4.1 (Almost Complex Structures with Kodaira Dimension $-\infty$ on Tori) Complex tori are complex Lie groups of splitting type with $k = m$. In this case, when applying Theorem 3.1, we are not actually building a new structure.

However, the proof of Theorem 3.2 can be slightly modified to produce a large family of almost complex structures on tori of arbitrary dimension with Kodaira dimension $-\infty$. Indeed, to obtain such a family, it is enough to choose $f_m = 0$ and the f_q as in the proof of Theorem 3.2. In this case, the family is parametrized by $\mathcal{G} \times C^\infty(M)^{m-2}$. We point out that, alternatively, a different family of almost complex tori with Kodaira dimension $-\infty$ can be obtained by taking product of the 4-dimensional example built by Chen and Zhang, see Section 6.2 in [9], with tori endowed with an integrable structure.

Example 4.2 (The Iwasawa and the Nakamura Manifolds) The Iwasawa manifold and the Nakamura manifold are classical examples of solvmanifolds obtained as the quotient of 3-dimensional complex Lie groups by a lattice [20]. They admit the structure of T^2-bundle over T^4, resp. of T^4-bundle over T^2, induced by a complex structure of splitting type. In particular, they all admit non-invariant almost complex structures with Kodaira dimension 0 or $-\infty$.

Example 4.3 (A 6-dimensional Solvmanifold Without Invariant Complex Structures) The construction of the manifold of this example is due to Fernández, de León and Saralegui [15]. Consider the Lie group given by the semi-direct product $G = \mathbb{R}^2 \ltimes_\eta \mathbb{R}^4$. If we denote by (t, x) coordinates on $\mathbb{R}^2$, then $\eta(t, x)$ acts on vectors in $\mathbb{R}^4$ as left-multiplication by the matrix

$$\eta(t,x) = \begin{bmatrix} e^t & 0 & xe^t & 0 \\ & e^{-t} & 0 & xe^{-t} \\ & & e^t & 0 \\ & & & e^{-t} \end{bmatrix}.$$

If we denote by (y_1, y_2, z_1, z_2) coordinates on $\mathbb{R}^4$, a basis of left-invariant 1-forms of G is given by

$$\begin{aligned} &e^1 = dt, \quad e^2 = dx, \quad e^3 = e^{-t}dy^1 - xe^{-t}dz^1, \\ &e^4 = e^t dy^2 - xe^t dz^2, \quad e^5 = e^{-t}dz^1, \quad e^6 = e^t dz^2, \end{aligned}$$

with differentials

$$\begin{aligned} &de^1 = de^2 = 0, \quad de^3 = -e^{13} - e^{25}, \\ &de^4 = e^{14} - e^{26}, \quad de^5 = -e^{15} \quad \text{and} \quad de^6 = e^{16}. \end{aligned}$$

The Lie group G admits a lattice Γ and the quotient manifold $M = \Gamma\backslash G$ has the structure of a T^4-bundle over T^2. A prominent feature of M is that it does not admit integrable invariant complex structures, even though it admits several almost complex structures of splitting type induced by those on G. We refer to [15] for the details. We consider two invariant almost complex structures J_1 and J_2 on M induced by different choices of integrable almost complex structure on $\mathbb{R}^4$. Both define an almost complex structure of splitting type on G whose associated SU(3)-structure is pseudoholomorphic, so that we can apply Theorems 3.1 and 3.2 to obtain families of non-invariant structures. However, the pseudoholomorphic trivialization of K_{J_1} is invariant, while that of K_{J_2} is necessarily non-invariant.

Let J_1 be the invariant almost complex structure on M given by the co-frame of $(1,0)$-forms

$$\phi^1 = e^1 + ie^2, \quad \phi^2 = e^3 + ie^4 \quad \text{and} \quad \phi^3 = e^5 + ie^6,$$

with differentials

$$d\phi^1 = 0, \quad d\phi^2 = -\frac{1}{2}(\phi^{1\bar{2}} + \phi^{\bar{1}\bar{2}}) \quad \text{and} \quad d\phi^3 = -\frac{1}{2}(\phi^{1\bar{3}} + \phi^{\bar{1}\bar{3}}) + \frac{i}{2}(\phi^{12} - \phi^{\bar{1}2}).$$

Thus, we have that $\bar{\partial}\phi^{123} = 0$ and that the standard SU(3)-structure is pseudoholomorphic. By Theorems 3.1 and 3.2, we can find families of non-invariant almost complex structures on M, built starting from J_1, with Kodaira dimension 0 or $-\infty$.

Let J_2 be the invariant almost complex structure on M given by the co-frame of $(1,0)$-forms

$$\phi^1 = e^1 + ie^2, \quad \phi^2 = e^3 + ie^5 \quad \text{and} \quad \phi^3 = e^4 + ie^6,$$

with differentials

$$d\phi^1 = 0, \quad d\phi^2 = -\frac{1}{4}(\phi^{12} + \phi^{1\bar{2}} + 3\phi^{\bar{1}2} - \phi^{\bar{1}\bar{2}}) \quad \text{and}$$

$$d\phi^3 = \frac{1}{4}(3\phi^{13} - \phi^{1\bar{3}} + \phi^{\bar{1}3} - \phi^{\bar{1}\bar{3}}).$$

In this case, we have that $\bar{\partial}\phi^{123} = -\frac{1}{2}\phi^{\bar{1}123}$ and the canonical bundle is not trivialized by invariant forms. However, we now show that it is trivialized by a non-invariant form.

Let $f \in C^\infty(M)$ and consider the equation $\bar{\partial}(f\phi^{123}) = 0$. This is equivalent to the vanishing of the $(0,1)$-form $\alpha = \bar{\partial}f - \frac{1}{2}f\phi^{\bar{1}}$. If we denote by ξ_j a basis of $(1,0)$-vector fields dual to the ϕ^j, then $\alpha = 0$ if and only if f is a solution of the system

$$\begin{cases} \xi_{\bar{1}}(f) = \frac{1}{2}f, \\ \xi_{\bar{2}}(f) = 0, \\ \xi_{\bar{3}}(f) = 0. \end{cases}$$

The last two equations, together with the T^4-bundle structure over T^2 of M, and the ellipticity on the fibers of the operator $\xi_2\xi_{\bar{2}} + \xi_3\xi_{\bar{3}}$, allow to conclude that f depends only on the variables (t, x). By writing $f = u + iv$, where u and v are real functions defined on the basis T^2, and using that $\xi_{\bar{1}} = \frac{1}{2}(\partial_t + i\partial_x)$, we are left with the system of real equations

$$\begin{cases} u_t - v_x - u = 0, \\ u_x + v_t - v = 0. \end{cases} \tag{4.1}$$

Taking derivative with respect to t of the first equation and with respect to x of the second one, then summing the two equations, we get

$$u_{tt} + u_{xx} - u_t - v_x = 0,$$

and, by substituting the value of v_x obtained from the first equation of (4.1), we have the decoupled equation

$$u_{tt} + u_{xx} - 2u_t + u = 0. \tag{4.2}$$

To find a solution, let

$$u = \sum_{m,n\in\mathbb{Z}} u_{mn}\, e^{imt+inx}, \quad u_{mn} \in \mathbb{C},$$

be the Fourier series expansion of u. Equation (4.2) is equivalent to the vanishing of

$$u_{mn}(-m^2 - n^2 - 2im + 1),$$

for all m, n, so that the only non-zero coefficients of the Fourier series are u_{01} and u_{0-1}, giving a solution

$$u = Ae^{ix} + Be^{-ix}, \quad A, B \in C.$$

By the first equation of (4.1), we know that $v_x = -u$ and, by integration, we get

$$v = iAe^{ix} - iBe^{-ix} + c(t),$$

where $c(t)$ is a function depending only on t. Finally, from the second equation of (4.1), we must have

$$0 = u_x + v_t - v = \partial_t(c(t)) - c(t).$$

Since solutions of $c'(t) = c(t)$ are not periodic in t, we have that $c(t) = 0$ and that $f = 2Be^{-ix}$, $B \in \mathbb{C} \setminus \{0\}$, is a never-vanishing solution of the system that provides a pseudoholomorphic trivialization of the canonical bundle.

Example 4.4 (An 8-dimensional Solvmanifold) The construction of the manifold of this example is due to Sferruzza and the second author [23]. Consider the Lie group given by the semi-direct product $G = \mathbb{C} \ltimes_\eta N$, where N is the 3-dimensional complex Heisenberg group. If we denote by (z_1, z_2, z_3) coordinates on N, then $\eta(z_0)$ acts on (z_1, z_2, z_3) as left-multiplication by the matrix

$$\begin{bmatrix} e^{-z_0} & & \\ & e^{z_0} & \\ & & 1 \end{bmatrix}.$$

The almost complex structure of splitting type on G induced by dz^j, $j = 0, \dots, 3$, is integrable and it is given by the co-frame of left-invariant $(1, 0)$-forms

$$\phi^1 = dz^0, \quad \phi^2 = e^{z_0} dz^1, \quad \phi^3 = e^{-z_0} dz^2 \quad \text{and} \quad \phi^4 = dz^3 - \frac{1}{2} z_1 dz^2 + \frac{1}{2} z_2 dz^1,$$

with differentials

$$d\phi^1 = 0, \quad d\phi^2 = \phi^{12}, \quad d\phi^3 = -\phi^{13} \quad \text{and} \quad d\phi^4 = -\phi^{23}.$$

In particular, the associated SU(4)-structure is holomorphic. Since G admits a lattice Γ, the holomorphic SU(4)-structure descends to the quotient $M = \Gamma \backslash G$. We refer to Section 6 in [23] for the construction of the lattice. By Theorems 3.1 and 3.2, there exist families of non-invariant almost complex structures on M with Kodaira dimension 0 or $-\infty$.

Acknowledgments the authors are grateful to the referee for useful comments and remarks.

References

1. A. Andrada, A. Tolcachier. On the canonical bundle of complex solvmanifolds and applications to hypercomplex geometry. Transform. Groups (2024). https://doi.org/10.1007/s00031-024-09866-z
2. D. Angella, A. Otal, L. Ugarte, R. Villacampa. Complex structures of splitting type. Rev. Mat. Iberoam. **33**(4), 1309–1350 (2017)
3. M.L. Barberis, I.G. Dotti, M. Verbitsky, Canonical bundles of complex nilmanifolds, with applications to hypercomplex geometry. Math. Res. Lett. **16**(2), 331–347 (2009)
4. R.L. Bryant, Metrics with exceptional holonomy. Ann. Math. (2) **126**(3), 525–576 (1987)
5. A. Cattaneo, Almost complex manifolds from the point of view of Kodaira dimension. Riv. Mat. Univ. Parma (N.S.) **13**(2), 535–549 (2022)
6. A. Cattaneo, A. Nannicini, A. Tomassini, Kodaira dimension of almost Kähler manifolds and curvature of the canonical connection. Ann. Mat. Pura Appl. (4) **199**(5), 1815–1842 (2020)
7. A. Cattaneo, A. Nannicini, A. Tomassini, On Kodaira dimension of almost complex 4-dimensional solvmanifolds without complex structures. Int. J. Math. **32**(10), Paper no. 2150075, 41 pp. (2021)
8. A. Cattaneo, A. Nannicini, A. Tomassini, Almost complex parallelizable manifolds: Kodaira dimension and special structures. Manuscripta Math. **173**(3–4), 1123–1145 (2024)

9. H. Chen, W. Zhang, Kodaira dimensions of almost complex manifolds, I. Am. J. Math. **145**(2), 477–514 (2023)
10. H. Chen, W. Zhang, Kodaira dimensions of almost complex manifolds II. Commun. Anal. Geom. **32**(3), 751–790 (2024)
11. J. Cirici, S.O. Wilson, Dolbeault cohomology for almost complex manifolds. Adv. Math. **391**, Paper no. 107970, 52 pp. (2021)
12. P. de Bartolomeis, G. Tian, Stability of complex vector bundles. J. Differ. Geom. **43**(2), 231–275 (1996)
13. P. de Bartolomeis, A. Tomassini, On the Maslov index of Lagrangian submanifolds of generalized Calabi-Yau manifolds. Int. J. Math. **17**(8), 921–947 (2006)
14. P. de Bartolomeis, A. Tomassini, Exotic deformations of Calabi-Yau manifolds. Ann. Inst. Fourier (Grenoble) **63**(2), 391–415 (2013)
15. M. Fernández, M. de León, M. Saralegui, A six-dimensional compact symplectic solvmanifold without Kähler structures. Osaka J. Math. **33**(1), 19–35 (1996)
16. A. Fino, A. Otal, L. Ugarte, Six-dimensional solvmanifolds with holomorphically trivial canonical bundle. Int. Math. Res. Not. IMRN **2015**(24), 13757–13799 (2015)
17. T. Holt, R. Piovani, Left-invariant almost complex structures on the higher dimensional Kodaira-Thurston manifolds. Ann. Global Anal. Geom. **66**(1), Paper No. 2, 19 pp. (2024)
18. D. Huybrechts, *Complex Geometry: An Introduction*. Universitext (Springer-Verlag, Berlin, 2005)
19. H. Kasuya, Techniques of computations of Dolbeault cohomology of solvmanifolds. Math. Z. **273**(1–2), 437–447 (2013)
20. I. Nakamura, Complex parallelisable manifolds and their small deformations. J. Differ. Geom. **10**(1), 85–112 (1975)
21. A. Otal, L. Ugarte, Six dimensional homogeneous spaces with holomorphically trivial canonical bundle. J. Geom. Phys. **194**, Paper no. 105014, 27 pp. (2023)
22. S. Salamon, *Riemannian Geometry and Holonomy Groups*. Pitman Research Notes in Mathematics Series, vol. 201 (Longman Scientific & Technical, Harlow; copublished in the United States with John Wiley & Sons, Inc., New York, 1989)
23. T. Sferruzza, A. Tomassini, Bott–Chern formality and massey products on strong kähler with torsion and kähler solvmanifolds. J. Geom. Anal. **34**(11), Paper No. 348, 42 pp. (2024)
24. A. Tolcachier, Six-dimensional complex solvmanifolds with non-invariant trivializing sections of their canonical bundle. Preprint (2024). arXiv:2412.02325 http://arxiv.org/abs/2412.02325v1
25. H.-C. Wang, Complex parallelisable manifolds. Proc. Am. Math. Soc. **5**, 771–776 (1954)

A Cheng-Yau Type Estimate for the Symplectic Calabi-Yau Equation

Valentino Tosatti

To Paul Gauduchon on the occasion of his 80th birthday

Abstract In the setting of Donaldson's conjecture on the Calabi-Yau equation on symplectic 4-manifolds, we prove an a priori estimate which in the Kähler case resembles a classical estimate of Cheng-Yau.

Keywords Symplectic Calabi-Yau equation · Cheng-Yau estimate · Symplectic 4-manifolds

1 Introduction

The Calabi Conjecture, solved by Yau in 1976 [20], says that given a smooth positive volume form σ on a compact Kähler manifold (M, ω) with total mass $\int_M \sigma = \int_M \omega^n$, we can find a unique Kähler metric $\tilde{\omega}$ in the same cohomology class as ω whose volume element is pointwise equal to σ, i.e $\tilde{\omega}^n = \sigma$. The uniqueness statement was proved by Calabi in 1954 [3], and the existence follows via a continuity method from a priori C^∞ estimates for a Kähler metric which only depend on its volume form, its cohomology class, and the underlying complex manifold.

In 2006, Donaldson [6] proposed a conjectural generalization of this result to symplectic 4-manifolds. More precisely, suppose (M^4, J) is a closed almost-complex manifold and ω is a symplectic form taming J, which means that

$$\omega(X, JX) > 0, \quad \text{for all } X \neq 0.$$

V. Tosatti (✉)
Courant Institute of Mathematical Sciences, New York University, New York, NY, USA
e-mail: tosatti@cims.nyu.edu

L. Ornea (ed.), *Real and Complex Geometry*,
https://doi.org/10.1007/978-3-031-92297-8_15

The $(1, 1)$-part of ω with respect to J is denoted by $\omega^{(1,1)}$. Let $\tilde{\omega}$ be another symplectic form, cohomologous to ω and compatible with J, which means that it tames J and furthermore

$$\tilde{\omega}(X, Y) = \tilde{\omega}(JX, JY).$$

There are associated Hermitian metrics $g, \tilde{g}$ which are defined by

$$g(X, Y) = \frac{1}{2}\left(\omega(X, JY) + \omega(Y, JX)\right), \quad \tilde{g}(X, Y) = \tilde{\omega}(X, JY)$$

with corresponding volume forms $(\omega^{(1,1)})^2$ and $\tilde{\omega}^2$. Donaldson then conjectured the following:

Conjecture 1.1 Let (M^4, J) be a closed almost-complex 4-manifold, ω a symplectic form taming J, $\tilde{\omega}$ a cohomologous symplectic form compatible with J, and σ a smooth positive volume form with $\int_M \sigma = \int_M \omega^2$. If $\tilde{\omega}$ satisfies the Calabi-Yau equation

$$\tilde{\omega}^2 = \sigma, \tag{1.1}$$

then for any $k \geqslant 0$, we can bound $\|\tilde{\omega}\|_{C^k(M,g)}$ by a constant that depends only on k and on bounds on σ, ω, (M, J).

When J is integrable and ω is compatible with J, we are on a Kähler surface and Conjecture 1.1 follows from the aforementioned theorem of Yau [20]. If solved, Donaldson's conjecture would have striking applications in symplectic geometry, see [6, 14]. Donaldson's conjecture was first investigated by Weinkove [19] in the case when ω is also compatible with J, where he showed that it holds provided J is close to being integrable. In [17], Tosatti-Weinkove-Yau showed that Conjecture 1.1 in general would follow if one could prove a bound

$$\mathrm{tr}_g \tilde{g} \leqslant C, \tag{1.2}$$

where C depends only on σ, ω, (M, J) (we will call such constants uniform), and that (1.2) can indeed be established when g has nonnegative curvature in a suitable sense. Further progress on Donaldson's conjecture has since proceeded in two main directions: proving it on explicit examples [1, 2, 7, 13, 16, 18], and for general manifolds reducing the bound (1.2) to bounding an "almost-Kähler potential" function. For this, following [17, 19], one defines a function $\varphi \in C^\infty(M, \mathbb{R})$ by

$$\Delta_{\tilde{g}} \varphi = 2 - \mathrm{tr}_{\tilde{g}} g, \quad \sup_M \varphi = 0, \tag{1.3}$$

which is uniquely determined and in the Kähler case would satisfy $\tilde{\omega} = \omega + i\partial\overline{\partial}\varphi$, i.e. φ would be a familiar Kähler potential. Estimate (1.2) was then successively reduced to proving uniform bounds for $\int_M e^{-\alpha\varphi}\omega^2$ (for some $\alpha > 0$) in [17] in

2007, for $\int_M |\varphi|\omega^2$ in [12, Rmk 3.1] in 2009 (with a very recent new proof in [8]), for *any* integral bound for φ in [16] in 2016, and lastly to a uniform positive lower bound for the Lebesgue measure of $\{\varphi \geqslant -C\}$ for any given uniform C in [16].

Nevertheless, none of these results use the crucial assumption that $\tilde{\omega}$ and ω are cohomologous: if this is not the case then one has to replace the constant 2 in (1.3) with another suitable constant, namely

$$\Delta_{\tilde{g}}\varphi = \frac{2\int_M \omega \wedge \tilde{\omega}}{\int_M \tilde{\omega}^2} - \mathrm{tr}_{\tilde{g}} g, \quad \sup_M \varphi = 0, \tag{1.4}$$

but there is no essential difference in all the above results, which still show that to obtain uniform C^∞ bounds for $\tilde{\omega}$ (and a uniform lower bound $\tilde{\omega} \geqslant C^{-1}\omega$) it suffices to have for example any uniform integral bound for φ (or just a uniform lower bound for the measure of a superlevel set), provided the constant $\frac{2\int_M \omega\wedge\tilde{\omega}}{\int_M \tilde{\omega}^2}$ is uniformly bounded above. Consider the Kähler case, where J is integrable and ω is compatible with it, and degenerate $[\tilde{\omega}]$ to a limiting class $[\alpha]$ on the boundary of the Kähler cone with $\int_M \alpha^2 > 0$ (there are many such examples, see e.g. [11]). In this case the constant in (1.4) remains uniformly bounded, but the corresponding solution metrics $\tilde{\omega}$ cannot converge smoothly since $[\alpha]$ does not contain any Kähler metric, hence in this case no uniform integral/measure bound for φ solving (1.4) can hold. In other words, if one wants to use the almost-Kähler potential φ to prove (1.2), then some new idea has to be used which makes crucial use of the cohomological assumption.

In this note we take a different approach. Because of the cohomological assumption, we can write

$$\tilde{\omega} = \omega + da, \tag{1.5}$$

for some real 1-form a. We are free to modify a by adding df for any smooth function f, and if we choose f solving the elliptic equation $d^*_{\tilde{g}}df = -d^*_{\tilde{g}}a$, we can then assume without loss that

$$d^*_{\tilde{g}}a = 0. \tag{1.6}$$

The 1-form a satisfying (1.5) and (1.6) is then uniquely determined modulo the addition of a $\tilde{g}$-harmonic 1-form, and we can fix this ambiguity for example by requiring that a be $L^2(\tilde{g})$-orthogonal to the space of such forms. The Calabi-Yau equation (1.1) is then a nonlinear elliptic system for the 1-form a (when $\dim M = 4$). Our main result is then the following:

Theorem 1.2 *In the above setting, we have the estimate*

$$\mathrm{tr}_g \tilde{g} \leqslant C(1 + \sup_M |a|^2_{\tilde{g}}), \tag{1.7}$$

where C is a uniform constant.

We make a few remarks about this result.

1. It is important for our arguments that the gauge-fixing condition (1.6) and the norm on the RHS of (1.7) are both with respect to $\tilde{g}$ as opposed to g. Changing (1.6) to $d^*_g a = 0$ does not seem completely out of the question, but replacing $|a|^2_{\tilde{g}}$ with $|a|^2_g$ in (1.7) does not seem feasible.
2. In the Kähler case, i.e. when J is integrable and ω is compatible with J, we actually have that $a = d^c\varphi$ where φ is the usual Kähler potential so that $\tilde{\omega} = \omega + i\partial\overline{\partial}\varphi$. Indeed $dd^c\varphi = i\partial\overline{\partial}\varphi$, and

$$d^*_{\tilde{g}} d^c\varphi = - *_{\tilde{g}} d *_{\tilde{g}} d^c\varphi = *_{\tilde{g}} d(d\varphi \wedge \tilde{\omega}) = 0,$$

and for any $\tilde{g}$-harmonic 1-form α,

$$\int_M d^c\varphi \wedge *_{\tilde{g}}\alpha = \int_M d^c\varphi \wedge J(\alpha) \wedge \tilde{\omega} = \int_M d\varphi \wedge \alpha \wedge \tilde{\omega} = 0.$$

Thus, in the Kähler case, estimate (1.7) becomes

$$\mathrm{tr}_g \tilde{g} \leqslant C(1 + \sup_M |\nabla\varphi|^2_{\tilde{g}}), \tag{1.8}$$

which in the local setting of pseudoconvex domains in $\mathbb{C}^n$ was proved by Cheng-Yau [4, Proposition 7.1]. Tracing through our arguments easily shows that in the Kähler case (1.8) holds in all dimensions, and it does not use any knowledge about L^∞ a priori bounds for φ (which were famously proved by Yau [20] using Moser iteration).
3. An estimate of the different but related form

$$\mathrm{tr}_g \tilde{g} \leqslant C(1 + \sup_M |\nabla\varphi|^2_g), \tag{1.9}$$

is known for different geometric PDEs on Kähler manifolds, including for example the complex Hessian equation [9] and the Monge-Ampère equation for $(n-1)$-psh functions [15], see also [10]. In these settings, (1.9) is then used together with a blowup argument and a Liouville theorem [5, 10, 15] to show that $\sup_M |\nabla\varphi|^2_g \leqslant C$ and hence $\mathrm{tr}_g \tilde{g} \leqslant C$. However, the fact that $\tilde{g}$ appears on the RHS of (1.7) instead of g makes it ill-suited to blowup arguments.

The proof of Theorem 1.2 is done by a maximum principle argument, and interestingly it uses crucially that the dimension of M is 4 (while the aforementioned results in [16, 17] apply in all even dimensions $2n \geqslant 4$), except in the Kähler case where our argument works in all dimensions.

2 Proof of Theorem 1.2

We work in the setting described in the Introduction. Define a smooth function F on M by

$$\sigma = e^F (\omega^{(1,1)})^2,$$

so that in any local chart the Calabi-Yau equation (1.1) can be written as

$$\det(\tilde{g}) = e^F \det(g). \tag{2.1}$$

Since $g, \tilde{g}$ are both Hermitian with respect to J, and $\dim M = 4$, it follows from (2.1) that we have

$$\mathrm{tr}_{\tilde{g}} g = e^{-F} \mathrm{tr}_g \tilde{g}, \tag{2.2}$$

which we will use repeatedly, often without mention.

As in [17], we will use covariant derivatives with respect the Chern connection ∇ of g (also known as "canonical connection", see [17, §2]). We recall two estimates proved in [17]. The first one is the differential inequality from [17] (which can be extracted from the proof of Lemma 3.2 there)

$$\Delta_{\tilde{g}} \mathrm{tr}_g \tilde{g} \geqslant |\nabla \tilde{g}|^2_{g,\tilde{g}} - C \mathrm{tr}_g \tilde{g}\, \mathrm{tr}_{\tilde{g}} g - C \geqslant |\nabla \tilde{g}|^2_{g,\tilde{g}} - C(\mathrm{tr}_g \tilde{g})^2 - C, \tag{2.3}$$

using (2.2), where the norm $|\cdot|^2_{g,\tilde{g}}$ uses both g and $\tilde{g}$, see (2.22) below for the definition. The second one is the following Cauchy-Schwarz type inequality [17, (3.20)]

$$|\nabla \tilde{g}|^2_{g,\tilde{g}} \geqslant \frac{|\nabla \mathrm{tr}_g \tilde{g}|^2_{\tilde{g}}}{\mathrm{tr}_g \tilde{g}}. \tag{2.4}$$

Given these preliminaries, the main claim is then the following:

Proposition 2.1 *There is a uniform $C > 0$ such that for any small $\varepsilon > 0$ we have the differential inequality*

$$\Delta_{\tilde{g}} |a|^2_{\tilde{g}} \geqslant \frac{1}{2} |\tilde{\nabla} a|^2_{\tilde{g}} + \frac{1}{2} |\overline{\tilde{\nabla}} a|^2_{\tilde{g}} + \frac{1}{C} (\mathrm{tr}_g \tilde{g})^2 - \varepsilon |\nabla \tilde{g}|^2_{g,\tilde{g}} - \frac{C}{\varepsilon^2} |a|^4_{\tilde{g}} - C. \tag{2.5}$$

In fact, as follows from the proof, the constants $\frac{1}{2}$ on the RHS of (2.5) can be taken to be $1 - \delta$ for any given $\delta > 0$, at the expense of making C larger.

Proof of Theorem 1.2 Assuming Proposition 2.1 Let us first assume that Proposition 2.1 holds, and use it to prove (1.7). For this, combining (2.3) and (2.5), and throwing away two positive terms, we have

$$\begin{aligned}\Delta_{\tilde{g}}(\mathrm{tr}_g\tilde{g} + A|a|^2_{\tilde{g}}) \geqslant |\nabla\tilde{g}|^2_{g,\tilde{g}} - C_0(\mathrm{tr}_g\tilde{g})^2 - C \\ + \frac{A}{C_1}(\mathrm{tr}_g\tilde{g})^2 - \varepsilon A|\nabla\tilde{g}|^2_{g,\tilde{g}} - \frac{CA}{\varepsilon^2}|a|^4_{\tilde{g}} - CA,\end{aligned}$$

where C_0, C_1 are uniform constants, and now if we choose $A = C_1(C_0 + 1)$ and $\varepsilon = \frac{1}{A}$, we obtain

$$\Delta_{\tilde{g}}(\mathrm{tr}_g\tilde{g} + A|a|^2_{\tilde{g}}) \geqslant (\mathrm{tr}_g\tilde{g})^2 - C|a|^4_{\tilde{g}} - C,$$

and the maximum principle concludes the proof of (1.7).

Beginning of Proof of Proposition 2.1 We now proceed to prove Proposition 2.1. As in [17] we also use the Chern connection $\tilde{\nabla}$ of $\tilde{g}$, and the formalism of moving frames. Thus, we work in a local $\tilde{g}$-unitary frame $\tilde{\theta}^i, i = 1, 2$, which are $(1, 0)$-forms, and we have the first structure equations

$$d\tilde{\theta}^i = -\tilde{\theta}^i_j \wedge \tilde{\theta}^j + \tilde{\Theta}^i,$$

where $\tilde{\theta}^i_j$ are the connection 1-forms and $\tilde{\Theta}^i$ are the torsion forms, which are of type $(0, 2)$ and equal to

$$\tilde{\Theta}^i = \tilde{N}^i_{\overline{jk}}\overline{\tilde{\theta}^j} \wedge \overline{\tilde{\theta}^k},$$

where $\tilde{N}^i_{\overline{jk}}$ are the components of the Nijenhuis tensor of J (and are skew-symmetric in j, k). The second structure equations read

$$d\tilde{\theta}^i_j = -\tilde{\theta}^i_k \wedge \tilde{\theta}^k_j + \tilde{\Omega}^i_j,$$

where $\tilde{\Omega}^i_j$ are the curvature 2-forms. They can be written as

$$\tilde{\Omega}^i_j = \tilde{R}^i_{jk\bar{\ell}}\tilde{\theta}^k \wedge \overline{\tilde{\theta}^\ell} + \tilde{K}^i_{jk\ell}\tilde{\theta}^k \wedge \tilde{\theta}^\ell + \tilde{K}^i_{j\overline{k\ell}}\overline{\tilde{\theta}^k} \wedge \overline{\tilde{\theta}^\ell},$$

and the Ricci curvature form is defined to be

$$\mathrm{Ric}(\tilde{g}) = \frac{\sqrt{-1}}{2\pi}\tilde{\Omega}^i_i,$$

where here and in the following we will mostly omit summation signs. It is well-known that $\mathrm{Ric}(\tilde{g})$ is a closed real 2-form which represents the first Chern class

$c_1(M, J)$ of the complex vector bundle $(T^{\mathbb{R}}M, J)$. Its $(1,1)$-part is given by

$$\frac{\sqrt{-1}}{2\pi}\tilde{R}_{k\bar{\ell}}\tilde{\theta}^k \wedge \overline{\tilde{\theta}^\ell}, \quad \tilde{R}_{k\bar{\ell}} = \tilde{R}^i_{\ ik\bar{\ell}}.$$

Its relation with the Calabi-Yau equation is this [17, (3.16)]: if $g, \hat{g}$ are two J-Hermitian metrics then we have

$$\frac{1}{2\pi}dd^c \log\frac{\det(g)}{\det(\hat{g})} = \mathrm{Ric}(\hat{g}) - \mathrm{Ric}(g), \tag{2.6}$$

where for a real-valued function f we define $(1,0)$ and $(0,1)$ forms $\partial f, \overline{\partial} f$ by $df = \partial f + \overline{\partial} f$, and define $d^c f = \frac{\sqrt{-1}}{2}(\overline{\partial} f - \partial f)$. A different sign convention was used in [17], so the term $dd^c f$ here is equal to the term $-\frac{1}{2}d(Jdf)$ in [17]. The trace of the $(1,1)$-part of $dd^c f$ with respect to $\tilde{g}$ equals its Laplacian $\Delta_{\tilde{g}} f$, see [17, Lemma 2.5], which is equal to the usual Laplace-Beltrami operator of $\tilde{g}$ (up to a factor of $\frac{1}{2}$). It can be written as

$$\Delta_{\tilde{g}} f = 2\frac{\tilde{\omega} \wedge dd^c f}{\tilde{\omega}^2}.$$

If we have a tensor T then we can consider its covariant derivatives $\tilde{\nabla} T$ and $\overline{\tilde{\nabla}} T$. For example, if in our frame T has components $\tilde{T}^i_{j\bar{k}}$ (as an example) then its covariant derivatives $\tilde{\nabla} T$ and $\overline{\tilde{\nabla}} T$ have components $\tilde{T}^i_{j\bar{k},p}$ and $\tilde{T}^i_{j\bar{k},\bar{p}}$ respectively, which can be obtained as follows:

$$\tilde{T}^i_{j\bar{k},p}\tilde{\theta}^p + \tilde{T}^i_{j\bar{k},\bar{p}}\overline{\tilde{\theta}^p} = d\tilde{T}^i_{j\bar{k}} + \tilde{T}^q_{j\bar{k}}\tilde{\theta}^i_q - \tilde{T}^i_{q\bar{k}}\tilde{\theta}^q_j - \tilde{T}^i_{j\bar{q}}\overline{\tilde{\theta}^q_k},$$

and similarly for tensors of other types. We can easily compute [17, (2.28)] that if f is a real-valued function then we have

$$dd^c f = \tilde{f}_{\bar{i}}\overline{\tilde{\Theta}^i} + \tilde{f}_{i\bar{j}}\tilde{\theta}^i \wedge \overline{\tilde{\theta}^j} + \tilde{f}_i\tilde{\Theta}^i. \tag{2.7}$$

To simplify computations, denote by $\alpha = a^{(1,0)}$, so that

$$a = \alpha + \overline{\alpha},$$

and we have

$$|a|^2_{\tilde{g}} = 2|\alpha|^2_{\tilde{g}},$$

where we are using $\tilde{g}$ here as a Riemannian metric on $T^{\mathbb{R}}M$ on the LHS and as a Hermitian metric on $T^{1,0}M$ on the RHS, and the equality follows from the fact that the Riemannian metric is J-invariant. In our frame we can write $\alpha = \tilde{\alpha}_i \tilde{\theta}^i$, so that

$$|\alpha|^2_{\tilde{g}} = |\tilde{\alpha}_i|^2.$$

We then compute

$$d|\alpha|^2_{\tilde{g}} = \tilde{\alpha}_{i,j}\overline{\tilde{\alpha}_i}\tilde{\theta}^j + \tilde{\alpha}_{i,\bar{j}}\overline{\tilde{\alpha}_i}\overline{\tilde{\theta}^j} + \tilde{\alpha}_i\overline{\tilde{\alpha}_{i,\bar{j}}}\tilde{\theta}^j + \tilde{\alpha}_i\overline{\tilde{\alpha}_{i,j}}\overline{\tilde{\theta}^j},$$

hence

$$\partial|\alpha|^2_{\tilde{g}} = \tilde{\alpha}_{i,j}\overline{\tilde{\alpha}_i}\tilde{\theta}^j + \tilde{\alpha}_i\overline{\tilde{\alpha}_{i,\bar{j}}}\tilde{\theta}^j,$$

$$(d\partial|\alpha|^2_{\tilde{g}})^{(1,1)} = (\tilde{\alpha}_{i,j\bar{k}}\overline{\tilde{\alpha}_i} + \tilde{\alpha}_{i,j}\overline{\tilde{\alpha}_{i,k}} + \tilde{\alpha}_{i,\bar{k}}\overline{\tilde{\alpha}_{i,\bar{j}}} + \tilde{\alpha}_i\overline{\tilde{\alpha}_{i,\bar{j}k}})\overline{\tilde{\theta}^k} \wedge \tilde{\theta}^j$$

and since $dd^c|\alpha|^2_{\tilde{g}} = -\sqrt{-1}d\partial|\alpha|^2_{\tilde{g}}$, this implies that

$$\Delta_{\tilde{g}}|\alpha|^2_{\tilde{g}} = |\tilde{\alpha}_{i,j}|^2 + |\tilde{\alpha}_{i,\bar{j}}|^2 + \tilde{\alpha}_{i,k\bar{k}}\overline{\tilde{\alpha}_i} + \tilde{\alpha}_i\overline{\tilde{\alpha}_{i,\bar{k}k}}. \tag{2.8}$$

Next, we need a commutation relation for covariant derivatives of α. We start with the definition

$$d\tilde{\alpha}_i = \tilde{\alpha}_j\tilde{\theta}^j_i + \tilde{\alpha}_{i,j}\tilde{\theta}^j + \tilde{\alpha}_{i,\bar{j}}\overline{\tilde{\theta}^j},$$

and applying d again

$$\begin{aligned}0 = dd\tilde{\alpha}_i &= \tilde{\alpha}_j\tilde{\Omega}^j_i + \tilde{\alpha}_{i,jp}\tilde{\theta}^p \wedge \tilde{\theta}^j + \tilde{\alpha}_{i,j\bar{p}}\overline{\tilde{\theta}^p} \wedge \tilde{\theta}^j + \tilde{\alpha}_{i,j}\tilde{\Theta}^j + \tilde{\alpha}_{i,\bar{j}p}\tilde{\theta}^p \wedge \overline{\tilde{\theta}^j} \\ &+ \tilde{\alpha}_{i,\bar{j}\bar{p}}\overline{\tilde{\theta}^p} \wedge \overline{\tilde{\theta}^j} + \tilde{\alpha}_{i,\bar{j}}\overline{\tilde{\Theta}^j},\end{aligned} \tag{2.9}$$

and the (1, 1)-part of (2.9) says

$$\tilde{\alpha}_j\tilde{R}^j_{ik\bar{\ell}}\tilde{\theta}^k \wedge \overline{\tilde{\theta}^\ell} - \tilde{\alpha}_{i,k\bar{\ell}}\tilde{\theta}^k \wedge \overline{\tilde{\theta}^\ell} + \tilde{\alpha}_{i,\bar{\ell}k}\tilde{\theta}^k \wedge \overline{\tilde{\theta}^\ell} = 0,$$

and so we obtain the commutation relation

$$\tilde{\alpha}_{i,k\bar{\ell}} = \tilde{\alpha}_{i,\bar{\ell}k} + \tilde{\alpha}_j\tilde{R}^j_{ik\bar{\ell}}, \tag{2.10}$$

exactly like in the Kähler case. Taking the (0, 2)-part of (2.9) gives

$$\tilde{\alpha}_j\tilde{K}^j_{i\bar{k}\bar{\ell}}\overline{\tilde{\theta}^k} \wedge \overline{\tilde{\theta}^\ell} + \tilde{\alpha}_{i,j}\tilde{N}^j_{\bar{k}\bar{\ell}}\overline{\tilde{\theta}^k} \wedge \overline{\tilde{\theta}^\ell} + \tilde{\alpha}_{i,\bar{\ell}\bar{k}}\overline{\tilde{\theta}^k} \wedge \overline{\tilde{\theta}^\ell} = 0,$$

which, after skew-symmetrizing $\tilde{\alpha}_{i,\overline{\ell k}}$ in k, ℓ, gives the commutation relation

$$\tilde{\alpha}_{i,\overline{\ell k}} = \tilde{\alpha}_{i,\overline{k\ell}} - 2\tilde{\alpha}_j \tilde{K}^j_{\overline{ik\ell}} - 2\tilde{\alpha}_{i,j}\tilde{N}^j_{\overline{k\ell}}. \tag{2.11}$$

Using (2.10) in (2.8) gives

$$\Delta_{\tilde{g}}|\alpha|^2_{\tilde{g}} = |\tilde{\alpha}_{i,j}|^2 + |\tilde{\alpha}_{i,\bar{j}}|^2 + 2\mathrm{Re}(\tilde{\alpha}_{i,\bar{k}k}\overline{\tilde{\alpha}_i}) + \tilde{\alpha}_j\overline{\tilde{\alpha}_i}\tilde{R}^j_{ik\bar{k}},$$

and recalling from [17, (2.21)] that

$$\tilde{R}^j_{ik\bar{k}} = \tilde{R}_{i\bar{j}} - 4\tilde{N}^q_{\bar{p}\bar{j}}\overline{\tilde{N}^p_{\bar{q}\bar{i}}} - 4\tilde{N}^p_{\bar{q}\bar{j}}\overline{\tilde{N}^i_{\bar{p}\bar{q}}}, \tag{2.12}$$

we obtain

$$\Delta_{\tilde{g}}|\alpha|^2_{\tilde{g}} = |\tilde{\alpha}_{i,j}|^2 + |\tilde{\alpha}_{i,\bar{j}}|^2 + 2\mathrm{Re}(\tilde{\alpha}_{i,\bar{k}k}\overline{\tilde{\alpha}_i}) + \tilde{R}_{i\bar{j}}\tilde{\alpha}_j\overline{\tilde{\alpha}_i} - 4\tilde{\alpha}_j\overline{\tilde{\alpha}_i}\tilde{N}^q_{\bar{p}\bar{j}}\overline{\tilde{N}^p_{\bar{q}\bar{i}}} - 4\tilde{\alpha}_j\overline{\tilde{\alpha}_i}\tilde{N}^p_{\bar{q}\bar{j}}\overline{\tilde{N}^i_{\bar{p}\bar{q}}}.$$

To deal with the term $\tilde{R}_{i\bar{j}}\tilde{\alpha}_j\overline{\tilde{\alpha}_i}$, observe that differentiating the PDE (2.1) and using (2.6) we have that $\mathrm{Ric}(\tilde{g}) = \mathrm{Ric}(g) - \frac{1}{2\pi}dd^cF$, which is a fixed background tensor (independent of $\tilde{g}$), hence we can write $\tilde{R}_{i\bar{j}} = \tilde{T}_{i\bar{j}}$ where $T = \sqrt{-1}\tilde{T}_{i\bar{j}}\tilde{\theta}^i \wedge \overline{\tilde{\theta}^j}$ is some fixed tensor. This implies that

$$\tilde{R}_{i\bar{j}}\tilde{\alpha}_j\overline{\tilde{\alpha}_i} \geqslant -|T|_{\tilde{g}}|\alpha|^2_{\tilde{g}} \geqslant -C\mathrm{tr}_g\tilde{g}|\alpha|^2_{\tilde{g}}.$$

Likewise, if we let now $\tilde{T}_{i\bar{j}} = \tilde{N}^q_{\bar{p}\bar{j}}\overline{\tilde{N}^p_{\bar{q}\bar{i}}}$, then $T = \sqrt{-1}\tilde{T}_{i\bar{j}}\tilde{\theta}^i \wedge \overline{\tilde{\theta}^j}$ is some fixed tensor (independent of $\tilde{g}$), and so by the same logic we can bound

$$-4\tilde{\alpha}_j\overline{\tilde{\alpha}_i}\tilde{N}^q_{\bar{p}\bar{j}}\overline{\tilde{N}^p_{\bar{q}\bar{i}}} \geqslant -C\mathrm{tr}_g\tilde{g}|\alpha|^2_{\tilde{g}}.$$

Unfortunately this doesn't work directly for the last term

$$-4\tilde{\alpha}_j\overline{\tilde{\alpha}_i}\tilde{N}^p_{\bar{q}\bar{j}}\overline{\tilde{N}^i_{\bar{p}\bar{q}}}, \tag{2.13}$$

since the corresponding tensor $T = \sqrt{-1}\tilde{N}^p_{\bar{q}\bar{j}}\overline{\tilde{N}^i_{\bar{p}\bar{q}}}\tilde{\theta}^i \wedge \overline{\tilde{\theta}^j}$ depends also on $\tilde{g}$. To understand more precisely how to bound this, denote by θ^i a local unitary frame for the background metric g, so that we can write

$$\tilde{\theta}^i = a^i_j\theta^j, \quad \theta^i = b^i_j\tilde{\theta}^j,$$

where the local matrices (a^i_j) and (b^i_j) are inverses of each other, i.e. $a^i_j b^k_i = \delta_{jk}$. We can express the components of the Nijenhuis tensor with respect to the frame θ^i as

$$N^i_{\overline{jk}} = \tilde{N}^p_{\overline{qr}} \overline{a^q_j} \overline{a^r_k} b^i_p, \quad \text{hence } \tilde{N}^i_{\overline{jk}} = N^p_{\overline{qr}} \overline{b^q_j} \overline{b^r_k} a^i_p,$$

and the components $N^i_{\overline{jk}}$ are all uniformly bounded. Similarly the components of $(1,0)$-form α are given by $\alpha_i = a^j_i \tilde{\alpha}_j$, and so we can write

$$4\tilde{\alpha}_j \overline{\tilde{\alpha}_i} \tilde{N}^p_{\overline{qj}} \overline{\tilde{N}^i_{\overline{pq}}} = 4\alpha_w \overline{\alpha_h} b^w_j \overline{b^h_i} N^k_{\overline{\ell r}} \overline{N^t_{\overline{uv}}} \overline{b^\ell_q} \overline{b^r_j} a^p_k b^u_p b^v_q \overline{a^i_t} = 4\alpha_w \overline{\alpha_h} b^w_j N^k_{\overline{\ell r}} \overline{N^h_{\overline{kv}}} \overline{b^\ell_q} \overline{b^r_j} b^v_q.$$

and working at an arbitrary point we may choose our unitary frames so that at this point we have

$$a^i_j = \sqrt{\lambda_i}\delta_{ij},$$

where $\lambda_1, \lambda_2 > 0$ are the eigenvalues of the Hermitian metric $\tilde{g}$ with respect to g. This implies that

$$b^i_j = \frac{1}{\sqrt{\lambda_j}}\delta_{ij},$$

and so at our point our term (2.13) simplifies to

$$-4\sum_{j,q=1}^{2} \lambda_j^{-1}\lambda_q^{-1}\alpha_j \overline{\alpha_h} N^k_{\overline{qj}} \overline{N^h_{\overline{kq}}},$$

and since $N^k_{\overline{qj}}$ is skew-symmetric in q, j, only the terms in the sum with $j \neq q$ survive, and so this equals

$$-4\lambda_1^{-1}\lambda_2^{-1}\alpha_1 \overline{\alpha_h} N^k_{\overline{21}} \overline{N^h_{\overline{k2}}} - 4\lambda_2^{-1}\lambda_1^{-1}\alpha_2 \overline{\alpha_h} N^k_{\overline{12}} \overline{N^h_{\overline{k1}}},$$

but from the Calabi-Yau equation (1.1) we have that

$$\lambda_1\lambda_2 = e^F, \tag{2.14}$$

which is uniformly bounded, and so our term (2.13) can be bounded below by

$$-C\sum_h |\alpha_h|^2 = -C|\alpha|^2_g \geqslant -C\mathrm{tr}_g\tilde{g}|\alpha|^2_{\tilde{g}},$$

and combining all of the above gives

$$\Delta_{\tilde{g}}|\alpha|^2_{\tilde{g}} \geqslant |\tilde{\alpha}_{i,j}|^2 + |\tilde{\alpha}_{i,\bar{j}}|^2 + 2\mathrm{Re}(\tilde{\alpha}_{i,\bar{k}k}\overline{\tilde{\alpha}_i}) - C\mathrm{tr}_g\tilde{g}|\alpha|^2_{\tilde{g}}.$$

Observe that here we have used crucially that $\dim M = 4$ so that there are only two eigenvalues. This fact will be also used further below.

The Main Claim (2.15) If we denote by $\Delta^H_{\tilde{g}} a = -(dd^*_{\tilde{g}} a + d^*_{\tilde{g}} da)$ the $\tilde{g}$-Hodge Laplacian of $a = \alpha + \overline{\alpha}$, then the main claim is that

$$2\mathrm{Re}(\tilde{\alpha}_{i,\bar{k}k}\overline{\tilde{\alpha}_i}) \geqslant \frac{1}{2}\tilde{g}(\Delta^H_{\tilde{g}} a, a) - \frac{1}{2}|\tilde{\alpha}_{i,j}|^2 - C\mathrm{tr}_g\tilde{g}|\alpha|^2_{\tilde{g}} - C|\nabla\tilde{g}|_{g,\tilde{g}}|\alpha|^2_{\tilde{g}}. \quad (2.15)$$

End of Proof of Proposition 2.1 Assuming (2.15) Assuming that (2.15) holds, let us complete the proof of Proposition 2.1. Plugging in, we get

$$\Delta_{\tilde{g}}|\alpha|^2_{\tilde{g}} \geqslant \frac{1}{2}|\tilde{\alpha}_{i,j}|^2 + |\tilde{\alpha}_{i,\bar{j}}|^2 + \frac{1}{2}\tilde{g}(\Delta^H_{\tilde{g}} a, a) - C\mathrm{tr}_g\tilde{g}|\alpha|^2_{\tilde{g}} - C|\nabla\tilde{g}|_{g,\tilde{g}}|\alpha|^2_{\tilde{g}}.$$

We compute

$$d\alpha = d(\tilde{\alpha}_i\tilde{\theta}^i) = \tilde{\alpha}_{i,j}\tilde{\theta}^j \wedge \tilde{\theta}^i + \tilde{\alpha}_{i,\bar{j}}\overline{\tilde{\theta}^j} \wedge \tilde{\theta}^i + \tilde{\alpha}_i\tilde{\Theta}^i,$$

and recall that

$$da = d\alpha + \overline{d\alpha} = \tilde{\omega} - \omega,$$

and taking the (1, 1)-part

$$(da)^{(1,1)} = \tilde{\omega} - \omega^{(1,1)},$$

hence

$$|(da)^{(1,1)}|^2_{\tilde{g}} = 2 - 2\mathrm{tr}_{\tilde{g}} g + |g|^2_{\tilde{g}} \geqslant 2 - 2\mathrm{tr}_{\tilde{g}} g + \frac{(\mathrm{tr}_{\tilde{g}} g)^2}{2} \geqslant \frac{(\mathrm{tr}_{\tilde{g}} g)^2}{4} - 2,$$

but we also have

$$(da)^{(1,1)} = (\tilde{\alpha}_{i,\bar{j}} - \overline{\tilde{\alpha}_{j,\bar{i}}})\overline{\tilde{\theta}^j} \wedge \tilde{\theta}^i,$$

$$|(da)^{(1,1)}|^2_{\tilde{g}} = |\tilde{\alpha}_{i,\bar{j}} - \overline{\tilde{\alpha}_{j,\bar{i}}}|^2 \leqslant 4|\tilde{\alpha}_{i,\bar{j}}|^2,$$

and so

$$|\tilde{\alpha}_{i,\bar{j}}|^2 \geqslant \frac{(\mathrm{tr}_{\tilde{g}} g)^2}{16} - \frac{1}{2} \geqslant \frac{(\mathrm{tr}_g \tilde{g})^2}{C} - \frac{1}{2}.$$

This gives

$$\Delta_{\tilde{g}}|\alpha|^2_{\tilde{g}} \geqslant \frac{1}{2}|\tilde{\alpha}_{i,j}|^2 + \frac{1}{2}|\tilde{\alpha}_{i,\bar{j}}|^2 + \frac{(\mathrm{tr}_g\tilde{g})^2}{C}$$
$$+ \frac{1}{2}\tilde{g}(\Delta^H_{\tilde{g}} a, a) - C\mathrm{tr}_g\tilde{g}|\alpha|^2_{\tilde{g}} - C|\nabla\tilde{g}|_{g,\tilde{g}}|\alpha|^2_{\tilde{g}} - C,$$

and we also have

$$\frac{1}{2}|\tilde{\alpha}_{i,j}|^2 + \frac{1}{2}|\tilde{\alpha}_{i,\bar{j}}|^2 \geqslant \frac{1}{4}|\tilde{\nabla}a|^2_{\tilde{g}} + \frac{1}{4}|\bar{\tilde{\nabla}}a|^2_{\tilde{g}},$$

and recalling that $|a|^2_{\tilde{g}} = 2|\alpha|^2_{\tilde{g}}$, we obtain

$$\Delta_{\tilde{g}}|a|^2_{\tilde{g}} \geqslant \frac{1}{2}|\tilde{\nabla}a|^2_{\tilde{g}} + \frac{1}{2}|\bar{\tilde{\nabla}}a|^2_{\tilde{g}} + \frac{(\mathrm{tr}_g\tilde{g})^2}{C} + \tilde{g}(\Delta^H_{\tilde{g}} a, a) - C\mathrm{tr}_g\tilde{g}|a|^2_{\tilde{g}}$$
$$- C|\nabla\tilde{g}|_{g,\tilde{g}}|a|^2_{\tilde{g}} - C. \tag{2.16}$$

We now deal with the term with the Hodge Laplacian. Using the gauge-fixing condition (1.6) we have

$$\Delta^H_{\tilde{g}} a = -dd^*_{\tilde{g}}a - d^*_{\tilde{g}}da = -d^*_{\tilde{g}}da = d^*_{\tilde{g}}\omega - d^*_{\tilde{g}}\tilde{\omega} = d^*_{\tilde{g}}\omega,$$

since $*_{\tilde{g}}\,\tilde{\omega} = \tilde{\omega}$ and so $d^*_{\tilde{g}}\tilde{\omega} = 0$. We also have the well-known formula

$$*_{\tilde{g}}\,\omega = \frac{2\omega\wedge\tilde{\omega}}{\tilde{\omega}^2}\tilde{\omega} - \omega + 2\omega^{(2,0)+(0,2)} = \mathrm{tr}_{\tilde{g}}g\,\tilde{\omega} - \omega + 2\omega^{(2,0)+(0,2)},$$

$$d *_{\tilde{g}}\omega = d\mathrm{tr}_{\tilde{g}}g\wedge\tilde{\omega} + 2d(\omega^{(2,0)+(0,2)}),$$

$$*_{\tilde{g}}\, d *_{\tilde{g}}\omega = d^c\mathrm{tr}_{\tilde{g}}g + 2 *_{\tilde{g}} d(\omega^{(2,0)+(0,2)}),$$

and so

$$\Delta^H_{\tilde{g}} a = d^*_{\tilde{g}}\omega = -d^c\mathrm{tr}_{\tilde{g}}g - 2 *_{\tilde{g}} d(\omega^{(2,0)+(0,2)}),$$

$$\begin{aligned}|\tilde{g}(\Delta^H_{\tilde{g}} a, a)| &\leqslant |\tilde{g}(d^c\mathrm{tr}_{\tilde{g}}g, a)| + 2|\tilde{g}(*_{\tilde{g}}d(\omega^{(2,0)+(0,2)}), a)|\\
&\leqslant C|\nabla\mathrm{tr}_{\tilde{g}}g|_{\tilde{g}}|a|_{\tilde{g}} + 2|*_{\tilde{g}}d(\omega^{(2,0)+(0,2)})|_{\tilde{g}}|a|_{\tilde{g}}\\
&= C|\nabla(\mathrm{tr}_g\tilde{g}e^{-F})|_{\tilde{g}}|a|_{\tilde{g}} + 2|d(\omega^{(2,0)+(0,2)})|_{\tilde{g}}|a|_{\tilde{g}}\\
&\leqslant C|\nabla\mathrm{tr}_g\tilde{g}|_{\tilde{g}}|a|_{\tilde{g}} + C\mathrm{tr}_g\tilde{g}|\nabla(e^{-F})|_{\tilde{g}}|a|_{\tilde{g}} + C(\mathrm{tr}_g\tilde{g})^{\frac{3}{2}}|a|_{\tilde{g}}\end{aligned}$$

$$
\begin{aligned}
&\leqslant C|\nabla \mathrm{tr}_g\tilde{g}|_{\tilde{g}}|a|_{\tilde{g}} + C(\mathrm{tr}_g\tilde{g})^{\frac{3}{2}}|a|_{\tilde{g}}\\
&\leqslant \varepsilon\frac{|\nabla \mathrm{tr}_g\tilde{g}|^2_{\tilde{g}}}{\mathrm{tr}_g\tilde{g}} + \frac{C}{\varepsilon}\mathrm{tr}_g\tilde{g}|a|^2_{\tilde{g}} + \varepsilon(\mathrm{tr}_g\tilde{g})^2 + \frac{C}{\varepsilon}|a|^4_{\tilde{g}}\\
&\leqslant \varepsilon\frac{|\nabla \mathrm{tr}_g\tilde{g}|^2_{\tilde{g}}}{\mathrm{tr}_g\tilde{g}} + 2\varepsilon(\mathrm{tr}_g\tilde{g})^2 + \frac{C}{\varepsilon^2}|a|^4_{\tilde{g}}\\
&\leqslant \varepsilon|\nabla\tilde{g}|^2_{g,\tilde{g}} + 2\varepsilon(\mathrm{tr}_g\tilde{g})^2 + \frac{C}{\varepsilon^2}|a|^4_{\tilde{g}},
\end{aligned}
$$

using (2.4), and so assuming without loss that ε is small, and substituting in (2.16) gives

$$
\Delta_{\tilde{g}}|a|^2_{\tilde{g}} \geqslant \frac{1}{2}|\tilde{\nabla}a|^2_{\tilde{g}} + \frac{1}{2}|\overline{\tilde{\nabla}}a|^2_{\tilde{g}} + \frac{(\mathrm{tr}_g\tilde{g})^2}{C} - \varepsilon|\nabla\tilde{g}|^2_{g,\tilde{g}} - \frac{C}{\varepsilon^2}|a|^4_{\tilde{g}} - C,
$$

which is exactly (2.5).

Proof of the Main Claim (2.15) ***Modulo*** (2.18) ***and*** (2.19) We now need to prove the main claim (2.15). Since the Hodge Laplacian is a real operator, we have

$$
\Delta^H_{\tilde{g}}a = \Delta^H_{\tilde{g}}\alpha + \Delta^H_{\tilde{g}}\overline{\alpha} = \Delta^H_{\tilde{g}}\alpha + \overline{\Delta^H_{\tilde{g}}\alpha},
$$

and

$$
(\Delta^H_{\tilde{g}}\overline{\alpha})^{(0,1)} = \overline{(\Delta^H_{\tilde{g}}\alpha)^{(1,0)}},
$$

and so

$$
\begin{aligned}
\tilde{g}(\Delta^H_{\tilde{g}}a, a) &= \tilde{g}(\Delta^H_{\tilde{g}}\alpha + \Delta^H_{\tilde{g}}\overline{\alpha}, \alpha + \overline{\alpha})\\
&= \tilde{g}(\Delta^H_{\tilde{g}}\alpha, \overline{\alpha}) + \tilde{g}(\Delta^H_{\tilde{g}}\alpha, \alpha) + \tilde{g}(\Delta^H_{\tilde{g}}\overline{\alpha}, \alpha) + \tilde{g}(\Delta^H_{\tilde{g}}\overline{\alpha}, \overline{\alpha})\\
&= \tilde{g}((\Delta^H_{\tilde{g}}\alpha)^{(1,0)}, \overline{\alpha}) + \tilde{g}((\Delta^H_{\tilde{g}}\alpha)^{(0,1)}, \alpha) + \tilde{g}((\Delta^H_{\tilde{g}}\overline{\alpha})^{(0,1)}, \alpha)\\
&\quad + \tilde{g}((\Delta^H_{\tilde{g}}\overline{\alpha})^{(1,0)}, \overline{\alpha})\\
&= \tilde{g}((\Delta^H_{\tilde{g}}\alpha)^{(1,0)}, \overline{\alpha}) + \tilde{g}((\Delta^H_{\tilde{g}}\alpha)^{(0,1)}, \alpha) + \tilde{g}(\overline{(\Delta^H_{\tilde{g}}\alpha)^{(1,0)}}, \alpha)\\
&\quad + \tilde{g}(\overline{(\Delta^H_{\tilde{g}}\alpha)^{(0,1)}}, \overline{\alpha})\\
&= 2\mathrm{Re}\tilde{g}((\Delta^H_{\tilde{g}}\alpha)^{(1,0)}, \overline{\alpha}) + 2\mathrm{Re}\tilde{g}((\Delta^H_{\tilde{g}}\alpha)^{(0,1)}, \alpha).
\end{aligned}
\tag{2.17}
$$

Next, we claim that

$$
(\Delta^H_{\tilde{g}}\alpha)^{(1,0)} = 2\tilde{\alpha}_{k,\bar{i}i}\tilde{\theta}^k, \tag{2.18}
$$

$$(\Delta^H_{\tilde{g}}\alpha)^{(0,1)} = \left(-4\tilde{\alpha}_{i,j}\tilde{N}^j_{\overline{ki}} - 2\tilde{\alpha}_j\tilde{K}^j_{\overline{iki}} + 2\tilde{\alpha}_i\tilde{N}^i_{\overline{jk},j}\right)\overline{\tilde{\theta}^k}. \tag{2.19}$$

We will prove these by long direct computations, but first let us assume we have these and complete the proof of the main claim (2.15). Combining (2.17), (2.18), and (2.19) gives

$$\frac{1}{2}\tilde{g}(\Delta^H_{\tilde{g}}a, a) = 2\mathrm{Re}(\tilde{\alpha}_{k,\bar{i}i}\overline{\tilde{\alpha}_k}) + \mathrm{Re}\left(-4\tilde{\alpha}_{i,j}\tilde{\alpha}_k\tilde{N}^j_{\overline{ki}} - 2\tilde{\alpha}_j\tilde{\alpha}_k\tilde{K}^j_{\overline{iki}} + 2\tilde{\alpha}_i\tilde{\alpha}_k\tilde{N}^i_{\overline{jk},j}\right), \tag{2.20}$$

and we can bound the last 3 terms in (2.20) using the same method as before. For the first term, using the same notation as earlier, we can write

$$\begin{aligned}\tilde{\alpha}_{i,j}\tilde{\alpha}_k\tilde{N}^j_{\overline{ki}} &= \sum_{i,k=1}^{2}\lambda_i^{-1}\lambda_k^{-1}\alpha_{i,j}\alpha_k N^j_{\overline{ki}} = \sum_{i\neq k}\lambda_i^{-1}\lambda_k^{-1}\alpha_{i,j}\alpha_k N^j_{\overline{ki}}\\ &= e^{-F}\sum_{i\neq k}\alpha_{i,j}\alpha_k N^j_{\overline{ki}},\end{aligned}$$

using that $N^j_{\overline{ki}}$ is skew-symmetric in i, k as well as the Calabi-Yau equation (2.14). Using the Calabi-Yau equation again, we can bound

$$\begin{aligned}-\mathrm{Re}(4\tilde{\alpha}_{i,j}\tilde{\alpha}_k\tilde{N}^j_{\overline{ki}}) &\leqslant C\sum_j\sum_{i\neq k}|\alpha_{i,j}||\alpha_k| \leqslant C\sum_j\sum_{i\neq k}\lambda_i^{-\frac{1}{2}}\lambda_k^{-\frac{1}{2}}|\alpha_{i,j}||\alpha_k|\lambda_j^{-\frac{1}{2}}\lambda_j^{\frac{1}{2}}\\ &= C\sum_j\sum_{i\neq k}|\tilde{\alpha}_{i,j}||\tilde{\alpha}_k|\lambda_j^{\frac{1}{2}} \leqslant C|\alpha|_{\tilde{g}}\sqrt{\mathrm{tr}_g\tilde{g}}\sum_{i,j}|\tilde{\alpha}_{i,j}|\\ &\leqslant \frac{1}{2}|\tilde{\alpha}_{i,j}|^2 + C\mathrm{tr}_g\tilde{g}|\alpha|^2_{\tilde{g}}.\end{aligned}$$

The second terms in the parenthesis in (2.20) is bounded by $C\mathrm{tr}_g\tilde{g}|\alpha|^2_{\tilde{g}}$ arguing exactly as we did to bound (2.13): using the skew-symmetry of $K^j_{\overline{iki}}$ in k, i and the Calabi-Yau equation (2.14) we have

$$\tilde{\alpha}_j\tilde{\alpha}_k\tilde{K}^j_{\overline{iki}} = \sum_{i,k=1}^{2}\lambda_i^{-1}\lambda_k^{-1}\alpha_j\alpha_k K^j_{\overline{iki}} = \sum_{i\neq k}\lambda_i^{-1}\lambda_k^{-1}\alpha_j\alpha_k K^j_{\overline{iki}} = e^{-F}\sum_{i\neq k}\alpha_j\alpha_k K^j_{\overline{iki}},$$

and so

$$-\mathrm{Re}(2\tilde{\alpha}_j\tilde{\alpha}_k\tilde{K}^j_{\overline{iki}}) \leqslant C|\alpha|^2_g \leqslant C\mathrm{tr}_g\tilde{g}|\alpha|^2_{\tilde{g}}.$$

The last term in the parenthesis in (2.20)

$$\tilde{\alpha}_i\tilde{\alpha}_k\tilde{N}^i_{\overline{jk},j}, \tag{2.21}$$

requires more work. First, we explain in detail the term $|\nabla\tilde{g}|^2_{g,\tilde{g}}$ in (2.3). As in [17, (3.9)] there are functions $a^i_{k\ell}$ defined by

$$da^i_m - a^i_j\theta^j_m + a^k_m\tilde{\theta}^i_k = a^i_{k\ell}a^k_m\tilde{\theta}^\ell,$$

which are the components of $\nabla\tilde{g}$ in the frame $\{\tilde{\theta}^i\}$, so that [17, Lemma 4.2]

$$|\nabla\tilde{g}|^2_{\tilde{g}} = |a^i_{k\ell}|^2.$$

The mixed norm $|\nabla\tilde{g}|^2_{g,\tilde{g}}$ that appears in (2.3) is given by

$$|\nabla\tilde{g}|^2_{g,\tilde{g}} = |a^i_{p\ell}a^p_k|^2. \tag{2.22}$$

We can now go back to the term in (2.21) and using [17, Lemma 4.4 (i)] we can write it as

$$\tilde{\alpha}_i\tilde{\alpha}_k\tilde{N}^i_{\overline{jk},j} = \alpha_\ell\alpha_m b^m_k\overline{b^s_k}b^u_j\overline{b^r_j}N^\ell_{\overline{rs},u} + \alpha_\ell\alpha_m b^\ell_i b^m_k\overline{b^s_k}\overline{b^r_j}N^t_{\overline{rs}}a^i_{uj}a^u_t,$$

and using the skew-symmetry of $N^i_{\overline{jk},j}$ in j,k and the Calabi-Yau equation we have

$$\begin{aligned}\alpha_\ell\alpha_m b^m_k\overline{b^s_k}b^u_j\overline{b^r_j}N^\ell_{\overline{rs},u} &= \sum_{j,k=1}^{2}\lambda_j^{-1}\lambda_k^{-1}\alpha_i\alpha_k N^i_{\overline{jk},j} = \sum_{j\neq k}^{2}\lambda_j^{-1}\lambda_k^{-1}\alpha_i\alpha_k N^i_{\overline{jk},j}\\ &= e^{-F}\sum_{j\neq k}\alpha_i\alpha_k N^i_{\overline{jk},j},\end{aligned}$$

whose absolute value is bounded by $C|\alpha|^2_g \leqslant C\mathrm{tr}_g\tilde{g}|\alpha|^2_{\tilde{g}}$. Similarly, using the skew-symmetry of $N^t_{\overline{jk}}$ in j,k and the Calabi-Yau equation we have

$$\begin{aligned}\alpha_\ell\alpha_m b^\ell_i b^m_k\overline{b^s_k}\overline{b^r_j}N^t_{\overline{rs}}a^i_{uj}a^u_t &= \sum_{i,j,k=1}^{2}\alpha_i\alpha_k\lambda_i^{-\frac{1}{2}}\lambda_k^{-1}\lambda_j^{-\frac{1}{2}}N^t_{\overline{jk}}a^i_{uj}a^u_t\\ &= \sum_i\sum_{j\neq k}\alpha_i\alpha_k\lambda_i^{-\frac{1}{2}}\lambda_k^{-1}\lambda_j^{-\frac{1}{2}}N^t_{\overline{jk}}a^i_{uj}a^u_t\\ &= e^{-\frac{F}{2}}\sum_i\sum_{j\neq k}\alpha_i\alpha_k\lambda_i^{-\frac{1}{2}}\lambda_k^{-\frac{1}{2}}N^t_{\overline{jk}}a^i_{uj}a^u_t\\ &= e^{-\frac{F}{2}}\sum_i\sum_{j\neq k}\tilde{\alpha}_i\tilde{\alpha}_k N^t_{\overline{jk}}a^i_{uj}a^u_t,\end{aligned}$$

whose absolute value is bounded by

$$C|\alpha|^2_{\tilde{g}}|a^i_{uj}a^u_t| = C|\alpha|^2_{\tilde{g}}|\nabla\tilde{g}|_{g,\tilde{g}},$$

and so

$$-\operatorname{Re}(2\tilde{\alpha}_i\tilde{\alpha}_k\tilde{N}^i_{\overline{jk},j}) \leqslant C\mathrm{tr}_g\tilde{g}|\alpha|^2_{\tilde{g}} + C|\alpha|^2_{\tilde{g}}|\nabla\tilde{g}|_{g,\tilde{g}}.$$

Putting all these together we obtain

$$\frac{1}{2}\tilde{g}(\Delta^H_{\tilde{g}}a, a) \leqslant 2\mathrm{Re}(\tilde{\alpha}_{k,\bar{i}i}\overline{\tilde{\alpha}_k}) + C\mathrm{tr}_g\tilde{g}|\alpha|^2_{\tilde{g}} + C|\alpha|^2_{\tilde{g}}|\nabla\tilde{g}|_{g,\tilde{g}} + \frac{1}{2}|\tilde{\alpha}_{i,j}|^2,$$

which is exactly the main claim (2.15).

Proof of (2.18) ***and*** (2.19) To complete the proof of the main claim (2.15), we are left with showing (2.18) and (2.19), which are Bochner-Kodaira-Weitzenböck type formulas. For these, recall that by definition of the Hodge star $*_{\tilde{g}}$, for any two (p, q)-forms β, γ we have

$$\begin{aligned}\beta \wedge \overline{*_{\tilde{g}}\gamma} &= \tilde{g}(\beta, \gamma)\tilde{\omega}^2 = \tilde{g}(\beta, \gamma)\sqrt{-1}\tilde{\theta}^1 \wedge \overline{\tilde{\theta}^1} \wedge \sqrt{-1}\tilde{\theta}^2 \wedge \overline{\tilde{\theta}^2}\\ &= -\tilde{g}(\beta, \gamma)\tilde{\theta}^1 \wedge \overline{\tilde{\theta}^1} \wedge \tilde{\theta}^2 \wedge \overline{\tilde{\theta}^2},\end{aligned}$$

and also $*_{\tilde{g}}\,\overline{\alpha} = \overline{*_{\tilde{g}}\alpha}$ (i.e. $*_{\tilde{g}}$ is a real operator), from which we can compute the Hodge star on basic combinations of our frame elements, using the following notation: for $i \in \{1, 2\}$ we let $\hat{\imath} \in \{1, 2\}$ be such that $\{i, \hat{\imath}\} = \{1, 2\}$ (unordered), i.e. $\hat{1} = 2, \hat{2} = 1$. Then from the definition we have

$$*_{\tilde{g}}\,(\tilde{\theta}^i \wedge \overline{\tilde{\theta}^i} \wedge \tilde{\theta}^{\hat{\imath}} \wedge \overline{\tilde{\theta}^{\hat{\imath}}}) = -1,$$

$$*_{\tilde{g}}\,\tilde{\theta}^i = \tilde{\theta}^i \wedge \tilde{\theta}^{\hat{\imath}} \wedge \overline{\tilde{\theta}^{\hat{\imath}}}, \quad *_{\tilde{g}}\overline{\tilde{\theta}^i} = -\overline{\tilde{\theta}^i} \wedge \tilde{\theta}^{\hat{\imath}} \wedge \overline{\tilde{\theta}^{\hat{\imath}}}$$

$$*_{\tilde{g}}\,(\overline{\tilde{\theta}^i} \wedge \tilde{\theta}^{\hat{\imath}} \wedge \overline{\tilde{\theta}^{\hat{\imath}}}) = -\tilde{\theta}^i, \quad *_{\tilde{g}}(\tilde{\theta}^i \wedge \tilde{\theta}^{\hat{\imath}} \wedge \overline{\tilde{\theta}^i}) = \tilde{\theta}^{\hat{\imath}},$$

$$*_{\tilde{g}}\,(\tilde{\theta}^{\hat{\imath}} \wedge \overline{\tilde{\theta}^{\hat{\imath}}} \wedge \overline{\tilde{\theta}^i}) = \overline{\tilde{\theta}^i}, \quad *_{\tilde{g}}(\tilde{\theta}^i \wedge \overline{\tilde{\theta}^{\hat{\imath}}} \wedge \overline{\tilde{\theta}^i}) = -\overline{\tilde{\theta}^{\hat{\imath}}},$$

$$*_{\tilde{g}}\,(\tilde{\theta}^i \wedge \tilde{\theta}^j) = \tilde{\theta}^i \wedge \tilde{\theta}^j, \quad *_{\tilde{g}}(\overline{\tilde{\theta}^i} \wedge \overline{\tilde{\theta}^j}) = \overline{\tilde{\theta}^i} \wedge \overline{\tilde{\theta}^j},$$

$$*_{\tilde{g}}\,(\tilde{\theta}^i \wedge \overline{\tilde{\theta}^j}) = \begin{cases} \tilde{\theta}^{\hat{\imath}} \wedge \overline{\tilde{\theta}^{\hat{\imath}}} & \text{if } i = j,\\ -\tilde{\theta}^i \wedge \overline{\tilde{\theta}^j} & \text{if } i = \hat{\jmath}.\end{cases}$$

With these, we can now start the computation of $\Delta^H_{\tilde{g}}\alpha = -dd^*_{\tilde{g}}\alpha - d^*_{\tilde{g}}d\alpha$, recalling that

$$d^*_{\tilde{g}}\alpha = - *_{\tilde{g}} d *_{\tilde{g}} \alpha,$$

we compute

$$*_{\tilde{g}}\, \alpha = \tilde{\alpha}_i \tilde{\theta}^i \wedge \tilde{\theta}^{\hat{\imath}} \wedge \overline{\tilde{\theta}^{\hat{\imath}}},$$

$$d *_{\tilde{g}} \alpha = \tilde{\alpha}_{i,\bar{j}} \overline{\tilde{\theta}^j} \wedge \tilde{\theta}^i \wedge \tilde{\theta}^{\hat{\imath}} \wedge \overline{\tilde{\theta}^{\hat{\imath}}} = \tilde{\alpha}_{i,\bar{i}} \overline{\tilde{\theta}^i} \wedge \tilde{\theta}^i \wedge \tilde{\theta}^{\hat{\imath}} \wedge \overline{\tilde{\theta}^{\hat{\imath}}},$$

$$d^*_{\tilde{g}}\alpha = -\tilde{\alpha}_{i,\bar{i}},$$

$$dd^*_{\tilde{g}}\alpha = -\tilde{\alpha}_{i,\bar{i}k}\tilde{\theta}^k - \tilde{\alpha}_{i,\bar{i}\bar{k}}\overline{\tilde{\theta}^k}. \tag{2.23}$$

$$d\alpha = \tilde{\alpha}_{i,j}\tilde{\theta}^j \wedge \tilde{\theta}^i + \tilde{\alpha}_{i,\bar{j}}\overline{\tilde{\theta}^j} \wedge \tilde{\theta}^i + \tilde{\alpha}_i \tilde{N}^i_{\bar{j}\bar{k}} \overline{\tilde{\theta}^j} \wedge \overline{\tilde{\theta}^k}$$

$$*_{\tilde{g}}\, d\alpha = \tilde{\alpha}_{i,j}\tilde{\theta}^j \wedge \tilde{\theta}^i - \tilde{\alpha}_{i,\bar{i}}\tilde{\theta}^{\hat{\imath}} \wedge \overline{\tilde{\theta}^{\hat{\imath}}} + \tilde{\alpha}_{i,\bar{\hat{\imath}}}\tilde{\theta}^i \wedge \overline{\tilde{\theta}^{\hat{\imath}}} + \tilde{\alpha}_i \tilde{N}^i_{\bar{j}\bar{k}} \overline{\tilde{\theta}^j} \wedge \overline{\tilde{\theta}^k}. \tag{2.24}$$

We first prove (2.18). We are thus interested in the (1, 0)-part of $d^*_{\tilde{g}}d\alpha$, so we first take the exterior derivative of (2.24) and take its (2, 1)-part (which becomes (1, 0) after taking the Hodge star)

$$\begin{aligned}(d *_{\tilde{g}} d\alpha)^{(2,1)} = {} & \tilde{\alpha}_{i,j\bar{k}}\overline{\tilde{\theta}^k} \wedge \tilde{\theta}^j \wedge \tilde{\theta}^i - \tilde{\alpha}_{i,\bar{i}k}\tilde{\theta}^k \wedge \tilde{\theta}^{\hat{\imath}} \wedge \overline{\tilde{\theta}^{\hat{\imath}}} + \tilde{\alpha}_{i,\bar{\hat{\imath}}k}\tilde{\theta}^k \wedge \tilde{\theta}^i \wedge \overline{\tilde{\theta}^{\hat{\imath}}} \\ & + 2\tilde{\alpha}_i \tilde{N}^i_{\bar{j}\bar{k}} \overline{\tilde{N}^j_{\bar{p}\bar{q}}} \tilde{\theta}^p \wedge \tilde{\theta}^q \wedge \overline{\tilde{\theta}^k},\end{aligned} \tag{2.25}$$

and in the first term in (2.25) we must have $j = \hat{\imath}$ and we can write it as

$$\tilde{\alpha}_{i,\hat{\imath}\bar{i}}\overline{\tilde{\theta}^i} \wedge \tilde{\theta}^{\hat{\imath}} \wedge \tilde{\theta}^i + \tilde{\alpha}_{i,\hat{\imath}\bar{\hat{\imath}}}\overline{\tilde{\theta}^{\hat{\imath}}} \wedge \tilde{\theta}^{\hat{\imath}} \wedge \tilde{\theta}^i,$$

and its Hodge star equals

$$- \tilde{\alpha}_{i,\hat{\imath}\bar{i}}\tilde{\theta}^{\hat{\imath}} + \tilde{\alpha}_{i,\hat{\imath}\bar{\hat{\imath}}}\tilde{\theta}^i = (-\tilde{\alpha}_{\hat{k},k\bar{\hat{k}}} + \tilde{\alpha}_{k,\hat{k}\bar{\hat{k}}})\tilde{\theta}^k = (-\tilde{\alpha}_{i,k\bar{i}} + \tilde{\alpha}_{k,i\bar{i}})\tilde{\theta}^k.$$

In the second term we must have $k = i$ and in the third term $k = \hat{\imath}$ so we can write them as

$$- \tilde{\alpha}_{i,\bar{i}i}\tilde{\theta}^i \wedge \tilde{\theta}^{\hat{\imath}} \wedge \overline{\tilde{\theta}^{\hat{\imath}}} + \tilde{\alpha}_{i,\bar{\hat{\imath}}\hat{\imath}}\tilde{\theta}^{\hat{\imath}} \wedge \tilde{\theta}^i \wedge \overline{\tilde{\theta}^{\hat{\imath}}},$$

and their Hodge star equals

$$(\tilde{\alpha}_{i,\bar{i}i} + \tilde{\alpha}_{i,\bar{i}\hat{i}})\tilde{\theta}^i = \tilde{\alpha}_{k,\bar{i}i}\tilde{\theta}^k.$$

In the last term in (2.25) we must have $q = \hat{p}$ so it equals

$$2\tilde{\alpha}_i \tilde{N}^i_{\bar{j}\bar{p}} \overline{\tilde{N}^j_{\bar{p}\bar{\hat{p}}}} \tilde{\theta}^p \wedge \tilde{\theta}^{\hat{p}} \wedge \overline{\tilde{\theta}^p} + 2\tilde{\alpha}_i \tilde{N}^i_{\bar{j}\bar{\hat{p}}} \overline{\tilde{N}^j_{\bar{p}\bar{\hat{p}}}} \tilde{\theta}^p \wedge \tilde{\theta}^{\hat{p}} \wedge \overline{\tilde{\theta}^{\hat{p}}},$$

and its Hodge star equals

$$\begin{aligned} 2\tilde{\alpha}_i \tilde{N}^i_{\bar{j}\bar{p}} \overline{\tilde{N}^j_{\bar{p}\bar{\hat{p}}}} \tilde{\theta}^{\hat{p}} - 2\tilde{\alpha}_i \tilde{N}^i_{\bar{j}\bar{\hat{p}}} \overline{\tilde{N}^j_{\bar{p}\bar{\hat{p}}}} \tilde{\theta}^p &= \left(2\tilde{\alpha}_i \tilde{N}^i_{\bar{j}\bar{\hat{k}}} \overline{\tilde{N}^j_{\bar{\hat{k}}\bar{k}}} - 2\tilde{\alpha}_i \tilde{N}^i_{\bar{j}\bar{\hat{k}}} \overline{\tilde{N}^j_{\bar{k}\bar{\hat{k}}}}\right) \tilde{\theta}^k \\ &= \left(2\tilde{\alpha}_i \tilde{N}^i_{\bar{j}\bar{\ell}} \overline{\tilde{N}^j_{\bar{\ell}\bar{k}}} - 2\tilde{\alpha}_i \tilde{N}^i_{\bar{j}\bar{\ell}} \overline{\tilde{N}^j_{\bar{k}\bar{\ell}}}\right) \tilde{\theta}^k \\ &= -4\tilde{\alpha}_i \tilde{N}^i_{\bar{j}\bar{\ell}} \overline{\tilde{N}^j_{\bar{k}\bar{\ell}}} \tilde{\theta}^k, \end{aligned}$$

using the skew-symmetry of $\tilde{N}^j_{\bar{k}\bar{\ell}}$ in k, ℓ. Putting these together gives

$$\begin{aligned} -(d^*_{\tilde{g}} d\alpha)^{(1,0)} &= (*_{\tilde{g}} d *_{\tilde{g}} d\alpha)^{(1,0)} = *_{\tilde{g}} \left((d *_{\tilde{g}} d\alpha)^{(2,1)}\right) \\ &= \left(-\tilde{\alpha}_{i,k\bar{i}} + \tilde{\alpha}_{k,i\bar{i}} + \tilde{\alpha}_{k,\bar{i}i} - 4\tilde{\alpha}_i \tilde{N}^i_{\bar{j}\bar{\ell}} \overline{\tilde{N}^j_{\bar{k}\bar{\ell}}}\right) \tilde{\theta}^k, \end{aligned}$$

and combining this with (2.23) gives

$$(\Delta^H_{\tilde{g}} \alpha)^{(1,0)} = \left(\tilde{\alpha}_{i,\bar{i}k} - \tilde{\alpha}_{i,k\bar{i}} + \tilde{\alpha}_{k,i\bar{i}} + \tilde{\alpha}_{k,\bar{i}i} - 4\tilde{\alpha}_i \tilde{N}^i_{\bar{j}\bar{\ell}} \overline{\tilde{N}^j_{\bar{k}\bar{\ell}}}\right) \tilde{\theta}^k.$$

From the commutation relation (2.10) we obtain

$$\tilde{\alpha}_{i,\bar{i}k} - \tilde{\alpha}_{i,k\bar{i}} = -\tilde{\alpha}_j \tilde{R}^{\,j}_{ik\bar{i}},$$

$$\tilde{\alpha}_{k,i\bar{i}} - \tilde{\alpha}_{k,\bar{i}i} = \tilde{\alpha}_j \tilde{R}^{\,j}_{ki\bar{i}},$$

while [17, (2.16)] gives

$$\tilde{R}^{\,j}_{ki\bar{i}} - \tilde{R}^{\,j}_{ik\bar{i}} = 4\tilde{N}^j_{\bar{p}\bar{i}} \overline{\tilde{N}^p_{\bar{k}\bar{i}}},$$

and so

$$(\Delta^H_{\tilde{g}} \alpha)^{(1,0)} = \left(2\tilde{\alpha}_{k,\bar{i}i} + 4\tilde{\alpha}_j \tilde{N}^j_{\bar{p}\bar{i}} \overline{\tilde{N}^p_{\bar{k}\bar{i}}} - 4\tilde{\alpha}_i \tilde{N}^i_{\bar{j}\bar{\ell}} \overline{\tilde{N}^j_{\bar{k}\bar{\ell}}}\right) \tilde{\theta}^k = 2\tilde{\alpha}_{k,\bar{i}i} \tilde{\theta}^k,$$

which proves (2.18). The proof of (2.19) is similar. First, from (2.23) we have

$$(dd_{\tilde{g}}^*\alpha)^{(0,1)} = -\tilde{\alpha}_{i,\overline{ik}}\overline{\tilde{\theta}^k}. \tag{2.26}$$

We are then interested in the (0, 1)-part of $d_{\tilde{g}}^* d\alpha$, so we take the exterior derivative of (2.24) and then take the (1, 2)-part to obtain

$$\begin{aligned}(d *_{\tilde{g}} d\alpha)^{(1,2)} &= \tilde{\alpha}_{i,j}\tilde{N}^j_{\overline{pq}}\overline{\tilde{\theta}^p} \wedge \overline{\tilde{\theta}^q} \wedge \tilde{\theta}^i - \tilde{\alpha}_{i,j}\tilde{N}^i_{\overline{pq}}\tilde{\theta}^j \wedge \overline{\tilde{\theta}^p} \wedge \overline{\tilde{\theta}^q} - \tilde{\alpha}_{i,\overline{ik}}\overline{\tilde{\theta}^k} \wedge \tilde{\theta}^{\hat{\imath}} \wedge \overline{\tilde{\theta}^{\hat{\imath}}} \\ &\quad + \tilde{\alpha}_{i,\overline{\hat{\imath}k}}\overline{\tilde{\theta}^k} \wedge \tilde{\theta}^i \wedge \overline{\tilde{\theta}^{\hat{\imath}}} + \tilde{\alpha}_{i,p}\tilde{N}^i_{\overline{jk}}\tilde{\theta}^p \wedge \overline{\tilde{\theta}^j} \wedge \overline{\tilde{\theta}^k} + \tilde{\alpha}_i \tilde{N}^i_{\overline{jk},p}\tilde{\theta}^p \wedge \overline{\tilde{\theta}^j} \wedge \overline{\tilde{\theta}^k} \\ &= \tilde{\alpha}_{i,j}\tilde{N}^j_{\overline{pq}}\overline{\tilde{\theta}^p} \wedge \overline{\tilde{\theta}^q} \wedge \tilde{\theta}^i - \tilde{\alpha}_{i,\overline{ik}}\overline{\tilde{\theta}^k} \wedge \tilde{\theta}^{\hat{\imath}} \wedge \overline{\tilde{\theta}^{\hat{\imath}}} + \tilde{\alpha}_{i,\overline{\hat{\imath}k}}\overline{\tilde{\theta}^k} \wedge \tilde{\theta}^i \wedge \overline{\tilde{\theta}^{\hat{\imath}}} \\ &\quad + \tilde{\alpha}_i \tilde{N}^i_{\overline{jk},p}\tilde{\theta}^p \wedge \overline{\tilde{\theta}^j} \wedge \overline{\tilde{\theta}^k},\end{aligned} \tag{2.27}$$

and in the first term in (2.27) we must have $q = \hat{p}$ so we can write it as

$$\tilde{\alpha}_{p,j}\tilde{N}^j_{\overline{p\hat{p}}}\overline{\tilde{\theta}^p} \wedge \overline{\tilde{\theta}^{\hat{p}}} \wedge \tilde{\theta}^p + \tilde{\alpha}_{\hat{p},j}\tilde{N}^j_{\overline{p\hat{p}}}\overline{\tilde{\theta}^p} \wedge \overline{\tilde{\theta}^{\hat{p}}} \wedge \tilde{\theta}^{\hat{p}},$$

and its Hodge star equals

$$\tilde{\alpha}_{p,j}\tilde{N}^j_{\overline{p\hat{p}}}\overline{\tilde{\theta}^{\hat{p}}} - \tilde{\alpha}_{\hat{p},j}\tilde{N}^j_{\overline{p\hat{p}}}\overline{\tilde{\theta}^p} = \tilde{\alpha}_{p,j}\tilde{N}^j_{\overline{pi}}\overline{\tilde{\theta}^i} - \tilde{\alpha}_{i,j}\tilde{N}^j_{\overline{pi}}\overline{\tilde{\theta}^p} = -2\tilde{\alpha}_{i,j}\tilde{N}^j_{\overline{pi}}\overline{\tilde{\theta}^p}.$$

In the second and third terms in (2.27) we must have $k = i$ so we can write them as

$$-\tilde{\alpha}_{i,\overline{ii}}\overline{\tilde{\theta}^i} \wedge \tilde{\theta}^{\hat{\imath}} \wedge \overline{\tilde{\theta}^{\hat{\imath}}} + \tilde{\alpha}_{i,\overline{\hat{\imath}i}}\overline{\tilde{\theta}^i} \wedge \tilde{\theta}^i \wedge \overline{\tilde{\theta}^{\hat{\imath}}},$$

and their Hodge star equals

$$-\tilde{\alpha}_{i,\overline{ii}}\overline{\tilde{\theta}^i} - \tilde{\alpha}_{i,\overline{\hat{\imath}i}}\overline{\tilde{\theta}^{\hat{\imath}}} = -\tilde{\alpha}_{i,\overline{ki}}\overline{\tilde{\theta}^k}.$$

In the last term in (2.27) we must have $k = \hat{\jmath}$ and so we can write it as

$$\tilde{\alpha}_i \tilde{N}^i_{\overline{j\hat{\jmath}},j}\tilde{\theta}^j \wedge \overline{\tilde{\theta}^j} \wedge \overline{\tilde{\theta}^{\hat{\jmath}}} + \tilde{\alpha}_i \tilde{N}^i_{\overline{j\hat{\jmath}},\hat{\jmath}}\tilde{\theta}^{\hat{\jmath}} \wedge \overline{\tilde{\theta}^j} \wedge \overline{\tilde{\theta}^{\hat{\jmath}}},$$

and its Hodge star equals

$$\tilde{\alpha}_i \tilde{N}^i_{\overline{j\hat{\jmath}},j}\overline{\tilde{\theta}^{\hat{\jmath}}} - \tilde{\alpha}_i \tilde{N}^i_{\overline{j\hat{\jmath}},\hat{\jmath}}\overline{\tilde{\theta}^j} = \tilde{\alpha}_i \tilde{N}^i_{\overline{jk},j}\overline{\tilde{\theta}^k} - \tilde{\alpha}_i \tilde{N}^i_{\overline{kj},j}\overline{\tilde{\theta}^k} = 2\tilde{\alpha}_i \tilde{N}^i_{\overline{jk},j}\overline{\tilde{\theta}^k}.$$

Putting these together gives

$$-(d_{\tilde{g}}^{*} d\alpha)^{(0,1)} = (*_{\tilde{g}} d *_{\tilde{g}} d\alpha)^{(0,1)} = *_{\tilde{g}} \left((d *_{\tilde{g}} d\alpha)^{(1,2)} \right)$$
$$= \left(-2\tilde{\alpha}_{i,j} \tilde{N}^{j}_{\overline{ki}} - \tilde{\alpha}_{i,\overline{ki}} + 2\tilde{\alpha}_i \tilde{N}^{i}_{\overline{jk},j} \right) \overline{\tilde{\theta}^k},$$

and combining this with (2.26) gives

$$(\Delta_{\tilde{g}}^{H} \alpha)^{(0,1)} = \left(\tilde{\alpha}_{i,\overline{ik}} - 2\tilde{\alpha}_{i,j} \tilde{N}^{j}_{\overline{ki}} - \tilde{\alpha}_{i,\overline{ki}} + 2\tilde{\alpha}_i \tilde{N}^{i}_{\overline{jk},j} \right) \overline{\tilde{\theta}^k}.$$

From the commutation relation (2.11) we obtain

$$\tilde{\alpha}_{i,\overline{ik}} - \tilde{\alpha}_{i,\overline{ki}} = -2\tilde{\alpha}_j \tilde{K}^{j}_{\overline{iki}} - 2\tilde{\alpha}_{i,j} \tilde{N}^{j}_{\overline{ki}},$$

and so

$$(\Delta_{\tilde{g}}^{H} \alpha)^{(0,1)} = \left(-4\tilde{\alpha}_{i,j} \tilde{N}^{j}_{\overline{ki}} - 2\tilde{\alpha}_j \tilde{K}^{j}_{\overline{iki}} + 2\tilde{\alpha}_i \tilde{N}^{i}_{\overline{jk},j} \right) \overline{\tilde{\theta}^k},$$

which proves (2.19) and concludes the proof of Proposition 2.1, and hence of Theorem 1.2.

Acknowledgments The author owes many thanks to Guido De Philippis for many related discussions. The author was partially supported by NSF grant DMS-2404599. This note is dedicated to our friend Paul Gauduchon, whose pioneering work in complex and almost-complex geometry has been a constant source of inspiration. *Buon compleanno Paul!*

References

1. E. Buzano, A. Fino, L. Vezzoni, The Calabi-Yau equation for T^2-bundles over T^2: the non-Lagrangian case. Rend. Semin. Mat. Univ. Politec. Torino **69**(3), 281–298 (2011)
2. E. Buzano, A. Fino, L. Vezzoni, The Calabi-Yau equation on the Kodaira-Thurston manifold, viewed as an S^1-bundle over a 3-torus. J. Differential Geom. **101**(2), 175–195 (2015)
3. E. Calabi, The space of Kähler metrics, in *Proceedings of the International Congress of Mathematicians, Amsterdam, 1954*, vol. 2 (North-Holland, Amsterdam, 1956), pp. 206–207
4. S.Y. Cheng, S.-T. Yau, On the existence of a complete Kähler metric on noncompact complex manifolds and the regularity of Fefferman's equation. Comm. Pure Appl. Math. **33**(4), 507–544 (1980)
5. S. Dinew, S. Kołodziej, Liouville and Calabi-Yau type theorems for complex Hessian equations. Amer. J. Math. **139**(2), 403–415 (2017)
6. S.K. Donaldson, Two-forms on four-manifolds and elliptic equations, in *Inspired by S. S. Chern*, Nankai Tracts Math., vol. 11 (World Sci. Publ., Hackensack, NJ, 2006), pp. 153–172
7. A. Fino, Y.Y. Li, S. Salamon, L. Vezzoni, The Calabi-Yau equation on 4-manifolds over 2-tori. Trans. Amer. Math. Soc. **365**(3), 1551–1575 (2013)

8. B. Guo, D.H. Phong, On L^∞ estimates for fully non-linear partial differential equations. Ann. Math. (2) **200**(1), 365–398 (2024)
9. Z. Hou, X.-N. Ma, D. Wu, A second order estimate for complex Hessian equations on a compact Kähler manifold. Math. Res. Lett. **17**(3), 547–561 (2010)
10. G. Székelyhidi, Fully non-linear elliptic equations on compact Hermitian manifolds. J. Differential Geom. **109**(2), 337–378 (2018)
11. V. Tosatti, Limits of Calabi-Yau metrics when the Kähler class degenerates. J. Eur. Math. Soc. (JEMS) **11**(4), 755–776 (2009)
12. V. Tosatti, B. Weinkove, Estimates for the complex Monge-Ampère equation on Hermitian and balanced manifolds. Asian J. Math. **14**(1), 19–40 (2010)
13. V. Tosatti, B. Weinkove, The Calabi-Yau equation on the Kodaira-Thurston manifold. J. Inst. Math. Jussieu **10**(2), 437–447 (2011)
14. V. Tosatti, B. Weinkove, The Calabi-Yau equation, symplectic forms and almost complex structures, in *Geometry and analysis. No. 1*, Adv. Lect. Math. (ALM), vol. 17 (Int. Press, Somerville, MA, 2011), pp. 475–493
15. V. Tosatti, B. Weinkove, The Monge-Ampère equation for $(n-1)$-plurisubharmonic functions on a compact Kähler manifold. J. Amer. Math. Soc. **30**(2), 311–346 (2017)
16. V. Tosatti, B. Weinkove, The Aleksandrov-Bakelman-Pucci estimate and the Calabi-Yau equation, in *Nonlinear Analysis in Geometry and Applied Mathematics. Part 2*, Harv. Univ. Cent. Math. Sci. Appl. Ser. Math., vol. 2 (Int. Press, Somerville, MA, 2018), pp. 147–158
17. V. Tosatti, B. Weinkove, S.-T. Yau, Taming symplectic forms and the Calabi-Yau equation. Proc. Lond. Math. Soc. (3) **97**(2), 401–424 (2008)
18. L. Vezzoni, On the Calabi-Yau equation in the Kodaira-Thurston manifold. Complex Manifolds **3**(1), 239–251 (2016)
19. B. Weinkove, The Calabi-Yau equation on almost-Kähler four-manifolds. J. Differential Geom. **76**(2), 317–349 (2007)
20. S.-T. Yau, On the Ricci curvature of a compact Kähler manifold and the complex Monge-Ampère equation, I. Comm. Pure Appl. Math. **31**, 339–411 (1978)

The manufacturer's authorised representative in the EU is Springer
Nature Customer Service Centre GmbH, Europaplatz 3, 69115 Heidelberg,
Germany. If you have any concerns regarding our products, please
contact ProductSafety@springernature.com

Printed and bound by CPI Group (UK) Ltd, Croydon, CR0 4YY
30/09/2026
02225572-0013